LINEAR PROGRAMMING

An Introduction to Finite Improvement Algorithms

Second Edition

Daniel Solow
Department of Operations
Weatherhead School of Management
Case Western Reserve University

Dover Publications, Inc.
Mineola, New York

Bibliographical Note

This Dover edition, first published in 2014, is an unabridged republication of the work originally published in 1984 by North-Holland Publishing Company, New York. For this second edition, the author has added an Introduction and an Appendix C.

Library of Congress Cataloging-in-Publication Data

Solow, Daniel, author.
Linear programming : an introduction to finite improvement algorithms / Daniel Solow. — Dover edition.
pages cm
"Second edition."
Reprint of: New York : North-Holland, 1984. With introduction and an appendix C.
Summary: "Suitable for undergraduate students of mathematics and graduate students of operations research and engineering, this text covers the basic theory and computation for a first course in linear programming. In addition to substantial material on mathematical proof techniques and sophisticated computation methods, the treatment features numerous examples and exercises. An introductory chapter offers a systematic and organized approach to problem formulation. Subsequent chapters explore geometric motivation, proof techniques, linear algebra and algebraic steps related to the simplex algorithm, standard phase 1 problems, and computational implementation of the simplex algorithm. Additional topics include duality theory, issues of sensitivity and parametric analysis, techniques for handling bound constraints, and network flow problems. Helpful appendixes conclude the text, including a new addition that explains how to use Excel to solve linear programming problems"—Provided by publisher.
Includes bibliographical references and index.
ISBN 978-0-486-49376-3 (paperback)
ISBN 0-486-49376-8 (paperback)
1. Linear programming. I. Title.
T57.74.S653 2014
519.7'2—dc23

2014016975

Manufactured in the United States by Courier Corporation
49376801 2014
www.doverpublications.com

To All of My Students

Contents

Introduction to the Dover Edition

When I first published *Linear Programming: An Introduction to Finite Improvement Algorithms* in 1984, my objective was to teach not only the simplex algorithm for solving linear programming (LP) problems, but also, as the subtitle suggests, the principle by which the simplex algorithm works. Specifically, a ***finite improvement algorithm*** attempts to solve a problem by moving from one of a finite number of "possible" solutions to another possible solution while strictly improving a *measure of goodness,* as determined by an objective function.

It was my belief that learning about a finite improvement algorithm was important not only because of the simplex algorithm, but also because many other algorithms for solving optimization problems are based on the same principle. To teach that principle, the approach I chose was to use insights from solving an LP geometrically to introduce the concept of a finite improvement algorithm and then to describe the steps of the simplex algorithm as a geometric finite improvement algorithm. The algebraic version of the simplex algorithm is then a translation of each step of the geometric algorithm to a corresponding computation using linear algebra. Although the latter is a non-traditional approach that is different from the usual "tableaux method," to date, I have found no better way to describe the simplex algorithm in a lasting way to my students. Furthermore, while certain aspects of linear programing have changed since 1984, the importance of a finite-improvement algorithm remains as strong, or stronger, today than in 1984.

While there have been a number of computational improvements to the simplex algorithm since 1984, perhaps the single greatest advancement in our ability to solve large real-world problems is due to technological improvements in computers. As a result, since I wrote this book, many efficient computer packages are now available. For this reason, I am including with this reprint, an appendix that explains how to use Solver in Excel to solve linear programming problems. This appendix also describes how to interpret the output reports that Solver provides.

A second major change in linear programming was due to theoretical research that led to a whole new approach to solving such problems—the ***interior point methods.*** These methods find the solution to an LP by traversing the interior of the feasible region rather than by following the boundary of the feasible region, as does the simplex algorithm. While the interior point methods have not replaced the simplex algorithm, they have become an accepted approach to solving certain large problems with special structure in the constraints. A description of these methods is beyond the scope of this reprint but details can be found in many other books and by searching the web.

Finally, I would like to thank John Grafton at Dover who believed enough in this project to make it happen.

DANIEL SOLOW
Department of Operations
Weatherhead School of Management
Case Western Reserve University
Cleveland, OH

Preface

Linear Programming: An Introduction to Finite Improvement Algorithms is designed for use by M.S. and Ph.D. students in operations research and engineering and for undergraduate mathematics majors. This book is a teaching text, not a comprehensive research reference book. It covers the basic theory and computation for a first course in linear programming.

It is natural to ask, "Why another text on linear programming?" The reason is that almost every chapter of this book contains or presents material in a manner significantly different from all other existing textbooks. This book makes three major contributions:

1. It includes, in Chapter 3, material on mathematical proof techniques so as to enable students to understand all of the proofs. These techniques are then used consistently in an outline that precedes each proof.
2. The tableau method is replaced with a more natural approach that remains closely related, at each step, to the geometry of linear programming. The advantage of this new approach is that it allows the student to focus on the principles underlying the simplex algorithm, rather than on the purely mechanical (and algebraically burdensome) steps of the tableau method. Indeed, with the ready availability of computer codes, there seems to be no reason for requiring students to solve any but the smallest problems by hand. In this spirit, this text solves problems having only 2×2 basis matrices.
3. Modern and sophisticated computational techniques are included.

To be more specific, the major contribution of each chapter will be outlined.

Chapter 1 presents a systematic and organized approach to the "art" of problem formulation. Specific steps are given as guidelines, and they are demonstrated in detail.

Chapter 2 not only presents the standard geometry of linear programming but also describes, in very elementary terms, the entire essence of a finite improvement algorithm—an algorithm that moves from one object to another getting "better" at each step. This concept is fundamental not only to the simplex algorithm but also to many other optimization procedures.

Chapter 3 provides unique material enabling the student to read and to understand all of the subsequent mathematical proofs. In essence, this chapter identifies, categorizes, and describes—*at the student's level*—the various techniques that will be used in the proofs. Furthermore, each proof in the text is preceded by an outline of the proof that details which techniques are going to be used and how they are to be applied. It should, of course, be pointed out to the better students that they need only read the "condensed" proof and not the detailed outline.

Chapter 4 provides the linear algebra (and only that portion) that is needed for the development of the simplex algorithm. For example, the concept of the rank of a matrix is not needed in the development of the simplex algorithm, and it has therefore been omitted. At each step, the relevance of the linear algebra is specifically related to its role in linear programming. One of the essential contributions of this chapter is the development of rules for using the linear programming data to specify conditions for detecting unboundedness.

Chapter 5 presents the algebraic steps of the simplex algorithm without using a tableau. Instead, all computations are illustrated with problems having a 2×2 basis matrix whose inverse can be computed quickly and easily by hand. Also, each algebraic step is related back to the geometric and conceptual motivation that was provided in Chapter 2. Degeneracy is resolved by Bland's rule, and the epsilon perturbation techniques are omitted.

Chapter 6 describes the standard phase 1 problem. However, when artificial variables remain in the optimal phase 1 basis at zero value, no attempt is made to remove them from the basis. Rather, an artificial constraint is added to ensure that all artificial variables remain at zero value.

Chapter 7 addresses many of the issues relating to a numerically accurate and efficient implementation of the simplex algorithm. Among the topics included are sparsity, numerical stability, reinversion, the geometry of phase 1, and crashing techniques.

Chapter 8 presents duality theory and stresses how duality theory can be used to determine whether a linear programming problem is infeasible, optimal, or unbounded without using a computer. This chapter also explains the more general role of duality theory in other finite improvement algorithms.

Chapter 9 presents, in a fairly standard way, the issues of sensitivity and postoptimality analysis. Parametric analysis is presented from the point of view of finite improvement.

Chapter 10 describes the handling of bound constraints, including unrestricted and superbasic variables. A phase 1 procedure that can be initiated from any given starting point is also developed. Generalized upper bounding is described, too.

Chapter 11 presents a general, primal network simplex algorithm. No attempt is made to specialize the algorithm to transportation or assignment problems, for example.

Appendix A correlates that tableau method with the finite improvement approach of this textbook, thus enabling students to understand both methods. Appendix B presents the fundamentals of Khacian's ellipsoid algorithm. No proofs, however, have been included.

Exercises have been designed to emphasize those topics presented in the text. Simply stated, it is advisable to assign all exercises for homework.

This textbook is decidedly weak in the area of advanced linear programming topics. This is in part because of the time, effort, and detail spent on the first nine chapters. It was felt that if the main body of linear programming could be made completely understandable, then students could easily learn on their own the more advanced topics. Perhaps those topics will, someday, be presented in a sequel to this book.

Acknowledgments

This book is truly a work of the students, for the students, and by the students. Its existence is directly attributable to them. Here is a partial list of what they accomplished. They prepared and tested numerical examples and homework exercises. They typed all the chapters into the computer. They proofread, edited, and printed out four entire revisions of the manuscript, complete with subscripts and superscripts. They made substantive suggestions for improving the readability. They made sketches of the figures. They prepared preliminary solutions to the exercises. Some students even wrote initial versions of parts of the book. And all of this work was done on a voluntary basis.

Since there were more than 50 students who contributed to this effort over three years, I cannot name them all. They know who they are and what they have done. I shall, however, mention the two students who undertook the enormously difficult task of coordinating the entire project—Reddy Dondeti and Sudhansu Baksi. Also, I acknowledge Ravi Kumar for supervising the whole computer operation and Robert Wenig for helping to develop subscripting and superscripting capabilities on the Case computer. Of course, I am grateful to Case for the use of their facilities.

I gratefully acknowledge John Wiley and Sons, Inc., for allowing me to use the material from my book *How to Read and Do Proofs* as Chapter 3 of this book. Finally, I thank my wife, Audrey, whose patience and understanding made it possible to complete this work.

Chapter 1

Problem Formulation

The success and survival of any organization depend on the quality of the decisions made by its managers. The central issue in many problems arising in business, industry, and government is that various activities compete for limited resources. It is the responsibility of the decision maker to allocate the resources among the competing activities so as to achieve the best results for the organization. The decision maker needs tools and techniques that can help in the quest for the optimal allocation of resources. Linear programming is the study of a large and important class of these problems together with the procedures currently available for obtaining their solutions. In this chapter, you will learn what a linear programming problem is, how to identify and to formulate one, and what distinguishes it from other types of problems.

1.1. WHAT IS A LINEAR PROGRAMMING PROBLEM?

To understand what a linear programming problem is, consider the dilemma of Midas M. Miner. One day he was traveling through the galaxy in his luxury spaceship when he developed minor engine trouble and was forced to land on a nearby asteroid. After making the necessary repairs, he was astounded (and delighted) to discover that the asteroid contained large quantities of silver ore and gold ore. Naturally, Midas wanted to take the whole asteroid with him, but practical considerations dictated otherwise. He knew that his ship could carry up to 100 pounds of extra cargo as long as it did not take up more than 150 cubic feet of space. After some measurements, Midas calculated that each pound of gold ore requires 2 cubic feet but each pound of silver ore requires only 1 cubic foot. Given that each pound of gold ore is worth 3 intergalactic credits (each credit being equivalent to approximately $10,000) and each pound of silver ore is worth 2 credits, how many pounds of each type of ore should Midas put on his ship so as to maximize his profit and still be able to take off?

In this problem, it is necessary to determine the quantities of gold and silver ore that Midas should take so as to achieve the maximum possible profit. Since it is not known in advance what these quantities should be, let x_1 and x_2 represent the number of pounds, respectively, of each of the two ores to be loaded on the ship. The variables x_1 and x_2 are sometimes called *decision variables*.

Since each pound of gold ore is worth 3 intergalactic credits, the worth of x_1 pounds of gold ore would be $3x_1$ credits. Similarly, the worth of x_2 pounds of silver ore would be $2x_2$ credits since each pound is worth 2 credits. Therefore, Mr. Miner's total profit would be $3x_1 + 2x_2$ credits. The function $3x_1 + 2x_2$ is called the *objective function* because the objective of the problem is to find values for x_1 and x_2 that maximize this function. Suppose, for instance, that x_1 and x_2 are chosen to be 40 pounds and 50 pounds, respectively. The corresponding profit would be $3(40) + 2(50) = 220$ credits. On the other hand, if x_1 is 60 pounds and x_2 is 80 pounds, then the total profit would be $3(60) + 2(80) = 340$ credits. Evidently, as x_1 and x_2 are increased, so are the profits. What is it, then, that prevents Midas from making an infinite profit? Evidently, it is the limited capacity of his cargo space.

If Midas decides to take x_1 pounds of gold ore and x_2 pounds of silver ore, then the total weight of the cargo will be $x_1 + x_2$ pounds. Similarly, since each pound of gold ore requires 2 cubic feet of space and each pound of silver ore requires 1 cubic foot, the total volume of the cargo will be $2x_1 + x_2$ cubic feet. For the specific case in which x_1 and x_2 are chosen to be 40 and 50 pounds, respectively, the corresponding total weight of the cargo would be $x_1 + x_2 = 40 + 50 = 90$ pounds, and the volume of this cargo would be $2x_1 + x_2 = 2(40) + (50) = 130$ cubic feet. Since the ship can indeed hold 100 pounds and 150 cubic feet of cargo, Mr. Miner will have no trouble taking off with 40 pounds of gold ore and 50 pounds of silver ore. For this reason, the choice of $x_1 = 40$, $x_2 = 50$ is said to be *feasible*.

In contrast, consider the case in which x_1 is 60 and x_2 is 80. The weight and volume of the resulting cargo are 140 pounds and 200 cubic feet, respectively. Since these quantities exceed the ship's capacity, it is not possible to take 60 pounds of gold ore and 80 pounds of silver ore aboard the ship. For this reason, the choice $x_1 = 60$, $x_2 = 80$ is said to be *infeasible*.

The conclusion is that Midas is *constrained* by the limited cargo capacity, and so his profits are limited also. The constraint on the total weight of the cargo can be expressed mathematically by the inequality

$$x_1 + x_2 \leq 100$$

Similarly, the total volume of the cargo should be less than or equal to 150 cubic feet, so

$$2x_1 + x_2 \leq 150$$

These are two constraints that x_1 and x_2 should satisfy, but are there any other restrictions on the variables? Although there are none explicitly stated in the problem, there is an implied understanding that the values of x_1 and x_2 should be ≥ 0 because it is not possible to have negative quantities of rocks. These implicit nonnegativity constraints can be made explicit by writing the inequalities

$$x_1 \geq 0, \qquad x_2 \geq 0$$

Such constraints should be included in the formulation of any real world problem whenever appropriate.

In mathematical terms, Mr. Miner's problem can be stated as follows:

$$\begin{aligned} \text{Maximize} \quad & 3x_1 + 2x_2 \\ \text{subject to} \quad & x_1 + x_2 \leq 100 \\ & 2x_1 + x_2 \leq 150 \\ & x_1, \quad x_2 \geq 0 \end{aligned}$$

The above problem is a linear programming problem (hereafter referred to as an LP). In an its final form, an LP has

1. a set of decision variables whose values need to be determined,
2. an objective function that has to be maximized (or minimized), and
3. a set of constraints that must be satisfied by the decision variables.

In addition to these features, there are three other characteristics that distinguish an LP from other mathematical problems: (1) proportionality, (2) additivity, and (3) divisibility. Each of these will be described briefly.

1. Proportionality

Proportionality means that if the value of a variable is multiplied by a constant, then its contribution to the objective and constraint functions is also multiplied by the same constant. In Mr. Miner's problem, this assumption is satisfied because if he were to double the amount of gold ore, say, from 20 to 40 pounds, then the profit from the gold ore would also double, from 60 to 120 credits. Similarly, the volume of the gold ore would double from 40 to 80 cubic feet. In other words, the values of the objective and constraint functions are proportional to the values of the decision variables.

2. Additivity

Additivity means that the total value of the objective function as well as the constraint functions are equal to the sum of the individual contributions of the decision variables. Once again, this assumption is satisfied in Mr. Miner's problem. If x_1 is 40 pounds and x_2 is 50 pounds, then the value of the objective function would be 220 credits, which is the sum of the individual contributions of the decision variables, namely, 120 and 100 credits, respectively. Similarly, the value of the first constraint function is 90 pounds, which is the sum of 40 and 50 pounds, the individual contributions of the decision variables. Finally, the value of the second constraint is equal to 130 cubic feet, which again is the sum of the individual contributions of the decision variables. In other words there is no "interaction" between the decision variables.

To understand the concept of additivity better, consider two examples involving the blending of materials. For instance, cattle feed is made by blending components such as corn, limestone, soybean, and fish meal. No matter how the ingredients are blended, the total weight of the mixture is equal to the sum of the individual weights, and the total volume of the mixture is equal to the sum of the individual volumes. Again there is no interaction between the different components of the mixture, and hence the weight and volume of the mixture are additive.

On the other hand, consider the situation where a certain volume of water is to be mixed with a certain volume of alcohol. Since water and alcohol are partially miscible, the resulting volume will actually be much less than the sum of the volumes of the water and alcohol. In this case the decision variables interact with each other, and thus the volume of the mixture is not additive. An LP model always requires that the property of additivity hold for the objective function and the constraints.

3. Divisibility

Divisibility means that the decision variables can, theoretically, assume any value, including fractions, in some interval between $-\infty$ and $+\infty$. Mr. Miner's problem satisfies this assumption because x_1 and x_2 can assume values such as 50.4763 or 28.9348 since the weights of the ores can be measured on a continuous scale. In contrast, consider a production facility that manufactures airplanes. In this case, the decision variables representing the number of airplanes to be produced have to be integers, for it is not possible to produce one-third of an airplane. Integer restrictions on the values of the decision variables are not permitted in an LP model. When it is necessary to have integer variables, a technique known as integer programming is required, but that topic will not be discussed here.

Now that you know what an LP is, it is time to learn how to formulate such problems mathematically. This process will be illustrated in the next two sections with several examples.

EXERCISES

1.1.1. For each of the following LPs, identify the objective function and the constraints. Also, determine if the given values of the variables are feasible or not.

(a)

$$\begin{array}{lrrrcr} \text{Minimize} & 3x_1 & +5x_2 & +6x_3 & & \\ \text{subject to} & x_1 & +\ x_2 & +\ x_3 & \leq & 8 \\ & 2x_1 & +3x_2 & +\ x_3 & = & 11 \\ & x_1 & +4x_2 & -3x_3 & \geq & -3 \\ & 5x_1 & -3x_2 & -4x_3 & \leq & 2 \\ & 3x_1 & +\ x_2 & -\ x_3 & \geq & 2 \\ & x_1, & x_2, & x_3 & \geq & 0 \end{array}$$

given the values $x_1 = 0$, $x_2 = 3$, $x_3 = 2$.

(b)

$$\begin{array}{lrrrrcr} \text{Maximize} & 4x_1 & +5x_2 & +3x_3 & +2x_4 & & \\ \text{subject to} & 7x_1 & +2x_2 & +4x_3 & +3x_4 & \leq & 600 \\ & 4x_1 & +\ x_2 & +2x_3 & +\ x_4 & \leq & 750 \\ & 2x_1 & +3x_2 & +\ x_3 & +2x_4 & \leq & 200 \\ & x_1 & +\ x_2 & +\ x_3 & +\ x_4 & = & 30 \\ & x_1 & & +\ x_3 & & \geq & 10 \\ & & x_2 & & +\ x_4 & \geq & 10 \\ & x_1, & x_2, & x_3, & x_4 & \geq & 0 \end{array}$$

given the values $x_1 = 5$, $x_2 = 12$, $x_3 = 6$, $x_4 = 15$.

(c)

$$\begin{array}{lrrrrrcr} \text{Maximize} & x_1 + & 10x_2 + & 8x_3 + & 15x_4 + & 2x_5 & & \\ \text{subject to} & 0.5x_1 + & 0.4x_2 + & 0.6x_3 + & 0.5x_4 + & 0.4x_5 & \leq & 250 \\ & 0.3x_1 + & 0.2x_2 + & 0.2x_3 + & 0.1x_4 + & 0.1x_5 & \leq & 120 \\ & 0.1x_1 + & 0.2x_2 + & 0.1x_3 + & 0.2x_4 + & 0.1x_5 & \leq & 100 \\ & & & & x_4 + & x_5 & \leq & 250 \end{array}$$

$$\begin{array}{l} 0 \leq x_1 \leq 200 \\ 0 \leq x_2 \leq 150 \\ 0 \leq x_3 \leq 200 \\ 0 \leq x_4 \\ 0 \leq x_5 \leq 100 \end{array}$$

given the values $x_1 = 150$, $x_2 = 150$, $x_3 = 0$, $x_4 = 150$, $x_5 = 100$.

1.1.2. For each of the following problems, indicate whether the properties of proportionality, additivity, and divisibility hold.

(a)

$$\begin{aligned}
\text{Minimize}\quad & 3x_1 + 2x_2 + 7x_3 + 0.2x_4 \\
\text{subject to}\quad & x_1 + x_2 + x_3 + x_4 \geq 2 \\
& 2x_1 - x_2 + x_3 - x_4 \leq 3 \\
& x_2 + x_3 + x_4 \leq 7 \\
& x_1 \leq 0 \\
& x_2 \geq 0 \\
& x_3 \geq 0 \\
& x_4 = 0 \text{ or } 1
\end{aligned}$$

(b)

$$\begin{aligned}
\text{Minimize}\quad & 2x_1x_2^2/x_3 \\
\text{subject to}\quad & x_1x_2 \geq 4 \\
& x_3/x_2 \leq 2 \\
& 2x_1x_3/x_2^2 \geq 1 \\
& x_1, x_2, x_3 \geq 0
\end{aligned}$$

(c) Convert the problem in part (b) into an LP by taking the natural logarithm of the objective function and the constraints and then defining new variables by $y_i = \ln(x_i)$.

1.2. HOW TO FORMULATE AN LP

Problem formulation is the process of transforming the verbal description of a decision problem into a mathematical form that can then be solved, most likely by a computer. In a complicated problem, it is not always easy to identify the relationship between the different decision variables. Formulating a mathematical model of the problem is the first (and probably the most important) step in the solution of any real world problem.

Problem formulation is an art in itself, and there are no simple methods or rules available. In solving real world problems, several approximations may have to be made before a problem can be modeled mathematically. Only with adequate practice can one learn to formulate problems correctly. In general, however, the following steps are involved.

Step 1. Identification of the Variables

The variables are those quantities whose values you wish to determine. The variables are closely linked to the objectives and activities of the problem. In a chemical mixing problem, for example, suppose that there are ten different chemicals to be mixed to obtain four different mixtures. If the desired amounts of the final mixtures are specified, then the variables would be the quantities of the different chemicals to be mixed. On the other hand, if the available quantities of the chemicals are specified, then the variables would be the amounts of the different mixtures to be produced. Thus, in a given problem, there could be different sets of variables at different times. In some instances, it may even be necessary to create auxiliary variables to simplify the mathematical representation or to make the problem more amenable to solution, for example. In any event, the variables should be identified as a first step in the development of an LP model.

Step 2. Identification of the Objective Function

The decision maker is the one who specifies the overall objective of the problem. For an LP model, the objective function might take the form of maximizing profits or minimizing costs, for example. In general there can be several (often conflicting) objectives, and the decision maker has to strike a compromise among them. Linear programming allows for the optimization of only one objective function. Thus the decision maker must indicate a single specific objective to be optimized. There are other techniques, notably goal programming, that attempt to optimize multiple objective functions, but the topic of multiobjective optimization will not be addressed here.

The objective function of an LP always has the form

$$c_1x_1 + c_2x_2 + \cdots + c_nx_n$$

where $x_1, x_2, \ldots, x_n$ are the variables and $c_1, c_2, \ldots, c_n$ are the known coefficients of the variables; these can be positive, negative, or even zero. The objective of the problem is to find values for the variables $x_1, \ldots, x_n$ that either maximize or minimize the value of $c_1x_1 + \cdots + c_nx_n$. The second step in formulating an LP is to identify the values of $c_1, \ldots, c_n$ and the type of optimization, i.e., maximization or minimization.

Step 3. Identification of the Constraints

The constraints of an LP can be divided into four basic groups.

1. *Resource (or less-than-or-equal-to) constraints.* These constraints usually reflect the limited availability of a resource, such as the cargo space in Mr. Miner's problem. Each such constraint has the form

   $$a_1x_1 + a_2x_2 + \cdots + a_nx_n \leq b$$

   where $a_1, a_2, \ldots, a_n$, and b are known constants. The coefficients $a_1, a_2, \ldots, a_n$, often referred to as *technological coefficients*, can be positive, negative, or zero. The value of b might represent the available quantity of a raw material, the production capacity of a machine, or the total funds available for investment.

2. *Requirement (or greater-than-or-equal-to) constraints.* These constraints usually reflect an imposed requirement on the problem. Each such constraint has the form

$$a_1x_1 + a_2x_2 + \cdots + a_nx_n \geq b$$

Here again, the coefficients $a_1, a_2, \ldots, a_n$ can be positive, negative, or zero. The value of b might represent the required quantity of a nutrient stipulated by a consumer organization, the quantity of material demanded by a customer, or a quality standard set by professional and governmental organizations.

3. *Structural (or equality) constraints.* These constraints usually reflect the structural or technological relationships among the variables. Each such constraint has the form

$$a_1x_1 + a_2x_2 + \cdots + a_nx_n = b$$

For instance, consider a production process in which the amount x_1 of product 1 and the amount x_2 of product 2 must always be produced in the ratio of 3 to 2; i.e., $x_1/x_2 = \frac{3}{2}$ or, equivalently, $2x_1 - 3x_2 = 0$.

4. *Bound constraints.* These are constraints that involve only one variable—a nonnegativity constraint, for instance. Alternatively, some problems might require a variable to be nonpositive (e.g., $x_1 \leq 0$), and other problems might require a variable to be between a given *lower bound* and *upper bound* (e.g., $-10 \leq x_1 \leq 20$). It is also possible that certain variables are *unrestricted*; i.e., they can assume any value—positive, negative, or zero.

An Example

To demonstrate the art of problem formulation, consider the energy problem faced by the country of Lilliput. Lilliput is a small country with limited energy resources. The Department of Energy of Lilliput is currently in the

TABLE 1.1. Data on Generation Capacities and Costs

Energy source	Total capacity[a] (MW-hr)	Unit cost of generation ($/MW-hr)
Coal	45,000	6.0
Natural gas	15,000	5.5
Nuclear materials	45,000	4.5
Hydroelectric projects	24,000	5.0
Petroleum[b]	48,000	7.0

[a]Capacity is measured in megawatt-hours, a unit of energy.
[b]Petroleum is used for producing gasoline and fuel oils, but for convenience it is given here in terms of equivalent electrical energy units.

TABLE 1.2. Data on Allowable Pollution Levels

Effluent	Limit (g)
Sulfur dioxide	75,000
Carbon monoxide	60,000
Dust particles	30,000
Solid waste	25,000

process of developing a national energy plan for the next year. Lilliput can generate energy from any one of five sources: coal, natural gas, nuclear materials, hydroelectric projects, and petroleum. The data on the energy resources, generation capacities, and unit costs of generation are given in Table 1.1.

Lilliput needs 100,000 MW-hr of energy for domestic use, but the country would also like to produce at least 25,000 MW-hr for export. Furthermore, to conserve the energy resources and also to protect the environment, the House and Senate have passed the following regulations:

1. The generation of energy from nuclear materials should not exceed 20% of the total energy generated by Lilliput.
2. At least 80% of the capacity of the coal plants should be utilized.
3. The effluents let off into the atmosphere should not exceed the limits specified in Table 1.2.
4. The generation of energy from natural gas should be at least 30% of the energy generated from petroleum.

The EPAL (Environmental Protection Agency of Lilliput) has collected data on the various amounts of pollutants given off by each of the energy sources. Those data are summarized in Table 1.3.

The three steps of problem formulation will be applied in order to formulate a plan for the Department of Energy that would minimize the total cost of generating the electricity while meeting all of the requirements.

Step 1. Identification of the Variables

Since it is necessary to determine the quantities of energy to be generated from the individual energy sources, let the variables x_1, x_2, x_3, x_4, x_5 represent the number of megawatt-hours of energy to be generated from coal, natural gas, nuclear materials, hydroelectric projects, and petroleum, respectively.

Step 2. Identification of the Objective Function

The objective of the problem is to minimize the total cost. The cost coefficients of the decision variables (namely, $c_1, \ldots, c_5$) are given in the

TABLE 1.3. Rates of Pollutant Emissions

Energy source	Sulfur dioxide (g/MW-hr)	Carbon monoxide (g/MW-hr)	Dust particles (g/MW-hr)	Solid waste (g/MW-hr)
Coal	1.5	1.2	0.7	0.4
Natural gas	0.2	0.5	—	—
Nuclear materials	0.5	0.2	0.4	0.5
Hydroelectric projects	—	—	—	—
Petroleum	0.4	0.8	0.5	0.1

third column of Table 1.1. Therefore, the objective function is

$$6.0x_1 + 5.5x_2 + 4.5x_3 + 5.0x_4 + 7.0x_5$$

Step 3. Identification of the Constraints

1. The entire demand of Lilliput should be met. Specifically, 100,000 MW-hr are needed for domestic use, and at least 25,000 MW-hr are needed for export. Therefore, the total number of MW-hr required is at least 125,000. In terms of the variables $x_1, \ldots, x_5$ the quantity of megawatt-hours to be generated is $x_1 + x_2 + x_3 + x_4 + x_5$. Hence the constraint for meeting the total demand is

$$x_1 + x_2 + x_3 + x_4 + x_5 \geq 125{,}000$$

2. The generation from nuclear materials (i.e., x_3) should be at most 20% of the total energy generated (i.e., 20% of $x_1 + x_2 + x_3 + x_4 + x_5$), so

$$x_3 \leq 0.20(x_1 + x_2 + x_3 + x_4 + x_5)$$

or, equivalently, on multiplying both sides by 5 and then subtracting $5x_3$,

$$x_1 + x_2 - 4x_3 + x_4 + x_5 \geq 0$$

3. At least 80% of the capacity of the coal plants should be utilized. From the second column of Table 1.1, the capacity of the coal plants is 45,000 MW-hr. Since x_1 should exceed 80% of this,

$$x_1 \geq 0.80(45{,}000)$$

or, equivalently,

$$x_1 \geq 36{,}000$$

4. The environmental constraints should be met. From Table 1.3, the total quantity of sulfur dioxide let into the atmosphere would be $1.5x_1 + 0.2x_2 + 0.5x_3 + 0.4x_5$. This quantity should be less than or equal to 75,000, as specified in Table 1.2:

$$1.5x_1 + 0.2x_2 + 0.5x_3 + 0.4x_5 \leq 75{,}000$$

Similarly, the constraint for carbon monoxide is

$$1.2x_1 + 0.5x_2 + 0.2x_3 + 0.8x_5 \le 60{,}000$$

The constraint for dust particles is

$$0.7x_1 + 0.4x_3 + 0.5x_5 \le 30{,}000$$

The constraint for solid waste is

$$0.4x_1 + 0.5x_3 + 0.1x_5 \le 25{,}000$$

5. The amount of energy generated from natural gas should be at least 30% of the energy generated from petroleum, so

$$x_2 \ge 0.3x_5$$

or, equivalently,

$$x_2 - 0.3x_5 \ge 0$$

6. The values of the decision variables cannot exceed the capacities of their respective sources as listed in the second column of Table 1.1, so

$$\begin{aligned} x_1 &\le 45{,}000 \\ x_2 &\le 15{,}000 \\ x_3 &\le 45{,}000 \\ x_4 &\le 24{,}000 \\ x_5 &\le 48{,}000 \end{aligned}$$

7. All of the decision variables should be nonnegative, so

$$x_j \ge 0 \qquad \text{for} \quad j = 1, 2, 3, 4, 5$$

After some rearrangement, the LP can be stated mathematically as follows:

$$\begin{aligned} \text{Minimize} \quad & 6.0x_1 + 5.5x_2 + 4.5x_3 + 5.0x_4 + 7.0x_5 \\ \text{subject to} \quad & x_1 + x_2 + x_3 + x_4 + x_5 \ge 125{,}000 \\ & x_1 + x_2 - 4.0x_3 + x_4 + x_5 \ge 0 \\ & 1.5x_1 + 0.2x_2 + 0.5x_3 + 0.0x_4 + 0.4x_5 \le 75{,}000 \\ & 1.2x_1 + 0.5x_2 + 0.2x_3 + 0.0x_4 + 0.8x_5 \le 60{,}000 \\ & 0.7x_1 + 0.0x_2 + 0.4x_3 + 0.0x_4 + 0.5x_5 \le 30{,}000 \\ & 0.4x_1 + 0.0x_2 + 0.5x_3 + 0.0x_4 + 0.1x_5 \le 25{,}000 \\ & 0.0x_1 + x_2 + 0.0x_3 + 0.0x_4 - 0.3x_5 \ge 0 \\ & 36{,}000 \le x_1 \le 45{,}000 \\ & 0 \le x_2 \le 15{,}000 \\ & 0 \le x_3 \le 45{,}000 \\ & 0 \le x_4 \le 24{,}000 \\ & 0 \le x_5 \le 48{,}000 \end{aligned}$$

Observe that this problem does satisfy the properties of proportionality, additivity, and divisibility.

This section has described and illustrated the process of problem formulation. The next section demonstrates the process again, but with a more difficult problem.

EXERCISES

1.2.1. For each of the LPs in Exercise 1.1.1, identify the resource, requirement, structural, and bound constraints.

1.2.2. An oil refinery can buy two types of oil: light crude oil and heavy crude oil. The cost per barrel is \$40 and \$34, respectively. The following quantities (in barrels) of gasoline, kerosene, and jet fuel can be produced per barrel of each type of crude oil:

	Gasoline	Kerosene	Jet fuel
Light crude oil	0.45	0.18	0.32
Heavy crude oil	0.34	0.36	0.22

The refinery has contracted to deliver 1,200,000 barrels of gasoline, 500,000 barrels of kerosene, and 300,000 barrels of jet fuel. Formulate, as an LP, the problem of determining how much of each type of crude oil to purchase so as to minimize the cost while meeting the appropriate demands. Are the properties of proportionality, additivity, and divisibility satisfied? Explain.

1.2.3. The Carmac company manufactures a compact and a subcompact car. The production of each car requires a certain amount of raw material and labor as specified in the following table:

	Raw material	Labor
Compact	20	18
Subcompact	15	20
Unit cost (\$)	100	70
Total available	8000	9000

The marketing division has estimated that at most 1500 compacts can be sold at \$6000 each and that at most 200 subcompacts can be sold at \$5000 each. Formulate, as an LP, the problem of determining how many of each type of car to manufacture so as to maximize the total profit. Are the properties of proportionality, additivity, and divisibility satisfied? Explain.

1.2.4. Each gallon of milk, pound of cheese, and pound of apple produces a known number of milligrams of protein and vitamins A, B, and C, as given in the table below. The minimum weekly requirements of the nutritional ingredients and the costs of the foods are also supplied.

	Milk (gal)	Cheese (lb)	Apples (lb)	Minimum weekly requirements
Protein	40	20	10	80
Vitamin A	5	40	30	60
Vitamin B	20	30	40	50
Vitamin C	30	50	60	30
Unit cost ($)	1	2.5	0.75	

Formulate, as an LP, the problem of determining how much of each food should be consumed so as to meet the weekly nutritional requirements while minimizing the total costs. Are the properties of proportionality, additivity, and divisibility satisfied? Explain.

1.2.5. An oil company near Cleveland transports gasoline to its distributors by trucks. The company has recently received a contract to begin supplying gasoline distributors in Cincinnati and has $500,000 available to spend on the necessary expansion of its fleet of gasoline tank trucks. Three types of gasoline tank trucks are available.

Truck type	Capacity (gal)	Purchase cost ($)	Operating cost/month ($)	Maximum trips/month
1	6000	50,000	800	20
2	3000	40,000	650	25
3	2000	25,000	500	30

The company estimates that the monthly demand for the region will be a total of 800,000 gallons of gasoline. Based on maintenance and driver availability, the firm does not want to add more than 20 new vehicles to its fleet. In addition, the company would like to make sure it purchases at least three of the type-3 trucks to use on the short-run/low-demand routes. Finally, the company does not want more than half of the new models to be type-1 trucks. Formulate a linear programming model to calculate the number of units of each truck type to be purchased in such a way that the monthly operating costs are minimized and the demand is satisfied. Are the properties of proportionality, additivity, and divisibility satisfied? Explain.

1.2.6. An airline refuels its aircraft regularly at the four airports it serves, and its purchasing agent must decide on the amounts of jet fuel to buy from three possible vendors. The vendors have said that they can furnish up to the following amounts during the coming month: 300,000 gallons from vendor 1, 600,000 gallons from vendor 2, and 700,000 gallons from vendor 3. The required amount of fuel is 150,000 gallons at airport 1, 250,000 gallons at airport 2, 350,000 gallons at airport 3, and 480,000 gallons at airport 4. The combination of the transportation costs and the bid price per gallon supplied from each vendor furnishing a specific airport is shown in the following table:

	Vendor 1	Vendor 2	Vendor 3
Airport 1	9	8	9
Airport 2	9	12	13
Airport 3	8	13	5
Airport 4	10	14	10

Formulate the decision problem as a linear programming model. Are the properties of proportionality, additivity, and divisibility satisfied? Explain.

1.3. ADVANCED PROBLEM FORMULATION

As you have seen in the previous sections, some problems lend themselves readily to formulation as linear programs. Here you will see that other problems can require a certain amount of cleverness in order to be put in the form of an LP.

Consider the production planning problem of the National Steel Corporation (NSC), which produces a special-purpose steel that is used in the aircraft and aerospace industries. The marketing department of NSC has received orders for 2400, 2200, 2700, and 2500 tons of steel during each of the next four months; NSC can meet these demands by producing the steel, by drawing from its inventory, or by any combination thereof.

The production costs per ton of steel during each of the next four months are projected to be \$7400, \$7500, \$7600, and \$7800. Because of these inflationary costs, it might be advantageous for NSC to produce more steel than it needs in a given month and store the excess, although production capacity can never exceed 4000 tons in any month. All production takes place at the beginning of the month, and immediately thereafter the demand is met. The remaining steel is then stored in inventory at a cost of \$120/ton for each month that it remains there.

In addition, if the production level is increased or decreased from one month to the next, then the company incurs a cost for implementing these changes. Specifically, for each ton of increased or decreased production over the previous month, the cost is \$50. The production of the first month, however, is exempt from this cost.

If the inventory at the beginning of the first month is 1000 tons of steel and the inventory level at the end of the fourth month should be at least 1500 tons, formulate a production plan for NSC that will minimize the total cost over the next four months.

Step 1. Identification of the Variables

Since the objective of NSC is to minimize its total cost, the variables should consist of all those items that are related to or that incur costs. For instance, there are costs associated with producing steel, so let x_1, x_2, x_3, x_4 represent the number of tons of steel to be produced during each of the next four months. Also, the inventory of steel incurs costs; thus it would seem natural to let I_1, I_2, I_3, I_4 be the number of tons of steel in inventory (after meeting the demand) for each of the next four months.

Finally, there are costs associated with changes in the production levels from one month to the next. At first glance, it would appear that these changes can be computed from $x_1, \ldots, x_4$. For instance, if $x_1 = 1500$ and $x_2 = 2000$, then the amount of increase in the production from the first month to the second would be $x_2 - x_1 = 2000 - 1500 = 500$. Hence one might conjecture that no new variables are needed. As will be seen shortly, this conjecture is false.

Step 2. Identification of the Objective Function

The total cost to NSC over the next four months is the sum of the costs for production, inventory, and changes in production. If $x_1, \ldots, x_4$ are the production quantities and the corresponding costs per ton are \$7400, \$7500, \$7600, and \$7800, then the total production cost is

$$7400x_1 + 7500x_2 + 7600x_3 + 7800x_4$$

Similarly, if $I_1, \ldots, I_4$ are the inventory quantities during each month, then they incur a cost of \$120/ton for that month. In other words, the total inventory cost is

$$120(I_1 + I_2 + I_3 + I_4)$$

Finally, the cost associated with changes in the production levels must be calculated. Consider, for instance, the change from the first month to the second, namely, $x_2 - x_1$. Since the cost per ton of change is \$50, one is tempted to claim that $50(x_2 - x_1)$ is the associated cost; however, if $x_1 = 1500$ tons and $x_2 = 1000$ tons, for example, then $x_2 - x_1 = -500$, and the associated cost would be computed as $50(-500) = -25{,}000$. This of course cannot be correct since it is impossible to incur a negative cost. One way to correct the problem would be to modify the formula for computing the cost from $50(x_2 - x_1)$ to $50|x_2 - x_1|$. If this is done, then the total cost associated with changes in the production levels over the next four months would be

$$50(|x_2 - x_1| + |x_3 - x_2| + |x_4 - x_3|)$$

Unfortunately, this cost is not in the form $c_1x_1 + c_2x_2 + c_3x_3 + c_4x_4$, as required in linear programming problems. However, by creating certain "auxiliary" variables, this cost can be rewritten in the appropriate form. Specifically, let S_1, S_2, and S_3 be the amount of increase in production during the second, third, and fourth months, respectively (recall that the first month is exempt from this cost). In the event that the production level decreases, let D_1, D_2, and D_3 represent the amount of decrease during the appropriate months. Observe that if production increases from one month to the next, then the corresponding S variable will be positive and the D variable will be zero; if production decreases, then the D variable will be positive and the S variable will be zero. The relationships between all of the variables will be described subsequently in the constraints of the problem.

In any event, with the aid of these new variables, the total cost for changing the production levels can be written as

$$50(S_1 + S_2 + S_3 + D_1 + D_2 + D_3)$$

which does have the appropriate form for a linear programming problem.

In summary, the objective of NSC can be stated mathematically as follows:

$$\begin{array}{llr} \text{Minimize} & 7400x_1 + 7500x_2 + 7600x_3 + 7800x_4 & \text{(production)} \\ & + 120I_1 + 120I_2 + 120I_3 + 120I_4 & \text{(inventory)} \\ & + 50S_1 + 50S_2 + 50S_3 & \text{(changes in} \\ & + 50D_1 + 50D_2 + 50D_3 & \text{production)} \end{array}$$

Step 3. Identification of the Constraints

The first set of constraints to be formulated mathematically are those that relate the production and inventory variables. To help visualize the sequence of events that take place during each month, consider Figure 1.1. The box represents the month. At the beginning of the first month, for example, there are 1000 tons of steel in inventory that can be used to meet the demand. Also, at the beginning of the month, x_1 tons of steel will be produced. One constraint is that the total amount of steel available (namely, $1000 + x_1$) should be adequate to meet the demand for that month (namely, 2400), so

$$1000 + x_1 \geq 2400$$

and similarly for each of the other months.

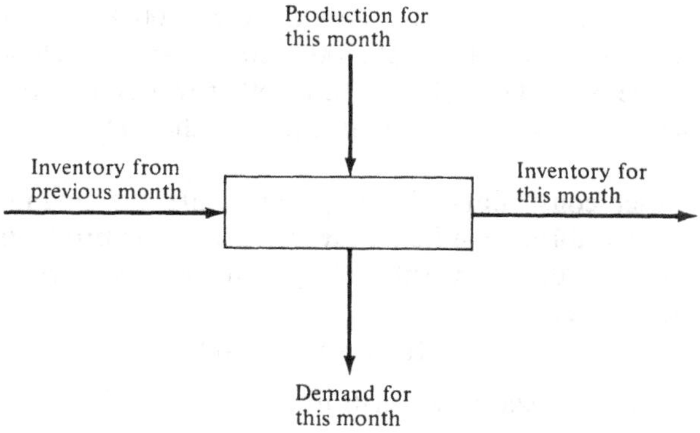

FIGURE 1.1. Events during a month of production.

Let us return to Figure 1.1. Once the demand has been met, the remaining steel (namely, $1000 + x_1 - 2400$) becomes the inventory for the next month, so

$$I_1 = 1000 + x_1 - 2400$$

Similarly, for the other months the structural constraints would be

$$I_2 = I_1 + x_2 - 2200$$
$$I_3 = I_2 + x_3 - 2700$$
$$I_4 = I_3 + x_4 - 2500$$

Since the desired inventory at the end of the fourth month should be at least 1500, one has

$$I_4 \geq 1500$$

Turning now to the constraints relating the production variables to the change-in-production variables, it is perhaps surprising to discover that these constraints can be written

$$x_2 - x_1 = S_1 - D_1$$
$$x_3 - x_2 = S_2 - D_2$$
$$x_4 - x_3 = S_3 - D_3$$

To see why these constraints are correct, imagine that $x_1 = 1500$ and $x_2 = 2000$. In this case, $x_2 - x_1 = 500$, $S_1 = 500$, and $D_1 = 0$. Thus, indeed, $x_2 - x_1 = S_1 - D_1$. Alternatively, if $x_1 = 1500$ and $x_2 = 1000$, then $x_2 - x_1 = -500$, $S_1 = 0$, and $D_1 = 500$, so again, $x_2 - x_1 = S_1 - D_1$.

It is important to note that if $x_1 = 1500$ and $x_2 = 2000$, then there are many different values for S_1 and D_1 that will make $x_2 - x_1 = S_1 - D_1$: for instance, $S_1 = 500$ and $D_1 = 0$, or $S_1 = 700$ and $D_1 = 200$. In the former case, however, the associated cost would be $50(500 + 0) = 25{,}000$; in the latter case, the cost would be $50(700 + 200) = 45{,}000$. Hence it is less expensive to have $S_1 = 500$ and $D_2 = 0$. In other words, it is the costs in the objective function that will ensure that if $S_1 > 0$, then $D_1 = 0$, and if $D_1 > 0$, then $S_1 = 0$.

The only other constraints of this problem are the bound constraints. Specifically, all variables must be nonnegative, and the production variables can never exceed 4000. It is worth noting that the inventory constraint for the first month, namely,

$$I_1 = 1000 + x_1 - 2400$$

together with the nonnegativity constraint

$$I_1 \geq 0$$

automatically ensures that $1000 + x_1 \geq 2400$, and hence the "redundant" constraints that were used to guarantee that the demand is met are not needed.

After some rewriting, the LP formulation of the production planning problem of NSC can be restated mathematically:

$$\begin{aligned}
\text{Minimize} \quad & 7400x_1 + 7500x_2 + 7600x_3 + 7800x_4 \\
& + 120I_1 + 120I_2 + 120I_3 + 120I_4 \\
& + 50S_1 + 50S_2 + 50S_3 \\
& + 50D_1 + 50D_2 + 50D_3 \\
\text{subject to} \quad & x_1 - I_1 = 1400 \\
& I_1 + x_2 - I_2 = 2200 \\
& I_2 + x_3 - I_3 = 2700 \\
& I_3 + x_4 - I_4 = 2500 \\
& x_2 - x_1 - S_1 + D_1 = 0 \\
& x_3 - x_2 - S_2 + D_2 = 0 \\
& x_4 - x_3 - S_3 + D_3 = 0 \\
& 1400 \leq x_1 \leq 4000 \\
& 0 \leq x_2 \leq 4000 \\
& 0 \leq x_3 \leq 4000 \\
& 0 \leq x_4 \leq 4000 \\
& 1500 \leq I_4 \\
& I_1, I_2, I_3, I_4 \geq 0 \\
& S_1, S_2, S_3 \geq 0 \\
& D_1, D_2, D_3 \geq 0
\end{aligned}$$

It is important to note that the properties of proportionality, additivity, and divisibility are satisfied in the problem.

The problems that have been presented in this chapter are considered to be very small in terms of the number of variables and constraints. It is not uncommon to have problems containing hundreds (or thousands) of variables and hundreds (or thousands) of constraints. In order to solve such problems, it is necessary to use computers. The remainder of this text will present the theory and algorithms currently available for solving these large-scale problems. The final section of this chapter will present a brief history of linear programming and a discussion of the subsequent material.

EXERCISES

1.3.1. A factory manufactures a product with each complete unit consisting of four units of component A and three units of component B. The two components (A and B) are manufactured from two different raw materials of which 100 units and 200 units are available, respectively. Each of three departments uses a different method for manufacturing the components. The table below gives the raw material requirements per production run and the resulting number of each component produced by that run.

Formulate a linear programming problem to determine the number of production runs for each department so as to maximize the total number of completed units of the final product. Ignore the divisibility requirement.

	Input/run (units)		Output/run (units)	
Department	Raw Material 1	Raw Material 2	Component A	Component B
1	8	6	7	5
2	5	9	6	9
3	3	8	8	4

1.3.2. A company wishes to plan its production of two items with seasonal demands over a six-month period. The demands, in thousands of units, are as follows:

Item	October	November	December	January	February	March
1	80	80	80	20	20	20
2	60	60	60	60	60	10

The unit production costs of items 1 and 2 are \$6 and \$10, respectively, provided that they are manufactured prior to January. Starting in January, the unit costs are reduced to \$5 and \$8 because of the installation of an improved manufacturing system. The total units of items 1 and 2 that can be manufactured during any particular month cannot exceed 140,000. Furthermore, each unit of item 1 occupies 3 cubic feet and each unit of item 2 occupies 6 cubic feet of inventory. The maximum inventory space is 220,000 cubic feet, and the inventory holding cost per cubic foot during any month is \$0.15. Formulate a linear programming model to minimize production and inventory costs. Assume that all production takes place at the beginning of the month and, immediately thereafter, demand is met and also that there is no initial inventory.

1.3.3. A chemical company makes two products that yield excessive amounts of three different types of pollutants. The state government has ordered the company to install and to employ antipollution devices. The following table shows the daily emission, in pounds, of each pollutant for every 1000 gallons of product manufactured.

Type of pollutant	Product 1	Product 2
A	25	40
B	10	15
C	80	60

The company is prohibited from emitting more than 300, 250, and 500 pounds of pollutant type A, B, and C, respectively. The profits per 1000 gallons of products 1 and 2 manufactured are \$100 and \$80, respectively. The production manager has approved the installation of two antipollution devices. Each unit of each product must go through exactly one antipollution device. The first device removes 40%, 60%, and 55% of pollutant types A, B, and C, respectively, regardless of the product made. The second device removes 30%, 0%, and 65% of pollutants A, B, and C, respectively, for product 1 and 20%, 0%, and 80% of pollutant types A, B, and C, respectively, for product 2. The first device reduces profit per thousand gallons manufactured daily by \$5, regardless of the product; similarly, the second device reduces profit by \$7 per thousand gallons manufactured daily, regardless of the product. A sales commitment dictates that at least 8000 gallons of product 1 and 5000 gallons of product 2 be produced per day. Formulate the appropriate LP model.

1.3.4. A manufacturing company produces, at n plants, a small component and distributes the component to m wholesalers. Sales forecasts indicate that monthly deliveries will be D_j units to wholesaler j. The monthly production capacity at plant i is u_i $(i = 1, \ldots, n)$. The direct cost of producing each unit is w_i at plant i. The transportation cost of shipping a unit from plant i to wholesaler j is c_{ij}. Formulate a linear programming model to indicate optimal production amounts at each plant and to show how many components each plant supplies each wholesaler.

1.4. THE DEVELOPMENT OF LINEAR PROGRAMMING

Prior to World War II, optimization models were solved with the methods of differential calculus. These techniques had been successfully applied to problems arising in economics, engineering, and the pure sciences for over 200 years. However, these methods were found to be inadequate for solving certain optimization problems (such as linear programming problems) that often arise in present-day government, military, and industrial organizations.

It was during World War II that a systematic research program was initiated by the military and government to develop new methods for the optimization of these large scale problems. George B. Dantzig, a member of the U.S. Air Force at that time, developed the theory of linear programming to address some of those problems. The specific solution procedure that he developed is called the *simplex algorithm*. It is a step-by-step procedure for solving all linear programming problems in a finite number of arithmetic operations.

Since Dantzig's initial work in 1951, progress in the development and application of linear programming has been very rapid. Much work has been done to improve the computational efficiency of the simplex algorithm, thus enabling organizations to solve larger and larger problems. Indeed, models containing thousands of variables are solved routinely. Also, many variants of the simplex algorithm have been developed. Each has its inherent advantages, disadvantages, and special uses. Yet none of these new procedures has replaced the straightforward approach of the simplex algorithm.

In addition, much effort has been directed toward developing and extending the theory of linear programming. This in turn has facilitated the solution of mathematical models that are far more complex than the linear programming ones. For instance, the property of divisibility can be replaced by the requirement that the variables assume integer values. The resulting type of problem is known as an integer programming problem. Another class of problems arises when the properties of proportionality and additivity are relaxed. In this case, the resulting objective function and constraints can be *nonlinear*, thus giving rise to the study of nonlinear programming problems. Collectively, the study of linear, integer, nonlinear, and similar problems is referred to as *mathematical programming*.

The major emphasis of this text is on the theory of linear programming, the computational implementation of the simplex algorithm (and many of its variants), and the methodology by which it works. Specifically, Chapter 2 provides the geometric and conceptual motivation behind the simplex algorithm. Chapters 3 and 4 present the mathematical thought process and background material that are needed for understanding the subsequent theory. Chapters 5–9 develop the algebraic, computational, and theoretical details of the simplex algorithm and several of its variants. Then, Chapters 10 and 11 illustrate how certain linear programming problems have special structure that can be exploited to improve the efficiency of the simplex algorithm.

DISCUSSION

On the History of Linear Programming and Related Topics

As was mentioned in the text, George Dantzig [4, 5] was responsible for identifying the structure and nature of linear programming problems and for developing the simplex algorithm. No better account of that development exists than that written by Dantzig himself [7]. Only after numerous years of research did Dantzig finally write a comprehensive reference book on linear programming [6].

Since its inception, enormous effort has gone into extending the theory, applications, and uses of linear programming to several other areas of optimization. A partial list of these areas together with one or two appropriate references will be supplied here:

1. nonlinear programming [1, 2],
2. integer programming [15],
3. goal programming [3],
4. large-scale programming [12],
5. network flow theory [11],
6. applications of linear programming [14],
7. economics [9],
8. complementary pivot theory [8],
9. stochastic programming [16],
10. game theory [10, 13].

REFERENCES

1. M. Avriel, *Nonlinear Programming Analysis and Methods*, Prentice-Hall, Englewood, NJ, 1976.
2. M. S. Bazaraa and C. M. Shetty, *Nonlinear Programming Theory and Algorithms*, Wiley, New York, 1977.
3. V. Chankong and Y. Haimes, *Multiobjective Decision Making: Theory and Methodology*, North-Holland, New York, 1983.
4. G. B. Dantzig, "Programming in a Linear Structure," Report of the Comptroller, U.S. Air Force, Washington, DC, 1948.
5. G. B. Dantzig, "Maximization of a Linear Function of Variables Subject to Linear Inequalities," in *Activity Analysis of Production and Allocation* (T. C. Koopmans, ed.), Cowles Commission Monograph No. 13, Wiley, New York, 1951, pp. 339–347.

6. G. B. Dantzig, *Linear Programming and Extensions*, Princeton University Press, Princeton, NJ, 1963.
7. G. Dantzig, "Reminiscences About the Origins of Linear Programming," Technical Report, SOL 81-5, Department of Operations Research, Stanford University, Stanford, CA, 1981.
8. B. C. Eaves, "A Short Course in Solving Equations with PL Homotopies," *SIAM-AMS Proc.* 9, 73–143 (1976).
9. M. D. Intriligator, *Mathematical Optimization and Economic Theory*, Prentice-Hall, Englewood Cliffs, NJ, 1971.
10. E. L. Kaplan, *Mathematical Programming and Games*, Wiley, New York, 1982.
11. J. Kennington and R. Helgason, *Algorithms for Network Programming*, Wiley, New York, 1981.
12. L. Lasdon, *Optimization Theory for Large Systems*, Macmillan, New York, 1970.
13. A. Rapoport, *N-Person Game Theory: Concepts and Applications*, University of Michigan Press, Ann Arbor, 1970.
14. J. Saha and H. Salkin, *Studies in Linear Programming*, North-Holland, Amsterdam, 1975.
15. H. Salkin, *Integer Programming*, Addison-Wesley, Reading, MA, 1975.
16. J. K. Sengupta, *Stochastic Programming: Methods and Applications*, North-Holland, New York, 1972.

Chapter 2

Geometric Motivation

The purpose of this chapter is to examine the geometry of linear programming problems having only two or three variables. This will provide insight into designing a finite algorithm for solving all linear programming problems. Any such algorithm must eventually be described in algebraic terms because computers work with algebra rather than geometry. The simplex algorithm is an algebraic translation of the geometry that will be presented here.

2.1. A NORMAL EXAMPLE

In this section, the geometry of an LP having two variables will be examined in detail, and a solution will be found by geometric means. Consider the following example.

EXAMPLE 2.1

$$\begin{array}{llr}
\text{Minimize} & 3x_1 + x_2 & \\
\text{subject to} & 5x_1 - 4x_2 \leq 14 & \text{(a)} \\
 & x_1 - 4x_2 \leq -2 & \text{(b)} \\
 & 2x_1 + x_2 \geq 5 & \text{(c)} \\
 & 6x_1 - x_2 \geq 3 & \text{(d)} \\
 & 4x_1 + 9x_2 \leq 60 & \text{(e)} \\
 & x_1, \quad x_2 \geq 0 & \text{(f)}
\end{array}$$

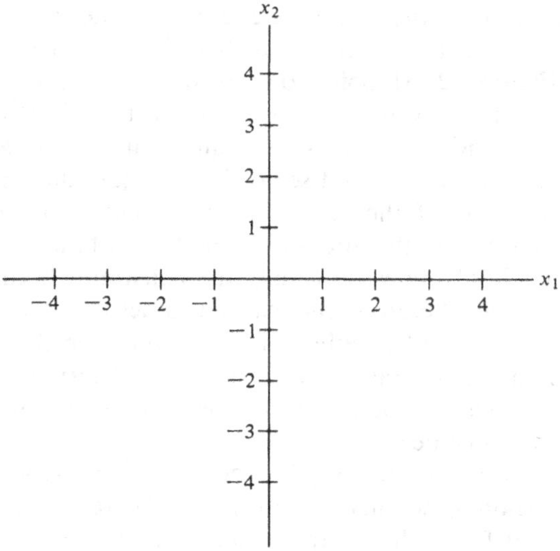

FIGURE 2.1. Graph of the x_1 and x_2 axes.

In the above LP, the variables x_1 and x_2 can assume many conceivable values. To "geometrize" all of these possible values, a graph can be drawn in the plane. The graph consists of a horizontal line (the x_1 axis) that corresponds to the values for x_1 and a vertical line (the x_2 axis) that corresponds to the values for x_2 as in Figure 2.1. The point at which the two axes meet is called the *origin*, and it corresponds to the values of $x_1 = x_2 = 0$. For any (numerical) values of x_1 and x_2, a (geometric) point corresponding to these values can be located in the graph. Similarly, for any point in the graph, the corresponding values for x_1 and x_2 can be found.

Recall that, in order to solve the LP, you must select values for x_1 and x_2 that

1. satisfy all of the constraints and
2. yield the smallest possible value of the objective function.

The geometric method for doing this is best understood by concentrating first on the constraints and subsequently on the objective function.

Graphing the Constraints

To determine which values of x_1 and x_2 satisfy all of the constraints, consider one constraint at a time. Each constraint by itself will permit certain feasible values for x_1 and x_2 (those that satisfy the constraint), while ruling out other infeasible values (those that do not satisfy the constraint). To see geometrically which values are feasible for the first constraint

$5x_1 - 4x_2 \leq 14$, plot on the graph the equation of the straight line $5x_1 - 4x_2 = 14$ obtained by replacing the inequality sign by an equality sign, as is illustrated in Figure 2.2. All points on this line give rise to feasible values for x_1 and x_2, as will every point on one of the two sides of this line. The question is which side. To answer the question, choose any point in the graph that is not on the line and see if the corresponding x_1 and x_2 values satisfy the constraint. If they do, then this point is on the feasible side; otherwise the point is on the infeasible side. For instance, consider the point $x_1 = 0$, $x_2 = 0$. These values do satisfy the constraint because $5x_1 - 4x_2 = 5(0) - 4(0) = 0 \leq 14$. Therefore, the feasible values for this constraint consist of the line $5x_1 - 4x_2 = 14$ together with everything on the same side of this line as the origin. In general, the easiest way to determine the feasible side of the line is to consider the origin. However, if the origin lies on the line, some other point must be selected.

Remember that you are looking for those values of x_1 and x_2 that satisfy all of the constraints, not just the first one. Therefore, the above process must be repeated for each constraint, and then those values of x_1 and x_2 that simultaneously satisfy all of the constraints must be identified. The resulting set of points constitutes what is called the *feasible region*. The feasible region corresponding to constraints (a)–(f) in Example 2.1 is depicted in Figure 2.3. Any point lying inside the feasible region is called a feasible point and gives rise to values for x_1 and x_2 that satisfy all of the constraints.

FIGURE 2.2. Feasible values for $5x_1 - 4x_2 \leq 14$.

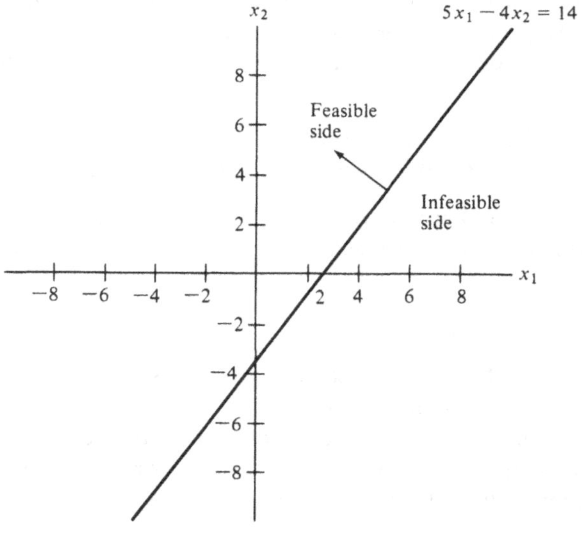

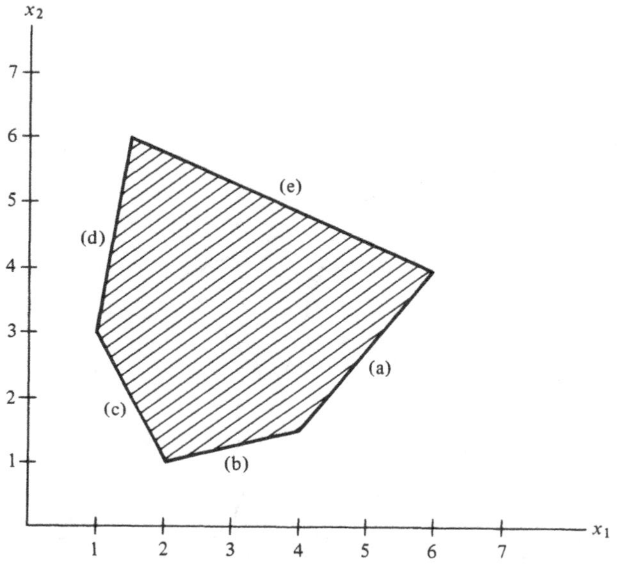

FIGURE 2.3. Feasible region of Example 2.1.

Graphing the Objective Function

Now a feasible point that minimizes the objective function must be found. Begin by choosing any point inside the feasible region and computing the value of the objective function at that point. In Example 2.1, suppose that the point $x_1 = 4$, $x_2 = 4$ is chosen, and hence the objective function value at this point is $3x_1 + x_2 = 16$. It is necessary to determine if there is any other point inside the feasible region that provides a smaller value of the objective function. To find out, first check if there are any other feasible points that yield the same objective value as the current one. In Example 2.1, this means that you must see if there are any other feasible values of x_1 and x_2 for which $3x_1 + x_2 = 16$. Hence you are led to draw the line $3x_1 + x_2 = 16$ through the (already chosen) point $x_1 = 4$, $x_2 = 4$, as is done in Figure 2.4.

Any point on this line will produce an objective value of 16. By experimenting, you can see that on one side of the line the values for x_1 and x_2 produce smaller values of the objective function, whereas on the other side they produce larger values. Since the objective function of Example 2.1 is to be minimized, you would like to determine which side of the line produces the smaller values. As before, this can be accomplished by choosing a point that is not on the line. If the new point produces a smaller value for the objective function, then all points on this same side of the line yield smaller

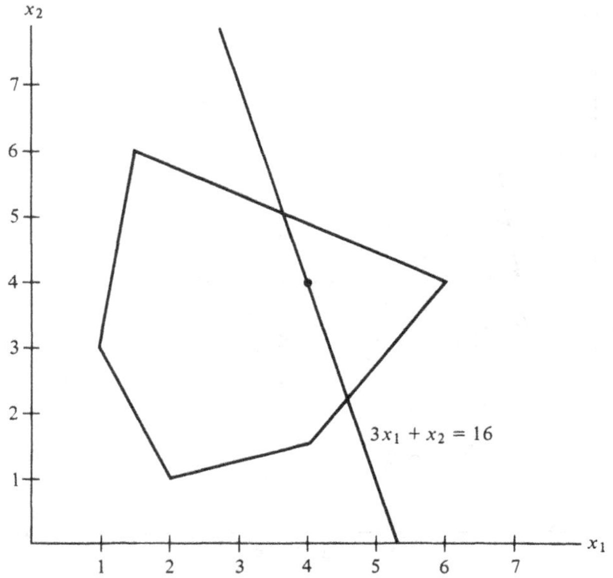

FIGURE 2.4. Objective function through $x_1 = 4$, $x_2 = 4$.

objective values. Otherwise, the points on the opposite side of the line produce the smaller values.

Once the desirable side of the objective function line has been determined, you must look on that side for a feasible point. If such a point can be found, then its x_1 and x_2 values yield a smaller value for the objective function. In Example 2.1, the point $x_1 = 3$, $x_2 = 2$ lies inside the feasible region and produces an objective value of 11, which is smaller than the previous objective value of 16. The entire process can now be repeated by drawing the new line $3x_1 + x_2 = 11$ through the point $x_1 = 3$, $x_2 = 2$ as is done in Figure 2.5.

From Figure 2.5 you can see that the new line is parallel to the old one. In other words, by moving the original line parallel to itself until reaching the second line, every feasible point on the second line will yield a smaller objective value than every point on the first line. In fact, there is no reason to stop at the second line! Why not continue moving the original line parallel to itself until it is just about to leave the feasible region? When you can no longer move the line, any feasible point lying on this last line will minimize the objective function and hence solve the LP! Such a point is said to be *optimal.* For Example 2.1, this graphical solution procedure is illustrated in Figure 2.6. The optimal point is $x_1 = 1$, $x_2 = 3$, and the value of the objective function is 6.

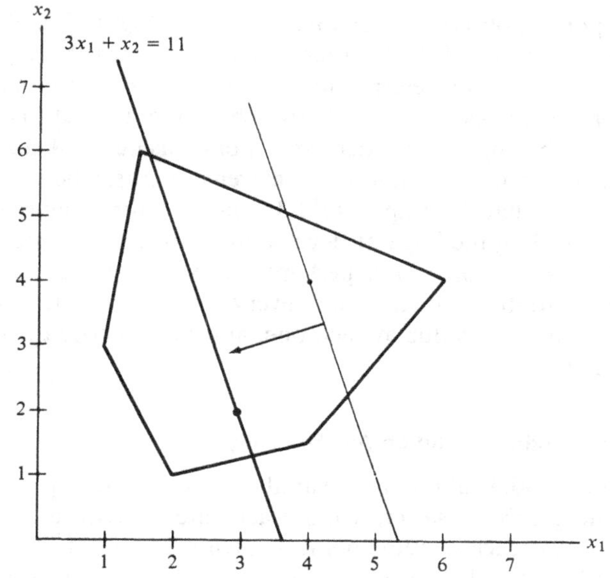

FIGURE 2.5. Objective function through $x_1 = 3$, $x_2 = 2$.

FIGURE 2.6. Graphical solution of Example 2.1.

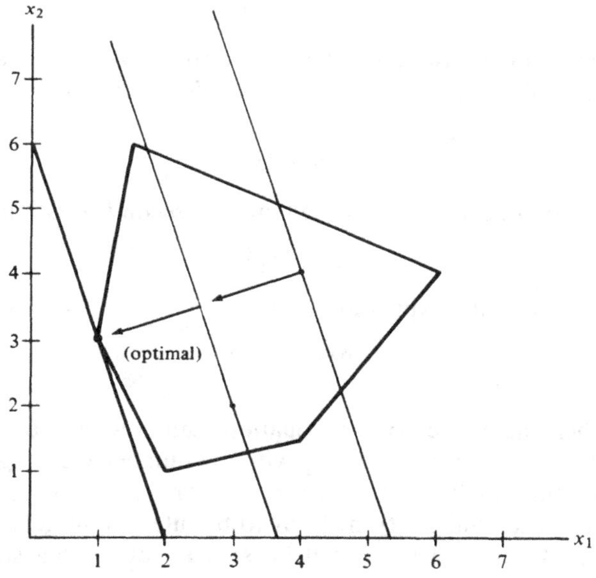

A most important observation can be made from Figure 2.6. The solution to this LP occurs at one of the "corner" points of the feasible region. These corner points are formally referred to as *extreme points*. If every LP has a solution at one of the extreme points (which may not always be the case), then, instead of having to consider every point inside the feasible region, you could restrict your attention to the extreme points. The importance of this observation is that there appears to be only a finite number of extreme points, thus providing the hope of developing a finite procedure for solving an LP. In particular, you could, perhaps, devise a method for listing the values of the variables at each and every extreme point, compute the corresponding objective value at each one, and then choose the one having the minimum value.

Computing the Optimal Solution Algebraically

One way to obtain the values of the variables at an extreme point is to read them from the graph; however, a more accurate procedure can be developed. Observe that each extreme point of Figure 2.6 lies at the intersection of two lines of the feasible region, i.e., the x_1 and x_2 values of the extreme point satisfy two linear equations. Hence the exact values for x_1 and x_2 can be found by solving a simultaneous system of two linear equations (corresponding to the two intersecting constraints) in the two unknowns x_1 and x_2. For instance, from Figure 2.6, the values for x_1 and x_2 at the optimal extreme point are found by solving

$$2x_1 + x_2 = 5$$

$$6x_1 - x_2 = 3$$

One way to do this is to use the first equation to express x_2 in terms of x_1. Specifically, on subtracting $2x_1$ from both sides of the first equation, it becomes

$$x_2 = 5 - 2x_1$$

Now this value for x_2 can be substituted into the second equation, obtaining

$$6x_1 - (5 - 2x_1) = 3$$

After collecting terms the expression becomes

$$8x_1 = 8$$

so $x_1 = 1$.

Finally, either one of the original equations can now be used to obtain a numerical value for x_2 by replacing x_1 with the number 1. Doing this in the first equation, one has $2x_1 + x_2 = 2 + x_2 = 5$, or $x_2 = 3$. Although in this case the values of x_1 and x_2 turned out to be integers, in general this will not happen. In Chapter 4 more will be said about solving simultaneous

TABLE 2.1. Extreme Points and Objective Values for Example 2.1

x_1	x_2	Objective value
4	$\frac{3}{2}$	$\frac{27}{2}$
2	1	7
1	3	6 (optimal)
$\frac{3}{2}$	6	$\frac{21}{2}$
6	4	22

linear equations, as that topic plays a central role in the simplex algorithm. For now, the above procedure will suffice, and it has been used in Example 2.1 to obtain the x_1 and x_2 values for all of the extreme points. The results are reported in Table 2.1.

The approach of listing all of the extreme points is not used in practice simply because, when a large number of variables and constraints are involved, the number of extreme points is so vast that even high-speed computers cannot begin to find them all in a reasonable amount of time. Fortunately, a better method has been found in the form of the simplex algorithm, and the next section introduces the basic idea.

EXERCISES

Solve each of the following problems by the graphical method. Compute the optimal solution and the objective function value algebraically.

2.1.1. Maximize $-x_1 + x_2$

subject to

$$\begin{aligned} 6x_1 - 2x_2 &\le 3 \\ -2x_1 + 3x_2 &\le 6 \\ 2x_1 + 3x_2 &\le 24 \\ x_1, \quad x_2 &\ge 0 \end{aligned}$$

2.1.2. Maximize $-4x_1 + 6x_2$

subject to

$$\begin{aligned} 6x_1 - 2x_2 &\le 3 \\ -2x_1 + 3x_2 &\le 6 \\ 2x_1 + 3x_2 &\le 24 \\ x_1 \qquad &\le 13 \\ x_2 &\le 9 \\ x_1, \quad x_2 &\ge 0 \end{aligned}$$

2.1.3. Minimize $x_1 + 2x_2$

subject to

$$\begin{aligned} 4x_1 + 3x_2 &\le 12 \\ -5x_1 + 4x_2 &\le 20 \\ x_1 + x_2 &\ge -3 \\ x_1, \quad x_2 &\ge 0 \end{aligned}$$

2.1.4. Minimize $3x_1 + 5x_2$

subject to

$$\begin{aligned} -3x_1 + 2x_2 &\le 6 \\ -x_1 + x_2 &\le 5 \\ -3x_1 + 8x_2 &\ge 12 \\ 3x_1 + 2x_2 &\ge 18 \\ x_1, \quad x_2 &\ge 0 \end{aligned}$$

2.1.5. Minimize $4x_1 + 5x_2$

subject to

$$\begin{aligned} -5x_1 + 4x_2 &\le 0 \\ x_1 + x_2 &\ge -3 \\ x_1 \qquad &\le 0 \end{aligned}$$

2.1.6. Algebraically compute all of the extreme points of the feasible region in Exercise 2.1.1.

2.2. FINITE IMPROVEMENT ALGORITHMS

In the previous section you observed that sometimes the solution to an LP occurs at an extreme point of the feasible region; this particular point was referred to as an optimal extreme point. If we assume, for the moment, that this is always the case, a simple yet effective approach to finding this optimal extreme point can be developed. It forms the very basis of the simplex algorithm and every other method to be studied in this text.

Imagine that each extreme point of the feasible region is a colored ball lying inside a closed box. Suppose further that the "optimal" balls (and only these) are colored red. How would you proceed to find a red one?

A reasonable way to start is by reaching into the box and taking out a ball. The next step is to see whether the ball is red (assuming that you are not color-blind). If it is, then of course you are done, so suppose that it is not. Cleverly now, you throw away the ball and reach into the box to pull out another one, and the whole process is repeated. Since there are only a finite number of balls in the box, eventually you must find a red ball, provided that there is one. As simple minded as this idea may seem, it is the very essence of how the simplex algorithm works. There are, however, several issues that complicate the situation in linear programming.

Perhaps the most serious complication is caused by the fact that, in linear programming, you are not allowed to throw away any ball (i.e., extreme point). Instead, each time you examine a ball, it must be returned to the box. The danger that this causes is serious, for it might happen that, at some later time, you reexamine the same ball over and over again. Thus the entire finiteness argument is gone. Unless this problem can be overcome, all appears lost. Fortunately, there is a solution, and it involves the objective function. To see how this can be done, remember that each ball is really an extreme point of the feasible region that gives rise to certain values for the variables. These values, in turn, can be used to produce a particular value for the objective function. Thus you can imagine that on each ball is painted a single number corresponding to the value of the objective function at the extreme point. These numbers are used to guarantee that, even though each ball is returned to the box, none is examined twice! The trick? Simple—you just have to make sure that each time a new ball is taken out of the box, its number is strictly less than the one being put back in. Now you should convince yourself that no ball is ever examined twice. Note that a strict decrease is necessary.

A second complication in linear programming is caused by the fact that, in reality, the balls are not colored! This minor change has created a major difficulty, for now you are going to have a very hard time determining which ball is the "red" one. Recall that the problem of finding the "red" ball is really that of finding the one with the smallest number (i.e., objective value) painted on it. Once you take a ball out of the box, how will you be able to tell whether the number painted on it is the smallest one? Comparing the number with that of every other ball is simply impractical due to the vast number of balls that are usually found in the box. Fortunately, an algebraic test involving only the current ball has been designed to circumvent this difficulty. The details will be presented in Chapter 5. For now, assume that such a test for optimality is available.

A third complication in linear programming is caused by the fact that the box containing the balls is actually the feasible region of the LP, so the question is exactly how to go about reaching into the box and taking out a ball. The answer lies in exploiting the geometry of the feasible region. Look again at Figure 2.3, and imagine that you are standing at one of the extreme points of the feasible region and that you have somehow tested to see that it is not the one with the smallest objective value. The simplex algorithm suggests that you follow one of the "edges" of the feasible region emanating from the current extreme point until you reach another extreme point. The question is which edge to follow. Recall that the edge must be chosen in such a way that the new extreme point will have a strictly smaller objective value associated with it. Surprisingly, the very fact that the current extreme point does not pass the test for optimality will indicate which edge to follow! Moreover, there is a simple algebraic formula that can be used to determine the values of the variables at the new extreme point. The details of the formula will be given in Chapter 5. A summary of the proposed algorithm follows.

Step 0. Initialization: Select an initial extreme point and go to step 1.

Step 1. Test for optimality: Perform a (it is hoped easy) test to determine if the current extreme point has the smallest possible objective value. If so, stop. Otherwise, go to step 2.

Step 2. Moving: Use the fact that the current extreme point is not an optimal one to find an edge of the feasible region that leads to another extreme point having a strictly smaller value of the objective function. By means of an algebraic formula, move along this edge to the new extreme point and determine the new values of the variables. Now return to step 1.

These steps embody the central idea behind the simplex algorithm and many other methods that move from one object (i.e., potential solution) to another getting better and better at each step. Such a procedure is called a *finite improvement algorithm.* It is worth noting that the key to a finite improvement algorithm lies in the ability to design a test for optimality with the property that, if the current object is not optimal, then that fact will indicate how to select a new object with a strictly better value. As long as there are a finite number of objects, the finite improvement algorithm must eventually find the best object.

It is now time to see if there are any geometric examples of linear programming problems that for some reason prevent you from finding a solution at an extreme point of the feasible region. Can you think of any? How do they affect the algorithm proposed above?

EXERCISES

2.2.1. Restate the finite improvement algorithm for a maximization problem. What changes must be made?

2.2.2. For a general finite improvement algorithm, explain what difficulties may arise that could conceivably cause the algorithm to fail in each step.

2.2.3. Consider the following algorithm for solving an LP:

Step 0. Initialization: Select an initial extreme point and go to step 1.

Step 1. Moving: Determine whether there is an edge of the feasible region leading from the current extreme point to another extreme point having a strictly smaller value of the objective function. If so, then move along this edge to the new extreme point and repeat the process.

(a) If no such edge exists, then what would you like to conclude about the current extreme point?

(b) Compare and contrast the algorithm of Exercise 2.2.3 with the one given in Section 2.2. What advantages and/or disadvantages does the algorithm of this exercise have?

2.3. AN INFEASIBLE EXAMPLE

In Section 2.1 you learned that the first step in solving a linear programming problem graphically is to determine the feasible region corresponding to the constraints. Then, by drawing the objective function together with the feasible region for Example 2.1, you saw that the optimal solution occurred at one of the extreme points of the feasible region. The importance of extreme points in solving linear programming problems was emphasized in Section 2.2 where the concept of a finite improvement algorithm was presented as a means of finding an optimal extreme point. Unfortunately, there are linear programming problems for which this approach simply will not work. This section deals with one such type of problem.

To see what can go wrong, consider a modified version of Example 2.1, consisting of constraints (a), (d), and (f), and in which the inequalities (a) and (d) are reversed.

EXAMPLE 2.2

$$\begin{aligned} \text{Minimize} \quad & 3x_1 + x_2 & \\ \text{subject to} \quad & 5x_1 - 4x_2 \geq 14 & \text{(a)} \\ & 6x_1 - x_2 \leq 3 & \text{(d)} \\ & x_1, \ x_2 \geq 0 & \text{(f)} \end{aligned}$$

Once again, the constraints are graphed one at a time by first drawing the equation of the straight line obtained by replacing the inequality sign with an equality sign. These lines will be the same as the corresponding ones in Example 2.1 since only the inequality signs have been reversed in Example 2.2. As a result, the feasible sides are reversed for constraints (a) and (d). Each constraint of Example 2.2 is plotted in Figure 2.7, with an arrow indicating the feasible side. The shaded region satisfies both (a) and (d). However, remember the nonnegativity constraints $x_1, x_2 \geq 0$, and notice that the shaded region of Figure 2.7 consists entirely of negative values for x_1 and x_2. Thus there are no values for x_1 and x_2 that simultaneously satisfy all of the constraints; i.e., there is no feasible region! For this reason, Example 2.2 is said to be infeasible.

In general, an *infeasible problem* is one in which there are no points

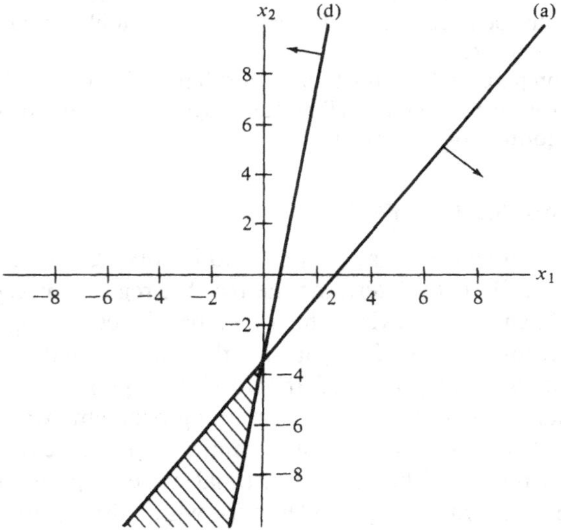

FIGURE 2.7. Constraints for Example 2.2.

simultaneously satisfying all of the constraints. Because such a problem has no feasible region, there are no extreme points. Hence it makes no sense to find the optimal solution since there is no feasible solution. In this case, the finite improvement algorithm breaks down at step 0 because an initial extreme point cannot be found. In the context of Section 2.2, an infeasible LP is analogous to reaching into the box for a ball (an extreme point) and finding the box empty! It is indeed fortunate that the simplex algorithm will detect this situation before attempting to find an optimal solution.

To illustrate another type of infeasible problem, consider a second variation of Example 2.1 in which the two inequalities (a) and (c) are reversed.

EXAMPLE 2.3

$$\begin{aligned}
\text{Minimize} \quad & 3x_1 + x_2 & \\
\text{subject to} \quad & 5x_1 - 4x_2 \geq 14 & \text{(a)} \\
& x_1 - 4x_2 \leq -2 & \text{(b)} \\
& 2x_1 + x_2 \leq 5 & \text{(c)} \\
& 6x_1 - x_2 \geq 3 & \text{(d)} \\
& 4x_1 + 9x_2 \leq 60 & \text{(e)} \\
& x_1, \quad x_2 \geq 0 & \text{(f)}
\end{aligned}$$

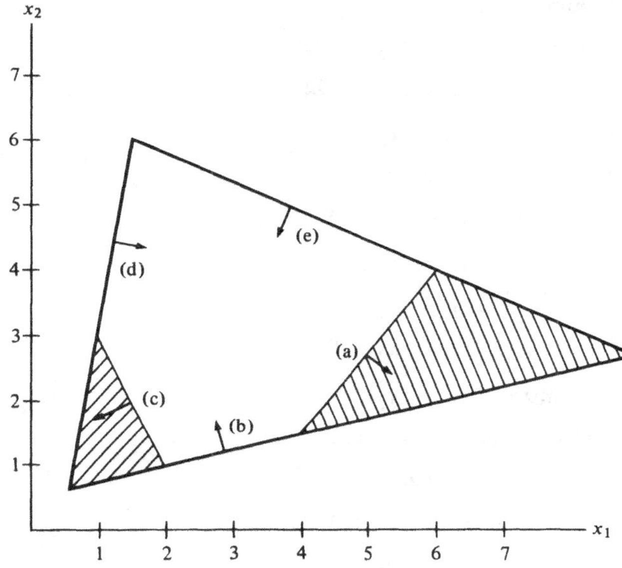

FIGURE 2.8. Constraints for Example 2.3.

For Example 2.3, the points that simultaneously satisfy the combination of constraints (b)–(e) constitute the shaded triangular region in the lower left portion of Figure 2.8. Likewise, the points that simultaneously satisfy the combination of constraints (a), (b), (d), and (e) constitute the shaded triangular region in the right portion of Figure 2.8. However, there are no points that satisfy all six constraints. Thus Example 2.3 is also an infeasible problem. Once again, there is no feasible region from which to select an initial extreme point for the finite improvement algorithm.

This section has presented infeasible linear programs. The next section describes the other type of linear programming problem for which you cannot find an optimal solution at an extreme point.

EXERCISES

Use the graphical method to indicate why each of the following LPs is infeasible.

2.3.1. Minimize $x_1 + x_2$

subject to
$$3x_1 - 5x_2 \geq 30$$
$$3x_1 + 2x_2 \leq 9$$
$$x_1, \quad x_2 \geq 0$$

2.3.2. Minimize $3x_1 + 4x_2$

$$\begin{aligned} \text{subject to } -6x_1 - 8x_2 &\geq -24 \\ 3x_1 + 4x_2 &\geq 24 \\ x_1, \quad x_2 &\geq 0 \end{aligned}$$

2.3.3. Minimize $3x_1 + 7x_2$

$$\begin{aligned} \text{subject to } x_1 - x_2 &\geq 4 \\ x_1 - 2x_2 &\leq 10 \\ -2x_1 - x_2 &\geq 2 \\ x_1, \quad x_2 &\geq 0 \end{aligned}$$

2.3.4. Maximize $2x_1 + 5x_2$

$$\begin{aligned} \text{subject to } 22x_1 - 6x_2 &\leq 21 \\ 5x_1 + 6x_2 &\geq 60 \\ -2x_1 + 3x_2 &\leq 3 \\ x_1, \quad x_2 &\geq 0 \end{aligned}$$

2.4. AN UNBOUNDED EXAMPLE

To illustrate the other type of problem for which an optimal solution cannot be found at an extreme point, return again to Example 2.1. This time, the original constraints are restored, with the exceptions that constraint (e) is omitted and that the objective function is changed. The new objective function is obtained by multiplying the objective function of Example 2.1 by -1.

EXAMPLE 2.4

$$\text{Minimize } -3x_1 - x_2$$

$$\begin{aligned} \text{subject to } 5x_1 - 4x_2 &\leq 14 \quad &(a) \\ x_1 - 4x_2 &\leq -2 \quad &(b) \\ 2x_1 + x_2 &\geq 5 \quad &(c) \\ 6x_1 - x_2 &\geq 3 \quad &(d) \\ x_1, \ x_2 &\geq 0 \quad &(f) \end{aligned}$$

There are no conflicts between any of these constraints, and the feasible region is shown in Figure 2.9.

To find the optimal solution by the graphical method, begin by choosing a feasible point and calculating the value of the objective function at this point, as was done for Example 2.1. For instance, at the feasible point $x_1 = 1$, $x_2 = 3$, the value of the objective function is -6. The objective

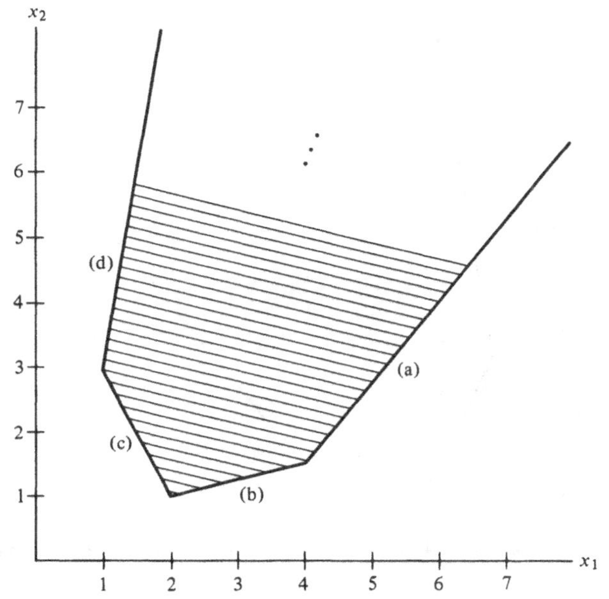

FIGURE 2.9. Feasible region for Example 2.4.

function $-3x_1 - x_2$ has been drawn through this point, as shown in Figure 2.10.

Points above this line produce smaller values of the objective function than points on the line. Therefore, the value of the objective function can be reduced by moving the line parallel to itself in the direction indicated by the arrow in Figure 2.10. This direction is exactly opposite to the direction of movement in Example 2.1 because the objective function has been multiplied by -1. Recall that the objective line should be moved as far as possible within the feasible region. However, in Example 2.4 there is nothing to limit the movement of the objective function line. In other words, no matter how small the value of the objective function, further reduction is always possible by moving the line further out into the feasible region. Example 2.4 is said to have an *unbounded solution*. The finite improvement algorithm breaks down at step 2 because movement along either edge (a) or edge (d) of the feasible region does not lead to a new extreme point. It is indeed fortunate that the simplex algorithm will detect this situation when it arises.

Linear programming problems with unbounded solutions represent the second type of problem for which a solution cannot be found at an extreme point of the feasible region. Thus you have seen three categories of linear programming problems:

1. those with an optimal solution,
2. those with infeasible constraints, and
3. those with an unbounded solution.

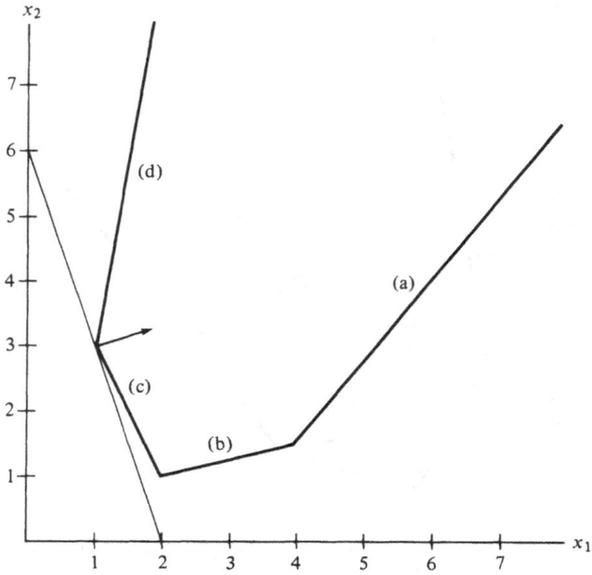

FIGURE 2.10. Objective function through $x_1 = 1$, $x_2 = 3$.

In fact, every LP falls into one of these three categories. The simplex algorithm is a finite algebraic method for determining which of these three conditions occurs and for producing an optimal solution when it exists. In problems arising from real world applications, an infeasible set of constraints or an unbounded solution is usually the result of poor problem formulation or the result of typing the wrong data into the computer.

The final section of this chapter will describe and illustrate the graphical solution procedure for an LP having three variables.

EXERCISES

Use the graphical method to illustrate that each of the following LPs is unbounded.

2.4.1. Maximize $3x_1 + 5x_2$

subject to

$$\begin{aligned} -3x_1 + 2x_2 &\leq 6 \\ -x_1 + x_2 &\leq 5 \\ -3x_1 + 8x_2 &\geq 12 \\ 3x_1 + 2x_2 &\geq 18 \\ x_1, \quad x_2 &\geq 0 \end{aligned}$$

2.4.2. Maximize $2x_1 + 2x_2$

subject to

$$-6x_1 + 10x_2 \geq 15$$
$$x_1 + x_2 \geq 7$$
$$x_1 \leq 7$$
$$x_2 \geq 0$$

2.4.3. Maximize $-6x_1 + 10x_2$

subject to

$$3x_1 - 5x_2 \leq 30$$
$$3x_1 + 2x_2 \geq 9$$
$$x_1, \quad x_2 \geq 0$$

2.5. THE GEOMETRY OF AN LP HAVING THREE VARIABLES

The geometric concepts introduced in Section 2.1 for an LP having two variables can be extended to problems having three variables. First, recall that the horizontal and vertical axes were used to represent the values of x_1 and x_2 when graphing the constraints and objective function. To represent a third variable x_3, another axis, which corresponds to the values of x_3, is added (as illustrated in Figure 2.11). A geometric point in this graph corresponds to numerical values for x_1, x_2, x_3 in the same manner as the graph for two variables.

FIGURE 2.11. Graph of the x_1, x_2, and x_3 axes.

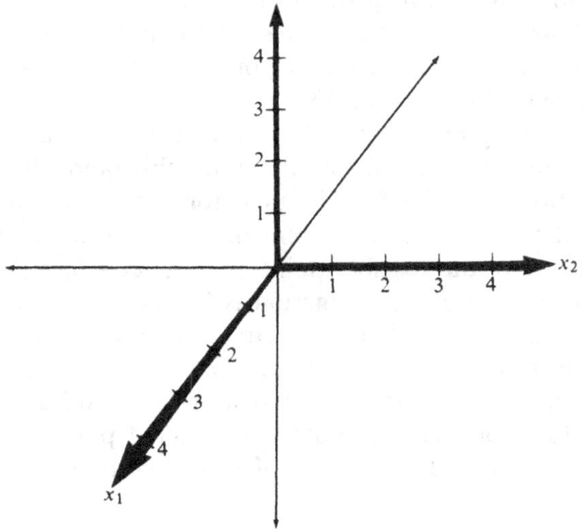

To illustrate the graphical solution to an LP having three variables, consider the following example.

EXAMPLE 2.5

$$\begin{array}{llr}
\text{Minimize} & x_1 - x_2 + x_3 & \\
\text{subject to} & 3x_2 + 2x_3 \le 27 & \text{(a)} \\
& 3x_1 + x_3 \le 24 & \text{(b)} \\
& 3x_2 - 2x_3 \ge 9 & \text{(c)} \\
& 3x_1 - x_3 \ge 15 & \text{(d)} \\
& x_3 \le 3 & \text{(e)} \\
& x_1, \quad x_2, \quad x_3 \ge 0 & \text{(f)}
\end{array}$$

As before, the feasible side of each constraint is determined one at a time. Recall that the feasible side for a constraint of the two-variable problem was represented graphically by a line together with all points on one side of the line. Similarly, the feasible side for a constraint of the three-variable problem is represented graphically by a plane together with all points on one side of the plane. For instance, the feasible side for constraint (a) of Example 2.5 consists of the plane $3x_2 + 2x_3 = 27$ (depicted in Figure 2.12) together with the side of the plane indicated by the arrow.

In Example 2.5, when the feasible sides of all of the constraints are determined, the resulting feasible region is now a solid three-dimensional region having depth, rather than the flat two-dimensional surface of the two-variable problems. For instance, the feasible region of a three-variable problem may look like a box, prism, or pyramid. In general, such a region is called a *polyhedron*. Each side, or *face*, of the polyhedron corresponds to one of the constraints. The feasible region defined by constraints (a)–(f) of Example 2.5 is shown in Figure 2.13.

Recall that any point inside the feasible region satisfies all of the constraints. To find an optimal solution, a feasible point that minimizes the objective function must be found. As before, the value of the objective function is first calculated at any point inside the feasible region. Suppose that in Example 2.5 the feasible point $x_1 = 7$, $x_2 = 6$, $x_3 = 1$ is selected; here the value of the objective function is 2. There are many other points that yield the same value of the objective function as this selected point. These points form the objective function plane, which is analogous to the objective function line for problems having two variables. In Example 2.5, the objective function plane through the selected point is defined by the equation $x_1 - x_2 + x_3 = 2$. The graph of this equation is the plane depicted in Figure 2.14.

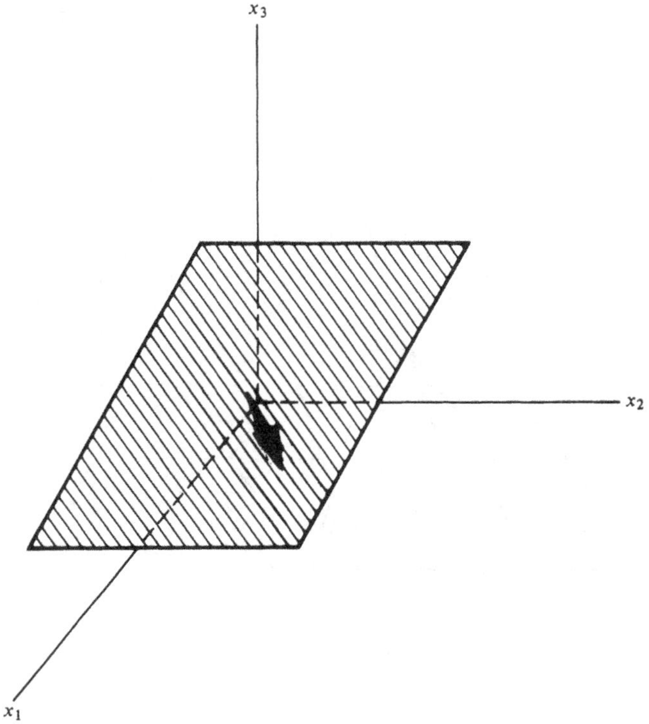

FIGURE 2.12. Feasible side of the constraint $3x_2 + 2x_3 \leq 27$.

FIGURE 2.13. Feasible region of Example 2.5.

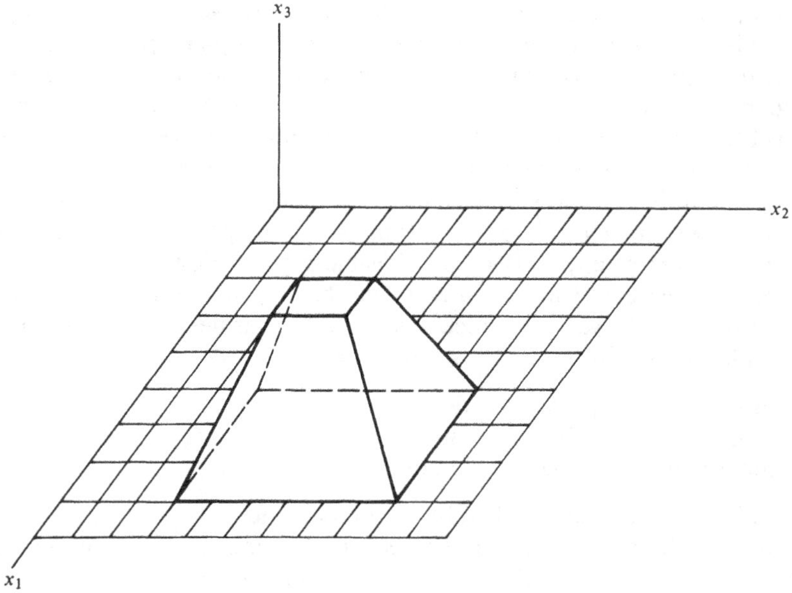

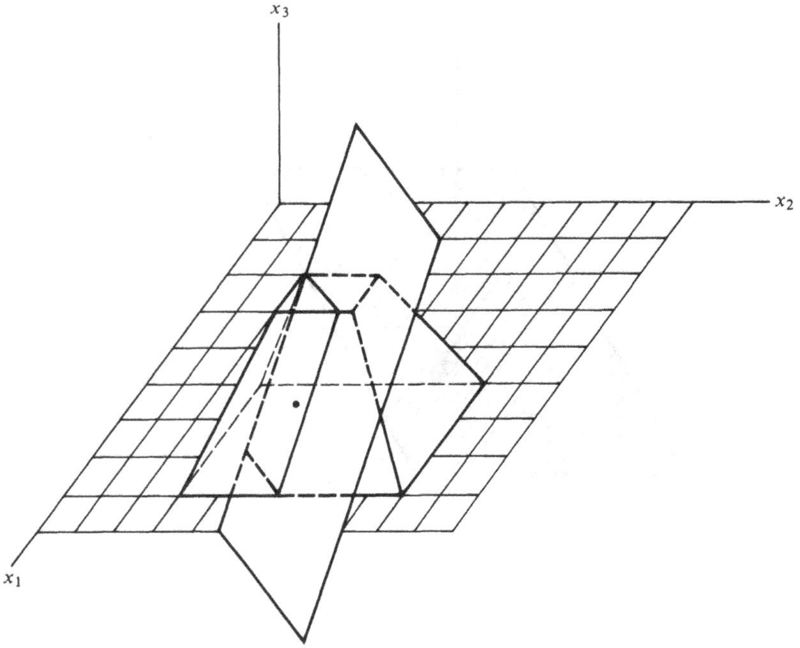

FIGURE 2.14. Objective function plane through the point $x_1 = 7$, $x_2 = 6$, $x_3 = 1$.

For a problem having two variables, the value of the objective function is reduced by moving the objective function line parallel to itself in the appropriate direction. Similarly, for the three-variable problem the objective value is reduced by moving the objective function plane parallel to itself in the appropriate direction until the plane is about to leave the feasible region. The point at which the objective function plane just touches the feasible region is the optimal solution. The graphical solution of the LP in Example 2.5 is shown in Figure 2.15. The optimal point is $x_1 = 5$, $x_2 = 9$, $x_3 = 0$, and the corresponding value of the objective function is -4.

As shown in Figure 2.15, an optimal solution again occurs at one of the corner, or extreme, points of the feasible region. However, an LP having three variables could also be infeasible or have an unbounded solution, as in the two-variable case. In fact, every LP, regardless of the number of variables, falls into one of three categories:

1. those with an optimal solution,
2. those with infeasible constraints, and
3. those with an unbounded solution,

as will be formally established in Chapters 5 and 6.

In both the two-variable and three-variable examples, the optimal point was an extreme point of the feasible region. In fact, this is true of LPs

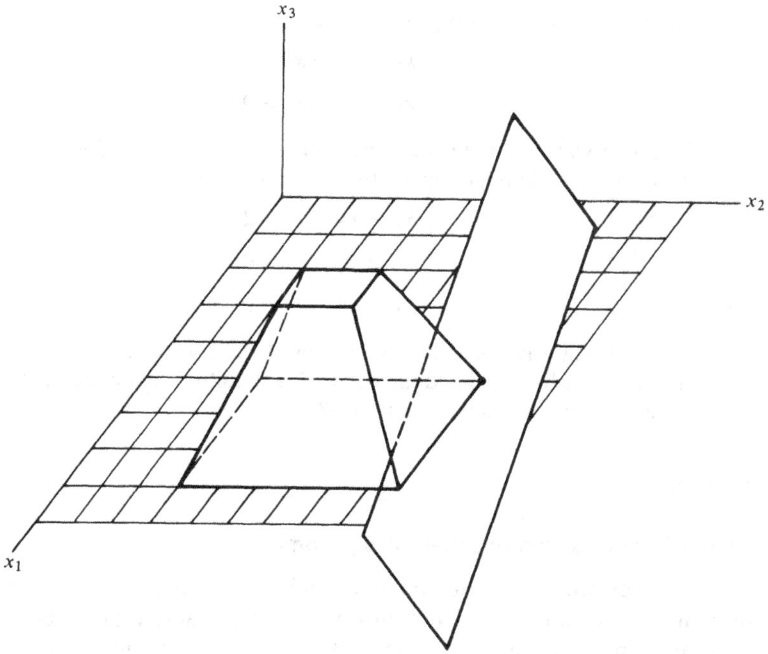

FIGURE 2.15. Graphical solution of Example 2.5.

having any number of variables. In other words, if an LP has an optimal solution, then it can be found by calculating the value of the objective function at every extreme point and then finding the one with the smallest objective value.

Although graphing the constraints and objective function is a bit more difficult in three dimensions than it is in two, a solution can still be obtained by the graphical method. For problems of higher dimension, however, the constraints and objective function cannot be drawn graphically. Also, the number of extreme points can become so large that finding all of them is impractical. Moreover, computers work with algebra, not geometry.

The simplex algorithm is a finite algebraic procedure for determining whether an LP is optimal, infeasible, or unbounded; in the first case, it produces an optimal solution. To "build" such a procedure, certain mathematical "tools" are needed. The purpose of the next two chapters is to provide these necessary tools.

EXERCISES

2.5.1. Compare and contrast the graphical solution procedures for solving LPs with two and three variables.

2.5.2. **(a)** Draw the feasible region given by the constraints

$$x_1 + x_2 + x_3 = 1$$
$$x_1, \quad x_2, \quad x_3 \geq 0$$

What can you say about the dimension of the feasible region?

(b) Draw the feasible region given by the constraints

$$x_1 + x_2 + x_3 = 2$$
$$x_1 - x_2 \qquad = 0$$
$$x_1, \quad x_2, \quad x_3 \geq 0$$

What can you say about the dimension of the feasible region?

(c) In general, what is the dimension of a feasible region defined by n variables and k equality constraints?

DISCUSSION

The Geometric Formalities of Linear Programming

In this chapter certain geometric concepts such as extreme points and edges were introduced. Mathematical definitions for these and related concepts can be found in Stoer and Witzgall [3]. Moreover, there is a formal correspondence between these geometric concepts and the algebraic steps of the simplex algorithm. For example, it can be shown that if an LP has an optimal solution, then there is an optimal solution at an extreme point of the feasible region. Similarly, it can be shown that there are a finite number of extreme points. However, a formal development of these geometric and algebraic relationships is beyond the scope of this text. Appropriate references would be the books by Bazaraa and Shetty [1] and by Stoer and Witzgall [3].

The mathematical notion of convexity and its role in optimization are described in detail by Rockafellar [2]. Simply stated, it is the convexity of the feasible region that ensures that, when an extreme point fails the test for optimality in the simplex algorithm, a suitable direction leading to a better extreme point can be found.

REFERENCES

1. M. S. Bazaraa and C. M. Shetty, *Foundations of Optimization*, Springer, New York, 1976.
2. R. T. Rockafellar, *Convex Analysis*, Princeton University Press, Princeton, NJ, 1970.
3. J. Stoer and C. Witzgall, *Convexity and Optimization in Finite Dimensions I*, Springer, New York, 1970.

Chapter 3

Proof Techniques

The objective of mathematicians is to discover and to communicate certain truths. Mathematics is the language of mathematicians, and a proof is a method of communicating a mathematical truth to another person who also "speaks" the language. To understand or present a proof, you must learn a new language, a new method of thought. This chapter explains much of the basic "grammar" you will need; however, no exercises on this material will be included here. For a further discussion of this material, see the reference at the end of this chapter.

The approach of this chapter is to categorize and to explain the various techniques that are used in proofs. One objective is to teach you how to read and how to understand a written proof by identifying the techniques that have been used. This chapter also describes when each technique is likely to be used and why. It is often the case that a correct technique can be chosen based on certain key words that appear in the problem under consideration.

3.1. THE TRUTH OF IT ALL

This section explains the type of relationships to which proofs can be applied. Given two statements A and B each of which may be either true or false, a fundamental problem of interest in mathematics is to show that

if A is true, then B is true

A proof is a formal method for accomplishing this task. As you will soon discover, the particular form of A and B can often indicate a way to proceed.

A mathematical proof is a convincing argument that is expressed in the language of mathematics. Thus a proof should contain enough mathematical detail to be convincing to the person(s) to whom the proof is addressed.

In order to do a proof, you must know exactly what it means to show that if A is true, then B is true. Statement A is often called the *hypothesis* and B the *conclusion*. For brevity, the statement "if A is true, then B is true" is shortened to "if A then B," or simply "A implies B." Mathematicians are often very lazy when it comes to writing. As such, they have developed a symbolic shorthand. For instance, a mathematician would often write "$A \Rightarrow B$" instead of "A implies B." This symbolic notation will not be used in this text.

It seems reasonable that the conditions under which "A implies B" is true should depend on whether A and B themselves are true. Consequently, there are four possible cases to consider:

1. A is true, and B is true.
2. A is true, and B is false.
3. A is false, and B is true.
4. A is false, and B is false.

The truth of "A implies B" in each of these four cases is given in Table 3.1. According to Table 3.1, when trying to show that "A implies B" is true, you can assume that the statement to the left of the word "implies" (namely, A) is true. Your goal is to conclude that the statement to the right (namely, B) is true. Note that a proof of the statement "A implies B" is not an attempt to verify whether A and B themselves are true, but rather to show that B is a logical result of having assumed that A is true.

In general, your ability to show that B is true will depend very heavily on the fact that you have assumed A to be true, and, ultimately, you will have to discover the linking relationship between A and B. Doing so will require

TABLE 3.1. The Truth of "A Implies B"

A	B	A implies B
True	True	True
True	False	False
False	True	True
False	False	True

a certain amount of creativity on your part. The proof techniques presented here are designed to get you started and to guide you along the path.

Hereafter A and B will be statements that are either true or false. The problem of interest will be that of showing "A implies B."

3.2. THE FORWARD–BACKWARD METHOD

The purpose of this section is to describe one of the fundamental proof techniques: the *forward–backward method*. Special emphasis is given to the material of this section because all of the other proof techniques will use the forward–backward method.

The first step in any proof requires recognizing statements A and B. Everything that you are assuming to be true (i.e., the hypothesis) is A; everything that you are trying to prove (i.e., the conclusion) is B. Consider the following example.

EXAMPLE 3.1: If the right triangle XYZ with sides of lengths x and y and hypotenuse of length z has an area of $z^2/4$, then the triangle XYZ is isosceles (see Figure 3.1).

OUTLINE OF PROOF. In this example one has the following statements:

A. The right triangle XYZ with sides of lengths x and y and hypotenuse of length z has an area of $z^2/4$.
B. The triangle XYZ is isosceles.

Recall that when proving "A implies B" you can assume that A is true, and you must somehow use this information to reach the conclusion that B is true. In attempting to figure out just how to reach the conclusion that B is true, you will be going through a *backward process*. On the other hand, when you make specific use of the information contained in A, you will be going through a *forward process*. Both of these processes will be described in detail.

In the backward process you begin by asking, "How or when can I conclude that statement B is true?" The very manner in which you phrase

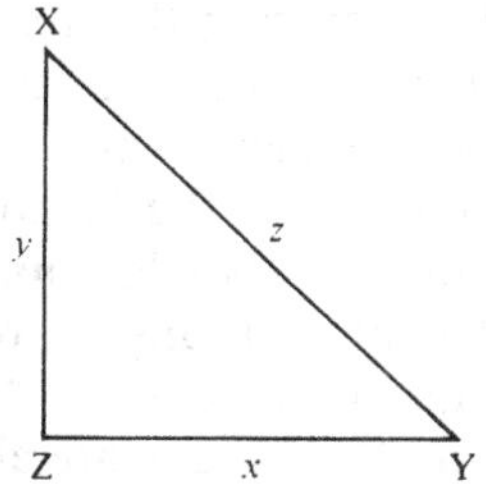

FIGURE 3.1. Right triangle XYZ.

this question is critical since you must eventually be able to answer it. The question should be posed in an abstract way. For Example 3.1 the correct abstract question is "How can I show that a triangle is isosceles?" Although it is true that you want to show that the particular triangle XYZ is isosceles, by asking the abstract question you call on your general knowledge of triangles, clearing away irrelevant details (such as the fact that the triangle is called XYZ instead of ABC), thus allowing you to focus on those aspects of the problem that really seem to matter. The question obtained from statement B in such problems will be called the *abstraction question*. A properly posed abstraction question should contain no symbols or other notation from the specific problem under consideration. The key to many proofs is formulating a correct abstraction question.

In any event, once you have posed the abstraction question, the next step in the backward process is to answer it. Returning to the example, how can you show that a triangle is isosceles? Certainly one way is to show that two of its sides have equal length. Referring to Figure 3.1, you should show that $x = y$. Observe that answering the abstraction question is a two-phase process. First you give an abstract answer: To show that a triangle is isosceles, show that two of its sides have equal length. Next you apply this answer to the specific situation: in this case, to show that two of its sides have equal length means to show that $x = y$, not that $x = z$ or $y = z$. The process of asking the abstraction question, answering it abstractly, and then applying that answer to the specific situation will be referred to as the *abstraction process*.

The abstraction process has given you a new statement, B1, with the property that, if you could show that B1 is true, then B would be true. For the example above, the new statement is

B1. $x = y$

Once you have the statement B1, all of your efforts must now be directed toward reaching the conclusion that B1 is true, for then it will follow that B is true. How can you show that B1 is true? Eventually you will have to make use of the assumption that A is true. If you were solving this problem, you would most likely do so now, but, for the moment, let us continue working backward by repeating the abstraction process on the statement B1. This will illustrate some of the difficulties that can arise in the backward process. Can you pose the new abstraction question?

Since x and y are the lengths of two sides of a triangle, a reasonable abstraction question would appear to be "How can I show that the lengths of two sides of a triangle are equal?" A second perfectly reasonable abstraction question would be "How can I show that two real numbers are equal?" After all, x and y are also real numbers. One of the difficulties that can arise in the abstraction process is the possibility of more than one abstraction question. Choosing the correct one is more of an art than a science. In fortunate circumstances there will be only one obvious abstraction question. In other cases you may have to proceed by trial and error. This is where your intuition, insight, creativity, experience, diagrams, and graphs can play an important role. One general guideline is to let the information in A (which you are assuming to be true) help you to choose the question, as will be done in this case.

Regardless of which question you finally settle on, the next step will be to answer it, first in the abstract and then in the specific situation. Can you do this for the two abstraction questions above? For the first one, you might show that two sides of a triangle have equal length by showing that the angles opposite them are equal. For triangle XYZ of Figure 3.1 this would mean that you have to show that angle X equals angle Y. A cursory examination of the contents of statement A does not seem to provide much information concerning the angles of triangle XYZ. For this reason the other abstraction question will be chosen.

Now one is faced with the question "How can I show that two real numbers (namely, x and y) are equal?" One answer to this question would be to show that the difference of the two numbers is 0. Applying this answer to the specific statement B1 means you would have to show that $x - y = 0$. Unfortunately, there is another perfectly acceptable answer: Show that the first number is less than or equal to the second number and also that the second number is less than or equal to the first number. Applying this answer to the specific statement B1, you would have to show that $x \leq y$ and $y \leq x$. Thus a second difficulty can arise in the backward process. Even if you choose the correct abstraction question, there may be more than one answer to it. Moreover, you might choose an answer that will not permit you to complete the proof. For instance, associated with the abstraction question "How can I show that a triangle is isosceles?" is the answer "Show that the triangle is equilateral." Of course it will be impossible to show that triangle XYZ of Example 3.1 is equilateral, since one of its angles is 90°.

Let us return to the abstraction question "How can I show that two real numbers (namely, x and y) are equal?" Suppose, for the sake of argument, that you choose the answer of showing that their difference is 0. Once again, the abstraction process has given you a new statement, B2, with the property that if you could show that B2 is true, then in fact B1 would be true, and hence so would B. Specifically, the new statement is

B2. $x - y = 0$

Now all of your efforts must be directed toward reaching the conclusion that B2 is true. You must ultimately make use of the information in A, but, for the moment, let us continue once more with the abstraction process applied to the new statement B2.

One abstraction question is "How can I show that the difference of two real numbers is 0?" At this point it may seem that there is no reasonable answer to this question. Yet another problem can arise in the abstraction process. The abstraction question might have no apparent answer! Do not despair—all is not lost. Remember that when proving "A implies B" you are allowed to assume that A is true. Nowhere have you made use of this fact. It is time to do so through the forward process.

The forward process involves starting with the statement A, which you assume to be true, and deriving from it some other statement, A1, that you know to be true as a result of A being true. It should be emphasized that the statements derived from A are not haphazard. Rather, they are directed toward linking up with the last statement derived in the backward process. This last statement should act as the guiding light in the forward process. Let us return to Example 3.1, keeping in mind that the last statement obtained in the backward process was "$x - y = 0$."

For Example 3.1 statement A is "The right triangle XYZ with sides of length x and y and hypotenuse of length z has an area of $z^2/4$." One fact that you know (or should know) as a result of A being true is that $xy/2 = z^2/4$, because the area of a right triangle is one-half the base times the height, in this case $xy/2$. You have obtained a new statement:

A1. $xy/2 = z^2/4$

Another useful statement follows from A by the Pythagorean theorem:

A2. $x^2 + y^2 = z^2$

The forward process can also combine and use the new statements to produce more true statements. For instance, it is possible to combine A1 and A2 by replacing z^2 in A1 with $x^2 + y^2$ from A2, obtaining

A3. $xy/2 = (x^2 + y^2)/4$

One of the problems with the forward process is that it is also possible to generate some useless statements: for instance, "angle X is less than 90°." There are no specific guidelines for producing new statements, but keep in mind that the forward process is directed toward obtaining B2, "$x - y = 0$," which was the last statement derived in the backward process. It is for this reason that z^2 was eliminated from A1 and A2.

Continuing with the forward process, you should attempt to rewrite A3 to make it look more like B2. For instance, you can multiply both sides of A3 by 4 and subtract $2xy$ from both sides to obtain

A4. $x^2 - 2xy + y^2 = 0$

By factoring you can obtain

A5. $(x-y)^2=0$

One of the most common steps of the forward process is to rewrite statements in different forms, as was done in obtaining A4 and A5. For Example 3.1, the final step in the forward process (and in the entire proof) is to note from A5 that the only way $(x-y)^2$ can be 0 is for $x-y$ to be 0, thus obtaining precisely the statement "$x-y=0$." The proof is now complete since you started with the assumption that A is true and used it to derive the conclusion that B2, and hence B, is true.

It is interesting to note that the forward process ultimately produced the elusive answer to the abstraction question associated with B2: "How can I show that the difference of two real numbers is 0?"—which was to show that the square of the difference is 0.

Finally, you should realize that, in general, it will not be practical to write down the entire thought process that goes into a proof, for this would require far too much time, effort, and space. Instead, a highly condensed version is usually presented that often makes little or no reference to the backward process. For the problem above it might go something like this.

PROOF OF EXAMPLE 3.1. From the hypothesis and the formula for the area of a right triangle, the area of XYZ is $xy/2=z^2/4$. By the Pythagorean theorem, $x^2+y^2=z^2$, and on substituting x^2+y^2 for z^2 and performing some algebraic manipulations one obtains $(x-y)^2=0$. Hence $x=y$, and triangle XYZ is isosceles. □ (The "□" or some equivalent symbol is usually used to indicate the end of the proof.)

Sometimes the shortened proof will be partly backward and partly forward. For example:

PROOF OF EXAMPLE 3.1. The statement will be proved by establishing that $x=y$, which in turn is done by showing that $(x-y)^2=x^2-2xy+y^2=0$. But the area of the triangle is $xy/2=z^2/4$, so $2xy=z^2$. By the Pythagorean theorem, $z^2=x^2+y^2$; hence $x^2+y^2=2xy$, or $x^2-2xy+y^2=0$, as required. □

Unfortunately, these shortened versions are typically given in mathematics books, which makes proofs so hard to read. You should strive for the ability to read and to dissect a condensed proof. To do so, you will have to figure out which proof techniques are being used (since the forward–backward method is not the only one available). Then from what is written you will have to discover the thought process that went into the proof and, finally, be able to verify all of the steps involved. The more condensed the proof, the harder this process will be. In this textbook your life will be made substantially easier because an outline describing the proof techniques, methodology, and reasoning that was involved will precede each condensed proof. Out of necessity, these outlines will be more succinct than the one given in Example 3.1.

As a general rule, the forward–backward method is probably the first technique to try on a problem unless you have reason to use a different approach based on the form of B, as will be described shortly. In any case, you will gain much insight into the relationship between A and B.

3.3. DEFINITIONS AND MATHEMATICAL TERMINOLOGY

In the previous section you learned the forward–backward method and saw the importance of formulating and answering the abstraction question. One of the simplest yet most effective ways of answering an abstraction question is through the use of a definition, as will be explained in this section. In addition, you will learn some of the "vocabulary" of the language of mathematics.

A *definition* is nothing more than a statement that is agreed on by all parties concerned. You have already come across a definition in Section 3.1. There, we defined what it means for the statement "A implies B" to be true. Specifically, we agreed that it is true in all cases except when A is true and B is false.

Definitions are not made randomly. Usually they are motivated by a mathematical concept that occurs repeatedly. In fact, a definition can be viewed as an abbreviation that is agreed on for a particular concept. Take, for example, the notion of a "positive integer greater than 1 that is not divisible by any positive integer other than 1 and itself," which is abbreviated (or defined) as a *prime*. Surely it is easier to say "prime" than "positive integer greater than 1...," especially if the concept comes up frequently. Several other examples of definitions follow:

Definition 3.1. An integer n *divides* an integer m (written $n|m$) if $m = kn$ for some integer k.

Definition 3.2. A positive integer $p > 1$ is *prime* if the only positive integers that divide p are 1 and p.

Definition 3.3. A triangle is *isosceles* if two of its sides have equal length.

Definition 3.4. Two pairs of real numbers (x_1, y_1) and (x_2, y_2) are *equal* if $x_1 = x_2$ and $y_1 = y_2$.

Definition 3.5. An integer n is *even* if and only if its remainder on division by 2 is 0.

Definition 3.6. An integer n is *odd* if and only if $n = 2k + 1$ for some integer k.

Definition 3.7. A real number r is a *rational number* if and only if r can be expressed as the ratio of two integers p and q in which the denominator q is not 0.

Definition 3.8. Two statements A and B are *equivalent* if and only if A implies B and B implies A.

Definition 3.9. The statement A AND B is true if and only if A is true and B is true.

Definition 3.10. The statement A OR B is true in all cases except when A is false and B is false.

Observe that the words "if and only if" have been used in some of the definitions, although "if" is often (incorrectly) used instead. Some terms, such as "set" and "point," are left undefined. One could possibly try to define a set as a collection of objects, but to do so is impractical because the concept of an object is too vague. One would then be led to ask for the definition of "object," and so on, and so on. Such philosophical issues are beyond the scope of this book.

In the proof of Example 3.1, a definition was already used to answer an abstraction question. Recall the very first one: "How can I show that a triangle is isosceles?" Using Definition 3.3, in order to show that a triangle is isosceles one shows that two of its sides have equal length. Definitions are equally useful in the forward process. For instance, if you know that an integer n is odd, then, by Definition 3.6 you know that $n = 2k + 1$ for some integer k. Using definitions to work forward and backward is a common occurrence in proofs.

Just as a definition can be used in the forward and backward processes, so can a (previously proven) proposition, as will be shown in the next example.

EXAMPLE 3.2: If a right triangle RST with legs of lengths r and s and hypotenuse of length t satisfies $t = \sqrt{2rs}$, then triangle RST is isosceles (see Figure 3.2).

OUTLINE OF PROOF. The forward–backward method gives rise to the abstraction question "How can I show that a triangle (namely, RST) is isosceles?" One answer is to use Definition 3.3, but a second answer is also provided by the conclusion of Example 3.1, which states that triangle XYZ is isosceles. Perhaps the current triangle, RST, is also isosceles for the same reason that triangle XYZ is. In order to find out, it is necessary to see if

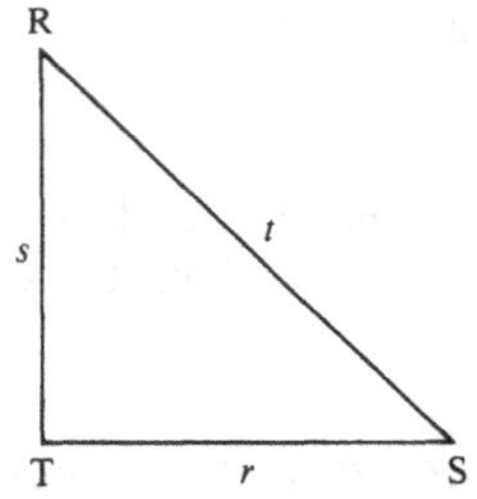

FIGURE 3.2. Right triangle RST.

RST also satisfies the hypothesis of Example 3.1, as triangle XYZ did, for then RST will also satisfy the conclusion and hence be isosceles.

In verifying the hypothesis of Example 3.1 for triangle RST, it is first necessary to "match up" the current notation with that of Example 3.1. To be specific, the corresponding lengths are $x = r$, $y = s$, and $z = t$. Thus, to check the hypothesis of Example 3.1 for the current problem, you must see if the area of triangle RST equals $t^2/4$, or, equivalently, since the area of triangle RST is $rs/2$, you must see if $rs/2 = t^2/4$.

The fact that $rs/2 = t^2/4$ will be established by working forward from the current hypothesis that $t = \sqrt{2rs}$. To be specific, on squaring both sides and dividing by 4, one obtains $rs/2 = t^2/4$, as desired. Do not forget to observe that the hypothesis of Example 3.1 also requires that triangle RST be a right triangle, which of course it is, as is stated in the current hypothesis.

Notice how much more difficult it would have been to match up the notation if the current triangle had been labeled WXY with legs of length w and x and hypotenuse of length y. Unfortunately, this overlapping notation can (and will) arise, and, when it does, it is particularly important to keep the symbols straight.

Terminology

Associated with a statement A is the statement NOT A. The statement NOT A is true when A is false, and vice versa. More will be said about the NOT of a statement in Section 3.10.

Three other statements related to "A implies B" are

1. "B implies A" (called the *converse*),
2. "NOT A implies NOT B" (called the *inverse*), and
3. "NOT B implies NOT A" (called the *contrapositive*).

Table 3.1 can be used to determine when each of these three statements is true. For instance, the contrapositive statement, "NOT B implies NOT A," is true in all cases except when the statement to the left of the word

TABLE 3.2. Truth Table for "NOT B implies NOT A"

A	B	NOT B	NOT A	A ⇒ B	NOT B ⇒ NOT A
True	True	False	False	True	True
True	False	True	False	False	False
False	True	False	True	True	True
False	False	True	True	True	True

"implies" (namely, NOT B) is true and the statement to the right of the word "implies" (namely, NOT A) is false. In other words, the contrapositive statement is true in all cases except when B is false and A is true, as shown in Table 3.2.

Note from Table 3.2 that the statement "NOT B implies NOT A" is true under the same conditions as "A implies B," that is, in all cases except when A is true and B is false. This observation gives rise to a new proof technique known as the contrapositive method, which will be described in Section 3.9. Truth tables similar to Table 3.2 can be derived for the converse and inverse statements but will not be given here.

A final statement of interest related to "A implies B" is the statement that A is equivalent to B, often written "A is true if and only if B is true" or, more simply, "A if and only if B." In mathematical notation one would write "A iff B" or "A ⇔ B." Whenever you are asked to show that "A if and only if B," you must show that "A implies B" and "B implies A."

This section has explained the meaning of many of the terms used in the language of mathematics. More importantly, it has shown how definitions and how previous propositions can often be used in the forward–backward method. Now it is time to learn more proof techniques.

3.4. QUANTIFIERS I: THE CONSTRUCTION METHOD

In the previous section you saw that a definition could successfully be used to answer an abstraction question. The next four sections provide you with several other techniques for formulating and answering an abstraction question that arises when B has a special form.

Two particular forms of B appear repeatedly throughout all branches of mathematics. They can always be identified by certain key words that appear in the statement. The first form has the words "there is" ("there are," "there exists"), whereas the second one has "for all" ("for each," "for every," "for any"). These two groups of words are referred to as *quantifiers*, and each one will give rise to its own proof technique. The remainder of this section deals with the *existential quantifier* "there is" and with the corresponding proof technique called the *construction method*. The *universal quantifier* "for all" and its associated proof technique is discussed in the next section.

The quantifier "there is" arises quite naturally in many mathematical statements. Recall that Definition 3.7 states that a rational number is a real number that can be expressed as the ratio of two integers in which the denominator is not zero. This definition could just as well have been written using the quantifier "there are."

Definition 3.11. A real number r is *rational* if and only if there are integers p and q with $q \neq 0$ such that $r = p/q$.

It is important to observe that the quantifier "there is" allows for the possibility of more than one such object, as is shown in the next definition.

Definition 3.12. An integer n is *square* if there is an integer k such that $n = k^2$.

Note that if an integer n (for example, $n = 9$) is square, then there are usually two values of k that satisfy $n = k^2$ (in this case, $k = 3$ and $k = -3$). More will be said in Section 3.11 about the issue of uniqueness (i.e., the existence of only one such object).

There are many other instances where an existential quantifier can and will be used, but from the examples above you can see that such statements always have the same basic structure. Each time the quantifier "there is," "there are," or "there exists" appears, the statement will have the following basic form:

There is an *object* with a *certain property*
such that *something happens*.

The italicized words depend on the particular statement under consideration, and you must learn to read, to identify, and to write each of the three components. For instance, consider the following statement:

There is an integer $x > 2$ such that $x^2 - 5x + 6 = 0$.

The object is the integer x;

the certain property is that $x > 2$;

the something that happens is $(x^2 - 5x + 6) = 0$.

Observe that the words "such that" (or equivalent words like "for which") always precede the something that happens.

During the backward process, if you ever come across a statement having the quantifier "there is," then one way in which you can proceed to show that the statement is true is through the construction method. The idea is to construct (guess, produce, devise an algorithm to produce, etc.) the desired object. Of course you must show that the object has the certain property and that the something happens. How you actually construct the desired object is not at all clear. Sometimes it will be by trial and error; sometimes an algorithm can be designed to produce the desired object. It all depends on the particular problem; nonetheless, the information in statement A will surely be used to help accomplish the task. Indeed, the appearance of the quantifier "there is" strongly suggests turning to the forward process to produce the desired object. Suppose, for instance, that you were faced with the following problem.

Proposition. *If a, b, c, d, e, and f are real numbers with the property that $ad - bc \neq 0$, then the two linear equations $ax + by = e$ and $cx + dy = f$ can be solved for x and y.*

OUTLINE OF PROOF. On starting the backward process, you should recognize that the statement B has the form discussed above, even though the quantifier "there are" does not appear explicitly. Observe that the statement B can be rewritten to contain the quantifier explicitly; for example, "There are real numbers x and y such that $ax + by = e$ and $cx + dy = f$." Statements containing hidden quantifiers occur frequently in problems, and you should watch for them.

Proceeding with the construction method, the issue is how to construct real numbers x and y such that $ax + by = e$ and $cx + dy = f$. If you are clever enough to guess that $x = (de - bf)/(ad - bc)$ and $y = (af - ce)/(ad - bc)$, then you are very fortunate, but you must still show that the something happens, in this case, that $ax + by = e$ and $cx + dy = f$. This, of course, is not hard to do. Also observe that, by guessing these values for x and y you have used the information in A, since the denominators are not 0.

While this guess and check approach is perfectly acceptable for producing the desired x and y, it is not very informative as to how these particular values were produced. A more instructive proof would be desirable.

The construction method is not the only technique available for dealing with statements having the quantifier "there is," but it often works and should be considered seriously. To be successful with the construction method, you must become a "builder" and use your creative ability to construct the desired object having the certain property. Also, you must not forget to show that the something happens. Your "building supplies" consist of the information contained in A.

3.5. QUANTIFIERS II: THE CHOOSE METHOD

This section develops the *choose method*, a proof technique for dealing with statements containing the quantifier "for all." Such statements arise quite naturally in many mathematical areas, one of which is set theory.

Recall that a *set* is nothing more than a collection of items. Each of the individual items is called a *member* or *element* of the set and each member of the set is said to *be in* or *belong to* the set. The set is usually denoted by enclosing the list of its members (separated by commas) in braces. Thus the set consisting of the numbers 1, 4, and 7 would be written $\{1, 4, 7\}$.

It is certainly desirable to make a list of all of the elements in a set; but when a set has an infinite number of elements (such as the set of real numbers that are greater than or equal to 0), it is actually impossible to make a complete list, even if you wanted to. To describe such large sets, one would write, for the example above, $S = \{\text{real numbers } x : x \geq 0\}$, where the ":" stands for the words "such that." Everything following the ":" is referred to as the defining property of the set although sometimes part of the defining property appears to the left of the ":".

From a proof theory point of view, the defining property plays the same role as a definition does: it is used to answer the abstraction question "How can I show that an item belongs to a particular set?" One answer is to check that the item satisfies the defining property.

While discussing sets, observe that it can happen that no item satisfies the defining property. Consider, for example,

$$\{\text{real numbers } x \geq 0 : x^2 + 3x + 2 = 0\}$$

Such a set is said to be *empty*, meaning that it has no members. The special symbol "$\varnothing$" is used to denote the empty set.

To motivate the use of the quantifier "for all," observe that it is usually possible to write a set in more than one way. Consider, for example, the sets

$$S = \{\text{real numbers } x : x^2 - 3x + 2 \leq 0\}$$

$$T = \{\text{real numbers } x : 1 \leq x \leq 2\}$$

where $1 \leq x \leq 2$ means that $1 \leq x$ and $x \leq 2$. Surely, for two sets S and T to be the same, each element of S should appear in T and vice versa. Using the quantifier "for all," we can now make the following definitions:

Definition 3.13. A set S is said to be a *subset* of a set T (written $S \subseteq T$) if and only if for each element x in S, x is in T.

Definition 3.14. Two sets S and T are said to be *equal* (written $S = T$) if and only if S is a subset of T and T is a subset of S.

Like any definition, these can be used to answer an abstraction question. The first one answers the question "How can I show that a set (namely, S) is a subset of another set (namely, T)?" by requiring you to show that for each element x in S, x is also in T. As you will see shortly, the choose method will enable you to show that for each element x in S, x is also in T.

In addition to set theory, there are many other instances where the quantifier "for all" can and will be used, but, from the above example, you can see that such statements appear to have the same consistent structure. When the quantifiers "for all," "for each," "for every," or "for any" appear, the statement will have the following basic form (which is similar to the one you saw in the previous section):

For every *object* with a *certain property*,
something happens.

The italicized words depend on the particular statement under consideration. For instance,

For all real numbers $y > 0$, there is a real number x such that $2^x = y$,

Object: the real numbers y;

Certain property: $y > 0$; and

Something that happens: There is a real number x such that $2^x = y$.

Observe that a comma always precedes the something that happens. Sometimes the quantifier is hidden; for example, the statement "The cosine of any angle strictly between 0 and $\pi/4$ is larger than the sine of the angle" could be phrased equally well "For every angle t with $0 < t < \pi/4$, $\cos(t) > \sin(t)$."

During the backward process, if you ever come across a statement having the quantifier "for all" in the form discussed above, then one way in which you might be able to show that the statement is true is to make a list of all of the objects having the certain property. Then, for each one, you could try to show that the something happens. When the list is finite, this might be a reasonable way to proceed. However, more often than not, this approach will not be practicable because the list is too long, or even infinite. You have already dealt with this type of obstacle in set theory, where the problem was overcome by using the defining property to describe the set. Here, the choose method will allow you to circumvent the difficulty.

The choose method can be thought of as a proof machine that, rather than actually checking that the something happens for each and every object having the certain property, has the *capability* of doing so. If you had such a machine, then there would be no need to check the whole (possibly infinite) list because you would know that the machine could always do so. The choose method shows you how to design the inner workings of the proof machine.

To understand the mechanics of the choose method, put yourself in the role of the proof machine and keep in mind that you need to have the capability of taking any object with the certain property and concluding that the something happens (see Figure 3.3). As such, pretend that someone gave you one of these objects, but remember that you do not know precisely which one. All you do know is that the particular object does have the certain property, and you must somehow be able to use the property to reach the conclusion that the something happens. This task is most easily accomplished by working forward from the certain property and backward from the something that happens. In other words, with the choose method, you choose one object that has the certain property. Then, by using the forward–backward method, you must conclude that, for the chosen object, the something happens. Your proof machine will then have the capability of repeating the proof for any of the objects having the certain property.

Suppose, for example, that in some proof you needed to show that, for all real numbers x with $x^2 - 3x + 2 \leq 0$, $1 \leq x \leq 2$. With the choose method you would choose one of these real numbers, say, x', that does have the certain property (in this case, $x'^2 - 3x' + 2 \leq 0$). Then, by working forward from the fact that $x'^2 - 3x' + 2 \leq 0$, you must reach the conclusion that for x' the something happens; that is, $1 \leq x' \leq 2$.

Here, the symbol x' has been used to distinguish the chosen object from the general object x. This distinction is often ignored (i.e., the same symbol is used for both the general object and the chosen one), and you must be careful to interpret the symbol correctly.

Observe that, when you use the choose method, you obtain additional information that is added to the assumption that A is true. Invariably, in the forward process, you will use the extra information.

The choose method is a viable approach for dealing with a statement that contains the quantifier "for all." Proceed by choosing an object having the certain property. Add this information to that in A and attempt to show that the something happens by using the forward–backward method.

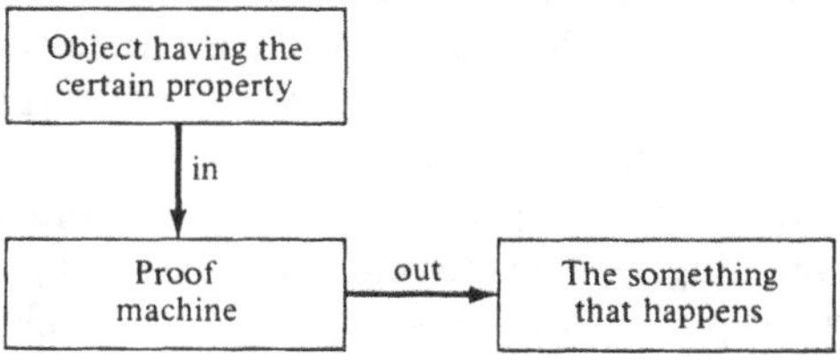

FIGURE 3.3. Proof machine for the choose method.

3.6. QUANTIFIERS III: INDUCTION

In the previous section you learned how to use the choose method when the quantifier "for all" appears in the statement B. There is one very special form of B for which a separate proof technique, known as *mathematical induction*, has been developed. Induction should seriously be considered (even before the choose method) when B has the following form:

For every integer $n \geq 1$, *something happens*,

where the something that happens is some statement that depends on the integer n. An example would be the following statement:

$$\text{For all integers } n \geq 1, \sum_{k=1}^{n} k = \frac{n(n+1)}{2}$$

where $\sum_{k=1}^{n} k = 1 + \cdots + n$. When considering induction, the key words to look for are "integer" and "≥ 1."

One way to attempt proving such statements would be to make an infinite list of problems, one for each of the integers starting from $n = 1$, and then prove each statement separately. The first few problems on the list are usually easy to verify, but the issue is how to check the nth one and beyond. For the example, the list is

$$\text{P}(1): \quad \sum_{k=1}^{1} k = \frac{1(1+1)}{2} \qquad \text{or} \quad 1 = 1$$

$$\text{P}(2): \quad \sum_{k=1}^{2} k = \frac{2(2+1)}{2} \qquad \text{or} \quad 1 + 2 = 3$$

$$\text{P}(3): \quad \sum_{k=1}^{3} k = \frac{3(3+1)}{2} \qquad \text{or} \quad 1 + 2 + 3 = 6$$

$$\vdots$$

$$\text{P}(n): \quad \sum_{k=1}^{n} k = \frac{n(n+1)}{2}$$

$$\text{P}(n+1): \quad \sum_{k=1}^{n+1} k = \frac{(n+1)[(n+1)+1]}{2} = \frac{(n+1)(n+2)}{2}$$

$$\vdots$$

Induction is a clever method for proving that each of the statements in the infinite list is true. As with the choose method, induction can be thought of as an automatic problem-solving machine that starts with P(1) and works its way progressively down the list proving each statement as it proceeds. Here is how it works. You start the machine by verifying that P(1) is true, as can easily be done for the example above. Then you feed P(1) into the machine. It uses the fact that P(1) is true and automatically proves that P(2) is true. You then take P(2) and put it into the machine. Once again, the machine uses the fact that P(2) is true to reach the conclusion that P(3) is true, and so on (see Figure 3.4). Observe that, by the time the machine is going to prove that $P(n+1)$ is true, it will already have shown that $P(n)$ is true (from the previous step). Thus, in designing the machine, you can assume that $P(n)$ is true, and your job is to make sure that $P(n+1)$ will also be true. Do not forget that, in order to start the whole process, you must also verify that P(1) is true.

To repeat, a proof by induction consists of two steps. The first step is to verify that the statement P(1) is true. To do so, you simply replace n everywhere by 1. Usually, to verify that the resulting statement is true, you will only have to do some minor rewriting.

The second step is much more challenging. It requires that you reach the conclusion that $P(n+1)$ is true by using the assumption that $P(n)$ is true. There is a standard way of doing this. Begin by writing down the statement $P(n+1)$. Since you are allowed to assume that $P(n)$ is true and you want to conclude that $P(n+1)$ is true, you should somehow try to rewrite the statement $P(n+1)$ in terms of $P(n)$—as will be illustrated in a moment—for

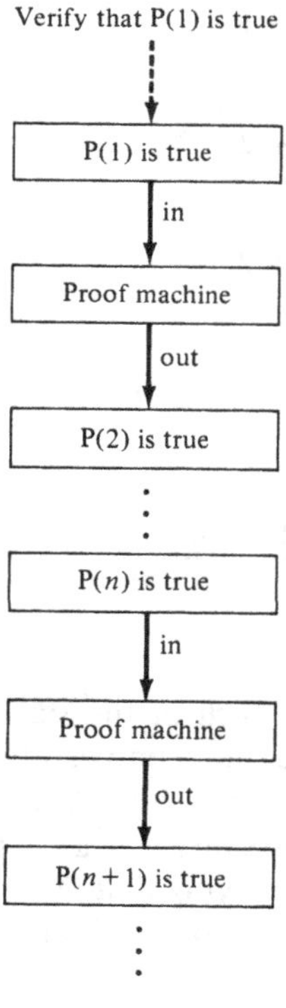

FIGURE 3.4. Proof machine for induction.

then you will be able to make use of the assumption that P(n) is true. On establishing that P($n+1$) is true, the proof will be complete.

To apply the second step of induction to the specific example above, you would assume that

$$P(n):\quad \sum_{k=1}^{n} k = \frac{n(n+1)}{2}$$

is true. Your goal would be to use this information to conclude that

$$P(n+1):\quad \sum_{k=1}^{n+1} k = \frac{(n+1)[(n+1)+1]}{2} = \frac{(n+1)(n+2)}{2}$$

is true.

To reach the conclusion that P($n+1$) is true, begin with the left-hand side of the equality in P($n+1$) and try to make it look like the right-hand side. In so doing, you should use the information in P(n) by relating the left-hand side of the equality in P($n+1$) to the left-hand side of the equality in P(n), for then you will be able to use the right-hand side of the equality in P(n). In this example,

$$\sum_{k=1}^{n+1} k = \sum_{k=1}^{n} k + (n+1)$$

Now you can use the assumption that P(n) is true by replacing $\sum_{k=1}^{n} k$ with $n(n+1)/2$, obtaining

$$\sum_{k=1}^{n+1} k = \frac{n(n+1)}{2} + (n+1)$$

All that remains is a bit of algebra to rewrite $n(n+1)/2+(n+1)$ as $(n+1)(n+2)/2$, thus obtaining the right-hand side of the equality in P($n+1$). In summary,

$$\begin{aligned} 1+\cdots+(n+1) &= (1+\cdots+n)+(n+1) \\ &= n(n+1)/2+(n+1) \\ &= (n^2+3n+2)/2 \\ &= (n+1)(n+2)/2 \end{aligned}$$

Your ability to relate P($n+1$) to P(n) so as to use the induction hypothesis that P(n) is true will determine the success of the proof by induction. If you are unable to relate P($n+1$) to P(n), then you might wish to consider a different proof technique.

Induction is a very powerful technique when applicable; however, it is important to realize that induction does not help you to discover the correct form of the statement P(n). Induction only verifies that a given statement P(n) is true for all integers n greater than or equal to some initial one. The key to its success rests in your ability to relate P($n+1$) to P(n), but do not forget to verify that the statement is also true for the very first possible value of n.

3.7. QUANTIFIERS IV: SPECIALIZATION

In the previous three sections you discovered how to proceed when a quantifier appeared in the statement B. This section will develop a method for exploiting quantifiers that appear in statement A. Suppose statement A contains the quantifier "there is" in the standard form

> There is an *object* with a *certain property*
> such that *something happens*.

You can use this information in a straightforward way. When showing that "A implies B" by the forward–backward method, you assume A is true, which in this case means you can assume that indeed there is an object with the certain property such that the something happens. In doing the proof, you would say, "Let x be an object with the certain property and for which the something happens...." The existence of this object will somehow be used in the forward process to obtain the conclusion that B is true.

The more interesting situation occurs when the statement A contains the quantifier "for all" in the standard form

> For all *objects* with a *certain property*,
> *something happens*.

To use this information, one typical method emerges, referred to as *specialization*. As a result of assuming that A is true, you know that, for all objects with the certain property, something happens. If, at some point in the backward process, you were to come across one of these objects that does have the certain property, then you can use the information in A to conclude that for this particular object the something does indeed happen, which should help you to conclude that B is true. In other words, you will have specialized statement A to one particular object having the certain property. For instance, if you know that, for every angle t, $\sin^2(t)+\cos^2(t) = 1$, then, in particular, for one angle, say $t=\pi/4$, you can conclude that $\sin^2(\pi/4)+\cos^2(\pi/4)=1$. An example demonstrates the proper use of specialization.

Definition 3.15. A real number u is an upper bound for a set of real numbers T if for all elements t in T, $t \leq u$.

EXAMPLE 3.3: If R is a subset of a set S of real numbers and u is an upper bound for S, then u is an upper bound for R.

OUTLINE OF PROOF. The forward–backward method gives rise to the abstraction question "How can I show that a real number (namely, u) is an upper bound for a set of real numbers (namely, R)?" Definition 3.15 is used to answer the question. Thus it must be shown that, for all elements r in R, $r \leq u$. The appearance of the quantifier "for all" in the backward process suggests proceeding with the choose method, whereby one chooses an element, say, r, in R for which it must be shown that $r \leq u$.

Turning now to the forward process, you will see how specialization is used to obtain the desired conclusion that $r \leq u$. From the hypothesis that R is a subset of S and by Definition 3.13, you know that each element in R is also in S. In the backward process you came across the particular element r in R, and hence you can use specialization to conclude that r is in S.

Also, from the hypothesis you know that u is an upper bound for S. By Definition 3.15 this means that, for every element s in S, $s \leq u$. Again, the appearance of the quantifier "for every" in the forward process suggests using specialization. In particular, r is an element of S, as was shown in the previous paragraph. So, by specialization, you can conclude that $r \leq u$. Since the statement "$r \leq u$" was the last one obtained in the backward process, the proof is now complete.

Note that when using specialization you must be very careful to keep your notation and symbols in order. Also, be sure that the particular object to which you are specializing does satisfy the certain property, for only then can you conclude that the something happens.

In the condensed proof that follows, note the lack of reference to the forward–backward, choose, and specialization methods.

PROOF OF EXAMPLE 3.3. To show that u is an upper bound for R, let r be an element of R (the word "let" indicates that the choose method has been used). By hypothesis, R is a subset of S, and so r is also an element of S (here is where specialization has been used). Furthermore, by hypothesis, u is an upper bound for S; thus every element in S is less than or equal to u. In particular, r is an element of S, so $r \leq u$ (again specialization has been used).

This and the previous three sections have provided various techniques for dealing with quantifiers that can appear in either A or B. As always, let the form of the statement guide you. When B contains the quantifier "there is," the construction method can be used to produce the desired object. The choose method is associated with the quantifier "for all," except when the statement B is supposed to be true for every integer starting from some initial one. In the latter case, induction is likely to be successful, provided that you can relate the statement for $n + 1$ to the one for n. Finally, if the quantifier "for all" appears in the statement A, specialization can often be exploited. When using specialization, be sure that the particular object under consideration does satisfy the certain property, for only then will the something happen.

All of the material thus far has been organized around the forward–backward method. Now it is time to see some other techniques for showing that "A implies B."

3.8. THE CONTRADICTION METHOD

As powerful as the forward–backward method is, you may well find yourself unable to complete a proof for one reason or another. Try using the forward–backward method, for instance, to prove the following proposition.

EXAMPLE 3.4: If n is an integer and n^2 is even, then n is even.

Fortunately, there are several other techniques that you might want to try before you give up. In this section, the *contradiction method* is described together with an indication of how and when it should be used. (Example 3.4 can easily be proved by this method; however, the details will not be given here.)

With the contradiction method, you begin by assuming that A is true, just as you would in the forward–backward method. However, to reach the desired conclusion that B is true, you proceed by asking yourself a simple question: "Why can't B be false?" After all, if B is supposed to be true, then there must be some reason why B cannot be false. The objective of the contradiction method is to discover that reason. In other words, the idea of a proof by contradiction is to assume that A is true and B is false and then see why this cannot happen. What does it mean to "see why this cannot happen?" Suppose, for example, that as a result of assuming that A is true and B is false (hereafter written as NOT B), you were somehow able to reach the conclusion that $0 = 1$. Would *that* not convince you that it is impossible for A to be true and B to be false simultaneously? Thus, in a proof by contradiction, you assume that A is true and that NOT B is true, and, somehow, you must use this information to reach a contradiction to something that you absolutely know to be true.

Another way of viewing the contradiction method is to recall that the statement "A implies B" is true in all cases except when A is true and B is false. In a proof by contradiction, you rule out this one unfavorable case by actually assuming that it does happen and then reaching a contradiction.

At this point, several very natural questions arise.

1. What contradiction should you be looking for?
2. Exactly how do you use the assumption that A is true and B is false to reach the contradiction?
3. Why and when should you use this approach instead of the forward–backward method?

The first question is by far the hardest to answer because there are no specific guidelines. Each problem gives rise to its own contradiction, and it usually takes creativity, insight, persistence, and luck to produce a contradiction.

As to the second question, one common approach to finding a contradiction is to work forward from the assumption that A and NOT B are true, as will be illustrated in a moment.

The discussion above also indicates why you might wish to use contradiction instead of the forward–backward method. With the forward–backward method you only assume that A is true, but in the contradiction method you can assume that both A and NOT B are true. Thus you get two statements from which to reason forward instead of just one (see Figure 3.5). On the other hand, the disadvantage is that you have no definite knowledge of where the contradiction will arise.

As a general rule, use contradiction when the statement NOT B gives you some useful information. In particular, the contradiction method is likely to be successful when the statement B contains the key word "no" or "not," as is shown in the next example.

EXAMPLE 3.5: If r is a real number such that $r^2 = 2$, then r is irrational.

OUTLINE OF PROOF. It is important to note that the conclusion of Example 3.5 can be rewritten to read "r is not rational." The appearance of the word "not" thus suggests using the contradiction method, whereby you can assume that A and NOT B are both true. In this case, that means you can assume that $r^2 = 2$ and that r is a rational number. Using this information, a contradiction must now be reached.

Working forward and using Definition 3.7 for a rational number, there are integers p and q with $q \neq 0$ such that $r = p/q$. There is still the unanswered question of where the contradiction arises, and to find it takes a lot of creativity. A crucial observation here will really help. It is possible to assume that p and q have no common divisor (i.e., no integer that divides both p and q), for if they did, you could divide this integer out of both the numerator p and the denominator q. Now a contradiction can be reached

FIGURE 3.5. Forward–backward method versus the contradiction method.

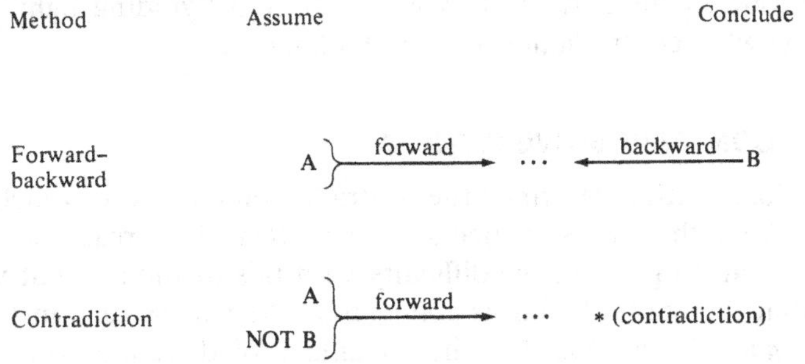

by showing that 2 is a common divisor of p and q. This will be done by showing that p and q are even, and hence 2 divides them both.

Working forward, since $r = p/q$, it follows that

A1. $r^2 = p^2/q^2$;
A2. $2 = p^2/q^2$ (substituting $r^2 = 2$);
A3. $2q^2 = p^2$ (multiplying A2 by q^2);
A4. p^2 is even (since $2q^2$ is even);
A5. p is even (see Example 3.4);
A6. there is an integer k such that $p = 2k$ (definition of "even integer");
A7. $2q^2 = (2k)^2 = 4k^2$ (substitute $p = 2k$ into A3);
A8. $q^2 = 2k^2$ (divide A7 by 2);
A9. q^2 is even (since $2k^2$ is even);
A10. q is even (see Example 3.4).

Thus A5 and A10 have resulted in the contradiction that p and q are both even.

There are several other valuable uses for the contradiction method. Recall that when statement B contains the quantifier "there is" the construction method is recommended in spite of the difficulty of actually producing the desired object. The contradiction method opens up a whole new approach. Instead of showing that there is an object with the certain property such that the something happens, you could proceed from the assumption that there is no such object. Now your job is to use this information to reach some kind of contradiction. How and where the contradiction arises is not at all clear, but it may be a lot easier than producing or constructing the object.

As you have seen, the contradiction method can be a very useful technique when the statement B contains the word "no" or "not" in it. You work forward from the assumption that A and NOT B are true to reach a contradiction. One of the disadvantages of the method is that you do not know exactly what the contradiction is going to be. The next section describes another proof technique in which you attempt to reach a very specific contradiction. As such, you will have a "guiding light" since you will know what contradiction you are looking for.

3.9. THE CONTRAPOSITIVE METHOD

The previous section described the contradiction method in which you work forward from the two statements A and NOT B to reach some kind of contradiction. In general, the difficulty with this method is that you do not know what the contradiction is going to be. As will be seen in this section, the *contrapositive method* has the advantage of directing you toward one specific type of contradiction.

The contrapositive method is similar to contradiction, in that you begin by assuming that A and NOT B are true. Unlike contradiction, however, you do not work forward from both A and NOT B. Instead, you work forward only from NOT B. Your objective is to reach the contradiction that A is false (hereafter written NOT A). Can you ask for a better contradiction than that? How can A be true and false at the same time? For example, if you were using the contrapositive method to prove that if n is an integer and n^2 is even, then n is even, then you would work forward from the assumption that n is not even (i.e., that n is odd), and backward from the desired conclusion that n^2 is not even (i.e., that n^2 is odd).

To repeat, in the contrapositive method, you assume that A and NOT B are true and you work forward from the statement NOT B to reach the contradiction that A is false. The contrapositive method can thus be thought of as a more passive form of contradiction in the sense that the assumption that A is true passively provides the contradiction. In the contradiction method, however, the assumption that A is true is actively used to reach a contradiction (see Figure 3.6).

From Figure 3.6 you can also see the advantages and disadvantages of the contrapositive method over the contradiction method. The disadvantage of the contrapositive method is that you work forward from only one statement (namely, NOT B) instead of two. On the other hand, the advantage is that you know precisely what you are looking for (namely, NOT A). Because of this, you can often apply the abstraction process to the statement NOT A in an attempt to work backward. The option of working backward is not available in the contradiction method because you do not know what contradiction you are looking for.

It is also quite interesting to compare the contrapositive and forward–backward methods. In the forward–backward method you work forward from A and backward from B, but in the contrapositive method, you work forward from NOT B and backward from NOT A (see Figure 3.7).

FIGURE 3.6. Contrapositive method versus the contradiction method.

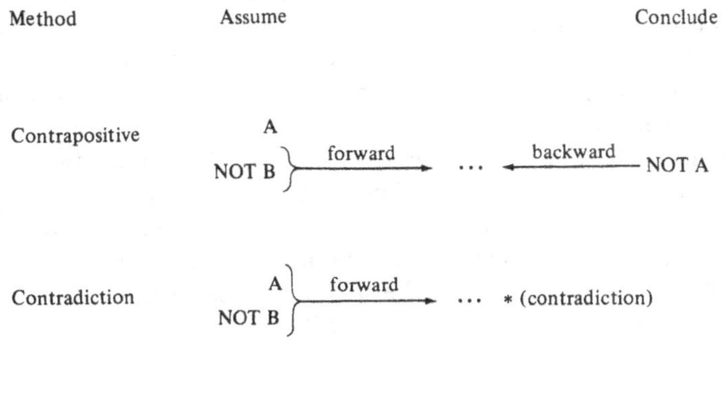

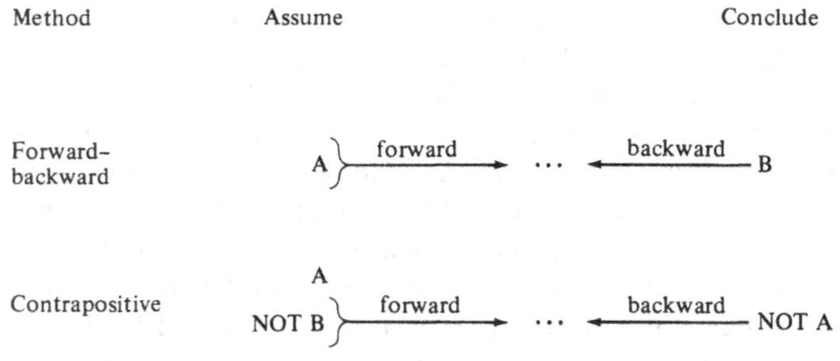

FIGURE 3.7. Forward–backward method versus the contrapositive method.

If you understand Figure 3.7, then it is not hard to see why the contrapositive method might be better than the forward–backward method. Perhaps you can obtain more useful information by working NOT B forward rather than A. Also, it might be easier to perform the abstraction process on the statement NOT A rather than on B, as would be done in the forward–backward method.

The forward–backward method arose from considering what happens to the truth of "A implies B" when A is true and when A is false (recall Table 3.1). The contrapositive method arises from similar considerations regarding B. Specifically, if B is true, then, according to Table 3.1, the statement "A implies B" is true. Hence there is no need to consider the case when B is true. Suppose B is false. In order to ensure that "A implies B" is true, according to Table 3.1, you would have to show that A is false. Thus the contrapositive method has you assume that B is false and try to conclude that A is false.

Indeed, the statement "A implies B" is logically equivalent to "NOT B implies NOT A" (see Table 3.2). The contrapositive method can thus be thought of as the forward–backward method applied to the statement "NOT B implies NOT A."

In general, it is difficult to know whether the forward–backward, contradiction, or contrapositive method will be most effective for a given problem without trying each one. However, there is one instance that often indicates that the contradiction or contrapositive method should be chosen, or at least considered seriously. This occurs when statement B contains the word "no" or "not" in it, for then you will usually find that the statement NOT B has some useful information.

In the contradiction method, you work forward from the two statements A and NOT B to obtain a contradiction. In the contrapositive method you also reach a contradiction, but you do so by working forward from NOT B to reach the conclusion NOT A. You should also, of course, work backward from NOT A. A comparison of the forward–backward, contradiction, and contrapositive methods is given in Figure 3.8.

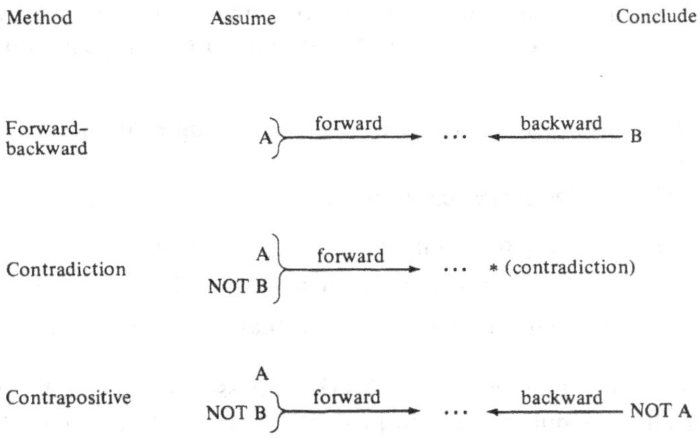

FIGURE 3.8. Comparison of the forward–backward, contradiction, and contrapositive methods.

Both the contrapositive and contradiction methods require that you be able to write the NOT of a statement. The next section shows you how to do so when the statements contain quantifiers.

3.10. NOTs OF NOTs LEAD TO KNOTS

As you saw in the previous section, the contrapositive method is a valuable proof technique. To use it, however, you must be able to write the statement NOT B so that you can work it forward. Similarly, you have to know exactly what the statement NOT A is so that you can apply the abstraction process to it. In some instances, the NOT of a statement is easy to find. For example, if A is the statement "the real number $x > 0$," then the NOT of A is "it is not the case that the real number $x > 0$," or equivalently, "the real number x is not greater than 0." In fact, the word "not" can be eliminated altogether by incorporating it into the statement to obtain "the real number $x \le 0$."

A more challenging situation arises when the statement contains quantifiers. In general, there are three easy steps to finding the NOT of a statement containing one or more quantifiers.

Step 1. Put the word NOT in front of the entire statement.

Step 2. If the word NOT appears to the left of a quantifier, then move the word NOT to the right of the quantifier and place it just before the something that happens. As you do so, you change the quantifier to its opposite, so that "for all" becomes "there is" and "there is" becomes "for all."

Step 3. When all of the quantifiers appear to the left of the NOT, eliminate the NOT by incorporating it into the statement that appears immediately to its right.

These steps will be demonstrated with the following examples.

EXAMPLE 3.6: For every real number $x \geq 2$, $x^2 + x - 6 \geq 0$.

Step 1. NOT for every real number $x \geq 2$, $x^2 + x - 6 \geq 0$.

Step 2. There is a real number $x \geq 2$ such that NOT $x^2 + x - 6 \geq 0$.

Step 3. There is a real number $x \geq 2$ such that $x^2 + x - 6 < 0$.

Note in step 2 that, when the NOT is passed from left to right, the quantifier changes but the certain property (namely, $x \geq 2$) does not. Also, since the quantifier "for every" is changed to "there exists," it becomes necessary to replace the "," by the words "such that." In a completely analogous manner, if the quantifier "there exists" is changed to "for all," then the words "such that" are removed and a "," is inserted, as illustrated in the next example.

EXAMPLE 3.7: There is a real number $x \geq 2$ such that $x^2 + x - 6 \geq 0$.

Step 1. NOT there is a real number $x \geq 2$ such that $x^2 + x - 6 \geq 0$.

Step 2. For all real numbers $x \geq 2$, NOT $x^2 + x - 6 \geq 0$.

Step 3. For all real numbers $x \geq 2$, $x^2 + x - 6 < 0$.

Finally, if the original statement contains more than one quantifier, then step 2 will have to be repeated until all of the quantifiers appear to the left of the NOT, as is demonstrated in the next example.

EXAMPLE 3.8: For every real number x between -1 and 1, there is a real number y between -1 and 1 such that $x^2 + y^2 \leq 1$.

Step 1. NOT for every real number x between -1 and 1, there is a real number y between -1 and 1 such that $x^2 + y^2 \leq 1$.

Step 2. There is a real number x between -1 and 1 such that, NOT there is a real number y between -1 and 1 such that $x^2 + y^2 \leq 1$.

Step 2 (again). There is a real number x between -1 and 1 such that, for all real numbers y between -1 and 1, NOT $x^2 + y^2 \leq 1$.

Step 3. There is a real number x between -1 and 1 such that, for all real numbers y between -1 and 1, $x^2 + y^2 \geq 1$.

Another situation where you must be careful is in taking the NOT of a statement containing the words AND or OR. Just as the quantifiers are interchanged when taking the NOT of the statement, so the words AND and OR interchange. Specifically, NOT [A AND B] becomes [NOT A] OR [NOT B]. Similarly, NOT [A OR B] becomes [NOT A] AND [NOT B].

EXAMPLE 3.9: NOT [$x \geq 3$ AND $y < 2$] becomes [$x < 3$] OR [$y \geq 2$].

EXAMPLE 3.10: NOT [$x \geq 3$ OR $y < 2$] becomes [$x < 3$] AND [$y \geq 2$].

Remember that when you use the contrapositive method of proof the first thing to do is to write the statements NOT B and NOT A.

3.11. SPECIAL PROOF TECHNIQUES

You now have three major proof techniques to help you in proving "A implies B": the forward–backward, the contrapositive, and the contradiction methods. In addition, when B has quantifiers, you have the choose and construction methods. There are several other special forms of B that have well-established and usually successful proof techniques associated with them. Three of these will be developed in this section.

Uniqueness Method

The first is referred to as the *uniqueness method*. It is associated with a statement B that wants you to show not only that there is an object with a certain property such that something happens, but also that the object is unique (i.e., it is the only such object). You will know to use the uniqueness method when the statement B contains the key word "unique" as well as the quantifier "there is."

In such a case, your first job is to show that the desired object does exist. This can be done by either the construction or the contradiction method. The next step will be to show uniqueness in one of two standard ways. The first approach, referred to as the *direct uniqueness method*, has you assume that there are two objects having the certain property and for which the something happens. If there really is only one such object, then, using the certain property, the something that happens, and perhaps the information in A, you must conclude that the two objects are one and the same (i.e., that they are really equal). The forward–backward method is usually the best way to prove that they are equal. This process is illustrated in the next example.

EXAMPLE 3.11: If a, b, c, d, e, and f are real numbers such that $ad - bc \neq 0$, then there are unique real numbers x and y such that $ax + by = e$ and $cx + dy = f$.

OUTLINE OF PROOF. The existence of the real numbers x and y was established in Section 3.4 by means of the construction method. Here the uniqueness will be established by the method described above. Thus you assume that (x_1, y_1) and (x_2, y_2) are two objects with the certain property and for which the something happens. Hence one obtains $a(x_1)+b(y_1)=e$ and $c(x_1)+d(y_1)=f$ and also that $a(x_2)+b(y_2)=e$ and $c(x_2)+d(y_2)=f$. Using these four equations and the assumption that A is true, you can show, by the forward–backward method, that the two objects (x_1, y_1) and (x_2, y_2) are equal. Specifically, the abstraction question is "How can I show that two pairs of real numbers [namely, (x_1, y_1) and (x_2, y_2)] are equal?" You can use the definition of equality of ordered pairs (see Definition 3.4) to show that both $x_1 = x_2$ and $y_1 = y_2$, or, equivalently, that $x_1 - x_2 = 0$ and $y_1 - y_2 = 0$. Both of these statements can be obtained from the forward process by applying some algebraic manipulations to the four equations and by using $ad - bc \neq 0$. The details will not be given here.

The second method of showing uniqueness has you assume that there are two different objects having the certain property and for which the something happens. Now, supposedly, this cannot happen, so, by using the certain property, the something that happens, the information in A, and especially the fact that the objects are different, you must then reach a contradiction. This process is demonstrated in the next example.

EXAMPLE 3.12: If r is a positive real number, then there is a unique real number x such that $x^3 = r$.

OUTLINE OF PROOF. The appearance of the quantifier "there is" in the conclusion suggests using the construction method to produce a real number x such that $x^3 = r$. This part of the proof will be omitted so that the issue of uniqueness can be addressed. To that end, suppose that x and y are two different real numbers such that $x^3 = r$ and $y^3 = r$. You can use this information together with the hypothesis that r is positive, and especially the fact that $x \neq y$, to show that $r = 0$, thus contradicting the hypothesis that r is positive.

To show that $r = 0$, one can work forward. In particular, since $x^3 = r$ and $y^3 = r$, it follows that $x^3 = y^3$. Thus $x^3 - y^3 = 0$, and on factoring, one obtains that $(x-y)(x^2+xy+y^2) = 0$. Here is where you can use the fact that $x \neq y$ to divide by $x - y$, obtaining $x^2 + xy + y^2 = 0$. This is a quadratic equation of the form $ax^2 + bx + c = 0$ in which $a = 1$, $b = y$, and $c = y^2$. The quadratic formula now states that

$$x = \frac{-y \pm \sqrt{(y^2 - 4y^2)}}{2} = \frac{-y \pm \sqrt{-3y^2}}{2}$$

Since x is real and the above formula for x requires taking the square root of $-3y^2$, it must be that $y=0$, and if $y=0$, then $r=y^3=0$, and the contradiction has been reached.

Either/Or Method

Another special proof technique, called the *either/or method*, arises when B is of the form "either C is true, or else D is true" (where C and D are statements). In other words, the either/or method is to be used when trying to show that the statement "A implies C OR D" is true. Applying the forward–backward method, you begin by assuming A is true and would like to conclude that either C is true or else D is true. Suppose that you were to make the additional assumption that C is not true. Clearly it had better turn out that in this case D is true. Thus, when using the either/or method to show that if $x^2-5x+6 \geq 0$, then $x \leq 2$ or $x \geq 3$, you would assume that $x^2-5x+6 \geq 0$ and $x > 2$. You would work forward from this information to show that $x \geq 3$.

It is worth noting that the either/or method could be used equally well by assuming that A is true and D is false and then concluding that C is true.

When the key words "either...or..." arise in the forward process, a technique known as a *proof by cases* is used. For instance, if you wanted to show that if either C or D is true, then B is true, you would use a proof by cases. By the forward–backward method, you would assume that either C is true or D is true, and your goal would be to conclude that B is true. However, since you do not know whether it is C that is true or D that is true, you should proceed by two cases.

Case 1. Assume that C is true and show that B is true. (Use the forward–backward method to accomplish this task.)

Case 2. Assume that D is true and show that B is true. (Again, use the forward–backward method to do so.)

Only by successfully completing both cases will you have shown that if either C or D is true, then B is true.

Max/Min Method

The final proof technique to be developed in this section is the *max/min method*, which applies to problems dealing with maxima and minima, such as in LPs. To understand the max/min method, suppose that S is a nonempty set of real numbers having both a largest and a smallest member. For a given real number x, you might be interested in the position of the set S relative to the number x. For instance, you might want to prove one of the following statements:

1. All of S is to the right of x (see Figure 3.9a).
2. Some of S is to the left of x (see Figure 3.9b).
3. All of S is to the left of x (see Figure 3.9c).
4. Some of S is to the right of x (see Figure 3.9d).

In mathematical problems these four statements are likely to appear, respectively, as

$$\min\{s : s \text{ is in } S\} \geq x \tag{a}$$

$$\min\{s : s \text{ is in } S\} \leq x \tag{b}$$

$$\max\{s : s \text{ is in } S\} \leq x \tag{c}$$

$$\max\{s : s \text{ is in } S\} \geq x \tag{d}$$

The proof techniques associated with the first two are discussed here; the remaining two are similar in spirit. The idea behind the max/min technique

FIGURE 3.9. **(a)** All of S to the right of x. **(b)** Some of S to the left of x. **(c)** All of S to the left of x. **(d)** Some of S to the right of x.

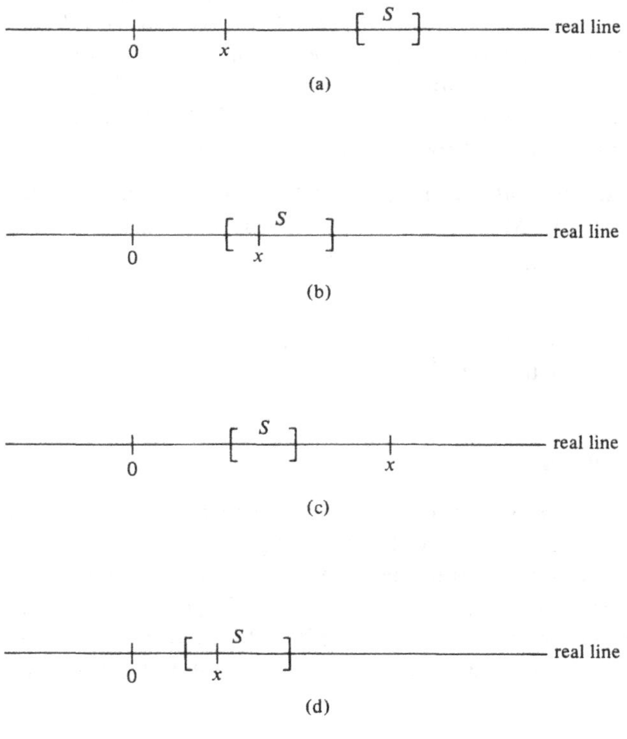

is to convert the given problem into an equivalent problem containing a quantifier. Then the appropriate choose or construction method can be used.

Consider, therefore, the problem of determining if the smallest member of S is greater than or equal to x. An equivalent problem containing a quantifier can be obtained by considering the corresponding statement 1 above. Since "all" of S should be to the right of x, you need to show that, for all elements s in S, $x \leq s$. For example, if a, b, and u are given real numbers and x is a variable, and if you need to prove that $\min\{cx : ax \leq b, x \geq 0\} \geq u$, then you really have to show that for all x with $ax \leq b$ and $x \geq 0$, $cx \geq u$. This statement might then be proved with the choose method.

Turning now to the problem of showing that the smallest member of S is less than or equal to x, the approach is slightly different. To proceed, consider the corresponding statement above. Since "some" of S should be to the left of x, an equivalent problem is to show that there is an element s in S such that $s \leq x$. Then, the construction or contradiction method could be used.

This section has described three special proof techniques that are appropriate when B has the corresponding special form. The final section provides a summary.

3.12. SUMMARY

The list of basic proof techniques is now complete. The techniques presented here are by no means the only ones, but they do constitute the basic set. Undoubtedly you will come across others as you are exposed to more mathematics. Perhaps you will develop some of your own. In any event, there are many fine points and tricks that you will pick up with experience. A final summary of how and when to use each of the various techniques for proving the proposition "A implies B" is in order.

With the forward–backward method, you can assume that A is true. Your job is to prove that B is true. Through the forward process, you will derive from A a sequence of statements A1, A2,... that are necessarily true as a result of A being assumed true. This sequence is not random. It is guided by the backward process, whereby, through asking and answering the abstraction question, you derive from B a statement, B1, with the property that if it is true, then so is B. This abstraction process can then be applied to B1, obtaining a new statement, B2, and so on. The objective is to link the forward sequence to the backward sequence by generating a statement in the forward sequence that is the same as the last statement obtained in the backward sequence. Then you can do the proof by going forward along the sequence, like a column of dominos, from A all the way to B. When obtaining the sequence of statements, watch for quantifiers to appear, for then the construction, choose, induction, or specialization method may be useful in doing the proof.

For instance, you should consider using the construction method to produce the desired object when the quantifier "there is" arises in the backward process in the standard form

> There is an *object* with a *certain property*
> such that *something happens*.

With the construction method, you work forward from the assumption that A is true to construct (produce, or devise an algorithm to produce, etc.) the object. Be sure to check that the object you construct satisfies the certain property and also that the something happens.

On the other hand, you should consider using the choose method when the quantifier "for all" arises in the backward process in the standard form

> For all *objects* with a *certain property*,
> *something happens*.

Here, your objective is to design a proof machine that is capable of taking any object with the certain property and proving that the something happens. To do so you select (or choose) an object that does have the certain property. You must conclude that, for that object, the something happens. Once you have chosen the object, it is best to proceed by working forward from the fact that the chosen object does have the certain property (together with the information in A if necessary) and backward from the something that happens.

The induction method should be considered (even before the choose method) when the statement B has the form

> For every integer n greater than or equal to
> some initial one, a statement $P(n)$ is true.

The first step of the induction method is to verify that the statement is true for the first possible value of n. The second step requires you to show that if $P(n)$ is true, then $P(n+1)$ is true. Remember that the success of a proof by induction rests on your ability to relate the statement $P(n+1)$ to $P(n)$ so that you can make use of the assumption that $P(n)$ is true. In other words, to perform the second step of the induction proof, you should write down the statement $P(n)$, replace n everywhere by $(n+1)$ to obtain $P(n+1)$, and then see if you can express $P(n+1)$ in terms of $P(n)$. Only then will you be able to use the assumption that $P(n)$ is true to reach the desired conclusion that $P(n+1)$ is also true.

Finally, you will probably want to use the specialization method when the quantifier "for all" arises in the forward process in the standard form

> For all *objects* with a *certain property*
> *something happens*.

To do so, watch for one of these objects to arise in the backward process, for then by using specialization you can conclude that the something does happen for the particular object that is under consideration. That fact should then be helpful in reaching the conclusion that B is true. When using specialization, be sure to verify that the particular object does satisfy the certain property, for only then will the something happen.

When the original statement B contains the word "no" or "not," or when the forward–backward method fails, you should consider the contrapositive or contradiction method, with the former being chosen first. To use the contrapositive approach, you must immediately write the statements NOT B and NOT A, using the techniques of Section 3.10 if necessary. Then, by beginning with the assumption that NOT B is true, your job is to conclude that NOT A is true. This is best accomplished by applying the forward–backward method, working forward from the statement NOT B and backward from the statement NOT A. Once again, remember to watch for quantifiers to appear in the forward or backward process, for if they do, then the corresponding construction, choose, induction, and/or specialization methods may be useful.

In the event that the contrapositive method fails, there is still hope with the contradiction method. With this approach, you are allowed to assume not only that A is true but also that B is false. This gives you two facts from which you must derive a contradiction to something that you know to be true. Where the contradiction arises is not always obvious, but it will be obtained by working the statements A and NOT B forward.

In trying to prove that "A implies B," let the form of B guide you as much as possible. Specifically, you should scan the statement B for certain key words, as they will often indicate how to proceed. For example, if you come across the quantifier "there is," then consider the construction method, whereas the quantifier "for all" suggests using the choose or induction method. When the statement B has the word "no" or "not" in it, you will probably want to use either the contrapositive or contradiction method. Other key words to look for are "uniqueness," "either...or...," and "maximum" and "minimum," for then you would use the corresponding uniqueness, either/or, and max/min methods. If you are unable to choose an approach based on the form of B, then you should proceed with the forward–backward method. Table 3.3 provides a complete summary.

You are now ready to "speak" mathematics. Your new "vocabulary" and "grammar" are complete. You have learned the three major proof techniques for proving propositions, theorems, lemmas, and corollaries. They are the forward–backward, contrapositive, and contradiction methods. You have come to know the quantifiers and the corresponding construction, choose, induction, and specialization methods. For special situations, your bag of proof techniques includes the uniqueness, either/or, and max/min methods. If all of these proof techniques fail, you may wish to stick to Greek—after all, it's all Greek to me.

TABLE 3.3. Summary of Proof Techniques

Proof technique	When to use it	What to assume	What to conclude	How to do it
Forward–backward	As a first attempt or when B does not have a recognizable form	A	B	Work forward from A and apply the abstraction process to B
Contrapositive	When B contains the word "no" or "not"	NOT B	NOT A	Work forward from NOT B and backward from NOT A
Contradiction	When B contains the word "no" or "not" or when the first two methods fail	A and NOT B	Some contradiction	Work forward from A and NOT B to reach a contradiction
Construction	When B has a term such as "there is" or "there exists"	A	There is the desired object	Guess, construct, etc., the object having the certain property and show that the something happens
Choose	When B has a term such as "for all" or "for each"	An object with the certain property and A	The something happens	Work forward from A and the fact that the object has the certain property; work backward from the something that happens
Induction	When B is true for each integer beginning with a particular one, say, n_0	The statement is true for n	The statement is true for $n+1$; also show it true for n_0	First substitute n_0 everywhere and show it true; second, invoke the induction hypothesis for n to prove it true for $n+1$

Specialization	When A has a term such as "for all" or "for each"	A	B	Work forward by specializing A to one particular object, the one obtained in the backward process
Uniqueness 1	When B contains the word "unique"	There are two such objects and A	The two objects are equal	Work forward using A and the properties of the two objects; also work backward to show the objects are equal
Uniqueness 2	When B contains the word "unique"	There are two different objects and A	Some contradiction	Work forward from A, the properties of the two objects, and the fact that they are different
Either/or	When B has the form "C OR D"	A and NOT C	D	Work forward from A and NOT C, and backward from D
Proof by cases	When A has the form "C OR D"	Case 1: C Case 2: D	B B	Work forward from C and backward from B, then work forward from D and backward from B
Max/min 1	When B has the form "max $S \leq x$" or "min $S \geq x$"	A and an s in S	$s \leq x$ or $s \geq x$	Work forward from A and the fact that s is in S; also work backward
Max/min 2	When B has the form "max $S \geq x$" or "min $S \leq x$"	A	Construct s in S so that $s \geq x$ or $s \leq x$	Use A and the construction method to produce the desired s in S

DISCUSSION

The material of this chapter is developed in a book by Solow [1], where further detail and extensive exercises can be found.

REFERENCES

1. D. Solow, *How to Read and Do Proofs*, Wiley, New York, 1982.

Chapter 4

Linear Algebra

Formal development of the simplex algorithm requires knowledge of some mathematical background material in linear algebra. The first four sections of this chapter develop the basic concepts of linear algebra; the remaining two sections show how linear algebra is useful in linear programming. The proof techniques of the previous chapter will be used to establish the needed propositions.

4.1. VECTORS

Recall that the objective of a linear programming problem is to determine the values of n variables that minimize (or maximize) an objective function while satisfying certain constraints. The resulting values $(x_1, \ldots, x_n)$ form what is known as an *n vector*. In other words, an n vector $\mathbf{x}$ is a list of n real numbers separated by commas. The number n is referred to as the *dimension* of the vector, and the ith number in the list is called the ith *component* of $\mathbf{x}$ and is denoted by x_i. Thus one writes $\mathbf{x} = (x_1, \ldots, x_n)$.

The remainder of this section will describe the various algebraic operations that can be performed on vectors, such as addition and multiplication. Corresponding to each algebraic operation is a geometric interpretation that can easily be visualized when the vectors have only two or three components. For instance, the vector $(1,2)$ is depicted geometrically in Figure 4.1 with its "tail" at the origin and its "head" at $(1,2)$. Throughout this section, let $\mathbf{x} = (x_1, \ldots, x_n)$, $\mathbf{y} = (y_1, \ldots, y_n)$, and $\mathbf{d} = (d_1, \ldots, d_n)$ be n vectors, and let t be a real number.

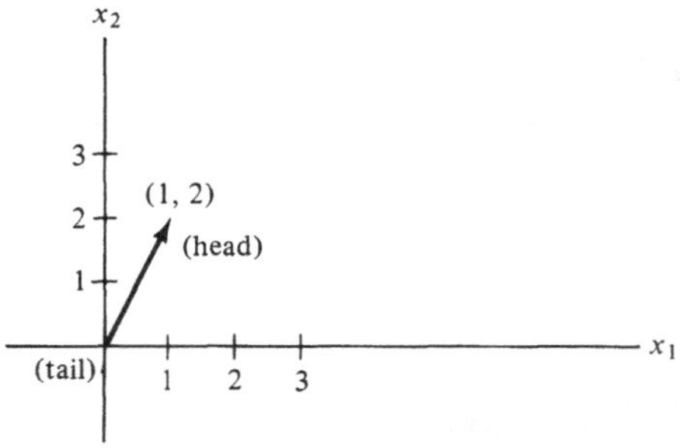

FIGURE 4.1. Geometry of the vector (1,2).

Vector Addition

Two n vectors $\mathbf{x}$ and $\mathbf{d}$ can be added to obtain a new n vector $\mathbf{x} + \mathbf{d}$. To add $\mathbf{x}$ and $\mathbf{d}$ algebraically, simply add the corresponding components: $\mathbf{x} + \mathbf{d} = (x_1 + d_1, \dots, x_n + d_n)$. For example, if $\mathbf{x} = (1,2)$ and $\mathbf{d} = (2,3)$, then $\mathbf{x} + \mathbf{d} = (1,2) + (2,3) = (1+2, 2+3) = (3,5)$. To add the two vectors $\mathbf{x}$ and $\mathbf{d}$ geometrically, first draw them in a plane. Then move the vector $\mathbf{d}$ parallel to itself until its tail coincides precisely with the head of $\mathbf{x}$ (see Figure 4.2). That is, $\mathbf{x} + \mathbf{d}$ is the vector whose tail is the origin and whose head is that of $\mathbf{d}$ (after it has been moved), as shown in Figure 4.3.

Observe that in order to find $\mathbf{x} + \mathbf{d}$ you could equally well have moved $\mathbf{x}$ parallel to itself until its tail coincides precisely with the head of $\mathbf{d}$. In other words, the vectors $\mathbf{x} + \mathbf{d}$ and $\mathbf{d} + \mathbf{x}$ are the same. Algebraically, two n vectors are equal if their corresponding components are equal.

Definition 4.1. Two n vectors $\mathbf{x} = (x_1, \dots, x_n)$ and $\mathbf{y} = (y_1, \dots, y_n)$ are equal (written $\mathbf{x} = \mathbf{y}$) if, for all i with $1 \le i \le n$, $x_i = y_i$.

In a similar manner it is possible to define $\mathbf{x} \le \mathbf{y}$ and $\mathbf{x} \ge \mathbf{y}$.

Definition 4.2. The n vector $\mathbf{x}$ is less than or equal to the n vector $\mathbf{y}$ (written $\mathbf{x} \le \mathbf{y}$) if, for all i with $1 \le i \le n$, $x_i \le y_i$.

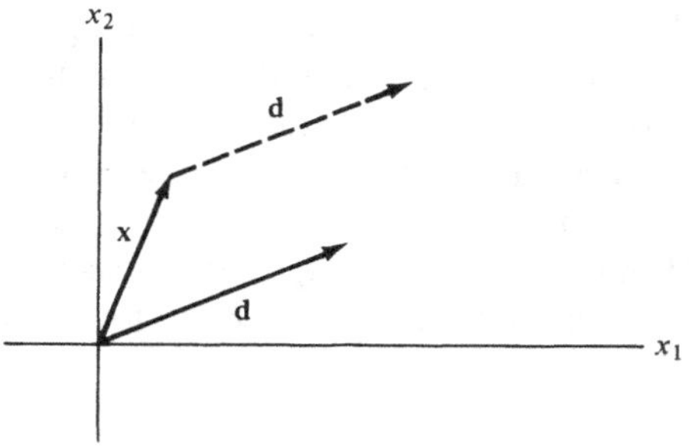

FIGURE 4.2. Moving a vector parallel to itself.

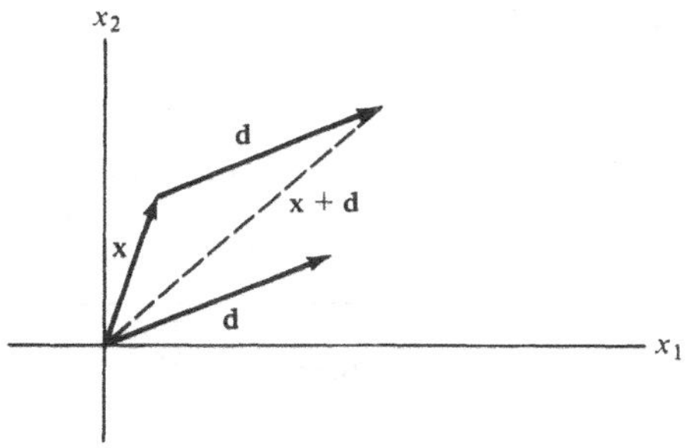

FIGURE 4.3. Geometry of vector addition.

Definition 4.3. The n vector **x** is greater than or equal to the n vector **y** (written $\mathbf{x} \geq \mathbf{y}$) if, for all i with $1 \leq i \leq n$, $x_i \geq y_i$.

In a linear programming problem for which the variables are restricted to be nonnegative, one could write either $x_1 \geq 0, \ldots, x_n \geq 0$ or, more concisely, $\mathbf{x} \geq \mathbf{0}$. Here, **0** is the n vector each of whose components is 0. In general, the dimension of the vector **0** is determined by the context in which it is used. For instance, if **x** is an n vector and $\mathbf{x} \geq \mathbf{0}$, then **0** is an n vector, and if **b** is an m vector and $\mathbf{b} \geq \mathbf{0}$, then **0** is an m vector. Another vector that occurs frequently is **e**. Each of its components is 1, and, as with the zero vector, its dimension is determined by the context in which it is used.

Return now to operations on vectors. It is common to multiply a vector **d** by a real number t. The result (written $t\mathbf{d}$) is an n vector obtained by multiplying each component of **d** by t, i.e., for each i with $1 \leq i \leq n$, $(t\mathbf{d})_i = t\mathbf{d}_i$. For example, if $\mathbf{d} = (2, 3)$ and $t = -2$, then $t\mathbf{d} = -2(2, 3) = (-4, -6)$. When $t > 0$, the vector $t\mathbf{d}$ points in the same direction as **d**. The vector $t\mathbf{d}$ is longer than **d** if $t > 1$ and shorter than **d** if $0 < t < 1$. Similarly, when $t < 0$, $t\mathbf{d}$ points in the direction "opposite" to **d**. It is longer than **d** if $t < -1$ and shorter than **d** if $-1 < t < 0$. Observe that if $t = 0$, then, no matter what **d** is, $t\mathbf{d}$ is the zero vector. Analogously, if $\mathbf{d} = \mathbf{0}$, then, no matter what t is, $t\mathbf{d}$ is the zero vector. The geometric interpretation of $t\mathbf{d}$ is illustrated in Figure 4.4.

In the simplex algorithm, it will often be necessary to "move" from a point **x** in the direction **d** by an "amount" t. Thus imagine that you are standing at **x** and that emanating from **x** is the vector **d** (see Figure 4.5). When $t = 0$, no movement is taking place. When $t = 1$, the new point is $\mathbf{x} + \mathbf{d}$; if $t = -1$, the new point is $\mathbf{x} - \mathbf{d}$. In general, for any value of t, the new point lies somewhere on the line that passes through **x** in the direction **d** (see Figure 4.6). Algebraically, the new point can be computed as $\mathbf{x} + t\mathbf{d}$. Observe that the line does not exist when $\mathbf{d} = \mathbf{0}$. In other words, if $\mathbf{d} = \mathbf{0}$, then no movement will take place, no matter what value t is.

In linear programming, the concept of moving from **x** an amount t in the direction $\mathbf{d} \neq \mathbf{0}$ is particularly useful. For instance, suppose that **x** is an

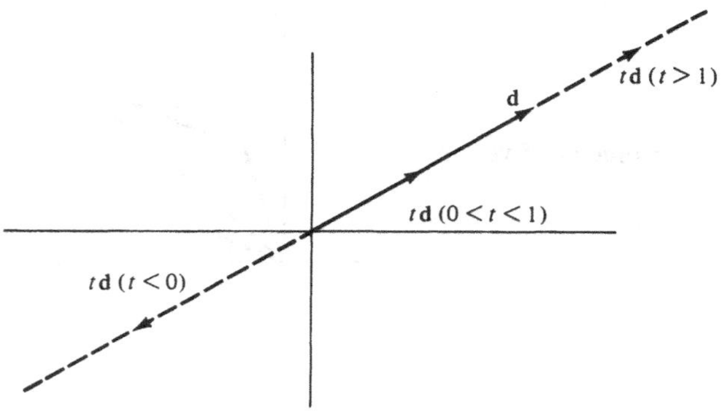

FIGURE 4.4. Geometry of $t\mathbf{d}$.

extreme point of the feasible region and $\mathbf{d}$ represents the direction of one of the "edges" emanating from $\mathbf{x}$ (see Figure 4.7). To move from $\mathbf{x}$ to the new extreme point indicated by $\mathbf{d}$, one would want to compute a value for t so that $\mathbf{x} + t\mathbf{d}$ is the new extreme point. This is precisely what the simplex algorithm does.

Vector Multiplication

Another common operation between two n vectors $\mathbf{x}$ and $\mathbf{y}$ is to take their inner product, a kind of vector multiplication. Unlike the previous operations defined in this chapter, it produces not a vector, but a number. Specifically

$$\mathbf{xy} = \sum_{i=1}^{n} x_i y_i = x_1 y_1 + \cdots + x_n y_n$$

(You may find this written in other books as $\mathbf{x} \cdot \mathbf{y}$.) For instance, if $\mathbf{x} = (1, 1)$

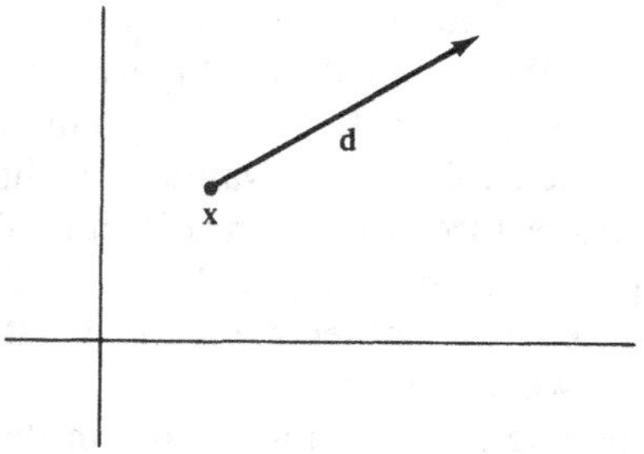

FIGURE 4.5. Vector $\mathbf{d}$ emanating from x.

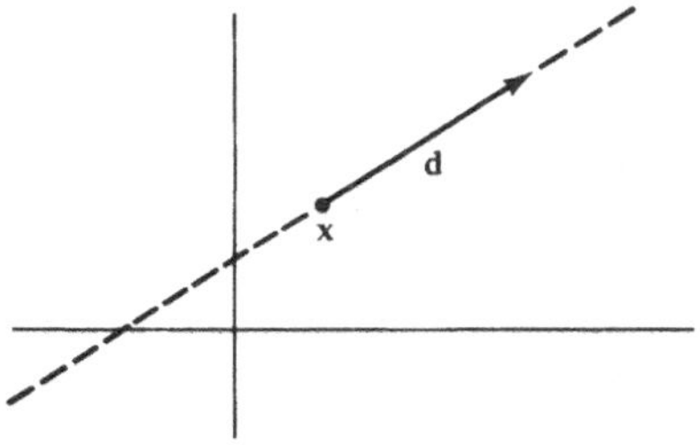

FIGURE 4.6. Line through x in the direction **d**.

and $\mathbf{y} = (-2, 3)$, then

$$\mathbf{xy} = (1,1)(-2,3) = 1(-2) + 1(3) = 1$$

To understand the geometric meaning of the sign of the number **xy**, consider the vector **x** emanating from the origin, and draw a line through the origin perpendicular to **x** (see Figure 4.8). Observe that the line divides the plane into two half spaces. As evident in Figure 4.8, the vector **y** must

1. lie in the same half space as **x** (when $\mathbf{xy} > 0$), or
2. lie in the opposite half space as **x** (when $\mathbf{xy} < 0$), or
3. lie precisely on the line (when $\mathbf{xy} = 0$).

In case 3 the vector **y** is perpendicular to **x**.

When the vectors have three components instead of two, the geometric interpretation of **xy** is similar, only this time one would draw a plane through the origin perpendicular to **x** instead of a line (see Figure 4.9). Once again, the plane gives rise to two half spaces. If $\mathbf{xy} > 0$, then **y** lies in the same half space as **x**; if $\mathbf{xy} < 0$, then **y** lies in the opposite half space. When $\mathbf{xy} = 0$, **y** lies exactly in the plane; i.e., **y** is perpendicular to **x**.

The inner product of two **n** vectors is useful in linear programming. For example, the objective function $c_1x_1 + \cdots + c_nx_n$ can be written as **cx**,

FIGURE 4.7. Moving from an extreme point along an edge.

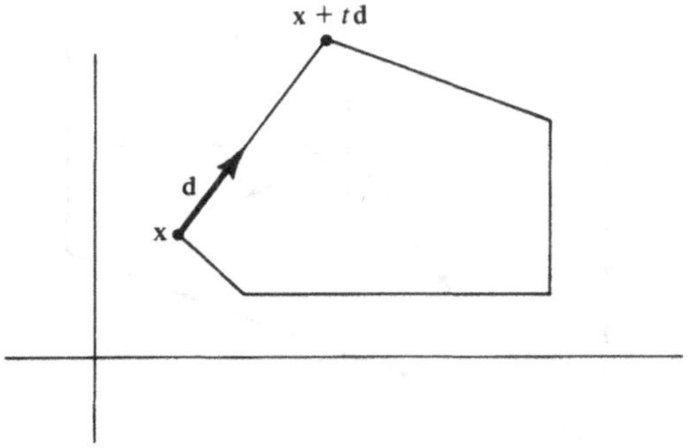

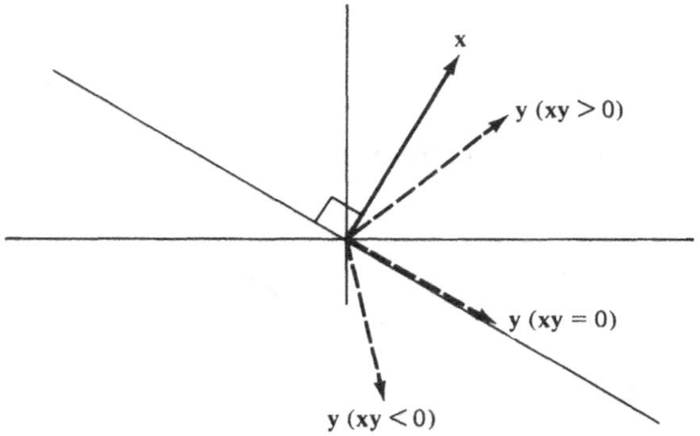

FIGURE 4.8. Geometry of vector multiplication.

where $\mathbf{c} = (c_1, \ldots, c_n)$ and $\mathbf{x} = (x_1, \ldots, x_n)$. The use of vector multiplication in the constraints of an LP will be discussed in the next section.

Just as $\mathbf{x} + \mathbf{d} = \mathbf{d} + \mathbf{x}$, so too $\mathbf{xy} = \mathbf{yx}$. In fact, there are several other useful properties of vector addition and multiplication. Table 4.1 summarizes the valid operations. The justification of the statements in Table 4.1 is left to Exercise 4.1.6.

FIGURE 4.9. Vector multiplication in three dimensions.

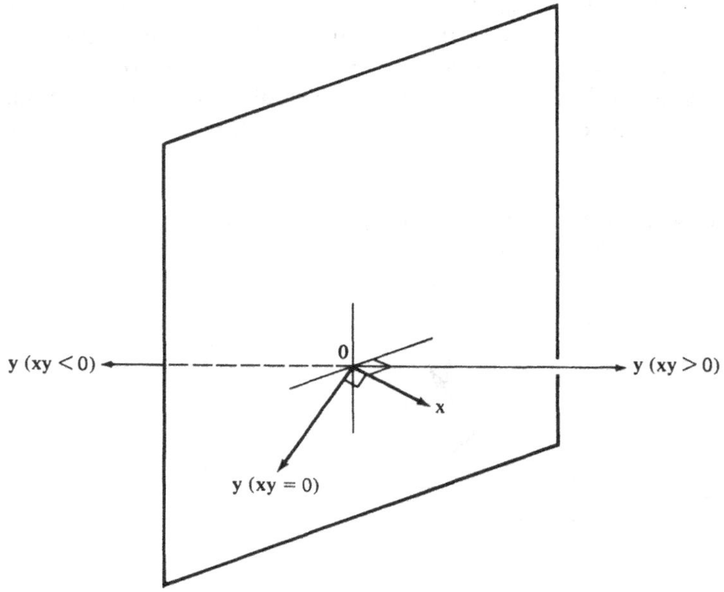

TABLE 4.1. Properties of Vector Operations

Vector addition	Vector multiplication
1. $\mathbf{x}+\mathbf{d}=\mathbf{d}+\mathbf{x}$	6. $\mathbf{xy}=\mathbf{yx}$
2. $\mathbf{x}+\mathbf{0}=\mathbf{0}+\mathbf{x}=\mathbf{x}$	7. $\mathbf{0x}=\mathbf{x0}=0$
3. $(\mathbf{x}+\mathbf{y})+\mathbf{d}=\mathbf{x}+(\mathbf{y}+\mathbf{d})$	8. $0\mathbf{x}=\mathbf{x}0=\mathbf{0}$
4. $\mathbf{x}-\mathbf{x}=\mathbf{0}$	10. $t(\mathbf{xy})=(t\mathbf{x})\mathbf{y}=\mathbf{x}(t\mathbf{y})$
5. $t(\mathbf{x}+\mathbf{d})=(t\mathbf{x})+(t\mathbf{d})$	11. $(\mathbf{x}+\mathbf{d})\mathbf{y}=(\mathbf{xy})+(\mathbf{dy})$

EXERCISES

4.1.1. Let $\mathbf{x}=(6,1)$ and $\mathbf{d}=(-4,3)$.
(a) Compute, algebraically and geometrically, the vectors $\mathbf{x}+\mathbf{d}$ and $\mathbf{x}-2\mathbf{d}$, as well as the real number $\mathbf{xd}$.
(b) Draw the line $\mathbf{x}+t\mathbf{d}$, where t can be any real number.

4.1.2. Given real numbers m and b with $m\neq 0$, consider the line $y=mx+b$. In terms of m and b, find vectors $\mathbf{x}=(x_1,x_2)$ and $\mathbf{d}=(d_1,d_2)$ so that the original line is the same as $\mathbf{x}+t\mathbf{d}$, where t can be any real number.

4.1.3. **(a)** Is the vector $\mathbf{x}=(-1,0,-2)\leq\mathbf{0}$ or $\geq\mathbf{0}$?
(b) Is the vector $\mathbf{x}=(-1,0,1)\leq\mathbf{0}$ or $\geq\mathbf{0}$?
(c) Use the techniques of Section 3.10 to define what it means to say that $\mathbf{x}$ is not greater than or equal to $\mathbf{0}$. Is this the same as $\mathbf{x}<\mathbf{0}$? Explain.

4.1.4. For each of the following pairs of vectors, determine if they lie in the same half space, opposite half spaces, or are perpendicular.
(a) $\mathbf{x}=(-1,0,5,3)$, $\mathbf{y}=(2,-2,-1,3)$.
(b) $\mathbf{x}=(2,0,-1,1)$, $\mathbf{y}=(0,5,3,3)$.
(c) $\mathbf{x}=(-2,2,3,-1)$, $\mathbf{y}=(2,1,-1,0)$.

4.1.5. Prove that if an n vector $\mathbf{y}\geq\mathbf{0}$ and $\mathbf{ey}=0$, then $\mathbf{y}=\mathbf{0}$.

4.1.6. Prove each of the properties in Table 4.1.

4.2. MATRICES

A *real matrix* is a rectangular table of real numbers; for instance,

$$\mathbf{A}=\begin{bmatrix}1&2&3\\4&5&6\end{bmatrix}$$

is a matrix. An $m\times n$ matrix $\mathbf{A}$ is one with m rows and n columns. The number of rows and columns of a matrix is called its *dimension*. Each individual number is called an *element* of the matrix. The element in row i and column j is denoted $(\mathbf{A})_{ij}$ or, more simply, $\mathbf{A}_{ij}$. Of course, it must be that $1 \le i \le m$ and $1 \le j \le n$.

Each of the m rows of an $m \times n$ matrix $\mathbf{A}$ is an n (row) vector. Row i of the matrix $\mathbf{A}$ will be denoted $\mathbf{A}_{i\cdot}$. For the 2×3 matrix above,

$$\mathbf{A}_{1\cdot} = (1,2,3), \qquad \mathbf{A}_{2\cdot} = (4,5,6)$$

In an analogous manner, each of the n columns of $\mathbf{A}$ is an m (column) vector, and column j of $\mathbf{A}$ will be denoted $\mathbf{A}_{\cdot j}$. For the 2×3 matrix above,

$$\mathbf{A}_{\cdot 1} = \begin{bmatrix} 1 \\ 4 \end{bmatrix}, \qquad \mathbf{A}_{\cdot 2} = \begin{bmatrix} 2 \\ 5 \end{bmatrix}, \qquad \mathbf{A}_{\cdot 3} = \begin{bmatrix} 3 \\ 6 \end{bmatrix}$$

Observe that an n vector $\mathbf{x}$ can be thought of as a matrix. The *row vector* $\mathbf{x} = (x_1, \ldots, x_n)$ is a $1 \times n$ matrix, and the *column vector*

$$\mathbf{x} = \begin{bmatrix} x_1 \\ \vdots \\ x_n \end{bmatrix}$$

is an $n \times 1$ matrix. The row vector is often referred to as the *transpose* of the column vector $\mathbf{x}$, and, in some texts, it is denoted $\mathbf{x}^{\mathrm{T}}$ to distinguish it from the column vector. Here, however, no such distinction will be made.

On the other hand, the transpose of an $m \times n$ matrix $\mathbf{A}$ will be denoted $\mathbf{A}^{\mathrm{T}}$, and is the $n \times m$ matrix in which column i is row i of $\mathbf{A}$. For the example above,

$$\mathbf{A}^{\mathrm{T}} = \begin{bmatrix} 1 & 4 \\ 2 & 5 \\ 3 & 6 \end{bmatrix}$$

As with vectors, it is possible to perform various algebraic operations on matrices. Throughout this section, $\mathbf{A}$ and $\mathbf{B}$ will be $m \times n$ matrices unless otherwise stated. Also, $\mathbf{x}$ will be an n vector and t a real number.

Matrix Addition

Two $m \times n$ matrices $\mathbf{A}$ and $\mathbf{B}$ can be added to obtain an $m \times n$ matrix $\mathbf{A} + \mathbf{B}$. To add $\mathbf{A}$ and $\mathbf{B}$ algebraically, simply add their corresponding elements, i.e., $(\mathbf{A} + \mathbf{B})_{ij} = \mathbf{A}_{ij} + \mathbf{B}_{ij}$. For example, if

$$\mathbf{A} = \begin{bmatrix} 1 & 2 & 3 \\ 4 & 5 & 6 \end{bmatrix}, \qquad \mathbf{B} = \begin{bmatrix} 3 & -1 & 0 \\ -5 & 1 & -4 \end{bmatrix}$$

then

$$\mathbf{A} + \mathbf{B} = \begin{bmatrix} 1+3 & 2-1 & 3+0 \\ 4-5 & 5+1 & 6-4 \end{bmatrix} = \begin{bmatrix} 4 & 1 & 3 \\ -1 & 6 & 2 \end{bmatrix}$$

Adding two matrices can also be accomplished by adding their corresponding rows (or columns) as vectors. For the example above,

$$\mathbf{A}+\mathbf{B}=\begin{bmatrix}(1, & 2, & 3)+(& 3, & -1, & 0)\\ (4, & 5, & 6)+(-5, & & 1, & -4)\end{bmatrix}=\begin{bmatrix}4 & 1 & 3\\ -1 & 6 & 2\end{bmatrix}$$

or, equivalently,

$$\mathbf{A}+\mathbf{B}=\left[\begin{bmatrix}1\\4\end{bmatrix}+\begin{bmatrix}3\\-5\end{bmatrix}, \begin{bmatrix}2\\5\end{bmatrix}+\begin{bmatrix}-1\\1\end{bmatrix}, \begin{bmatrix}3\\6\end{bmatrix}+\begin{bmatrix}0\\-4\end{bmatrix}\right]=\begin{bmatrix}4 & 1 & 3\\ -1 & 6 & 2\end{bmatrix}$$

From the example, it would appear that the matrix $\mathbf{A}+\mathbf{B}$ is the same as $\mathbf{B}+\mathbf{A}$. Algebraically, two $m\times n$ matrices are equal if all of their elements are equal.

Definition 4.4. Two $m\times n$ matrices are equal (written $\mathbf{A}=\mathbf{B}$) if, for all i with $1\le i\le m$, and for all j with $1\le j\le n$, $\mathbf{A}_{ij}=\mathbf{B}_{ij}$.

Equivalently, the matrices are equal if their corresponding rows are equal (i.e., for all i with $1\le i\le m$, $\mathbf{A}_{i\cdot}=\mathbf{B}_{i\cdot}$), or if their corresponding columns are equal (i.e., for all j with $1\le j\le n$, $\mathbf{A}_{\cdot j}=\mathbf{B}_{\cdot j}$).

Matrix Multiplication

To return to our discussion of operations on matrices, it is possible to multiply an $m\times n$ matrix $\mathbf{A}$ by a real number t. The resulting $m\times n$ matrix (written $t\mathbf{A}$) is obtained by multiplying each element of $\mathbf{A}$ by t; i.e., for all i with $1\le i\le m$, and for all j with $1\le j\le n$, $(t\mathbf{A})_{ij}=t(\mathbf{A}_{ij})$. For example, if

$$\mathbf{A}=\begin{bmatrix}1 & 2 & 3\\ 4 & 5 & 6\end{bmatrix}, \qquad t=2$$

then

$$t\mathbf{A}=\begin{bmatrix}2 & 4 & 6\\ 8 & 10 & 12\end{bmatrix}$$

Equivalently, the matrix $t\mathbf{A}$ can be obtained by multiplying each row (or column) of $\mathbf{A}$ by t. When $t=0$, $t\mathbf{A}$ is the matrix in which each element is 0. This particular matrix arises frequently and will be denoted by the same symbol as the zero vector, namely, $\mathbf{0}$. The meaning of the symbol will be clear from the context in which it is used.

The multiplication of two matrices requires more care than the corresponding operation for vectors. When multiplying the matrices $\mathbf{A}$ and $\mathbf{B}$, it is necessary to multiply each row of $\mathbf{A}$ by each column of $\mathbf{B}$. In order to be able to do so, each row of $\mathbf{A}$ must have the same dimension as each column of $\mathbf{B}$. In other words, to multiply the $m\times n$ matrix $\mathbf{A}$ by the matrix $\mathbf{B}$, it is necessary that $\mathbf{B}$ have the dimensions $n\times p$. Thus each row of $\mathbf{A}$ is an n vector, as is each column of $\mathbf{B}$. If this is the case, then the result of the multiplication (written **AB**) is an $m \times p$ matrix in which $(\mathbf{AB})_{ij}$ is row i of **A** multiplied by column j of **B**; i.e., $(\mathbf{AB})_{ij} = \mathbf{A}_{i.}\mathbf{B}_{.j}$. For example, if

$$\mathbf{A} = \begin{bmatrix} 1 & 2 & 3 \\ 4 & 5 & 6 \end{bmatrix}, \qquad \mathbf{B} = \begin{bmatrix} -1 & 2 \\ 1 & 0 \\ 0 & -1 \end{bmatrix}$$

then **AB** is the 2×2 matrix in which

$$\begin{aligned} (\mathbf{AB})_{11} &= \mathbf{A}_{1.}\mathbf{B}_{.1} = (1,2,3)(-1,1,0) = 1 \\ (\mathbf{AB})_{12} &= \mathbf{A}_{1.}\mathbf{B}_{.2} = (1,2,3)(2,0,-1) = -1 \\ (\mathbf{AB})_{21} &= \mathbf{A}_{2.}\mathbf{B}_{.1} = (4,5,6)(-1,1,0) = 1 \\ (\mathbf{AB})_{22} &= \mathbf{A}_{2.}\mathbf{B}_{.2} = (4,5,6)(2,0,-1) = 2 \end{aligned}$$

so

$$\mathbf{AB} = \begin{bmatrix} 1 & -1 \\ 1 & 2 \end{bmatrix}$$

Of particular interest in linear programming is the multiplication of an $m \times n$ matrix **A** by an $n \times 1$ column vector **x**. Consider, for example, the following constraints for an LP:

$$\begin{aligned} a_{11}x_1 + a_{12}x_2 + \cdots + a_{1n}x_n &= b_1 \\ a_{21}x_1 + a_{22}x_2 + \cdots + a_{2n}x_n &= b_2 \\ &\vdots \\ a_{m1}x_1 + a_{m2}x_2 + \cdots + a_{mn}x_n &= b_m \end{aligned}$$

By constructing the $m \times n$ matrix **A** in which $\mathbf{A}_{ij} = a_{ij}$ and the $n \times 1$ column vector

$$\mathbf{x} = \begin{bmatrix} x_1 \\ \vdots \\ x_n \end{bmatrix}$$

one can write the left-hand side of the constraints as the $m \times 1$ column vector **Ax**.

There are two instructive ways of performing the multiplication **Ax**. The first way is to multiply each row of **A** by **x** so, for each i with $1 \le i \le m$, $(\mathbf{Ax})_i = \mathbf{A}_{i.}\mathbf{x}$. This form of multiplication will be referred to as *row multiplication*.

The second method for computing **Ax** arises from observing that

$$\mathbf{Ax} = \begin{bmatrix} a_{11}x_1 + \cdots + a_{1n}x_n \\ \vdots \qquad\qquad \vdots \\ a_{m1}x_1 + \cdots + a_{mn}x_n \end{bmatrix} = \begin{bmatrix} a_{11} \\ \vdots \\ a_{m1} \end{bmatrix} x_1 + \cdots + \begin{bmatrix} a_{1n} \\ \vdots \\ a_{mn} \end{bmatrix} x_n$$

In other words, the vector **Ax** can be obtained by multiplying the first column of **A** by the first component of **x**, then adding the second column of **A** multiplied by the second component of **x**, and so on; i.e.,

$$\mathbf{Ax} = \sum_{j=1}^{n} \mathbf{A}_{.j}x_j = \mathbf{A}_{.1}x_1 + \mathbf{A}_{.2}x_2 + \cdots + \mathbf{A}_{.n}x_n$$

Observe that each vector $\mathbf{A}_{.j}x_j$ is an m vector. Computing **Ax** in this way will be referred to as *column multiplication*.

In a similar manner, it is possible to multiply an $m \times n$ matrix **A** on the left by a $1 \times m$ row vector, say, $\mathbf{u} = (u_1, \ldots, u_m)$. The result of the multiplication is the $1 \times n$ row vector **uA**. The multiplication can be performed by using either the columns of **A** (i.e., for all j with $1 \le j \le n$, $(\mathbf{uA})_j = \mathbf{uA}_{.j}$) or, equivalently, by using the rows of **A** (i.e., $\mathbf{uA} = \sum_{i=1}^{m} u_i \mathbf{A}_{i.}$).

Now that you know how to multiply an $m \times n$ matrix **A** by an $n \times p$ matrix **B**, it is instructive to compare the multiplication of matrices with the multiplication of real numbers. For instance, if a and b are real numbers, then it is always the case that $ab = ba$. On the other hand, for matrices, it may not always be possible to compute **BA** because the number of elements in a row of **B** might not be the same as the number of elements in a column of **A**. The one exception is when **A** and **B** are both *square matrices* (i.e., $n \times n$ matrices). Even in this case, **AB** might not be equal to **BA**, as the next example shows:

$$\mathbf{A} = \begin{bmatrix} 1 & 2 \\ 3 & 4 \end{bmatrix}, \qquad \mathbf{B} = \begin{bmatrix} 1 & -2 \\ 0 & 1 \end{bmatrix}$$

$$\mathbf{AB} = \begin{bmatrix} 1 & 0 \\ 3 & -2 \end{bmatrix}, \qquad \mathbf{BA} = \begin{bmatrix} -5 & -6 \\ 3 & 4 \end{bmatrix}$$

In the real number system, the number 1 has the property that, for all real numbers a, $(a)1 = 1(a) = a$. The corresponding notion for matrices is the $n \times n$ *identity matrix* **I** defined by

$$\mathbf{I} = \begin{bmatrix} 1 & 0 & \cdots & 0 \\ 0 & 1 & \cdots & 0 \\ & 0 & & \vdots \\ \vdots & \vdots & & 0 \\ 0 & 0 & \cdots & 1 \end{bmatrix}$$

It has the property that, for any $m \times n$ matrix **A**, **AI** = **A**, and, for any $n \times n$ matrix **B**, **IB** = **BI** = **B**. The symbol "**I**" will be used for the identity matrix, and its dimension will be determined by the context; e.g., **AI** means that **I** is $n \times n$, and **IA** means that **I** is $m \times m$. Other similarities between matrices

TABLE 4.2. Properties of Matrix Operations[a]

Matrix addition	Matrix multiplication
1. $\mathbf{A}+\mathbf{B}=\mathbf{B}+\mathbf{A}$	7. $\mathbf{AI}=\mathbf{A}$, $\mathbf{IB}=\mathbf{B}$
2. $\mathbf{A}+\mathbf{0}=\mathbf{0}+\mathbf{A}=\mathbf{A}$	8. $\mathbf{A}0=0\mathbf{A}=0$
3. $(\mathbf{A}+\mathbf{B})+\mathbf{C}=\mathbf{A}+(\mathbf{B}+\mathbf{C})$	9. $\mathbf{A0}=\mathbf{0}$
4. $t(\mathbf{A}+\mathbf{B})=(t\mathbf{A})+(t\mathbf{B})$	10. $(\mathbf{AB})\mathbf{C}=\mathbf{A}(\mathbf{BC})$
5. $(\mathbf{A}+\mathbf{B})\mathbf{x}=(\mathbf{Ax})+(\mathbf{Bx})$	11. $t(\mathbf{AB})=(t\mathbf{A})\mathbf{B}=\mathbf{A}(t\mathbf{B})$
6. $\mathbf{A}(\mathbf{x}+\mathbf{y})=(\mathbf{Ax})+(\mathbf{Ay})$	12. $\mathbf{A}(\mathbf{B}+\mathbf{C})=(\mathbf{AB})+(\mathbf{AC})$

[a]**A**, **B**, and **C** are matrices, **x** and **y** are vectors, and t is a scalar.

and real numbers are summarized in Table 4.2. The justification of the statements in Table 4.2 is left to the reader (see Exercise 4.2.3).

This section has introduced matrices and has shown the various operations that can be performed on them. The next two sections show how and when matrices can be used to solve n linear equations in n unknown variables. The need to do so arises repeatedly in solving linear programming problems.

EXERCISES

4.2.1. For each of the following LPs, identify the **A** matrix, the **c** vector (objective function), and the **b** vector (right-hand side), and indicate the dimension of each.

(a) Minimize

$$5x_1 + 3x_2 + x_3 - 9x_4 - 2x_5$$

subject to

$$\begin{aligned} 6x_1 - 2x_2 + 3x_3 - 7x_5 &= 23 \\ 11x_1 + x_2 + 13x_3 + 5x_4 + x_5 &= 32 \\ -8x_2 + 2x_3 + 3x_4 &= 17 \\ x_1,\ x_2,\ x_3,\ x_4,\ x_5 &\geq 0 \end{aligned}$$

(b) Maximize

$$-4x_1 + 6x_2 + 2x_3 - x_4$$

subject to

$$\begin{aligned} -x_2 + 13x_3 + 7x_4 &\leq 26 \\ 2x_1 + 3x_2 - 5x_3 + 6x_4 &\leq 43 \\ -7x_1 - 2x_2 + x_4 &\leq 3 \\ 12x_1 + x_3 + 7x_4 &\leq 30 \\ -3x_1 + 4x_2 + 3x_3 + 4x_4 &\leq 21 \\ x_1 + 6x_2 + 8x_3 &\leq 8 \\ x_1,\ x_2,\ x_3,\ x_4 &\geq 0 \end{aligned}$$

4.2.2. For the matrices

$$\mathbf{A} = \begin{bmatrix} 8 & 7 & -1 \\ -5 & -3 & 8 \\ 4 & 2 & 6 \end{bmatrix}, \quad \mathbf{B} = \begin{bmatrix} 0 & 3 & -6 \\ 2 & 9 & -5 \\ 3 & 4 & 2 \end{bmatrix}$$

and for the vector $\mathbf{x} = (3, -1, 2)$ and $t = -2$, compute

(a) $\mathbf{A} + \mathbf{B}$,

(b) $\mathbf{AB}$,

(c) $(\mathbf{BA})^{\mathrm{T}}$,

(d) $t\mathbf{A}$,

(e) $\mathbf{Ax}$ (by both row and column multiplication), and

(f) $\mathbf{xB}$ (by both row and column multiplication).

4.2.3. Prove each of the properties in Table 4.2.

4.3. SOLVING LINEAR EQUATIONS WITH MATRICES I

Given two real numbers a and b, it is an easy matter to determine a real number x such that $ax = b$. As long as $a \neq 0$, the unique value of x that satisfies $ax = b$ is b/a. In the development of the simplex algorithm, it will be necessary to solve a similar but more difficult problem, namely, that of solving linear equations. Given an $n \times n$ matrix $\mathbf{B}$ and an n vector $\mathbf{b}$, the problem is finding an $n \times 1$ vector $\mathbf{x}$ such that $\mathbf{Bx} = \mathbf{b}$.

Based on the corresponding problem for real numbers, one might conjecture that if $\mathbf{B}$ is not the zero matrix, then there is a vector $\mathbf{x}$ such that $\mathbf{Bx} = \mathbf{b}$. Unfortunately, this is not the case, but even if it were one would still need a method for computing the values for $\mathbf{x}$. Observe that it is not possible to set $\mathbf{x} = \mathbf{b}/\mathbf{B}$, for what does it mean to divide a vector by a matrix? This section will address the issue of developing conditions on the matrix $\mathbf{B}$ that guarantee the existence of a vector $\mathbf{x}$ such that $\mathbf{Bx} = \mathbf{b}$. The next section discusses methods for computing the values of $\mathbf{x}$ when $\mathbf{B}$ does satisfy the appropriate conditions.

The approach for solving linear equations can be understood best by considering again the corresponding problem for real numbers, namely, that of finding a value for x such that $ax = b$. Assuming that $a \neq 0$, one could divide both sides of the equation by a to find x. Alternatively, one could multiply both sides of the equation by $1/a$, thus

$$(1/a)(ax) = [(1/a)a]x = (1)x = x = (1/a)b$$

The advantage of computing x in this manner is that a similar approach can be developed for solving the linear system of equation $\mathbf{Bx} = \mathbf{b}$. Instead of trying to divide both sides by $\mathbf{B}$, suppose that it were possible to find an $n \times n$ matrix $\mathbf{C}$ for which $\mathbf{CB} = \mathbf{I}$ (the $n \times n$ identity matrix). In this case, one could multiply both sides of the equation $\mathbf{Bx} = \mathbf{b}$ by $\mathbf{C}$ to obtain

$$\mathbf{C}(\mathbf{Bx}) = (\mathbf{CB})\mathbf{x} = \mathbf{Ix} = \mathbf{x} = \mathbf{Cb}$$

In order to be sure that $\mathbf{x} = \mathbf{Cb}$ is a solution to the system of equations $\mathbf{Bx} = \mathbf{b}$, one must verify that $\mathbf{B(Cb)} = \mathbf{b}$, and this will be true provided that $\mathbf{BC} = \mathbf{I}$, for then

$$\mathbf{Bx} = \mathbf{B(Cb)} = \mathbf{(BC)b} = \mathbf{Ib} = \mathbf{b}$$

In other words, in order to solve the linear system of equations $\mathbf{Bx} = \mathbf{b}$, it would be desirable to find an $n \times n$ matrix $\mathbf{C}$ for which $\mathbf{CB} = \mathbf{BC} = \mathbf{I}$. The matrix $\mathbf{C}$ is called the *inverse* of $\mathbf{B}$ and will be written $\mathbf{B}^{-1}$ from now on. Unfortunately, the inverse matrix does not always exist, even when $\mathbf{B}$ is not the zero matrix. This discussion motivates the following definitions.

Definition 4.5. The $n \times n$ matrix $\mathbf{B}$ is *nonsingular* if there is an $n \times n$ matrix $\mathbf{B}^{-1}$ such that $\mathbf{B}^{-1}\mathbf{B} = \mathbf{BB}^{-1} = \mathbf{I}$.

Definition 4.6. The $n \times n$ matrix $\mathbf{B}$ is *singular* if it is not nonsingular.

One purpose of a nonsingular matrix is to ensure that, no matter what the values of $\mathbf{b}$ are, the linear system of equations $\mathbf{Bx} = \mathbf{b}$ has a unique solution, namely, $\mathbf{x} = \mathbf{B}^{-1}\mathbf{b}$, as shown in the next proposition.

Proposition 4.1. *If* $\mathbf{B}$ *is a nonsingular* $n \times n$ *matrix, then, for any* n *vector* $\mathbf{b}$, *there is a unique* n *vector* $\mathbf{x}$ *such that* $\mathbf{Bx} = \mathbf{b}$.

OUTLINE OF PROOF. The appearance of the quantifier "for any" in the conclusion suggests using the choose method to select a vector $\mathbf{b}$ for which it must be shown that there is a unique n vector $\mathbf{x}$ such that $\mathbf{Bx} = \mathbf{b}$. The word "unique" then suggests proceeding by one of the two uniqueness methods. Here, the first uniqueness method will be used, whereby one must first construct the desired $\mathbf{x}$ and subsequently show the uniqueness by assuming that there is a second solution $\mathbf{y}$ and then establishing that $\mathbf{x} = \mathbf{y}$.

To construct the desired solution $\mathbf{x}$, work forward from the hypothesis that $\mathbf{B}$ is nonsingular. By Definition 4.5, there is an $n \times n$ matrix $\mathbf{B}^{-1}$ for which $\mathbf{B}^{-1}\mathbf{B} = \mathbf{BB}^{-1} = \mathbf{I}$. The desired vector is $\mathbf{x} = \mathbf{B}^{-1}\mathbf{b}$. To see that this vector does indeed satisfy the linear system of equations, observe that

$$\mathbf{Bx} = \mathbf{B}(\mathbf{B}^{-1}\mathbf{b}) = (\mathbf{BB}^{-1})\mathbf{b} = \mathbf{Ib} = \mathbf{b}$$

It remains only to show that $\mathbf{x} = \mathbf{B}^{-1}\mathbf{b}$ is the unique solution to $\mathbf{Bx} = \mathbf{b}$. According to the first uniqueness method, one should assume that $\mathbf{x}$ and $\mathbf{y}$ are both solutions. Thus $\mathbf{Bx} = \mathbf{b}$ and $\mathbf{By} = \mathbf{b}$ or, equivalently, $\mathbf{Bx} = \mathbf{By}$. To reach the desired conclusion that $\mathbf{x} = \mathbf{y}$, the equation $\mathbf{Bx} = \mathbf{By}$ can be multiplied on both sides by $\mathbf{B}^{-1}$ to yield

$$
\begin{aligned}
(\mathbf{B}^{-1})\mathbf{Bx} &= (\mathbf{B}^{-1})\mathbf{By}\\
(\mathbf{B}^{-1}\mathbf{B})\mathbf{x} &= (\mathbf{B}^{-1}\mathbf{B})\mathbf{y}\\
\mathbf{Ix} &= \mathbf{Iy}\\
\mathbf{x} &= \mathbf{y}
\end{aligned}
$$

PROOF OF PROPOSITION 4.1. Let $\mathbf{b}$ be an n vector. First it will be shown that there is an n vector $\mathbf{x}$ such that $\mathbf{Bx} = \mathbf{b}$. By hypothesis, $\mathbf{B}$ is nonsingular, so, by the definition, there is an $n \times n$ matrix $\mathbf{B}^{-1}$ such that $\mathbf{B}^{-1}\mathbf{B} = \mathbf{BB}^{-1} = \mathbf{I}$. Let $\mathbf{x} = \mathbf{B}^{-1}\mathbf{b}$, for then

$$\mathbf{Bx} = \mathbf{B}(\mathbf{B}^{-1}\mathbf{b}) = (\mathbf{BB}^{-1})\mathbf{b} = \mathbf{Ib} = \mathbf{b}$$

To show that the linear system of equations has a unique solution, suppose that $\mathbf{x}$ and $\mathbf{y}$ are two solutions, so $\mathbf{Bx} = \mathbf{b}$ and $\mathbf{By} = \mathbf{b}$, or, equivalently, $\mathbf{Bx} = \mathbf{By}$. On multiplying both sides by $\mathbf{B}^{-1}$ one obtains $\mathbf{x} = \mathbf{y}$. □

The next task is to find conditions on the $\mathbf{B}$ matrix that ensure that it is nonsingular, and, subsequently, to compute the matrix $\mathbf{B}^{-1}$. Consider, for example, the 2×2 matrix

$$\mathbf{B} = \begin{bmatrix} a & b \\ c & d \end{bmatrix}$$

In order to construct the inverse matrix

$$\mathbf{B}^{-1} = \begin{bmatrix} e & f \\ g & h \end{bmatrix}$$

one would need e, f, g, h to satisfy the property that

$$\mathbf{BB}^{-1} = \begin{bmatrix} a & b \\ c & d \end{bmatrix}\begin{bmatrix} e & f \\ g & h \end{bmatrix} = \begin{bmatrix} ae + bg & af + bh \\ ce + dg & cf + dh \end{bmatrix} = \begin{bmatrix} 1 & 0 \\ 0 & 1 \end{bmatrix} = \mathbf{I}$$

In order words, given a, b, c, and d, the inverse matrix will exist if there are numbers e, f, g, and h such that

$$ae + bg = 1 \qquad (1)$$

$$ce + dg = 0 \qquad (2)$$

$$af + bh = 0 \qquad (3)$$

$$cf + dh = 1 \qquad (4)$$

Equations (1) and (2) might be used to solve for e and g. For instance, on multiplying (1) by d and (2) by b and then subtracting (2) from (1), one obtains $(ad - bc)e = d$. If $ad - bc \neq 0$, then it is possible to divide by $ad - bc$ to obtain $e = d/(ad - bc)$. In fact, if $ad - bc \neq 0$, it is possible to compute e, f, g, and h. Specifically,

$$e = d/(ad - bc)$$

$$f = -b/(ad - bc)$$

$$g = -c/(ad - bc)$$

$$h = a/(ad - bc)$$

In other words, if $ad - bc \neq 0$, then the inverse of $\mathbf{B}$ is

$$\mathbf{B}^{-1} = \begin{bmatrix} d/(ad-bc) & -b/(ad-bc) \\ -c(ad-bc) & a/(ad-bc) \end{bmatrix}$$

$$= \frac{1}{ad-bc}\begin{bmatrix} d & -b \\ -c & a \end{bmatrix}$$

The need to compute the inverse of a 2×2 matrix will arise frequently in this text. An easy way to remember the formula is to start with

$$\mathbf{B} = \begin{bmatrix} a & b \\ c & d \end{bmatrix}$$

First interchange a and d, then multiply b and c by -1, and finally multiply the resulting matrix by $1/(ad - bc)$. For example, if $a = 1$, $b = 2$, $c = 3$, and $d = 4$, then

$$\mathbf{B} = \begin{bmatrix} 1 & 2 \\ 3 & 4 \end{bmatrix}$$

and $ad - bc = -2$, and so

$$\mathbf{B}^{-1} = \frac{1}{-2}\begin{bmatrix} 4 & -2 \\ -3 & 1 \end{bmatrix} = \begin{bmatrix} -2 & 1 \\ \frac{3}{2} & -\frac{1}{2} \end{bmatrix}$$

As you have seen, the key to finding the inverse of the 2×2 matrix

$$\mathbf{B} = \begin{bmatrix} a & b \\ c & d \end{bmatrix}$$

is the number $ad - bc$. If this number is not 0, then the matrix $\mathbf{B}$ is nonsingular; otherwise, it is singular. In fact, associated with each $n \times n$ matrix is a number called the *determinant* of $\mathbf{B}$ (written det($\mathbf{B}$)). There is a general formula for using the elements of $\mathbf{B}$ to compute det($\mathbf{B}$). Moreover, when det($\mathbf{B}$) $\neq 0$, it is possible to use det($\mathbf{B}$) to find the inverse matrix that, in turn, can be used to solve a linear system of equations. More efficient methods have been found for solving a linear system of equations, as will be discussed in the next section. Therefore, the determinant will not be used to find the inverse of a matrix.

Nonetheless, the determinant of a matrix has a very important geometric interpretation. $|\det(\mathbf{B})|$ is the volume of the parallelepiped whose n edges are the n columns of $\mathbf{B}$, as shown in Figure 4.10.

Based on the geometry of Figure 4.10, another approach to determining if a matrix is nonsingular can be developed. The two vectors in Figure 4.10a "open up" properly, whereas those in Figure 4.10b do not. The property of "opening up" is formally referred to as linear independence.

Definition 4.7. Vectors $\mathbf{x}^1, \ldots, \mathbf{x}^k$ of dimension n are *linearly independent* if for every set of real numbers $t_1, \ldots, t_k$ for which $\mathbf{x}^1 t_1 + \cdots + \mathbf{x}^k t_k = \mathbf{0}$, it follows that $t_1 = t_2 = \cdots = t_k = 0$.

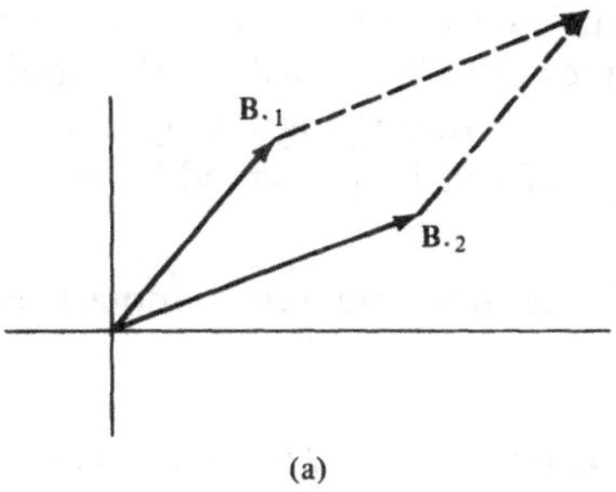

FIGURE 4.10. **(a)** Geometry of det(**B**) ≠ 0.
(b) Geometry of det(**B**) = 0.

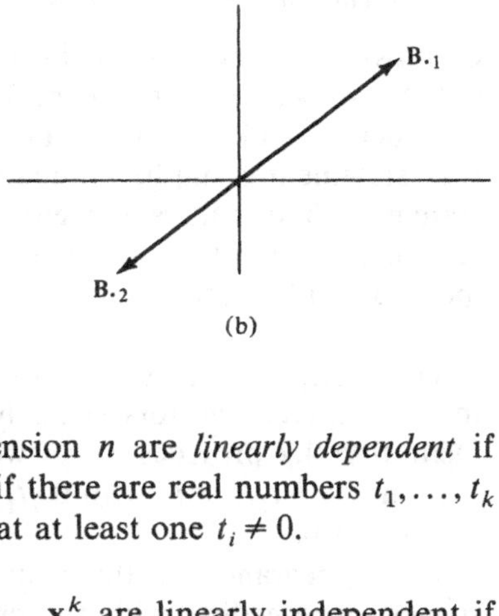

Definition 4.8. Vectors $\mathbf{x}^1,\ldots,\mathbf{x}^k$ of dimension n are *linearly dependent* if they are not linearly independent, i.e., if there are real numbers $t_1,\ldots,t_k$ for which $\mathbf{x}^1t_1+\cdots+\mathbf{x}^kt_k=\mathbf{0}$ such that at least one $t_i \neq 0$.

Definition 4.7 states that the vectors $\mathbf{x}^1,\ldots,\mathbf{x}^k$ are linearly independent if the only way to obtain the zero vector by adding multiples of $\mathbf{x}^1,\ldots,\mathbf{x}^k$ is to multiply each vector by 0.

The concept of linear independence provides yet another method for determining if a matrix is nonsingular, as shown in the next proposition.

Proposition 4.2. *If the columns of the $n \times n$ matrix* **B** *are linearly independent, then* **B** *is nonsingular.*

OUTLINE OF PROOF. The forward–backward method gives rise to the abstraction question "How can I show that a matrix (namely, **B**) is nonsingular?" According to Definition 4.5, one answer is to show that there is a matrix $\mathbf{B}^{-1}$ such that $\mathbf{B}\mathbf{B}^{-1}=\mathbf{B}^{-1}\mathbf{B}=\mathbf{I}$. The appearance of the quantifier "there is" now suggests using the construction method to produce the desired inverse matrix. To that end, a procedure will be suggested for attempting to construct $\mathbf{B}^{-1}$. To show that the proposed procedure does not fail, the contradiction method will be used. Specifically, if the procedure does fail, then it will be shown that the columns of **B** are linearly dependent, thus contradicting the hypothesis. The actual details of the method and the proof will be presented in the next section.

The next theorem summarizes the results of this section.

Theorem 4.1. *The following statements are equivalent for an $n \times n$ matrix* **B**:

1. **B** *is nonsingular.*
2. *For every n vector* **b**, *there is a unique n vector* **x** *such that* $\mathbf{Bx} = \mathbf{b}$.
3. *The columns of* **B** *are linearly independent.*

OUTLINE OF PROOF. To establish that the three statements are equivalent, it will be shown that statement 1 implies statement 2, statement 2 implies statement 3, and statement 3 implies statement 1. Proposition 4.1 establishes that statement 1 implies statement 2, and Proposition 4.2 establishes that statement 3 implies statement 1. Thus all that remains is to show that statement 2 implies statement 3. To that end, the forward–backward method will be used.

The abstraction question associated with statement 3 is "How can I show that n different vectors (namely, $\mathbf{B}_{.1}, \ldots, \mathbf{B}_{.n}$) are linearly independent?" Definition 4.7 provides an answer, whereby it must be shown that, for all real numbers $t_1, \ldots, t_n$ with $\mathbf{B}_{.1}t_1 + \cdots + \mathbf{B}_{.n}t_n = \mathbf{0}$, it follows that $t_1 = t_2 = \cdots = t_n = 0$.

The appearance of the quantifier "for all" in the backward process suggests proceeding with the choose method, so one should choose real numbers $t_1, \ldots, t_n$ for which $\mathbf{B}_{.1}t_1 + \cdots + \mathbf{B}_{.n}t_n = 0$ or, in matrix/vector notation, $\mathbf{Bt} = \mathbf{0}$, where $\mathbf{t} = (t_1, \ldots, t_n)$. It must be shown that $\mathbf{t} = \mathbf{0}$.

The desired conclusion can be obtained by applying specialization to statement 2. Specifically, specialize statement 2 to $\mathbf{b} = \mathbf{0}$, thus obtaining the conclusion that there is a unique n vector **x** such that $\mathbf{Bx} = \mathbf{0}$. Evidently, that unique vector is $\mathbf{x} = \mathbf{0}$. Consequently, since $\mathbf{Bt} = \mathbf{0}$, it must be that $\mathbf{t} = \mathbf{0}$, and the proof is complete.

PROOF OF THEOREM 4.1. Proposition 4.1 establishes that statement 1 implies statement 2, and Proposition 4.2 shows that statement 3 implies statement 1. Thus all that remains is to show that statement 2 implies statement 3. To that end, let $t_1, \ldots, t_n$ be real numbers with $\mathbf{B}_{.1}t_1 + \cdots + \mathbf{B}_{.n}t_n = \mathbf{0}$, or, equivalently, $\mathbf{Bt} = \mathbf{0}$, where $\mathbf{t} = (t_1, \ldots, t_n)$. It will be shown that $\mathbf{t} = \mathbf{0}$. By statement 2 it follows that there is a unique solution to $\mathbf{Bx} = \mathbf{0}$, namely, $\mathbf{x} = \mathbf{0}$. Since $\mathbf{Bt} = \mathbf{0}$, it must be that $\mathbf{t} = \mathbf{0}$. □

Another statement could have been added to the list in Theorem 4.1, namely,

4. $\det(\mathbf{B}) \neq 0$;

however, the necessary proofs will not be given here. Also, by replacing each statement in Theorem 4.1 with its negation, one obtains another useful theorem.

Theorem 4.2. *The following statements are equivalent for an* $n \times n$ *matrix* $\mathbf{B}$:

1. $\mathbf{B}$ *is singular.*
2. *There is an n vector* $\mathbf{b}$ *for which there does not exist a unique vector* $\mathbf{x}$ *such that* $\mathbf{Bx} = \mathbf{b}$.
3. *The columns of* $\mathbf{B}$ *are linearly dependent.*
4. $\det(\mathbf{B}) = 0$.

The final proposition of this section establishes that the product of two nonsingular matrices is nonsingular.

Proposition 4.3. *If* $\mathbf{A}$ *and* $\mathbf{B}$ *are nonsingular* $n \times n$ *matrices, then the matrix* $\mathbf{AB}$ *is nonsingular, and* $(\mathbf{AB})^{-1} = \mathbf{B}^{-1}\mathbf{A}^{-1}$.

OUTLINE OF PROOF. The forward–backward method gives rise to the abstraction question "How can I show that a matrix (namely, $\mathbf{AB}$) is nonsingular?" One answer is by Definition 4.5, whereby it must be shown that there is an $n \times n$ matrix $\mathbf{C}$ such that $\mathbf{C}(\mathbf{AB}) = (\mathbf{AB})\mathbf{C} = \mathbf{I}$.

The appearance of the quantifier "there is" suggests turning to the forward process to construct the desired matrix $\mathbf{C}$. Specifically, since $\mathbf{A}$ is nonsingular, by the definition, there is an $n \times n$ matrix $\mathbf{A}^{-1}$ such that $\mathbf{AA}^{-1} = \mathbf{A}^{-1}\mathbf{A} = \mathbf{I}$. Similarly, from the hypothesis that $\mathbf{B}$ is nonsingular, there is an $n \times n$ matrix $\mathbf{B}$ such that $\mathbf{BB}^{-1} = \mathbf{B}^{-1}\mathbf{B} = \mathbf{I}$. The desired matrix $\mathbf{C}$ is $\mathbf{B}^{-1}\mathbf{A}^{-1}$. To see that this construction is correct, it must be shown that $\mathbf{C}(\mathbf{AB}) = (\mathbf{AB})\mathbf{C} = \mathbf{I}$, but

$$\begin{aligned} \mathbf{C}(\mathbf{AB}) &= (\mathbf{B}^{-1}\mathbf{A}^{-1})(\mathbf{AB}) = \mathbf{B}^{-1}(\mathbf{A}^{-1}\mathbf{A})\mathbf{B} \\ &= \mathbf{B}^{-1}\mathbf{IB} = \mathbf{B}^{-1}\mathbf{B} \\ &= \mathbf{I} \end{aligned}$$

and

$$\begin{aligned} (\mathbf{AB})\mathbf{C} &= (\mathbf{AB})(\mathbf{B}^{-1}\mathbf{A}^{-1}) = \mathbf{A}(\mathbf{BB}^{-1})\mathbf{A}^{-1} \\ &= \mathbf{AIA}^{-1} = \mathbf{AA}^{-1} \\ &= \mathbf{I} \end{aligned}$$

PROOF OF PROPOSITION 4.3. By the hypothesis that $\mathbf{A}$ and $\mathbf{B}$ are nonsingular, there are $n \times n$ matrices $\mathbf{A}^{-1}$ and $\mathbf{B}^{-1}$ such that $\mathbf{AA}^{-1} = \mathbf{A}^{-1}\mathbf{A} = \mathbf{I}$ and

$\mathbf{BB}^{-1} = \mathbf{B}^{-1}\mathbf{B} = \mathbf{I}$. Let $\mathbf{C} = \mathbf{B}^{-1}\mathbf{A}^{-1}$; then

$$\mathbf{C}(\mathbf{AB}) = (\mathbf{B}^{-1}\mathbf{A}^{-1})(\mathbf{AB}) = \mathbf{B}^{-1}\mathbf{IB} = \mathbf{I}$$

and

$$(\mathbf{AB})\mathbf{C} = (\mathbf{AB})(\mathbf{B}^{-1}\mathbf{A}^{-1}) = \mathbf{AIA}^{-1} = \mathbf{I}$$

Thus

$$(\mathbf{AB})^{-1} = \mathbf{C} = \mathbf{B}^{-1}\mathbf{A}^{-1}$$ □

This section has provided various conditions under which an $n \times n$ matrix $\mathbf{B}$ has an inverse. The next section addresses the issues of computing the inverse matrix and solving linear equations.

EXERCISES

4.3.1. Determine which of the following matrices are nonsingular and, for each such matrix, find the inverse:

$$\begin{bmatrix} 1 & 2 \\ 3 & 4 \end{bmatrix}, \quad \begin{bmatrix} 2 & 3 \\ 1 & 5 \end{bmatrix}$$

$$\begin{bmatrix} 6 & 10 \\ 15 & 25 \end{bmatrix}, \quad \begin{bmatrix} 5 & 4 \\ 1 & -1 \end{bmatrix}$$

4.3.2. Use the inverse of the matrices in Exercise 4.3.1 to solve the following systems of equations.
(a) $x + 2y = 6$, $3x + 4y = 7$;
(b) $2x + 3y = 3$, $x + 5y = 0$;
(c) $5x + 4y = 3$, $x - y = 10$.

4.3.3. Determine, algebraically and geometrically, which of the following pairs of vectors are linearly independent.
(a) $(3,5)$, $(7,9)$;
(b) $(1,2)$, $(-2,-4)$;
(c) $(2,3,2)$, $(2,3,1)$;
(d) $(3,2,5)$, $(7,\frac{14}{3},\frac{35}{3})$.

4.3.4. Let $\mathbf{B}$ be an $n \times n$ matrix and y and n vector. Prove that if $\mathbf{By} = \mathbf{0}$ and $\mathbf{y} \neq \mathbf{0}$, then $\mathbf{B}$ is singular.

4.3.5. Prove that if $\mathbf{B}$ is a nonsingular $n \times n$ matrix, then $\mathbf{B}^{\mathrm{T}}$ is a nonsingular matrix and $(\mathbf{B}^{\mathrm{T}})^{-1} = (\mathbf{B}^{-1})^{\mathrm{T}}$. You can assume knowledge of the fact that $(\mathbf{AB})^{\mathrm{T}} = \mathbf{B}^{\mathrm{T}}\mathbf{A}^{\mathrm{T}}$.

4.3.6. Let $\mathbf{x}^1, \mathbf{x}^2, \ldots, \mathbf{x}^k$ be k linearly independent n vectors. Prove that any subset of these vectors is also linearly independent.

4.4. SOLVING LINEAR EQUATIONS WITH MATRICES II

In the previous section it was shown that if an $n \times n$ matrix $\mathbf{B}$ is nonsingular (or has linearly independent columns, or has a nonzero determinant), then $\mathbf{B}$ has an inverse $\mathbf{B}^{-1}$. The purpose of this section is to discuss several methods for computing $\mathbf{B}^{-1}$ and for using $\mathbf{B}$ to solve a linear system of equations. In general, computing the inverse of a matrix is somewhat complicated; however, for certain matrices, such as diagonal matrices, the inverse can be computed very easily.

Definition 4.9. An $n \times n$ matrix $\mathbf{D}$ is *diagonal* if all elements except possibly the diagonal ones are 0, i.e., if for all i with $1 \le i \le n$, and for all j with $1 \le j \le n$, with $i \ne j$, $\mathbf{D}_{ij} = 0$.

An example of a diagonal matrix is

$$\mathbf{D} = \begin{bmatrix} 1 & 0 & 0 & 0 \\ 0 & 2 & 0 & 0 \\ 0 & 0 & -3 & 0 \\ 0 & 0 & 0 & 4 \end{bmatrix}$$

In the event that none of the diagonal elements of $\mathbf{D}$ is 0, the inverse can be obtained by creating the diagonal matrix $\mathbf{D}^{-1}$ whose diagonal elements are the reciprocals of those of $\mathbf{D}$. For the example above

$$\mathbf{D}^{-1} = \begin{bmatrix} 1 & 0 & 0 & 0 \\ 0 & \frac{1}{2} & 0 & 0 \\ 0 & 0 & -\frac{1}{3} & 0 \\ 0 & 0 & 0 & \frac{1}{4} \end{bmatrix}$$

It is easy to verify that $\mathbf{D}\mathbf{D}^{-1} = \mathbf{D}^{-1}\mathbf{D} = \mathbf{I}$.

Another class of matrices for which it is easy to find the inverse is permutation matrices.

Definition 4.10. An $n \times n$ matrix $\mathbf{P}$ is a *permutation matrix* if in each row and each column all elements are 0 except for one element whose value is 1.

For example,

$$\mathbf{P} = \begin{bmatrix} 1 & 0 & 0 & 0 \\ 0 & 0 & 0 & 1 \\ 0 & 0 & 1 & 0 \\ 0 & 1 & 0 & 0 \end{bmatrix}$$

It is left as an exercise (see Exercise 4.4.6) to show that the inverse of a permutation matrix is its transpose, i.e., $\mathbf{P}^{-1} = \mathbf{P}^{\mathrm{T}}$.

A permutation matrix provides a convenient method for interchanging the columns (or rows) of a matrix, say, **B**. For instance, to interchange columns j and k of **B**, one need only multiply **B** on the right by the permutation matrix in which columns j and k of the identity matrix have been interchanged. The permutation matrix **P** given above would interchange columns 2 and 4 of a 4×4 matrix **B** when multiplied on the right; for example,

$$\mathbf{B} = \begin{bmatrix} 1 & 2 & 3 & 4 \\ 5 & 6 & 7 & 8 \\ 9 & 10 & 11 & 12 \\ 13 & 14 & 15 & 16 \end{bmatrix}$$

$$\mathbf{BP} = \begin{bmatrix} 1 & 4 & 3 & 2 \\ 5 & 8 & 7 & 6 \\ 9 & 12 & 11 & 10 \\ 13 & 16 & 15 & 14 \end{bmatrix}$$

In an analogous manner, multiplying **B** on the left by **P** will interchange rows j and k, so for the example

$$\mathbf{PB} = \begin{bmatrix} 1 & 2 & 3 & 4 \\ 13 & 14 & 15 & 16 \\ 9 & 10 & 11 & 12 \\ 5 & 6 & 7 & 8 \end{bmatrix}$$

Yet another class of matrices for which it is relatively easy to compute the inverse is eta matrices.

Definition 4.11. An $n \times n$ matrix **E** is an *eta matrix* if it differs from the identity matrix in only one column.

The matrix

$$\mathbf{E} = \begin{bmatrix} 1 & 0 & 0 \\ 0 & 2 & 0 \\ 0 & 4 & 1 \end{bmatrix}$$

is an eta matrix.

To find the inverse of an eta matrix **E** in which column k is different from that of the identity matrix, it is necessary that $\mathbf{E}_{kk} \neq 0$, for then $\mathbf{E}^{-1}$ is the eta matrix in which column k is

$$(\mathbf{E}^{-1})_{\cdot k} = \begin{bmatrix} -\mathbf{E}_{1k}/\mathbf{E}_{kk} \\ \vdots \\ 1/\mathbf{E}_{kk} \\ \vdots \\ -\mathbf{E}_{nk}/\mathbf{E}_{kk} \end{bmatrix}$$

so

$$\mathbf{E}^{-1} = \begin{bmatrix} 1 & -\mathbf{E}_{1k}/\mathbf{E}_{kk} & 0 \\ 0 & \vdots & \\ \vdots & 1/\mathbf{E}_{kk} & \vdots \\ & \vdots & 0 \\ 0 & -\mathbf{E}_{nk}/\mathbf{E}_{kk} & 1 \end{bmatrix}$$

Thus for the 3×3 matrix above,

$$\mathbf{E}^{-1} = \begin{bmatrix} 1 & 0 & 0 \\ 0 & \frac{1}{2} & 0 \\ 0 & -2 & 1 \end{bmatrix}$$

It is not hard to verify by direct computation that the formula for $\mathbf{E}^{-1}$ is correct (see Exercises 4.4.2).

Slightly more complicated classes of matrices whose inverse can be computed easily are the upper and lower triangular matrices.

Definition 4.12. An $n \times n$ matrix $\mathbf{U}$ is *upper triangular* if all elements below the diagonal are 0; i.e., if for all i with $1 \le i \le n$, and for all j with $1 \le j < i$, $\mathbf{U}_{ij} = 0$.

Definition 4.13. An $n \times n$ matrix $\mathbf{L}$ is *lower triangular* if all elements above the diagonal are 0; i.e., if for all i with $1 \le i \le n$, and for all j with $i < j \le n$, $\mathbf{L}_{ij} = 0$.

An example of an upper and a lower triangular matrix would be

$$\mathbf{U} = \begin{bmatrix} 1 & 2 & 3 & 4 \\ 0 & 5 & 6 & 7 \\ 0 & 0 & 8 & 9 \\ 0 & 0 & 0 & 10 \end{bmatrix}, \qquad \mathbf{L} = \begin{bmatrix} 1 & 0 & 0 & 0 \\ 2 & 3 & 0 & 0 \\ 4 & 5 & 6 & 0 \\ 7 & 8 & 9 & 10 \end{bmatrix}$$

respectively.

It is not difficult to find the inverse of a lower or upper triangular matrix provided that none of the diagonal elements is 0. One such method will be developed in Exercise 4.4.3. It is interesting to note that when a nonsingular lower or upper triangular matrix appears in a linear system of equations it is not necessary to compute the inverse matrix to solve the system. Instead, it is possible (and more efficient) to proceed by a process known as *back substitution*.

Consider, for example, the linear system of equations $\mathbf{Ux} = \mathbf{b}$:

$$\begin{aligned} 2x_1 + 2x_2 + x_3 &= 5 \\ 4x_2 + 5x_3 &= 1 \\ 6x_3 &= 6 \end{aligned}$$

To solve the system, one can use the last equation to solve for x_3, obtaining $x_3 = 1$. On substituting this value for x_3 in the second equation, one obtains $4x_2 + 5 = 1$ or, equivalently, $4x_2 = -4$. Thus the second equation can be used to solve for x_2, obtaining $x_2 = -1$. Finally, on substituting these values for x_2 and x_3 in the first equation, one obtains $2x_1 - 2 + 1 = 5$ or, equivalently, $2x_1 = 6$. Thus the first equation can be used to solve for x_1, obtaining $x_1 = 3$. A similar approach (known as *forward substitution*) can be used to solve the lower triangular system of equations $\mathbf{Lx} = \mathbf{b}$.

Finally, with the use of eta matrices and permutation matrices, a general method for attempting to compute the inverse of a matrix will be presented. When the method fails to produce the inverse, it stops by concluding that the matrix is singular, and hence there is no inverse. The method is referred to as *Gaussian elimination*.

The method of Gaussian elimination attempts to produce m eta matrices, $\mathbf{E}_1, \ldots, \mathbf{E}_m$, for which $(\mathbf{E}_m \cdots \mathbf{E}_1)\mathbf{B} = \mathbf{I}$, for then, $\mathbf{B}^{-1} = \mathbf{E}_m \cdots \mathbf{E}_1$. Of course, it must also be shown that $\mathbf{B}(\mathbf{E}_m \cdots \mathbf{E}_1) = \mathbf{I}$; however, if $\mathbf{CB} = \mathbf{I}$, then $\mathbf{BC} = \mathbf{I}$ (the proof will not be given here). The matrix $\mathbf{E}_1$ is chosen so that the first column of $\mathbf{E}_1\mathbf{B}$ is the first column of $\mathbf{I}$. Then, starting with $\mathbf{E}_1\mathbf{B}$, $\mathbf{E}_2$ is chosen so that the first two columns of $\mathbf{E}_2(\mathbf{E}_1\mathbf{B})$ are the first two columns of $\mathbf{I}$. In general, starting with $\mathbf{E}_{k-1} \cdots \mathbf{E}_1\mathbf{B}$, $\mathbf{E}_k$ is chosen so that the first k columns of $\mathbf{E}_k \cdots \mathbf{E}_1\mathbf{B}$ are the first k columns of $\mathbf{I}$. Thus, after m iterations, $\mathbf{E}_m \cdots \mathbf{E}_1\mathbf{B}$ will be the identity matrix.

The kth eta matrix, $\mathbf{E}_k$, is obtained by replacing column k of the identity matrix with a single column vector $\mathbf{w}$ that is derived from column k of $\mathbf{E}_{k-1} \cdots \mathbf{E}_1\mathbf{B}$. For notational purposes let $\mathbf{C} = \mathbf{E}_{k-1} \cdots \mathbf{E}_1\mathbf{B}$. The vector $\mathbf{w}$ is defined by

$$w_i = \begin{cases} 1/\mathbf{C}_{kk} & \text{if } i = k \\ -\mathbf{C}_{ik}/\mathbf{C}_{kk} & \text{if } i \neq k \end{cases}$$

provided, of course, that $\mathbf{C}_{kk} \neq 0$. Consider the next example.

EXAMPLE 4.1: Consider the following matrix $\mathbf{C}$, the first column of which is $\mathbf{I}_{.1}$:

$$\mathbf{C} = \begin{bmatrix} 1 & 2 & 1 & 1 & 1 \\ 0 & 5 & 5 & 0 & 1 \\ 0 & -5 & -5 & -1 & 2 \\ 0 & 9 & 10 & 2 & 4 \\ 0 & 5 & 7 & 6 & -1 \end{bmatrix}$$

From the formula the eta matrix is

$$\mathbf{E}_2 = \begin{bmatrix} 1 & -\frac{2}{5} & 0 & 0 & 0 \\ 0 & \frac{1}{5} & 0 & 0 & 0 \\ 0 & 1 & 1 & 0 & 0 \\ 0 & -\frac{9}{5} & 0 & 1 & 0 \\ 0 & -1 & 0 & 0 & 1 \end{bmatrix}$$

It can readily be seen that the first two columns of $\mathbf{E}_2\mathbf{C}$ are those of the identity matrix:

$$\mathbf{E}_2\mathbf{C} = \begin{bmatrix} 1 & -\frac{2}{5} & 0 & 0 & 0 \\ 0 & \frac{1}{5} & 0 & 0 & 0 \\ 0 & 1 & 1 & 0 & 0 \\ 0 & -\frac{9}{5} & 0 & 1 & 0 \\ 0 & -1 & 0 & 0 & 1 \end{bmatrix} \begin{bmatrix} 1 & 2 & 1 & 1 & 1 \\ 0 & 5 & 5 & 0 & 1 \\ 0 & -5 & -5 & -1 & 2 \\ 0 & 9 & 10 & 2 & 4 \\ 0 & 5 & 7 & 6 & -1 \end{bmatrix}$$

$$= \begin{bmatrix} 1 & 0 & -1 & 1 & \frac{3}{5} \\ 0 & 1 & 1 & 0 & \frac{1}{5} \\ 0 & 0 & 0 & -1 & 3 \\ 0 & 0 & 1 & 2 & \frac{11}{5} \\ 0 & 0 & 2 & 6 & -2 \end{bmatrix}$$

If $\mathbf{C}_{kk} = 0$, an alternative strategy must be found. The method consists of looking down column k of $\mathbf{C}$ below row k in search of a nonzero element. If one can be found (say, in row $r > k$), then row k and row r are interchanged by multiplying $\mathbf{C}$ on the left by the appropriate permutation matrix $\mathbf{P}$. The result is that, in the new matrix $\mathbf{PC}$, $(\mathbf{PC})_{kk} \neq 0$ and so the eta matrix $\mathbf{E}_k$ can then be found, as illustrated in the next example.

EXAMPLE 4.2: From Example 4.1, $(\mathbf{E}_2\mathbf{C})_{33} = 0$, so let

$$\mathbf{P}_3 = \begin{bmatrix} 1 & 0 & 0 & 0 & 0 \\ 0 & 1 & 0 & 0 & 0 \\ 0 & 0 & 0 & 1 & 0 \\ 0 & 0 & 1 & 0 & 0 \\ 0 & 0 & 0 & 0 & 1 \end{bmatrix}$$

On multiplying $\mathbf{E}_2\mathbf{C}$ on the left by $\mathbf{P}_3$ one obtains

$$\mathbf{P}_3\mathbf{E}_2\mathbf{C} = \begin{bmatrix} 1 & 0 & 0 & 0 & 0 \\ 0 & 1 & 0 & 0 & 0 \\ 0 & 0 & 0 & 1 & 0 \\ 0 & 0 & 1 & 0 & 0 \\ 0 & 0 & 0 & 0 & 1 \end{bmatrix} \begin{bmatrix} 1 & 0 & -1 & 1 & \frac{3}{5} \\ 0 & 1 & 1 & 0 & \frac{1}{5} \\ 0 & 0 & 0 & -1 & 3 \\ 0 & 0 & 1 & 2 & \frac{11}{5} \\ 0 & 0 & 2 & 6 & -2 \end{bmatrix}$$

$$= \begin{bmatrix} 1 & 0 & -1 & 1 & \frac{3}{5} \\ 0 & 1 & 1 & 0 & \frac{1}{5} \\ 0 & 0 & 1 & 2 & \frac{11}{5} \\ 0 & 0 & 0 & -1 & 3 \\ 0 & 0 & 2 & 6 & -2 \end{bmatrix}$$

Now observe that $(\mathbf{P}_3\mathbf{E}_2\mathbf{C})_{33} \neq 0$ and the next eta matrix $\mathbf{E}_3$ can be computed so that the first three columns of $\mathbf{E}_3(\mathbf{P}_3\mathbf{E}_2\mathbf{C})$ are those of the identity matrix.

From Example 4.2 it should be clear that if $\mathbf{C}_{kk} = 0$, then you want to look down column k of $\mathbf{C}$ only below row k. Only those rows below k can be exchanged with row k without destroying the property that the first k columns of $\mathbf{C}$ are those of the identity matrix.

The final issue to be addressed is what happens if $\mathbf{C}_{kk} = 0$ and, below row k of $\mathbf{C}$, column k contains all zeros. Now no row of $\mathbf{C}$ can be interchanged with row k. The next proposition formally establishes that, in this event, the columns of $\mathbf{B}$ are linearly dependent, and hence no inverse matrix can be found, as was stated in Proposition 4.2 of the previous section, which is repeated here.

Proposition 4.2. *If the columns of the $n \times n$ matrix $\mathbf{B}$ are linearly independent, then $\mathbf{B}$ is nonsingular.*

OUTLINE OF PROOF. The forward–backward method gives rise to the abstraction question "How can I show that a matrix (namely, $\mathbf{B}$) is nonsingular?" Using the definition, it must be shown that there is an $n \times n$ matrix $\mathbf{B}^{-1}$ such that $\mathbf{B}\mathbf{B}^{-1} = \mathbf{B}^{-1}\mathbf{B} = \mathbf{I}$. Because the key words "there is" appear, the construction method will be used to produce $\mathbf{B}^{-1}$. To that end, apply the procedure of Gaussian elimination to $\mathbf{B}$. To show that this procedure cannot fail to produce the inverse matrix, the key word "not" suggests proceeding with the contrapositive method. Specifically, it will be shown that if Gaussian elimination does fail to produce the inverse matrix, then the columns of $\mathbf{B}$ are not linearly independent; i.e., the columns of $\mathbf{B}$ are linearly dependent.

When trying to show that the columns of $\mathbf{B}$ are linearly dependent, one is led to the abstraction question "How can I show that a collection of vectors (namely, $\mathbf{B}_{.1}, \ldots, \mathbf{B}_{.n}$) is linearly dependent?" Using the definition provides the answer that one must construct scalars $t_1, \ldots, t_n$, not all 0, such that $\mathbf{B}_{.1}t_1 + \cdots + \mathbf{B}_{.n}t_n = \mathbf{0}$. Indeed, when the procedure of Gaussian elimination fails, the desired scalars will be produced.

Specifically, suppose that the nonsingular eta matrices $\mathbf{E}_{k-1}, \ldots, \mathbf{E}_1$ have been produced by Gaussian elimination. Thus the first $k-1$ columns of $\mathbf{C} = \mathbf{E}_{k-1} \cdots \mathbf{E}_1\mathbf{B}$ are those of the identity matrix. If Gaussian elimination fails to produce $\mathbf{E}_k$, then, for each $i \geq k$, $\mathbf{C}_{ik} = 0$. The desired scalars $t_1, \ldots, t_n$ are

$$t_i = \begin{cases} -\mathbf{C}_{ik} & \text{if} \quad i < k \\ 1 & \text{if} \quad i = k \\ 0 & \text{if} \quad i > k \end{cases}$$

To see that this construction is correct, it must be shown that

1. not all of these scalars are 0 and
2. $\mathbf{B}_{.1}t_1 + \cdots + \mathbf{B}_{.n}t_n = \mathbf{0}$ or, equivalently, $\mathbf{Bt} = \mathbf{0}$, where $\mathbf{t}$ is the column vector in which component i is t_i.

The first statement is true because $t_k = 1$, and so not all of the scalars are 0. To prove statement 2, it will be shown that $\mathbf{Ct} = \mathbf{0}$, for then, since $\mathbf{C} = \mathbf{E}_{k-1} \cdots \mathbf{E}_1\mathbf{B}$, it will follow that $\mathbf{E}_{k-1} \cdots \mathbf{E}_1\mathbf{Bt} = \mathbf{0}$. Then, by multiplying on the left by the inverses of $\mathbf{E}_{k-1}, \ldots, \mathbf{E}_1$, one will obtain precisely that $\mathbf{Bt} = \mathbf{0}$.

To see that $\mathbf{Ct} = \mathbf{0}$, consider $\mathbf{Ct}$ as column multiplication. Then

$$\begin{aligned}
\mathbf{Ct} &= \mathbf{C}_{\cdot 1}t_1 + \cdots + \mathbf{C}_{\cdot n}t_n \\
&= [\mathbf{C}_{\cdot 1}t_1 + \cdots + \mathbf{C}_{\cdot k-1}t_{k-1}] + \mathbf{C}_{\cdot k}t_k + [\mathbf{C}_{\cdot k+1}t_{k+1} + \cdots + \mathbf{C}_{\cdot n}t_n] \\
&= [-\mathbf{I}_{\cdot 1}\mathbf{C}_{1k} - \cdots - \mathbf{I}_{\cdot k-1}\mathbf{C}_{k-1,k}] + \mathbf{C}_{\cdot k}(1) \\
&= -\mathbf{C}_{\cdot k} + \mathbf{C}_{\cdot k} \\
&= \mathbf{0}
\end{aligned}$$

PROOF OF PROPOSITION 4.2. To show that $\mathbf{B}$ is nonsingular, Gaussian elimination will be used to produce the inverse matrix. Suppose that $\mathbf{E}_{k-1}, \ldots, \mathbf{E}_1$ are nonsingular eta matrices such that the first k columns of $\mathbf{C} = \mathbf{E}_{k-1} \cdots \mathbf{E}_1\mathbf{B}$ are those of the identity matrix. If Gaussian elimination cannot produce $\mathbf{E}_k$, then, for each $i \geq k$, $\mathbf{C}_{ik} = 0$. To see that this situation cannot happen, construct scalars $t_1, \ldots, t_n$ (not all of which are 0) by

$$t_i = \begin{cases} -\mathbf{C}_{ik} & \text{if} \quad i < k \\ 1 & \text{if} \quad i = k \\ 0 & \text{if} \quad i > k \end{cases}$$

It is not hard to see that $\mathbf{Ct} = \mathbf{0}$ (where component i of $\mathbf{t}$ is t_i), for

$$\begin{aligned}
\mathbf{Ct} &= [\mathbf{C}_{\cdot 1}t_1 + \cdots + \mathbf{C}_{\cdot k-1}t_{k-1}] + \mathbf{C}_{\cdot k}t_k + [\mathbf{C}_{\cdot k+1}t_{k+1} + \cdots + \mathbf{C}_{\cdot n}t_n] \\
&= [-\mathbf{I}_{\cdot 1}\mathbf{C}_{1k} - \cdots - \mathbf{I}_{\cdot k-1}\mathbf{C}_{k-1,k}] + \mathbf{C}_{\cdot k}(1) \\
&= -\mathbf{C}_{\cdot k} + \mathbf{C}_{\cdot k} \\
&= \mathbf{0}
\end{aligned}$$

But then it would follow that $\mathbf{E}_{k-1} \cdots \mathbf{E}_1\mathbf{Bt} = \mathbf{0}$ and so $\mathbf{Bt} = \mathbf{0}$. This would contradict the assumption that the columns of $\mathbf{B}$ are linearly independent.□

Many other improvements have been made to the procedure for solving linear systems of equations. Some of them will be described in Chapter 7. For now, however, it is time to see how linear algebra is useful in linear programming.

EXERCISES

4.4.1. Find the inverses of the following matrices:

(a) $\begin{bmatrix} 2 & 0 & 0 & 0 \\ 0 & -5 & 0 & 0 \\ 0 & 0 & 3 & 0 \\ 0 & 0 & 0 & -1 \end{bmatrix}$

(b) $\begin{bmatrix} 0 & 0 & 1 & 0 \\ 0 & 1 & 0 & 0 \\ 1 & 0 & 0 & 0 \\ 0 & 0 & 0 & 1 \end{bmatrix}$

(c) $\begin{bmatrix} 1 & 0 & 0 & 0 \\ 0 & -2 & 0 & 0 \\ 0 & 5 & 1 & 0 \\ 0 & 1 & 0 & 1 \end{bmatrix}$

4.4.2. Suppose that $\mathbf{E}$ is an $n \times n$ eta matrix that differs from the identity matrix in column k. Prove that if $\mathbf{E}_{kk} \neq 0$, then $\mathbf{E}^{-1}$ is the eta matrix in which

$$(\mathbf{E}^{-1})_{\cdot k} = \begin{bmatrix} -\mathbf{E}_{1k}/\mathbf{E}_{kk} \\ \vdots \\ 1/\mathbf{E}_{kk} \\ \vdots \\ -\mathbf{E}_{nk}/\mathbf{E}_{kk} \end{bmatrix}$$

4.4.3. Using the method of back substitution, describe a procedure for finding the inverse of a lower triangular matrix $\mathbf{L}$ and an upper triangular matrix $\mathbf{U}$.

4.4.4. Solve the following linear systems of equations by the method of back substitution:

(a)
$$\begin{aligned} 4x_1 \qquad\qquad\quad &= 4 \\ 2x_1 + 5x_2 \qquad\quad &= 17 \\ x_1 + 3x_2 + 2x_3 &= 13 \end{aligned}$$

(b)
$$\begin{aligned} 2x_1 + 5x_2 + 3x_3 &= 23 \\ 12x_2 + x_3 &= 16 \\ 3x_3 &= 12 \end{aligned}$$

4.4.5. Using the method of Gaussian elimination, find the inverse of the following matrices:

$$\begin{bmatrix} 1 & 0 & 0 & 1 \\ 0 & 1 & 1 & 0 \\ 0 & 1 & -1 & 0 \\ -1 & 0 & 0 & 1 \end{bmatrix}, \quad \begin{bmatrix} 1 & 0 & 1 & 0 \\ 0 & 0 & -1 & -1 \\ 1 & -1 & 1 & 1 \\ 0 & 1 & 0 & 1 \end{bmatrix}$$

4.4.6. Show that if $\mathbf{P}$ is a permutation matrix, then $\mathbf{P}^{-1} = \mathbf{P}^{\mathrm{T}}$.

4.5. LINEAR ALGEBRA IN LINEAR PROGRAMMING

The linear algebra of the previous four sections will be used in the development of the simplex algorithm. First, however, some of the fundamental relationships between linear algebra and linear programming will be established. For example, the LP,

$$\begin{array}{lrrrrrl}
\text{Minimize} & 3x_1 & -2x_2 & + \; x_3 & - \; x_4 & + \; 2x_5 & \\
\text{subject to} & x_1 & +2x_2 & +3x_3 & +4x_4 & + \; 5x_5 & = 15 \\
 & 6x_1 & +7x_2 & +8x_3 & +9x_4 & +10x_5 & = 18 \\
 & x_1, & x_2, & x_3, & x_4, & x_5 & \geq 0
\end{array}$$

can be written using matrices and vectors. To do so, one constructs

1. the cost vector $\mathbf{c} = (3, -2, 1, -1, 2)$,
2. the vector of variables $\mathbf{x} = (x_1, x_2, x_3, x_4, x_5)$,
3. the right-hand-side vector $\mathbf{b} = \begin{bmatrix} 15 \\ 18 \end{bmatrix}$, and
4. the 2×5 coefficient matrix

$$\mathbf{A} = \begin{bmatrix} 1 & 2 & 3 & 4 & 5 \\ 6 & 7 & 8 & 9 & 10 \end{bmatrix}$$

consisting of the coefficients in the constraint.

Recall that $\mathbf{cx}$ is the real number resulting from the multiplication of the two vectors $\mathbf{c}$ and $\mathbf{x}$; specifically,

$$\begin{aligned}
\mathbf{cx} &= (3, -2, 1, -1, 2)(x_1, x_2, x_3, x_4, x_5) \\
&= 3x_1 - 2x_2 + x_3 - x_4 + 2x_5 \\
&= \text{the objective function of the LP}
\end{aligned}$$

The vector $\mathbf{Ax}$ is the result of multiplying the matrix $\mathbf{A}$ by the (column) vector $\mathbf{x}$; specifically,

$$\begin{aligned}
\mathbf{Ax} &= \begin{bmatrix} 1 & 2 & 3 & 4 & 5 \\ 6 & 7 & 8 & 9 & 10 \end{bmatrix} \begin{bmatrix} x_1 \\ x_2 \\ x_3 \\ x_4 \\ x_5 \end{bmatrix} \\
&= \begin{bmatrix} x_1 + 2x_2 + 3x_3 + 4x_4 + \; 5x_5 \\ 6x_1 + 7x_2 + 8x_3 + 9x_4 + 10x_5 \end{bmatrix} \\
&= \text{the left-hand side of the constraints of the LP}
\end{aligned}$$

In general, for an LP in the form,

$$\begin{array}{lllll} \text{Minimize} & c_1x_1 + \cdots + & c_nx_n & & \\ \text{subject to} & a_{11}x_1 + \cdots + & a_{1n}x_n & = & b_1 \\ & \vdots & \vdots & & \vdots \\ & a_{m1}x_1 + \cdots + & a_{mn}x_n & = & b_m \\ & x_1 \quad ,\ldots, & x_n & \geq & 0 \end{array}$$

one would construct the n vectors $\mathbf{c} = (c_1, \ldots, c_n)$ and $\mathbf{x} = (x_1, \ldots, x_n)$, the m vector $\mathbf{b} = (b_1, \ldots, b_m)$, and the $m \times n$ matrix $\mathbf{A}$ in which $\mathbf{A}_{ij} = a_{ij}$. In matrix/vector notation, the LP becomes to

$$\begin{array}{ll} \text{minimize} & \mathbf{cx} \\ \text{subject to} & \mathbf{Ax} = \mathbf{b} \\ & \mathbf{x} \geq \mathbf{0} \end{array}$$

It is useful to be able to think of $\mathbf{Ax}$ in terms of column multiplication (i.e., $\mathbf{Ax} = \sum_{j=1}^{n} \mathbf{A}_{.j}x_j$) or, equivalently, in terms of row multiplication (i.e., for all i with $1 \leq i \leq m$, $(\mathbf{Ax})_i = \mathbf{A}_{i.}\mathbf{x}$). The constraints $\mathbf{Ax} = \mathbf{b}$ are referred to as the equality constraints. The nonnegativity constraints are $\mathbf{x} \geq \mathbf{0}$. Much of the remainder of this text will use matrix/vector notation.

In the development of the simplex algorithm to follow, it will be necessary to extract certain selected columns from the $\mathbf{A}$ matrix and place them together to form the $\mathbf{B}$ matrix. The remaining columns of $\mathbf{A}$ will form the matrix $\mathbf{N}$. For the numerical example above, if columns 2 and 4 are extracted from $\mathbf{A}$, then

$$\mathbf{B} = \begin{bmatrix} 2 & 4 \\ 7 & 9 \end{bmatrix}, \qquad \mathbf{N} = \begin{bmatrix} 1 & 3 & 5 \\ 6 & 8 & 10 \end{bmatrix}$$

Although it is technically incorrect to write $\mathbf{A} = [\mathbf{B}, \mathbf{N}]$, this notation will be used to indicate that the columns of $\mathbf{A}$ have been rearranged so that those columns corresponding to $\mathbf{B}$ appear first and are followed by those columns constituting $\mathbf{N}$. Observe that if $\mathbf{A}$ is an $m \times n$ matrix and $\mathbf{B}$ consist of q columns from $\mathbf{A}$, then $\mathbf{B}$ is $m \times q$ and $\mathbf{N}$ is $m \times (n-q)$. In the simplex algorithm the $\mathbf{B}$ matrix consists of m columns from $\mathbf{A}$, so $\mathbf{B}$ is $m \times m$ and $\mathbf{N}$ is $m \times (n-m)$.

Consider the column multiplication $\mathbf{Ax} = \sum_{j=1}^{n} \mathbf{A}_{.j}x_j$. If the columns of $\mathbf{A}$ are rearranged so that $\mathbf{A} = [\mathbf{B}, \mathbf{N}]$, then it is also necessary to rearrange the components of $\mathbf{x}$. Specifically, we let $\mathbf{x_B}$ denote those components of $\mathbf{x}$ that multiply the corresponding columns of $\mathbf{B}$. Analogously, $\mathbf{x_N}$ denotes those components of $\mathbf{x}$ that multiply the corresponding columns of $\mathbf{N}$, and is written $\mathbf{x} = (\mathbf{x_B}, \mathbf{x_N})$. For the present example

$$\mathbf{Ax} = \begin{bmatrix}1\\6\end{bmatrix}x_1 + \begin{bmatrix}2\\7\end{bmatrix}x_2 + \begin{bmatrix}3\\8\end{bmatrix}x_3 + \begin{bmatrix}4\\9\end{bmatrix}x_4 + \begin{bmatrix}5\\10\end{bmatrix}x_5$$

When

$$\mathbf{B} = [\mathbf{A}_{\cdot 2}, \mathbf{A}_{\cdot 4}] = \begin{bmatrix}2 & 4\\7 & 9\end{bmatrix}$$

$$\mathbf{x_B} = (x_2, x_4)$$

and

$$\mathbf{N} = [\mathbf{A}_{\cdot 1}, \mathbf{A}_{\cdot 3}, \mathbf{A}_{\cdot 5}] = \begin{bmatrix}1 & 3 & 5\\6 & 8 & 10\end{bmatrix}$$

$$\mathbf{x_N} = (x_1, x_3, x_5)$$

then

$$\begin{aligned}\mathbf{Ax} &= \begin{bmatrix}1\\6\end{bmatrix}x_1 + \begin{bmatrix}2\\7\end{bmatrix}x_2 + \begin{bmatrix}3\\8\end{bmatrix}x_3 + \begin{bmatrix}4\\9\end{bmatrix}x_4 + \begin{bmatrix}5\\10\end{bmatrix}x_5\\ &= \begin{bmatrix}2\\7\end{bmatrix}x_2 + \begin{bmatrix}4\\9\end{bmatrix}x_4 + \begin{bmatrix}1\\6\end{bmatrix}x_1 + \begin{bmatrix}3\\8\end{bmatrix}x_3 + \begin{bmatrix}5\\10\end{bmatrix}x_5\\ &= \begin{bmatrix}2 & 4\\7 & 9\end{bmatrix}\begin{bmatrix}x_2\\x_4\end{bmatrix} + \begin{bmatrix}1 & 3 & 5\\6 & 8 & 10\end{bmatrix}\begin{bmatrix}x_1\\x_3\\x_5\end{bmatrix}\\ &= \mathbf{Bx_B} + \mathbf{Nx_N}\\ &= [\mathbf{B}, \mathbf{N}]\begin{bmatrix}\mathbf{x_B}\\\mathbf{x_N}\end{bmatrix}\end{aligned}$$

In a similar manner, for the cost vector $\mathbf{c}$ it is possible to rearrange the components so that $\mathbf{c} = (\mathbf{c_B}, \mathbf{c_N})$. Thus the objective function $\mathbf{cx} = \sum_{j=1}^{n} c_j x_j$ of an LP can equally well be computed as

$$\mathbf{cx} = (\mathbf{c_B}, \mathbf{c_N})\begin{bmatrix}\mathbf{x_B}\\\mathbf{x_N}\end{bmatrix} = \mathbf{c_B x_B} + \mathbf{c_N x_N}$$

The need to work with the **B** and **N** components of matrices and vectors will arise frequently throughout the text.

As you saw in Chapter 1, an LP can come in variety of sizes and shapes. It can have

1. inequality or equality constraints,
2. nonnegative, nonpositive, or unrestricted variables, and
3. an objective function that is to be maximized or minimized.

Rather than develop an algorithm that can handle any and all combinations of properties 1–3, it is conceptually simpler to deal with only one particular form of an LP. For algebraic reasons that will become evident in the next chapter, the simplex algorithm is designed to solve an LP in the

Standard Form:

$$\begin{aligned} \text{Minimize} \quad & \mathbf{cx} \\ \text{subject to} \quad & \mathbf{Ax} = \mathbf{b} \\ & \mathbf{x} \geq \mathbf{0} \end{aligned}$$

Also, for reasons that will not be clear until Chapter 6, the standard form requires that the vector $\mathbf{b} \geq \mathbf{0}$. The question naturally arises as to how to proceed when the original LP is not in standard form. That issue will be addressed in Section 4.6.

As was seen from the geometry presented in Chapter 2, an LP can be infeasible, optimal, or unbounded. These three terms will be defined formally. First, however, a feasible and optimal solution will be defined.

Definition 4.14. A vector $\mathbf{x}$ is *feasible* for an LP if it satisfies the constraints. (In particular, $\mathbf{x}$ is feasible for an LP in standard form if $\mathbf{Ax} = \mathbf{b}$ and $\mathbf{x} \geq \mathbf{0}$.)

A vector that is feasible is sometimes called a *feasible solution*. An optimal solution is a feasible solution that produces the best value of the objective function.

Definition 4.15. A vector $\mathbf{x}^*$ is *optimal* for an LP if $\mathbf{x}^*$ is feasible and, for every feasible solution $\mathbf{x}$, the objective value at $\mathbf{x}^*$ is at least as good as that at $\mathbf{x}$. (For an LP in standard form, $\mathbf{x}^*$ is optimal if $\mathbf{Ax}^* = \mathbf{b}$ and $\mathbf{x}^* \geq \mathbf{0}$, and, for all $\mathbf{x}$ with $\mathbf{Ax} = \mathbf{b}$ and $\mathbf{x} \geq \mathbf{0}$, $\mathbf{cx}^* \leq \mathbf{cx}$.)

A vector $\mathbf{x}^*$ that is optimal is often called an *optimal solution* or, equivalently, is said to solve the LP. Now one can define what it means for an LP to be feasible or optimal.

Definition 4.16. An LP is *feasible* if there is a feasible solution $\mathbf{x}$.

Definition 4.17. An LP is *infeasible* if it is not feasible, i.e., if there is no feasible solution.

Definition 4.18. An LP is *optimal* if there is an optimal solution $\mathbf{x}^*$.

The concept of unboundedness is more complicated. For an LP to be unbounded, it should first be feasible, i.e., there should be a feasible solution $\mathbf{x}$. Then, according to Figure 4.11, the idea of unboundedness is to find a

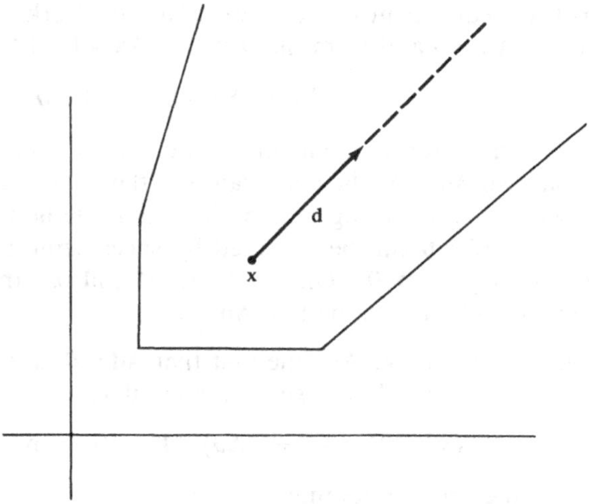

FIGURE 4.11. An unbounded LP.

direction in the form of an n vector $\mathbf{d}$ that should satisfy two properties:

1. No matter how far one moves from $\mathbf{x}$ in the direction $\mathbf{d}$, one remains inside the feasible region.
2. The objective function continually improves as one moves from $\mathbf{x}$ in the direction $\mathbf{d}$ (i.e., decreases when minimizing and increases when maximizing).

The issue that must be addressed is that of determining if a given direction $\mathbf{d}$ satisfies the above properties. For an LP in standard form, it will be shown that the $\mathbf{A}$ matrix and $\mathbf{c}$ vector can be used to test these properties. To that end, let $\mathbf{x}$ be a feasible solution (i.e., $\mathbf{Ax} = \mathbf{b}$ and $\mathbf{x} \geq \mathbf{0}$), and let $\mathbf{d}$ be an n vector. Recall from Section 4.1 that a point on the line depicted in Figure 4.11 can be represented algebraically as $\mathbf{x} + t\mathbf{d}$, where t is a nonnegative real number. In order for $\mathbf{d}$ to satisfy property 1, it should be that, for all $t \geq 0$, $\mathbf{x} + t\mathbf{d}$ is feasible, i.e., for all $t \geq 0$, (a) $\mathbf{A}(\mathbf{x} + t\mathbf{d}) = \mathbf{b}$ and (b) $\mathbf{x} + t\mathbf{d} \geq \mathbf{0}$. Proposition 4.4 provides the condition on $\mathbf{d}$ so that (a) is true, and Proposition 4.5 does the same for (b).

Proposition 4.4. *The vector* $\mathbf{Ad} = \mathbf{0}$ *if and only if for all* $t \geq 0$, $\mathbf{A}(\mathbf{x} + t\mathbf{d}) = \mathbf{b}$.

OUTLINE OF PROOF. Recognizing the words "if and only if," one sees that two proofs will be required (see Definition 3.8). Suppose first that $\mathbf{Ad} = \mathbf{0}$. It will be shown that, for all $t \geq 0$, $\mathbf{A}(\mathbf{x} + t\mathbf{d}) = \mathbf{b}$. The appearance of the quantifier "for all" in the conclusion suggests proceeding with the choose method, whereby one selects a $t \geq 0$ for which it must be shown that

$\mathbf{A}(\mathbf{x} + t\mathbf{d}) = \mathbf{b}$ or, equivalently, that $\mathbf{Ax} + t\mathbf{Ad} = \mathbf{b}$. Work forward from the hypothesis that $\mathbf{Ad} = \mathbf{0}$ and from the fact that $\mathbf{Ax} = \mathbf{b}$. It follows that

$$\mathbf{A}(\mathbf{x} + t\mathbf{d}) = \mathbf{Ax} + t\mathbf{Ad} = \mathbf{b} + t(\mathbf{0}) = \mathbf{b}$$

To complete the proof, it remains to be shown that if, for all $t \geq 0$, $\mathbf{A}(\mathbf{x} + t\mathbf{d}) = \mathbf{b}$, then $\mathbf{Ad} = \mathbf{0}$. The appearance of the quantifier "for all" in the forward process suggests using the specialization method. Indeed, the desired result that $\mathbf{Ad} = \mathbf{0}$ can be obtained by specializing to $t = 1$ (which is greater than or equal to 0). On so doing, it follows that $\mathbf{Ax} + \mathbf{Ad} = \mathbf{b}$. Finally, since $\mathbf{Ax} = \mathbf{b}$, it must be that $\mathbf{Ad} = \mathbf{0}$.

PROOF OF PROPOSITION 4.4. Assume first that $\mathbf{Ad} = \mathbf{0}$, and let $t \geq 0$. It will be shown that $\mathbf{A}(\mathbf{x} + t\mathbf{d}) = \mathbf{b}$, but since $\mathbf{x}$ is feasible,

$$\mathbf{A}(\mathbf{x} + t\mathbf{d}) = \mathbf{Ax} + t(\mathbf{Ad}) = \mathbf{b} + t(\mathbf{0}) = \mathbf{b}$$

To complete the proof, note that, for $t = 1$,

$$\mathbf{A}(\mathbf{x} + t\mathbf{d}) = \mathbf{Ax} + \mathbf{Ad} = \mathbf{b}$$

and since $\mathbf{Ax} = \mathbf{b}$, it follows that $\mathbf{Ad} = \mathbf{0}$. □

The next proposition establishes the needed property on $\mathbf{d}$ so that, for all $t \geq 0$, $\mathbf{x} + t\mathbf{d} \geq \mathbf{0}$.

Proposition 4.5. *The vector* $\mathbf{d} \geq \mathbf{0}$ *if and only if, for all* $t \geq 0$, $\mathbf{x} + t\mathbf{d} \geq \mathbf{0}$.

OUTLINE OF PROOF. Again the words "if and only if" require two proofs. First assume that $\mathbf{d} \geq \mathbf{0}$. It will be shown that, for all $t \geq 0$, $\mathbf{x} + t\mathbf{d} \geq \mathbf{0}$. The appearance of the quantifier "for all" in the conclusion suggests using the choose method to select a $t \geq 0$ for which it must be shown that $\mathbf{x} + t\mathbf{d} \geq \mathbf{0}$. But $\mathbf{d} \geq \mathbf{0}$ by hypothesis, and since $\mathbf{x}$ is feasible, $\mathbf{x} \geq \mathbf{0}$ (see Definition 4.14). Thus it must be that $\mathbf{x} + t\mathbf{d} \geq \mathbf{0}$.

To complete the proof, it is necessary to show that if, for all $t \geq 0$, $\mathbf{x} + t\mathbf{d} \geq \mathbf{0}$, then $\mathbf{d} \geq \mathbf{0}$. Here, the contrapositive method will be used. According to that technique, it can be assumed that $\mathbf{d} \geq \mathbf{0}$ is false, i.e., that there is a j with $1 \leq j \leq n$, such that $d_j < 0$. It must be shown that there is a $t \geq 0$ such that $\mathbf{x} + t\mathbf{d} \geq \mathbf{0}$ does not hold. In other words, a value for $t \geq 0$ must be produced for which one component of $\mathbf{x} + t\mathbf{d}$ is less than 0. The value for t will be chosen in such a way that $(\mathbf{x} + t\mathbf{d})_j < 0$ or, equivalently, $td_j < -x_j$. Since $d_j < 0$, t should be chosen so that $t > -x_j/d_j$. In order to verify that such a choice for t is correct, it had better be that $t \geq 0$. To see that it is, recall that $\mathbf{x}$ is feasible, so that $x_j \geq 0$. Furthermore, $d_j < 0$; thus $-x_j/d_j \geq 0$, and so $t > -x_j/d_j \geq 0$.

PROOF OF PROPOSITION 4.5. Assume first that $\mathbf{d} \geq \mathbf{0}$, and let $t \geq 0$. Then, since $\mathbf{x}$ is feasible, $\mathbf{x} \geq \mathbf{0}$, and so $\mathbf{x} + t\mathbf{d} \geq \mathbf{0}$.

To complete the proof, assume it is not the case that $\mathbf{d} \geq \mathbf{0}$, i.e., that there is a j with $1 \leq j \leq n$, such that $d_j < 0$. A value for $t \geq 0$ will be constructed with the property that $(\mathbf{x} + t\mathbf{d})_j = x_j + td_j < 0$. Specifically, choose $t > -x_j/d_j$. Since $\mathbf{x}$ is feasible, $x_j \geq 0$, and since $d_j < 0$, it follows that $t > -x_j/d_j \geq 0$. Finally, when $t > -x_j/d_j$, $td_j < -x_j$, or, equivalently, $x_j + td_j < 0$. □

The previous two propositions have provided conditions on $\mathbf{d}$ that guarantee that feasibility is maintained no matter how far one moves from $\mathbf{x}$ in the direction $\mathbf{d}$. In order for $\mathbf{d}$ to be a direction of unboundedness, the final task is to develop a condition on $\mathbf{d}$ that ensures that the objective function decrease in that direction. To determine the condition, observe that the value of the objective function at $\mathbf{x}$ is $\mathbf{cx}$, but the value at $\mathbf{x} + t\mathbf{d}$ is $\mathbf{c}(\mathbf{x} + t\mathbf{d}) = \mathbf{cx} + t\mathbf{cd}$. If the objective function is to decrease, then it had better be the case that $\mathbf{cx} + t\mathbf{cd} < \mathbf{cx}$ or, equivalently, by subtracting $\mathbf{cx}$ from both sides, that $t\mathbf{cd} < 0$. If $t > 0$, then one would like $\mathbf{d}$ to satisfy the property that $\mathbf{cd} < 0$, as formally established in the next proposition.

Proposition 4.6. *The real number* $\mathbf{cd} < 0$ *if and only if, for all* $t > 0$, $\mathbf{c}(\mathbf{x} + t\mathbf{d}) < \mathbf{cx}$.

OUTLINE OF PROOF. As in the previous propositions, two proofs are required. Assume first that $\mathbf{cd} < 0$. It must be shown that, for all $t > 0$, $\mathbf{c}(\mathbf{x} + t\mathbf{d}) < \mathbf{cx}$. The appearance of the quantifier "for all" suggests proceeding with the choose method to select a $t > 0$ for which it must be shown that $\mathbf{c}(\mathbf{x} + t\mathbf{d}) = \mathbf{cx} + t\mathbf{cd} < \mathbf{cx}$ or, equivalently, that $t\mathbf{cd} < 0$. But, since $t > 0$ and $\mathbf{cd} < 0$, $t\mathbf{cd} < 0$, and this part of the proof is done.

To complete the proof, it must be shown that if, for all $t > 0$, $\mathbf{c}(\mathbf{x} + t\mathbf{d}) < \mathbf{cx}$, then $\mathbf{cd} < 0$. To reach the desired conclusion, specialize the hypothesis to the value $t = 1$ (which is greater than 0). Then one has $\mathbf{cx} + \mathbf{cd} < \mathbf{cx}$ and, on subtracting $\mathbf{cx}$ from both sides, $\mathbf{cd} < 0$, as desired.

PROOF OF PROPOSITION 4.6. Assume first that $\mathbf{cd} < 0$, and let $t > 0$. It will be shown that $\mathbf{c}(\mathbf{x} + t\mathbf{d}) = \mathbf{cx} + t\mathbf{cd} < \mathbf{cx}$. But since $\mathbf{cd} < 0$ and $t > 0$, $t\mathbf{cd} < 0$. On adding $\mathbf{cx}$ to both sides, one obtains the desired conclusion that $\mathbf{cx} + t\mathbf{cd} < \mathbf{cx}$.

To complete the proof, let $t = 1$. Then, since $\mathbf{c}(\mathbf{x} + t\mathbf{d}) = \mathbf{cx} + \mathbf{cd} < \mathbf{cx}$, it follows that $\mathbf{cd} < 0$, as desired. □

As a result of the previous three propositions, it is possible to define formally what it means for an LP in standard form to be unbounded.

Definition 4.19. An LP in standard form is *unbounded* if it is feasible and there is a *direction of unboundedness* $\mathbf{d} \geq \mathbf{0}$ such that $\mathbf{Ad} = \mathbf{0}$ and $\mathbf{cd} < 0$.

To illustrate the definition of unboundedness, consider the following LP:

$$\begin{array}{lrrrl} \text{Minimize} & -x_1 & -x_2 & -x_3 & \\ \text{subject to} & -2x_1 & +3x_2 & & = 6 \\ & -10x_1 & +15x_2 & +3x_3 & = 45 \\ & x_1, & x_2, & x_3 & \geq 0 \end{array}$$

Observe that the vector $\mathbf{x} = (x_1, x_2, x_3) = (0, 2, 5)$ is feasible for the LP, the direction $\mathbf{d} = (d_1, d_2, d_3) = (3, 2, 0)$ satisfies $\mathbf{d} \geq \mathbf{0}$, and

$$\mathbf{cd} = (-1, -1, -1)(3, 2, 0) = -5 < 0$$

$$\mathbf{Ad} = \begin{bmatrix} -2 & 3 & 0 \\ -10 & 15 & 3 \end{bmatrix} \begin{bmatrix} 3 \\ 2 \\ 0 \end{bmatrix} = \begin{bmatrix} 0 \\ 0 \end{bmatrix}$$

Thus the LP is unbounded, and $\mathbf{d}$ is the direction of unboundedness.

For an LP that is not in standard form, the conditions for unboundedness of $\mathbf{d}$ must be modified appropriately. For instance, if the objective function were to be maximized instead of minimized, then $\mathbf{d}$ would have to satisfy the property that $\mathbf{cd} > 0$ instead of $\mathbf{cd} < 0$. Similar modifications would be needed for inequality constraints. For instance, for constraints of the form $\mathbf{Ax} \leq \mathbf{b}$, the appropriate condition for $\mathbf{d}$ would be $\mathbf{Ad} \leq \mathbf{0}$. The various conditions for unboundedness of $\mathbf{d}$ are summarized in Table 4.3.

This section has defined formally what it means for an LP to be feasible, optimal, or unbounded. The simplex algorithm to be developed in Chapter 5

TABLE 4.3. Conditions for Unboundedness

Objective function	Property of $\mathbf{d}$
Minimize $\mathbf{cx}$	$\mathbf{cd} < 0$
Maximize $\mathbf{cx}$	$\mathbf{cd} > 0$
Constraints	
$\mathbf{Ax} \leq \mathbf{b}$	$\mathbf{Ad} \leq \mathbf{0}$
$\mathbf{Ax} = \mathbf{b}$	$\mathbf{Ad} = \mathbf{0}$
$\mathbf{Ax} \geq \mathbf{b}$	$\mathbf{Ad} \geq \mathbf{0}$
Variables	
$\mathbf{x} \geq \mathbf{0}$	$\mathbf{d} \geq \mathbf{0}$
$\mathbf{x} \leq \mathbf{0}$	$\mathbf{d} \leq \mathbf{0}$
$\mathbf{x}$ unrestricted	$\mathbf{d}$ unrestricted

will determine which of these three conditions actually occurs for a given LP. First, however, it is necessary to know what to do when the original LP is not in standard form, as discussed in the next section.

EXERCISES

4.5.1. Consider the following LP:

$$
\begin{array}{lrrrrrrrl}
\text{Minimize} & 4x_1 & +5x_2 & -\ x_3 & -\ 2x_4 & +3x_5 & +x_6 & -\ x_7 & \\
\text{subject to} & x_1 & -3x_2 & & -\ 8x_4 & +\ x_5 & +x_6 & & =-10 \\
 & & -5x_2 & -2x_3 & -\ 2x_4 & & +x_6 & +3x_7 & =-12 \\
 & -x_1 & -4x_2 & & -10x_4 & & & +2x_7 & =-19 \\
 & x_1, & x_2, & x_3, & x_4, & x_5, & x_6, & x_7 \geq & 0
\end{array}
$$

(a) For the vector $\mathbf{x}=(1,2,0,1,3,0,0)$, identify the $\mathbf{B}$ matrix consisting of the columns of $\mathbf{A}$ corresponding to the positive components of $\mathbf{x}$. Also, identify the $\mathbf{N}$ matrix.

(b) Show that the vector $\mathbf{x}$ of part (a) is feasible for the LP by computing $\mathbf{Ax}$ using the formula $\mathbf{Bx_B}+\mathbf{Nx_N}$.

(c) Compute the value of the objective function at $\mathbf{x}$ by computing $\mathbf{c_B x_B}+\mathbf{c_N x_N}$.

4.5.2. For each of the following LPs, state precisely what it means for a vector $\mathbf{x}$ to be (i) feasible, and (ii) optimal.

(a)

$$
\begin{array}{lrrrl}
\text{Maximize} & -\ x_1 & +\ 3x_2 & -4x_3 & \\
\text{subject to} & -\ 2x_1 & +\ 3x_2 & -\ x_3 & \leq 6 \\
 & -10x_1 & +15x_2 & +3x_3 & \geq 45 \\
 & x_1 & +\ x_2 & +\ x_3 & =10 \\
 & x_1, & x_2, & x_3 & \geq 0
\end{array}
$$

(b)

$$
\begin{array}{ll}
\text{Minimize} & \mathbf{cx} \\
\text{subject to} & \mathbf{Ax} \leq \mathbf{b} \\
 & \mathbf{Gx} \geq \mathbf{b}' \\
 & \mathbf{x} \geq \mathbf{0}
\end{array}
$$

where $\mathbf{A}$ is an $m \times n$ matrix, $\mathbf{G}$ is a $p \times n$ matrix, $\mathbf{b}$ is an m vector, $\mathbf{b}'$ is a p vector, and $\mathbf{c}$ and $\mathbf{x}$ are n vectors.

4.5.3. In the proof of Proposition 4.5, explain why it is not possible to specialize to $t=1$ as was done in the proof of Proposition 4.4.

4.5.4. For each of the following LPs, use Table 4.3 to write precisely what it means for **d** to be a direction of unboundedness:

(a)
$$\begin{array}{llllll} \text{Minimize} & 4x_1 & +5x_2 & +3x_3 & +7x_4 & \\ \text{subject to} & 2x_1 & & +5x_3 & & \leq 2 \\ & & 2x_2 & & +6x_4 & \geq 5 \\ & x_1 + & x_2 + & x_3 + & x_4 & \geq 20 \\ & x_1, & x_2 & & & \geq 0 \\ & & & x_3, & x_4 & \leq 0 \end{array}$$

(b)
$$\begin{array}{lllll} \text{Maximize} & -3x_1 & +2x_2 & -5x_3 & \\ \text{subject to} & x_1 & +3x_2 & -2x_3 & = 10 \\ & & 2x_2 & +5x_3 & \geq -5 \\ & x_1 + & x_2 & & \leq 5 \\ & x_1, & x_2 & & \geq 0 \\ & & & x_3 \text{ unrestricted} & \end{array}$$

(c)
$$\begin{array}{ll} \text{Minimize} & \mathbf{cx} \\ \text{subject to} & \mathbf{Ax} \leq \mathbf{b} \\ & \mathbf{x} \geq \mathbf{0} \end{array}$$

(d)
$$\begin{array}{lll} \text{Maximize} & \mathbf{cx} & \\ \text{subject to} & \mathbf{Ax} \geq \mathbf{b} & (\mathbf{A} \text{ is } m \times n) \\ & \mathbf{Gx} \leq \mathbf{b}' & (\mathbf{G} \text{ is } p \times n) \\ & \mathbf{Qx} = \mathbf{q} & (\mathbf{Q} \text{ is } k \times n) \\ & \mathbf{x} & \text{unrestricted} \end{array}$$

4.5.5. Consider the following LP:

$$\begin{array}{llllll} \text{Minimize} & 4x_1 & -2x_2 & - x_3 & + x_4 & \\ \text{subject to} & 3x_1 & -2x_2 & -2x_3 & +4x_4 & \leq -2 \\ & & 2x_2 & +2x_3 & -4x_4 & \leq -1 \\ & x_1, & x_2, & x_3, & x_4 & \geq 0 \end{array}$$

(a) Show that the vector $\mathbf{d} = (0,1,1,1)$ is a direction of unboundedness in the sense that $\mathbf{d} \geq \mathbf{0}$, $\mathbf{Ad} \leq \mathbf{0}$, and $\mathbf{cd} < 0$.

(b) Explain why this LP is not unbounded.
[*Hint*: Add the two constraints.]

4.5.6. Prove that if $\mathbf{x}^*$ and $\mathbf{y}^*$ are optimal for an LP in standard form, then, for all $0 \leq t \leq 1$, $t\mathbf{x}^* + (1-t)\mathbf{y}^*$ is also optimal.

4.6. CONVERTING AN LP TO STANDARD FORM

The simplex algorithm is designed to solve an LP that is in standard form. This section will describe how to convert any LP into an equivalent one that is in standard form. The word "equivalent" is used in the sense that an optimal solution to the new LP can be used to create an optimal solution to the original LP and vice versa.

Consider, for example, an LP in the following form:

$$\begin{aligned} \text{Maximize} \quad & \mathbf{cx} \\ \text{subject to} \quad & \mathbf{Ax} = \mathbf{b} \\ & \mathbf{x} \geq \mathbf{0} \end{aligned} \qquad \text{(LP 1)}$$

which differs from standard form only in the form of the objective function. (Unless otherwise stated, it will be assumed that $\mathbf{b} \geq \mathbf{0}$.) The associated LP in standard form is to

$$\begin{aligned} \text{Minimize} \quad & (-\mathbf{c})\mathbf{x} \\ \text{subject to} \quad & \mathbf{Ax} = \mathbf{b} \\ & \mathbf{x} \geq \mathbf{0} \end{aligned} \qquad \text{(LP 2)}$$

as is shown in the next proposition.

Proposition 4.7. *If* $\mathbf{x}^*$ *is optimal for LP* 2, *then* $\mathbf{x}^*$ *is optimal for LP* 1.

OUTLINE OF PROOF. The forward–backward method gives rise to the abstraction question "How can I show that a vector (namely, $\mathbf{x}^*$) is optimal for an LP in the form of LP 1?" One answer is to use the definition, whereby it must be shown that

1. $\mathbf{x}^*$ is feasible for LP 1, and
2. for every feasible solution $\mathbf{x}$ to LP 1, $\mathbf{cx}^* \geq \mathbf{cx}$.

Both of these statements can be established by working forward from the hypothesis that $\mathbf{x}^*$ is optimal for LP 2. Specifically, by Definition 4.15,

a. $\mathbf{x}^*$ is feasible for LP 2 (i.e., $\mathbf{Ax}^* = \mathbf{b}$ and $\mathbf{x}^* \geq \mathbf{0}$), and
b. for every feasible solution $\mathbf{x}$ to LP 2, $(-\mathbf{c})\mathbf{x}^* \leq (-\mathbf{c})\mathbf{x}$, or, equivalently, $\mathbf{cx}^* \geq \mathbf{cx}$.

Statement (a) establishes property 1 and statement (b) establishes property 2.

PROOF OF PROPOSITION 4.7. To see that $\mathbf{x}^*$ is optimal for LP 1, observe first that $\mathbf{x}^*$ is feasible for LP 1 since it is optimal, and hence feasible, for LP 2. Furthermore, since $\mathbf{x}^*$ is optimal for LP 2, for every feasible solution $\mathbf{x}$ to LP 2, $(-\mathbf{c})\mathbf{x}^* \leq (-\mathbf{c})\mathbf{x}$, or, equivalently, $\mathbf{cx}^* \geq \mathbf{cx}$. Thus $\mathbf{x}^*$ is optimal for LP 1. □

In an analogous manner, it can be shown that if $\mathbf{x}^*$ is optimal for LP 1, then it is also optimal for LP 2. Note, however, that the objective value of LP 1 at $\mathbf{x}^*$ is $\mathbf{cx}^*$ but that the value of LP 2 is $-\mathbf{cx}^*$.

It is also possible to handle variations in the constraints. For example, consider an LP in the following form:

$$\begin{aligned} \text{Minimize} \quad & \mathbf{cx} \\ \text{subject to} \quad & \mathbf{Ax} \le \mathbf{b} \\ & \mathbf{x} \ge \mathbf{0} \end{aligned} \qquad \text{(LP 3)}$$

in which the equality constraints have been replaced by inequality constraints. In order to create the associated LP in standard form, it is necessary to add one additional (nonnegative) variable to each of the m inequality constraints. The new variables, $\mathbf{s} = (s_1, \ldots, s_m)$, are called *slack variables*. The associated LP in standard form is to

$$\begin{aligned} \text{Minimize} \quad & \mathbf{cx} \\ \text{subject to} \quad & \mathbf{Ax} + \mathbf{Is} = \mathbf{b} \\ & \mathbf{x}, \quad \mathbf{s} \ge \mathbf{0} \end{aligned} \qquad \text{(LP 4)}$$

where $\mathbf{I}$ is the $m \times m$ identity matrix.

The process of adding slack variables is illustrated with the following LP:

$$\begin{aligned} \text{Minimize} \quad & -3x_1 + 4x_2 \\ \text{subject to} \quad & -4x_1 + 5x_2 \le 11 \\ & 6x_1 + x_2 \le 8 \\ & x_1, \quad x_2 \ge 0 \end{aligned}$$

The associated LP in standard form is

$$\begin{aligned} \text{Minimize} \quad & -3x_1 + 4x_2 \\ \text{subject to} \quad & -4x_1 + 5x_2 + s_1 = 11 \\ & 6x_1 + x_2 + s_2 = 8 \\ & x_1, \quad x_2, \quad s_1, \quad s_2 \ge 0 \end{aligned}$$

The next proposition establishes that an optimal solution to LP 4 provides one for LP 3.

Proposition 4.8. *If* $(\mathbf{x}^*, \mathbf{s}^*)$ *is an optimal solution to LP* 4, *then* $\mathbf{x}^*$ *is optimal for LP* 3.

OUTLINE OF PROOF. The forward–backward method gives rise to the abstraction question "How can I show that a vector (namely, $\mathbf{x}^*$) is optimal for LP 3?" As usual, the definition can be used to provide an answer, whereby it must be shown that

1. $\mathbf{x}^*$ is feasible for LP 3 (i.e., $\mathbf{Ax}^* \le \mathbf{b}$ and $\mathbf{x}^* \ge \mathbf{0}$) and,
2. for all vectors $\mathbf{x}$ that are feasible for LP 3, $\mathbf{cx}^* \le \mathbf{cx}$.

Both of these facts can be established by working forward from the hypothesis that $(\mathbf{x}^*, \mathbf{s}^*)$ is optimal for LP 4. Specifically, by definition,

a. $(\mathbf{x}^*, \mathbf{s}^*)$ is feasible for LP 4 (i.e., $\mathbf{Ax}^* + \mathbf{Is}^* = \mathbf{b}$, $\mathbf{x}^* \ge \mathbf{0}$, and $\mathbf{s}^* \ge \mathbf{0}$), and
b. for all vectors $(\mathbf{x}, \mathbf{s})$ that are feasible for LP 4, $\mathbf{cx}^* \le \mathbf{cx}$.

From condition (a) it follows that $\mathbf{x}^* \ge 0$ and also that $\mathbf{Ax}^* = \mathbf{b} - \mathbf{Is}^*$. But since $\mathbf{s}^* \ge \mathbf{0}$, $-\mathbf{Is}^* \le \mathbf{0}$; and on adding $\mathbf{b}$ to both sides, one obtains $\mathbf{b} - \mathbf{Is}^* \le \mathbf{b}$. Consequently, $\mathbf{Ax}^* = \mathbf{b} - \mathbf{Is}^* \le \mathbf{b}$, and so property 1 has been proven.

To establish property 2, the appearance of the quantifier "for all" in the backward process suggests using the choose method to select a vector, say, $\mathbf{x}'$, that is feasible for LP 3, i.e., $\mathbf{Ax}' \le \mathbf{b}$ and $\mathbf{x}' \ge \mathbf{0}$. To reach the desired conclusion that $\mathbf{cx}^* \le \mathbf{cx}'$, it is necessary to use the fact that $(\mathbf{x}^*, \mathbf{s}^*)$ is optimal for LP 4. Specifically, the desired result will follow on specializing condition (b) to the vector $(\mathbf{x}, \mathbf{s}) = (\mathbf{x}', \mathbf{b} - \mathbf{Ax}')$, for then $\mathbf{cx}^* \le \mathbf{cx}'$, as desired. In order to be able to apply specialization to $(\mathbf{x}', \mathbf{b} - \mathbf{Ax}')$, it is necessary to verify that the certain property holds, in this case, that $(\mathbf{x}', \mathbf{b} - \mathbf{Ax}')$ is feasible for LP 4, i.e., that it satisfies the constraints of LP 4. But $\mathbf{x}'$ was chosen to be $\ge \mathbf{0}$, and since $\mathbf{Ax}' \le \mathbf{b}$, $\mathbf{s} = \mathbf{b} - \mathbf{Ax}' \ge \mathbf{0}$. Thus $(\mathbf{x}', \mathbf{b} - \mathbf{Ax}')$ satisfies the nonnegativity constraints of LP 4. Finally, to see that $(\mathbf{x}', \mathbf{b} - \mathbf{Ax}')$ satisfies the equality constraints of LP 4, note that

$$\mathbf{Ax}' + \mathbf{Is} = \mathbf{Ax}' + \mathbf{I}(\mathbf{b} - \mathbf{Ax}') = \mathbf{Ax}' + \mathbf{b} - \mathbf{Ax}' = \mathbf{b}$$

PROOF OF PROPOSITION 4.8. To see that $\mathbf{x}^*$ is optimal for LP 3, observe first that, since $(\mathbf{x}^*, \mathbf{s}^*)$ is optimal (and hence feasible) for LP 4, $\mathbf{Ax}^* + \mathbf{Is}^* = \mathbf{b}$, $\mathbf{x}^* \ge \mathbf{0}$, and $\mathbf{s}^* \ge \mathbf{0}$. Thus $\mathbf{x}^* \ge \mathbf{0}$ and $\mathbf{Ax}^* = \mathbf{b} - \mathbf{Is}^* \le \mathbf{b}$ (since $\mathbf{s}^* \ge \mathbf{0}$), and so $\mathbf{x}^*$ is feasible for LP 3.

To show that $\mathbf{x}^*$ is actually optimal for LP 3, let $\mathbf{x}$ be any feasible solution for LP 3, i.e., $\mathbf{Ax} \le \mathbf{b}$ and $\mathbf{x} \ge \mathbf{0}$. Then $\mathbf{x}$ and $\mathbf{s} = \mathbf{b} - \mathbf{Ax}$ are feasible for LP 4. Since $(\mathbf{x}^*, \mathbf{s}^*)$ is optimal for LP 4, it must be that $\mathbf{cx}^* \le \mathbf{cx}$, and so $\mathbf{x}^*$ is optimal for LP 3. □

In a similar manner, an optimal solution to LP 3 can be used to create one for LP 4. The actual construction will be developed in Exercise 4.6.2.

In LP 3, it was assumed that all of the inequality constraints were of the form $\le$. When this is not the case, one simply adds a slack variable to each $\le$ constraint.

When constraints of form $\geq$ appear in the original LP, one subtracts a slack variable from each constraint instead of adding it. Thus, if the original LP is in the following form:

$$\begin{aligned} \text{Minimize} \quad & \mathbf{cx} \\ \text{subject to} \quad & \mathbf{Ax} \geq \mathbf{b} \\ & \mathbf{x} \geq \mathbf{0} \end{aligned} \qquad \text{(LP 5)}$$

then the associated LP in standard form is to

$$\begin{aligned} \text{Minimize} \quad & \mathbf{cx} \\ \text{subject to} \quad & \mathbf{Ax} - \mathbf{Is} = \mathbf{b} \\ & \mathbf{x}, \quad \mathbf{s} \geq \mathbf{0} \end{aligned} \qquad \text{(LP 6)}$$

As usual, it is possible to show that an optimal solution to LP 5 provides one for LP 6 and vice versa. (See Exercise 4.6.3 for the details.) Once again, if only some of the constraints are of the form $\geq$, then one would subtract slack variables only from those constraints.

The next topic to be addressed is the issue of what to do if the original variables are not restricted to be nonnegative. Consider, for example, an LP in the following form:

$$\begin{aligned} \text{Minimize} \quad & \mathbf{cx} \\ \text{subject to} \quad & \mathbf{Ax} = \mathbf{b} \\ & \mathbf{x} \leq \mathbf{0} \end{aligned} \qquad \text{(LP 7)}$$

The associated LP in standard form is obtained by replacing the original $\mathbf{x}$ variables by the new variables $-\mathbf{x}'$:

$$\begin{aligned} \text{Minimize} \quad & \mathbf{c}(-\mathbf{x}') \\ \text{subject to} \quad & \mathbf{A}(-\mathbf{x}') = \mathbf{b} \\ & -\mathbf{x}' \leq \mathbf{0} \end{aligned}$$

or, equivalently,

$$\begin{aligned} \text{Minimize} \quad & (-\mathbf{c})\mathbf{x}' \\ \text{subject to} \quad & (-\mathbf{A})\mathbf{x}' = \mathbf{b} \\ & \mathbf{x}' \geq \mathbf{0} \end{aligned} \qquad \text{(LP 8)}$$

It is then possible to show that if $\mathbf{x}'$ is optimal for LP 8 then $\mathbf{x}^* = -\mathbf{x}'$ is optimal for LP 7, and if $\mathbf{x}^*$ is optimal for LP 7, then $\mathbf{x}' = -\mathbf{x}^*$ is optimal for LP 8 (see Exercises 4.6.5 and 4.6.6). As before, if some of the original variables are less than or equal to 0, then only those variables need to be replaced, as illustrated with the following LP:

$$\begin{array}{ll} \text{Minimize} & -3x_1 + 2x_2 - 4x_3 \\ \text{subject to} & -4x_1 - 5x_2 + x_3 = 11 \\ & 6x_1 - x_2 - 2x_3 = 8 \\ & x_1, \quad x_2 \geq 0 \\ & x_3 \leq 0 \end{array}$$

With the substitution $x_3 = -x_3'$, the associated LP in standard form is obtained:

$$\begin{array}{ll} \text{Minimize} & -3x_1 + 2x_2 + 4x_3' \\ \text{subject to} & -4x_1 - 5x_2 - x_3' = 11 \\ & 6x_1 - x_2 + 2x_3' = 8 \\ & x_1, \quad x_2, \quad x_3' \geq 0 \end{array}$$

The appearance of unrestricted variables requires a bit more care. Consider an LP in the following form:

$$\begin{array}{ll} \text{Minimize} & \mathbf{cx} \\ \text{subject to} & \mathbf{Ax} = \mathbf{b} \\ & \mathbf{x} \text{ unrestricted} \end{array} \qquad \text{(LP 9)}$$

The idea is to replace each unrestricted variable x_i by the difference $x_i^+ - x_i^-$ of two nonnegative variables, of which x_i^+ is positive (and $x_i^- = 0$) when x_i is positive and x_i^- is positive (and $x_i^+ = 0$) when x_i is negative. Specifically, the associated LP in standard form is to

$$\begin{array}{ll} \text{Minimize} & \mathbf{c}(\mathbf{x}^+ - \mathbf{x}^-) \\ \text{subject to} & \mathbf{A}(\mathbf{x}^+ - \mathbf{x}^-) = \mathbf{b} \\ & \mathbf{x}^+, \quad \mathbf{x}^- \geq \mathbf{0} \end{array}$$

or, equivalently, to

$$\begin{array}{ll} \text{Minimize} & \mathbf{cx}^+ - \mathbf{cx}^- \\ \text{subject to} & \mathbf{Ax}^+ - \mathbf{Ax}^- = \mathbf{b} \\ & \mathbf{x}^+, \quad \mathbf{x}^- \geq \mathbf{0} \end{array} \qquad \text{(LP 10)}$$

If $(\mathbf{x}^+, \mathbf{x}^-)$ is an optimal solution to LP 10, then the corresponding optimal solution to LP 9 is $\mathbf{x}^* = \mathbf{x}^+ - \mathbf{x}^-$, as will be established in Exercise 4.6.7. On the other hand, if $\mathbf{x}^*$ is an optimal solution to LP 9, then the optimal solution to LP 10 is

$$x_i^+ = \begin{cases} x_i^* & \text{if } x_i^* \geq 0 \\ 0 & \text{otherwise} \end{cases}$$

$$x_i^- = \begin{cases} -x_i^* & \text{if } x_i^* \leq 0 \\ 0 & \text{otherwise} \end{cases}$$

Furthermore, if only some of the variables in LP 9 are unrestricted, then only those variables need to be replaced.

Although it is theoretically possible to handle unrestricted variables in this manner, in practice it is not necessary to do so. A more efficient approach has been developed by making certain modifications to the simplex algorithm to enable it to handle unrestricted variables directly. These modifications are described in Chapter 10; for now unrestricted variables will be replaced as indicated above.

Finally, the issue of what to do if $\mathbf{b} \geq \mathbf{0}$ does not hold will be addressed. Consider an LP in the following form:

$$\begin{aligned} \text{Minimize} \quad & \mathbf{cx} \\ \text{subject to} \quad & \mathbf{Ax} = \mathbf{b} \\ & \mathbf{x} \geq \mathbf{0} \end{aligned}$$

in which some (or all) components of $\mathbf{b}$ are less than 0. On multiplication of both sides of those equations by -1, the corresponding components of $\mathbf{b}$ will be greater than 0. In other words, if $b_i < 0$, then one would replace row i of $\mathbf{A}$ (namely, $\mathbf{A}_{i\cdot}$) with $-\mathbf{A}_{i\cdot}$, and b_i with $-b_i$.

Observe that if an inequality constraint of the form $\leq$ is multiplied by -1 to make the corresponding component of $b_i > 0$, then the new constraint has the form $\geq$. Similarly, if the original constraint is of the form $\geq$, then the new one becomes $\leq$.

A flow chart provides a convenient summary of how to convert any LP into an equivalent one in standard form:

Step 1. If the objective function is to be maximized, then multiply the objective function by -1 and replace the word "maximize" with "minimize."

Step 2. If the constraints contain inequalities, then add (subtract) a nonnegative slack variable to the left side of each $\leq$ ($\geq$) constraint.

Step 3. If any variables are restricted to be nonpositive, then replace each such variable x_i in the objective function and the constraints with $-x_i'$. Then require that each x_i' be nonnegative.

Step 4. If any variables are unrestricted, then replace each such variable x_i in the objective function and the constraints with $x_i^+ - x_i^-$. Then require that each x_i^+ and x_i^- be nonnegative.

Step 5. If any component of the $\mathbf{b}$ vector is negative, then multiply the corresponding equation by -1.

To illustrate all the steps in the flow chart, consider the following LP:

$$\begin{array}{llrrrcr}
\text{Maximize} & -2x_1 + x_2 - 3x_3 & & & & & \\
\text{subject to} & x_1 + 2x_2 - x_3 \le & 11 \\
& x_1 - x_2 + x_3 = & -1 \\
& -2x_1 - 3x_2 + 4x_3 \ge & 8 \\
& x_1 \ge & 0 \\
& x_2 \le & 0 \\
& x_3 & \text{unrestricted}
\end{array}$$

By applying the steps in the order of the flow chart, one obtains the associated LP in standard form:

$$\begin{array}{ll}
\text{Minimize} & 2x_1 + x_2' + 3x_3^+ - 3x_3^- \\
\text{subject to} & x_1 - 2x_2' - x_3^+ + x_3^- + s_1 = 11 \\
& -x_1 - x_2' - x_3^+ + x_3^- = 1 \\
& -2x_1 + 3x_2' + 4x_3^+ - 4x_3^- - s_2 = 8 \\
& x_1,\ x_2',\ x_3^+,\ x_3^-,\ s_1,\ s_2 \ge 0
\end{array}$$

As a result of the work in this section, any LP can be converted to an equivalent LP in the standard form:

$$\begin{array}{ll}
\text{Minimize} & \mathbf{cx} \\
\text{subject to} & \mathbf{Ax} = \mathbf{b} \\
& \mathbf{x} \ge \mathbf{0}
\end{array}$$

in which $\mathbf{b} \ge \mathbf{0}$. Thus, when designing the simplex algorithm, one can assume that the LP is going to be in standard form. The reason for choosing this particular form arises from algebraic considerations that will become evident in the next chapter. From now on, unless otherwise stated, it will be assumed that the LP is in standard form.

EXERCISES

4.6.1. Prove that if $\mathbf{x}^*$ is optimal for LP 1, then $\mathbf{x}^*$ is optimal for LP 2.

4.6.2. Prove that if $\mathbf{x}^*$ is optimal for LP 3, then there is a vector $\mathbf{s}^*$ such that $(\mathbf{x}^*, \mathbf{s}^*)$ is optimal for LP 4.

4.6.3. Prove that if $(\mathbf{x}^*, \mathbf{s}^*)$ is optimal for LP 6, then $\mathbf{x}^*$ is optimal for LP 5.

4.6.4. State and prove the converse of Exercise 4.6.3.

4.6.5. Prove that if $\mathbf{x}'$ is optimal for LP 8, then $\mathbf{x}^* = -\mathbf{x}'$ is optimal for LP 7.

4.6.6. State and prove the converse of Exercise 4.6.5.

4.6.7. Prove that if $(\mathbf{x}^+, \mathbf{x}^-)$ is optimal for LP 10, then $\mathbf{x}^* = \mathbf{x}^+ - \mathbf{x}^-$ is optimal for LP 9.

4.6.8. If $\mathbf{x}^*$ is optimal for LP 9, how can one find an optimal solution for LP 10? Prove that your answer is correct.

4.6.9. For each of the following LPs write the associated LP in standard form.

(a)

$$\begin{array}{ll} \text{Maximize} & 3x_1 - 2x_2 \\ \text{subject to} & 6x_1 + x_2 \geq 4 \\ & -3x_1 + 2x_2 \leq -10 \\ & x_1 \leq 0 \\ & x_2 \geq 0 \end{array}$$

(b)

$$\begin{array}{ll} \text{Minimize} & 6x_1 - 9x_2 + 7x_3 \\ \text{subject to} & 5x_1 - 7x_2 - 4x_3 \geq 6 \\ & 4x_1 + 7x_2 + 9x_3 = 4 \\ & 6x_1 - 2x_2 \leq -10 \\ & x_1, \; x_2 \geq 0 \\ & x_3 \text{ unrestricted} \end{array}$$

(c)

$$\begin{array}{ll} \text{Maximize} & x_1 + 2x_2 - 4x_3 - 6x_4 \\ \text{subject to} & -2x_1 + 9x_2 + 9x_3 - x_4 \geq -7 \\ & 6x_1 - 9x_2 + 4x_3 - 7x_4 \leq 21 \\ & 6x_1 - 2x_2 - 6x_3 + x_4 = 3 \\ & 7x_1 + 11x_2 - 9x_3 - 21x_4 \geq 0 \\ & x_1, \; x_3 \geq 0 \\ & x_2, \; x_4 \leq 0 \end{array}$$

DISCUSSION

Linear Algebra

In this chapter only those topics of linear algebra that will be used in this book were presented. A more detailed discussion of topics in linear algebra that apply to linear programming can be found in a text by Hadley [2]. A unique contribution of this chapter is the development of rules for identifying the conditions for a vector to be a direction of unboundedness.

The ability to solve linear systems of equations is fundamental to the simplex algorithm and to many other branches of mathematics. Much study has gone into the development of efficient and numerically accurate methods for solving linear equations. Some of these concepts will be presented in Chapter 7, but a more extensive account is that of Forsythe and Moler [1].

REFERENCES

1. G. E. Forsythe and C. B. Moler, *Computer Solutions to Linear Algebraic Systems*, Prentice-Hall, Englewood Cliffs, NJ, 1967.
2. G. Hadley, *Linear Algebra*, Addison-Wesley, Reading, MA, 1961.

Chapter 5

The Simplex Algorithm

All of the preliminary work is now complete. This chapter will develop the simplex algorithm, which, in a finite number of steps, will determine if an LP in the standard form

$$\begin{aligned} \text{Minimize} \quad & \mathbf{cx} \\ \text{subject to} \quad & \mathbf{Ax} = \mathbf{b} \\ & \mathbf{x} \geq \mathbf{0} \end{aligned}$$

is infeasible, optimal, or unbounded. If the LP is unbounded, a direction of unboundness will be produced; if the LP is determined to be optimal, then an optimal solution will be produced.

5.1. BASIC FEASIBLE SOLUTIONS

Throughout this chapter, it will be helpful to recall the geometric and conceptual motivation of the simplex algorithm that was presented in Chapter 2. That information is summarized in Table 5.1.

From Table 5.1 you can see that the geometric approach suggests moving from one extreme point of the feasible region to another. Unfortunately, computers cannot solve problems geometrically, and so algebra becomes necessary. Consequently, the first step in designing an algebraic method is to develop an algebraic representation of an extreme point. That concept is referred to as a *basic feasible solution* (hereafter called a *bfs*). The simplex

TABLE 5.1. Geometric and Conceptual Approach of the Simplex Algorithm

	Conceptual	Geometric
The problem:		
Given	Colored balls in a box	Extreme points of the feasible region
Objective	Find a red ball	Find one with the smallest objective value
The algorithm:		
0. Initialization	Pick a ball from the box	Find an initial extreme point
1. Test for optimality	Is the ball red?	Does it have the smallest objective value?
2. Selecting a new object	Select a new ball from the box	Select a direction leading to a new extreme point

algorithm will move algebraically from one bfs to another while trying to decrease the objective value each time.

To motivate the formal definition of a bfs, consider the following LP in standard form that will be used throughout this chapter for demonstration purposes.

EXAMPLE 5.1:

$$\begin{array}{llllll} \text{Minimize} & -3x_1 & +x_2 & -x_3 & +2x_4 & \\ \text{subject to} & x_1 & & +x_3 & +2x_4 & = 2 \\ & & x_2 & +x_3 & +2x_4 & = 3 \\ & x_1, & x_2, & x_3, & x_4 & \geq 0 \end{array}$$

In this example

$$\mathbf{c} = (-3, 1, -1, 2)$$

$$\mathbf{b} = \begin{bmatrix} 2 \\ 3 \end{bmatrix}$$

$$\mathbf{A} = \begin{bmatrix} 1 & 0 & 1 & 2 \\ 0 & 1 & 1 & 2 \end{bmatrix}$$

A basic feasible solution is a feasible solution that is obtained by using the equality constraints to solve for some of the variables in terms of the remaining ones. For example, the first equality constraint can be used to express x_1 in terms of x_3 and x_4; $x_1 = 2 - x_3 - 2x_4$. Similarly, the second constraint can be used to express x_2 in terms of x_3 and x_4;

$x_2 = 3 - x_3 - 2x_4$. Thus the original equality constraints can be rewritten

$$x_1 = 2 - x_3 - 2x_4$$
$$x_2 = 3 - x_3 - 2x_4$$

To specify the actual bfs, values must be given to the variables. This is done by first assigning a value of 0 to all of the variables on the right-hand side of the equality sign. In this case that means $x_3 = 0$ and $x_4 = 0$. The variables on the right-hand side of the equality sign are referred to as the *nonbasic variables*, and they always have the value 0. Then, the values for the remaining variables can be determined. Specifically, on substituting 0 for x_3 and x_4 above, it must be that $x_1 = 2$ and $x_2 = 3$. The variables on the left-hand side of the equality sign are called *basic variables*. It is important to distinguish between the basic variables and their values. In this instance, x_1 and x_2 are the basic variables and their values are 2 and 3, respectively. Thus the vector $\mathbf{x} = (x_1, x_2, x_3, x_4) = (2, 3, 0, 0)$ is said to be a basic feasible solution in which x_1 and x_2 are the basic variables and x_3 and x_4 are the nonbasic ones. Also, note that the bfs $\mathbf{x} = (2, 3, 0, 0)$ satisfies the nonnegativity constraints.

Not surprisingly, there are usually many ways to choose which variables are to be basic and which nonbasic. For instance, the first equality constraint could just as well have been used to express x_3 in terms of x_1 and x_4; that is, $x_3 = 2 - x_1 - 2x_4$. On replacing this expression for x_3 in the second constraint, one obtains $x_2 + (2 - x_1 - 2x_4) + 2x_4 = 3$ and, after rearranging terms, $-x_1 + x_2 = 1$. Now one can solve for x_2 in terms of x_1; specifically, $x_2 = 1 + x_1$. Thus the original equality constraints can be rewritten

$$x_2 = 1 + x_1$$
$$x_3 = 2 - x_1 - 2x_4$$

Here, the nonbasic variables are x_1 and x_4, and the basic variables are x_2 and x_3. To specify the bfs, assign a value of 0 to the nonbasic variables x_1 and x_4. This in turn forces the basic variables x_2 and x_3 to assume the values 1 and 2, respectively. Thus the bfs is $\mathbf{x} = (x_1, x_2, x_3, x_4) = (0, 1, 2, 0)$. Notice once again that this vector satisfies the nonnegativity constraints.

Some choices for the basic and nonbasic variables can result in a vector that does not satisfy the nonnegativity constraints. To illustrate this possibility in Example 5.1, suppose that x_2 and x_3 are chosen to be nonbasic. This choice can be accomplished by using the second equality constraint to express x_4 in terms of x_2 and x_3; $x_4 = \frac{3}{2} - x_2/2 - x_3/2$. On substituting this expression for x_4 into the first constraint and rearranging terms, one obtains $x_1 - x_2 = -1$. Now x_1 can be expressed in terms of x_2; specifically, $x_1 = -1 + x_2$. Once again, the original equality constraints have been rewritten, this time as

$$x_1 = -1 + x_2$$
$$x_4 = \tfrac{3}{2} - x_2/2 - x_3/2$$

Here, x_2 and x_3 are the nonbasic variables, and x_1 and x_4 are the basic ones. As before, consider what happens when the nonbasic variables x_2 and x_3 are assigned a value of 0. This forces $x_1 = -1$, $x_4 = \frac{3}{2}$. The problem is that the vector $\mathbf{x} = (x_1, x_2, x_3, x_4) = (-1, 0, 0, \frac{3}{2})$ does not satisfy the nonnegativity constraints since $x_1 = -1 < 0$. Therefore $(-1, 0, 0, \frac{3}{2})$ is not a bfs.

When formally defining a bfs, it is not necessary to manipulate the original equality constraints so as to place all of the nonbasic variables on the right-hand side of the equality sign and all of the basic variables on the left-hand side. Since the nonbasic variables are going to be assigned a value of 0, all that is needed is a "mechanism" for computing the corresponding values of the basic variables. The mechanism manifests itself in the form of an $m \times m$ submatrix $\mathbf{B}$ of $\mathbf{A}$ consisting of those columns of $\mathbf{A}$ corresponding to the basic variables; e.g., if x_1 and x_4 are the basic variables, then $\mathbf{B} = [\mathbf{A}_{.1}, \mathbf{A}_{.4}]$. In order to compute the values of the basic variables, it is necessary for the $\mathbf{B}$ matrix to be nonsingular. Then, the basic variables $\mathbf{x_B}$ will be assigned the values $\mathbf{B}^{-1}\mathbf{b}$. In order for the nonnegativity constraints to be satisfied, it had better be the case that $\mathbf{B}^{-1}\mathbf{b} \geq \mathbf{0}$. The formal definition of a bfs follows.

Definition 5.1. An n vector $\mathbf{x}$ is a *basic feasible solution* if there is a nonsingular $m \times m$ submatrix $\mathbf{B}$ of $\mathbf{A}$ such that, together with the remaining $n - m$ columns of $\mathbf{A}$ (denoted by the $\mathbf{N}$ matrix),

1. $\mathbf{x_B} = \mathbf{B}^{-1}\mathbf{b}$,
2. $\mathbf{x_B} \geq \mathbf{0}$, and
3. $\mathbf{x_N} = \mathbf{0}$.

Identifying a bfs

Since the simplex algorithm is entirely dependent on the concept of a bfs, it is necessary to become thoroughly familiar with Definition 5.1. The first step in this direction is learning to identify the $\mathbf{B}$ matrix associated with a given bfs $\mathbf{x}$. The $\mathbf{N}$ matrix then consists of the remaining $n - m$ columns of $\mathbf{A}$.

The $\mathbf{B}$ matrix can be obtained from the bfs $\mathbf{x}$ by considering the multiplication $\mathbf{Ax}$ as column multiplication, that is, $\mathbf{Ax} = \sum_{k=1}^{n} \mathbf{A}_{.k} x_k$ (see Section 4.2). Under favorable circumstances, the $\mathbf{B}$ matrix will consist of precisely those columns $\mathbf{A}_{.k}$ of $\mathbf{A}$ for which $x_k > 0$. Of course, it must be shown that the $\mathbf{B}$ matrix thus obtained has all of the properties specified in Definition 5.1. Identifying the $\mathbf{B}$ matrix in this way is demonstrated in the next proposition.

Proposition 5.1. *If* $\mathbf{x} = (x_1, x_2, x_3, x_4) = (0, 1, 0, 1)$, *then* $\mathbf{x}$ *is a bfs for Example* 5.1.

OUTLINE OF PROOF. The forward–backward method gives rise to the abstraction question, "How can I show that a vector (namely, (0,1,0,1)) is a bfs?" which is answered by means of Definition 5.1. In so doing, the quantifier "there is" arises, and hence the construction method is used to produce the actual 2×2 basis matrix $\mathbf{B}$ having the desired properties. From the discussion preceding this proposition, since $x_2 = 1 > 0$ and $x_4 = 1 > 0$, $\mathbf{B}$ consists of columns 2 and 4 of the $\mathbf{A}$ matrix;

$$\mathbf{B} = [\mathbf{A}_{.2}, \mathbf{A}_{.4}] = \begin{bmatrix} 0 & 2 \\ 1 & 2 \end{bmatrix}$$

The $\mathbf{N}$ matrix then consists of the remaining columns of $\mathbf{A}$, so

$$\mathbf{N} = [\mathbf{A}_{.1}, \mathbf{A}_{.3}] = \begin{bmatrix} 1 & 1 \\ 0 & 1 \end{bmatrix}$$

It remains to show that $\mathbf{B}$ has all of the desired properties—that $\mathbf{B}$ is nonsingular, $\mathbf{x}_\mathbf{B} = \mathbf{B}^{-1}\mathbf{b}$, $\mathbf{x}_\mathbf{B} \geq \mathbf{0}$, and $\mathbf{x}_\mathbf{N} = \mathbf{0}$.

In trying to show that $\mathbf{B}$ is nonsingular, one is faced with the abstraction question "How can I show that a matrix (namely, $\mathbf{B}$) is nonsingular?" One answer is to produce the inverse matrix $\mathbf{B}^{-1}$ (see Definition 4.5). Recall that the inverse of a 2×2 matrix of the form

$$\begin{bmatrix} a & b \\ c & d \end{bmatrix}$$

is given by

$$\frac{1}{ad - bc}\begin{bmatrix} d & -b \\ -c & a \end{bmatrix}$$

provided, of course, that $ad - bc \neq 0$ (see Section 4.3). The formula for the inverse of a 2×2 nonsingular matrix will be used repeatedly throughout the chapter and is well worth memorizing. For the bfs of the current problem,

$$\mathbf{B} = \begin{bmatrix} 0 & 2 \\ 1 & 2 \end{bmatrix}$$

and so

$$\mathbf{B}^{-1} = \frac{1}{-2}\begin{bmatrix} 2 & -2 \\ -1 & 0 \end{bmatrix} = \begin{bmatrix} -1 & 1 \\ \frac{1}{2} & 0 \end{bmatrix}$$

Thus $\mathbf{B}$ is nonsingular.

The remaining properties of $\mathbf{B}$ are easily verified:

1. $\mathbf{B}^{-1}\mathbf{b} = \begin{bmatrix} -1 & 1 \\ \frac{1}{2} & 0 \end{bmatrix}\begin{bmatrix} 2 \\ 3 \end{bmatrix} = \begin{bmatrix} 1 \\ 1 \end{bmatrix} = \begin{bmatrix} x_2 \\ x_4 \end{bmatrix} = \mathbf{x}_\mathbf{B}$,
2. $\mathbf{x}_\mathbf{B} = \begin{bmatrix} 1 \\ 1 \end{bmatrix} \geq \mathbf{0}$, and
3. $\mathbf{x}_\mathbf{N} = \begin{bmatrix} x_1 \\ x_3 \end{bmatrix} = \begin{bmatrix} 0 \\ 0 \end{bmatrix}$.

PROOF OF PROPOSITION 5.1. Let **B** consist of columns 2 and 4 of the **A** matrix; thus

$$\mathbf{B}=[\mathbf{A}_{.2},\mathbf{A}_{.4}]=\begin{bmatrix}0 & 2\\ 1 & 2\end{bmatrix},\qquad \mathbf{N}=[\mathbf{A}_{.1},\mathbf{A}_{.3}]=\begin{bmatrix}1 & 1\\ 0 & 1\end{bmatrix}$$

It will be shown that **B** has all of the desired properties. **B** is nonsingular since its inverse is

$$\mathbf{B}^{-1}=\begin{bmatrix}-1 & 1\\ \frac{1}{2} & 0\end{bmatrix}$$

Moreover:

1. $\mathbf{B}^{-1}\mathbf{b}=\begin{bmatrix}-1 & 1\\ \frac{1}{2} & 0\end{bmatrix}\begin{bmatrix}2\\ 3\end{bmatrix}=\begin{bmatrix}1\\ 1\end{bmatrix}=\begin{bmatrix}x_2\\ x_4\end{bmatrix}=\mathbf{x_B}$,
2. $\mathbf{x_B}=\begin{bmatrix}1\\ 1\end{bmatrix}\geq \mathbf{0}$, and
3. $\mathbf{x_N}=\begin{bmatrix}x_1\\ x_3\end{bmatrix}=\begin{bmatrix}0\\ 0\end{bmatrix}$,

and the proof is complete. □

The standard method for checking whether a given vector **x** is a bfs is to identify the basic variables and the corresponding **B** matrix (often called the *basis matrix*), together with the nonbasic variables (and the **N** matrix). Any variable with a positive value must be a basic variable because, if it were nonbasic, then by property 3 of Definition 5.1 its value would have to be 0. Consequently, all variables whose values are positive must be basic variables, and the associated columns of the **A** matrix become part of the **B** matrix. If m such variables can be found, then the **B** matrix contains exactly m columns. It must then be verified that **B** is nonsingular, $\mathbf{x_B}=\mathbf{B}^{-1}\mathbf{b}$, $\mathbf{x_B}\geq \mathbf{0}$, and $\mathbf{x_N}=\mathbf{0}$. The next proposition demonstrates the process once again.

Proposition 5.2. *If* $\mathbf{x}=(2,3,0,0)$, *then* **x** *is a bfs for Example* 5.1.

OUTLINE OF PROOF. The approach is exactly the same as in the previous proposition. In constructing the **B** matrix, note that x_1 and x_2 have positive values of 2 and 3, respectively. Thus x_1 and x_2 are the basic variables, and so

$$\mathbf{B}=[\mathbf{A}_{.1},\mathbf{A}_{.2}]=\begin{bmatrix}1 & 0\\ 0 & 1\end{bmatrix}$$

This means that x_3 and x_4 are the nonbasic variables, so

$$\mathbf{N}=[\mathbf{A}_{.3},\mathbf{A}_{.4}]=\begin{bmatrix}1 & 2\\ 1 & 2\end{bmatrix}$$

The remaining properties of **B** are verified as in Proposition 5.1.

PROOF OF PROPOSITION 5.2. Let **B** consist of columns 1 and 2 of the **A** matrix, so

$$\mathbf{B} = \begin{bmatrix} 1 & 0 \\ 0 & 1 \end{bmatrix}, \quad \mathbf{N} = \begin{bmatrix} 1 & 2 \\ 1 & 2 \end{bmatrix}$$

It will be shown that **B** has all of the desired properties. **B** is nonsingular since its inverse is

$$\mathbf{B}^{-1} = \begin{bmatrix} 1 & 0 \\ 0 & 1 \end{bmatrix}$$

Moreover

$$\mathbf{B}^{-1}\mathbf{b} = \begin{bmatrix} 1 & 0 \\ 0 & 1 \end{bmatrix}\begin{bmatrix} 2 \\ 3 \end{bmatrix} = \begin{bmatrix} 2 \\ 3 \end{bmatrix} = \begin{bmatrix} x_1 \\ x_2 \end{bmatrix} = \mathbf{x_B}$$

$$\mathbf{x_B} = \begin{bmatrix} 2 \\ 3 \end{bmatrix} \geq \mathbf{0}$$

$$\mathbf{x_N} = \begin{bmatrix} x_3 \\ x_4 \end{bmatrix} = \begin{bmatrix} 0 \\ 0 \end{bmatrix}$$

and so the proof is complete. □

In trying to construct the **B** matrix for a given bfs **x**, difficulty can arise if you are unable to find m variables with positive values. The vector **x** is then called a *degenerate bfs*. The issues relating to degeneracy will be dealt with in Section 5.7.

Constructing a bfs

Propositions 5.1 and 5.2 were designed to help you to identify the basic variables and the basis matrix for a given bfs **x**. Further understanding of a bfs can be gained by attempting to construct one from scratch rather than simply checking if a given vector is a bfs. Indeed, Table 5.1 indicates that the very first step of the simplex algorithm requires finding an initial bfs.

One way to attempt constructing a bfs is to select any set of m variables as the basic ones and form the corresponding **B** matrix. The remaining $n - m$ variables must then be the nonbasic ones, and hence are assigned the value 0. One must then determine the values of the basic variables. According to the first property of Definition 5.1, one would like to compute the values of the basic variables by the formula $\mathbf{x_B} = \mathbf{B}^{-1}\mathbf{b}$. This is possible only if the selected **B** matrix is nonsingular. If it is not, then a bad choice has been made for the **B** matrix, and a new set of m basic variables must be sought. On the other hand, if **B** is nonsingular, then it is possible to compute $\mathbf{x_B} = \mathbf{B}^{-1}\mathbf{b}$, but unfortunately this is not the end of the story.

Because of the second property of Definition 5.1, if all the values of the basic variables thus computed are nonnegative, then indeed the vector

$\mathbf{x} = (\mathbf{x_B}, \mathbf{x_N}) = (\mathbf{B}^{-1}\mathbf{b}, \mathbf{0})$ is a bfs; otherwise, it is not, and again, a new set of basic variables should be sought.

For Example 5.1, suppose that one were to select x_2 and x_3 as basic variables, leaving x_1 and x_4 as the nonbasic ones. The corresponding **B** matrix would consist of columns 2 and 3 of **A**, so

$$\mathbf{B} = \begin{bmatrix} 0 & 1 \\ 1 & 1 \end{bmatrix}$$

B is nonsingular since

$$\mathbf{B}^{-1} = \begin{bmatrix} -1 & 1 \\ 1 & 0 \end{bmatrix}$$

The values of the basic variables can then be computed:

$$\mathbf{x_B} = \mathbf{B}^{-1}\mathbf{b} = \begin{bmatrix} -1 & 1 \\ 1 & 0 \end{bmatrix}\begin{bmatrix} 2 \\ 3 \end{bmatrix} = \begin{bmatrix} 1 \\ 2 \end{bmatrix} = \begin{bmatrix} x_2 \\ x_3 \end{bmatrix}$$

The values of the nonbasic variables x_1 and x_4 are both 0. Since $\mathbf{x_B} = (x_2, x_3) = (1,2) \geq \mathbf{0}$, the vector $\mathbf{x} = (0,1,2,0)$ is a bfs.

On the other hand, suppose that one were to select x_3 and x_4 as the basic variables. The corresponding basis matrix is

$$\mathbf{B} = [\mathbf{A}_{.3}, \mathbf{A}_{.4}] = \begin{bmatrix} 1 & 2 \\ 1 & 2 \end{bmatrix}$$

Unfortunately, this **B** matrix is singular because $\det(\mathbf{B}) = 1(2) - 1(2) = 0$, and hence x_3 and x_4 cannot be chosen as basic variables.

Suppose one were to list every $m \times m$ submatrix **B** of **A**. Some of them will produce a bfs (namely, those for which **B** is nonsingular and $\mathbf{B}^{-1}\mathbf{b} \geq \mathbf{0}$), but others will not. In any event, one thing is certain. There are only a finite number of possible choices for the **B** matrix. Indeed, there are

$$\binom{n}{m} = \frac{n!}{(n-m)!m!}$$

possibilities. Thus there are at most $\binom{n}{m}$ bfs, and it is this fact that will ultimately ensure that the simplex algorithm is finite. For Example 5.1, $n = 4$, $m = 2$, and so there are $4!/[(4-2)!2!] = 6$ possible **B** matrices, of which three lead to bfs, as shown in Table 5.2.

Recall from Table 5.1 that the very first step of the simplex algorithm requires finding a bfs (or determining that none exists). To do so, one could randomly choose a **B** matrix and proceed to see if it produces a bfs, as was done above. However, one might hope for a more systematic approach. Such a method has been found. It is called phase 1 and will be described in the next chapter.

The final proposition of this section formally establishes that a bfs is feasible for the LP. The proof also demonstrates a technique that will be used repeatedly when proving statements involving a bfs.

TABLE 5.2. All Possible **B** Matrices for Example 5.1

1. $\mathbf{B} = [\mathbf{A}_{.1}, \mathbf{A}_{.2}] = \begin{bmatrix} 1 & 0 \\ 0 & 1 \end{bmatrix}$ $\qquad \mathbf{N} = [\mathbf{A}_{.3}, \mathbf{A}_{.4}] = \begin{bmatrix} 1 & 2 \\ 1 & 2 \end{bmatrix}$

$$\mathbf{x_B} = \begin{bmatrix} x_1 \\ x_2 \end{bmatrix} = \mathbf{B}^{-1}\mathbf{b} = \begin{bmatrix} 1 & 0 \\ 0 & 1 \end{bmatrix}\begin{bmatrix} 2 \\ 3 \end{bmatrix} = \begin{bmatrix} 2 \\ 3 \end{bmatrix} \geq \mathbf{0} \qquad \mathbf{x_N} = \begin{bmatrix} x_3 \\ x_4 \end{bmatrix} = \begin{bmatrix} 0 \\ 0 \end{bmatrix}$$

$\mathbf{x} = (2, 3, 0, 0)$ is a bfs with objective value

$\mathbf{cx} = (-3, 1, -1, 2)(2, 3, 0, 0) = -3$ (optimal)

2. $\mathbf{B} = [\mathbf{A}_{.1}, \mathbf{A}_{.3}] = \begin{bmatrix} 1 & 1 \\ 0 & 1 \end{bmatrix}$ $\qquad \mathbf{N} = [\mathbf{A}_{.2}, \mathbf{A}_{.4}] = \begin{bmatrix} 0 & 2 \\ 1 & 2 \end{bmatrix}$

$$\mathbf{x_B} = \begin{bmatrix} x_1 \\ x_3 \end{bmatrix} = \mathbf{B}^{-1}\mathbf{b} = \begin{bmatrix} 1 & -1 \\ 0 & 1 \end{bmatrix}\begin{bmatrix} 2 \\ 3 \end{bmatrix} = \begin{bmatrix} -1 \\ 3 \end{bmatrix} \ngeq \mathbf{0} \qquad \mathbf{x_N} = \begin{bmatrix} x_2 \\ x_4 \end{bmatrix} = \begin{bmatrix} 0 \\ 0 \end{bmatrix}$$

$\mathbf{x} = (-1, 0, 3, 0)$ is not a bfs because $x_1 < 0$

3. $\mathbf{B} = [\mathbf{A}_{.1}, \mathbf{A}_{.4}] = \begin{bmatrix} 1 & 2 \\ 0 & 2 \end{bmatrix}$ $\qquad \mathbf{N} = [\mathbf{A}_{.2}, \mathbf{A}_{.3}] = \begin{bmatrix} 0 & 1 \\ 1 & 1 \end{bmatrix}$

$$\mathbf{x_B} = \begin{bmatrix} x_1 \\ x_4 \end{bmatrix} = \mathbf{B}^{-1}\mathbf{b} = \begin{bmatrix} 1 & -1 \\ 0 & \frac{1}{2} \end{bmatrix}\begin{bmatrix} 2 \\ 3 \end{bmatrix} = \begin{bmatrix} -1 \\ \frac{3}{2} \end{bmatrix} \ngeq \mathbf{0} \qquad \mathbf{x_N} = \begin{bmatrix} x_2 \\ x_3 \end{bmatrix} = \begin{bmatrix} 0 \\ 0 \end{bmatrix}$$

$\mathbf{x} = (-1, 0, 0, \frac{3}{2})$ is not a bfs because $x_1 < 0$

4. $\mathbf{B} = [\mathbf{A}_{.2}, \mathbf{A}_{.3}] = \begin{bmatrix} 0 & 1 \\ 1 & 1 \end{bmatrix}$ $\qquad \mathbf{N} = [\mathbf{A}_{.1}, \mathbf{A}_{.4}] = \begin{bmatrix} 1 & 2 \\ 0 & 2 \end{bmatrix}$

$$\mathbf{x_B} = \begin{bmatrix} x_2 \\ x_3 \end{bmatrix} = \mathbf{B}^{-1}\mathbf{b} = \begin{bmatrix} -1 & 1 \\ 1 & 0 \end{bmatrix}\begin{bmatrix} 2 \\ 3 \end{bmatrix} = \begin{bmatrix} 1 \\ 2 \end{bmatrix} \geq \mathbf{0} \qquad \mathbf{x_N} = \begin{bmatrix} x_1 \\ x_4 \end{bmatrix} = \begin{bmatrix} 0 \\ 0 \end{bmatrix}$$

$\mathbf{x} = (0, 1, 2, 0)$ is a bfs with objective value

$\mathbf{cx} = (-3, 1, -1, 2)(0, 1, 2, 0) = -1$

5. $\mathbf{B} = [\mathbf{A}_{.2}, \mathbf{A}_{.4}] = \begin{bmatrix} 0 & 2 \\ 1 & 2 \end{bmatrix}$ $\qquad \mathbf{N} = [\mathbf{A}_{.1}, \mathbf{A}_{.3}] = \begin{bmatrix} 1 & 1 \\ 0 & 1 \end{bmatrix}$

$$\mathbf{x_B} = \begin{bmatrix} x_2 \\ x_4 \end{bmatrix} = \mathbf{B}^{-1}\mathbf{b} = \begin{bmatrix} -1 & 1 \\ \frac{1}{2} & 0 \end{bmatrix}\begin{bmatrix} 2 \\ 3 \end{bmatrix} = \begin{bmatrix} 1 \\ 1 \end{bmatrix} \geq \mathbf{0} \qquad \mathbf{x_N} = \begin{bmatrix} x_1 \\ x_3 \end{bmatrix} = \begin{bmatrix} 0 \\ 0 \end{bmatrix}$$

$\mathbf{x} = (0, 1, 0, 1)$ is a bfs with objective value

$\mathbf{cx} = (-3, 1, -1, 2)(0, 1, 0, 1) = 3$

6. $\mathbf{B} = [\mathbf{A}_{.3}, \mathbf{A}_{.4}] = \begin{bmatrix} 1 & 2 \\ 1 & 2 \end{bmatrix}$ $\qquad \mathbf{N} = [\mathbf{A}_{.1}, \mathbf{A}_{.2}] = \begin{bmatrix} 1 & 0 \\ 0 & 1 \end{bmatrix}$

det($\mathbf{B}$) = 0; **B** is a singular and cannot be used to create a bfs

Proposition 5.3. *If* $\mathbf{x} = (\mathbf{x_B}, \mathbf{x_N}) = (\mathbf{B}^{-1}\mathbf{b}, \mathbf{0})$ *is a bfs, then* $\mathbf{x}$ *is feasible for the LP.*

OUTLINE OF PROOF. The forward–backward method gives rise to the abstraction question "How can I show that a vector (namely, $\mathbf{x}$) is feasible for an LP?" The answer is to show that $\mathbf{x}$ satisfies the constraints, i.e., $\mathbf{Ax} = \mathbf{b}$ and $\mathbf{x} \geq \mathbf{0}$.

When trying to prove that $\mathbf{x} \geq \mathbf{0}$, one is faced with the abstraction question "How can I show that a vector (namely, $\mathbf{x}$) is greater than or equal to another vector (namely, $\mathbf{0}$)?" Here, the answer is to show that the basic and nonbasic components of $\mathbf{x}$ are both nonnegative; i.e., $\mathbf{x_B} \geq \mathbf{0}$ and $\mathbf{x_N} \geq \mathbf{0}$. But this follows from the hypothesis that $\mathbf{x}$ is a bfs. Specifically, working forward from the second and third properties in the definition of a bfs yields $\mathbf{x_B} \geq \mathbf{0}$ and $\mathbf{x_N} = \mathbf{0}$, and thus $\mathbf{x} \geq \mathbf{0}$.

To see that $\mathbf{Ax} = \mathbf{b}$, one can work forward algebraically from the hypothesis that $\mathbf{x} = (\mathbf{x_B}, \mathbf{x_N}) = (\mathbf{B}^{-1}\mathbf{b}, \mathbf{0})$:

$$\begin{aligned}\mathbf{Ax} &= [\mathbf{B}, \mathbf{N}]\begin{bmatrix}\mathbf{x_B}\\ \mathbf{x_N}\end{bmatrix} = \mathbf{Bx_B} + \mathbf{Nx_N}\\ &= \mathbf{B}(\mathbf{B}^{-1}\mathbf{b}) + \mathbf{N}(\mathbf{0}) = \mathbf{Ib} + \mathbf{0}\\ &= \mathbf{b}\end{aligned}$$

PROOF OF PROPOSITION 5.3. To establish the conclusion, it must be shown that $\mathbf{Ax} = \mathbf{b}$ and $\mathbf{x} \geq \mathbf{0}$. Since $\mathbf{x}$ is a bfs, by Definition 5.1 $\mathbf{x_B} \geq \mathbf{0}$ and $\mathbf{x_N} = \mathbf{0}$, and so $\mathbf{x} \geq \mathbf{0}$. Also,

$$\begin{aligned}\mathbf{Ax} &= [\mathbf{B}, \mathbf{N}]\begin{bmatrix}\mathbf{x_B}\\ \mathbf{x_N}\end{bmatrix} = \mathbf{Bx_B} + \mathbf{Nx_N}\\ &= \mathbf{B}(\mathbf{B}^{-1}\mathbf{b}) + \mathbf{N}(\mathbf{0}) = \mathbf{Ib} + \mathbf{0}\\ &= \mathbf{b}\end{aligned}$$ □

In the proof of Proposition 5.3, $\mathbf{Ax} = \mathbf{b}$ was established by writing $\mathbf{Ax} = \mathbf{Bx_B} + \mathbf{Nx_N}$. Similarly, to show that $\mathbf{x} \geq \mathbf{0}$, the vector $\mathbf{x}$ was written $(\mathbf{x_B}, \mathbf{x_N})$. The approach of working with the $\mathbf{B}$ and $\mathbf{N}$ components of vectors and matrices will be used frequently throughout this chapter. Whenever $\mathbf{x}$ is a bfs. It will be written $\mathbf{x} = (\mathbf{x_B}, \mathbf{x_N}) = (\mathbf{B}^{-1}\mathbf{b}, \mathbf{0}) \geq \mathbf{0}$, where it is understood that $\mathbf{B}^{-1}\mathbf{b}$ comprises the values of the basic variables, $\mathbf{0}$ comprises the values of the nonbasic variables, and $\mathbf{B}^{-1}\mathbf{b} \geq \mathbf{0}$. The symbols "$\mathbf{x_B}$" and "$\mathbf{x_N}$" can refer to either the variables or their values.

This section has defined the concept of a basic feasible solution and has shown you how to (1) check if a given vector is a bfs and (2) go about creating a bfs from scratch. Imagine that, somehow, you have been successful in constructing an initial bfs. The next step of the simplex algorithm requires that you test the bfs for optimality (see Table 5.1). The test for optimality is developed in the next section.

EXERCISES

5.1.1. For the following LP, prepare a table similar to Table 5.2. Specifically, consider all possible **B** matrices and determine which ones result in a bfs. Supply all of the appropriate information.

$$\begin{array}{ll} \text{Minimize} & -x_1 + 2x_2 + x_3 - x_4 \\ \text{subject to} & x_1 + 3x_2 + 2x_3 = 5 \\ & x_2 + 2x_3 + x_4 = 6 \\ & x_1,\ x_2,\ x_3,\ x_4 \geq 0 \end{array}$$

5.1.2. For the LP

$$\begin{array}{ll} \text{Minimize} & -2x_1 - x_2 + x_3 + 5x_4 - 3x_5 \\ \text{subject to} & 4x_1 + 2x_2 + x_3 + x_4 = 3 \\ & 2x_1 + 2x_2 + 3x_4 + x_5 = 2 \\ & x_1,\ x_2,\ x_3,\ x_4,\ x_5 \geq 0 \end{array}$$

determine whether each of the following vectors is a bfs or not. Explain your answer.

(a) $\mathbf{x} = (0,0,3,0,2)$,
(b) $\mathbf{x} = (1,0,-1,0,0)$,
(c) $\mathbf{x} = (0,1,1,0,0)$,
(d) $\mathbf{x} = (\frac{1}{2},1,0,0,0)$.

5.1.3. Is the vector $\mathbf{x} = (2,3,1,0,1,0,4)$ a bfs for the following LP? Why or why not? Explain.

$$\begin{array}{ll} \text{Minimize} & x_1 - x_2 + 2x_3 + x_4 - x_5 + 3x_6 + x_7 \\ \text{subject to} & x_1 + 2x_3 + 2x_4 + x_6 = 4 \\ & x_2 + x_3 + 3x_4 + 4x_6 = 4 \\ & x_3 + 2x_4 + x_5 + 4x_6 = 2 \\ & x_3 + 3x_4 + x_6 + x_7 = 5 \\ & x_1,\ x_2,\ x_3,\ x_4,\ x_5,\ x_6,\ x_7 \geq 0 \end{array}$$

5.1.4. When the nonbasic variables are assigned a value of 0, the basic variables are computed by the formula $\mathbf{B}^{-1}\mathbf{b}$. Derive a formula for computing the values of the basic variables when the nonbasic variables are assigned arbitrary values. The basic variables should be computed in such a way that the resulting vector $\mathbf{x}$ satisfies $\mathbf{Ax} = \mathbf{b}$.

5.2. THE TEST FOR OPTIMALITY

Imagine, for the moment, that you have somehow found a basis matrix $\mathbf{B}$ and have constructed a bfs $\mathbf{x} = (\mathbf{x_B}, \mathbf{x_N}) = (\mathbf{B}^{-1}\mathbf{b}, \mathbf{0}) \geq \mathbf{0}$ for an LP in standard form. Using these values of the variables, you can compute the value of the objective function at $\mathbf{x}$ as $\mathbf{cx}$, but how will you know if the current bfs has the smallest possible objective value? The next proposition provides an algebraic test for optimality that uses the basic and nonbasic components of the cost vector $\mathbf{c}$ together with the matrices $\mathbf{B}^{-1}$ and $\mathbf{N}$. It is surprising that the test does not require looking at any other bfs.

Proposition 5.4. *If* $\mathbf{x}^* = (\mathbf{x}_\mathbf{B}^*, \mathbf{x}_\mathbf{N}^*) = (\mathbf{B}^{-1}\mathbf{b}, \mathbf{0}) \geq \mathbf{0}$ *is a bfs and has the property that* $\mathbf{c_N} - \mathbf{c_B}\mathbf{B}^{-1}\mathbf{N} \geq \mathbf{0}$, *then* $\mathbf{x}^*$ *is optimal for the LP.*

OUTLINE OF PROOF. The forward–backward method gives rise to the abstraction question "How can I show that a vector (namely, $\mathbf{x}^*$) is optimal for an LP?" This question is answered by Definition 4.15, whereby it must be shown that (1) $\mathbf{x}^*$ is feasible and (2) for all feasible vectors $\mathbf{x}$, $\mathbf{cx} \geq \mathbf{cx}^*$.

Working forward from the fact that $\mathbf{x}^*$ is a bfs, one notes that Proposition 5.3 ensures that $\mathbf{x}^*$ is feasible, thus establishing (1). To obtain (2), the appearance of the quantifier "for all" in the backward process suggests proceeding with the choose method. Consequently, $\mathbf{x}$ is chosen to be a feasible solution (so $\mathbf{Ax} = \mathbf{b}$ and $\mathbf{x} \geq \mathbf{0}$), and it must be shown that $\mathbf{cx} \geq \mathbf{cx}^*$, or, equivalently, that $\mathbf{c}(\mathbf{x} - \mathbf{x}^*) \geq 0$. To do so requires working forward from the fact that $\mathbf{x}$ is feasible together with the hypothesis that $\mathbf{x}^*$ satisfies the property that $\mathbf{c_N} - \mathbf{c_B}\mathbf{B}^{-1}\mathbf{N} \geq \mathbf{0}$. Specifically, since $\mathbf{x}$ is feasible, $\mathbf{x_N} \geq \mathbf{0}$ (note that $\mathbf{x_N}$ might not be $\mathbf{0}$ because $\mathbf{x}$ is not a bfs, it is merely a feasible solution). Then, by algebraic manipulation it will be shown that $\mathbf{c}(\mathbf{x} - \mathbf{x}^*) = (\mathbf{c_N} - \mathbf{c_B}\mathbf{B}^{-1}\mathbf{N})\mathbf{x_N}$; since $\mathbf{c_N} - \mathbf{c_B}\mathbf{B}^{-1}\mathbf{N} \geq \mathbf{0}$ and $\mathbf{x_N} \geq \mathbf{0}$, $\mathbf{c}(\mathbf{x} - \mathbf{x}^*) \geq 0$, and the proof is complete.

PROOF OF PROPOSITION 5.4. Since $\mathbf{x}^* = (\mathbf{x}_\mathbf{B}^*, \mathbf{x}_\mathbf{N}^*) = (\mathbf{B}^{-1}\mathbf{b}, \mathbf{0}) \geq \mathbf{0}$ is a bfs, Proposition 5.3 ensures that $\mathbf{x}^*$ is feasible for the LP. To show that it is optimal, let $\mathbf{x}$ be a feasible solution, for which it will be shown that $\mathbf{cx} \geq \mathbf{cx}^*$. Since $\mathbf{x}$ is feasible for the LP, $\mathbf{Ax} = \mathbf{Bx_B} + \mathbf{Nx_N} = \mathbf{b}$, or, equivalently, $\mathbf{x_B} = \mathbf{B}^{-1}(\mathbf{b} - \mathbf{Nx_N})$. Consequently,

$$\begin{aligned}
\mathbf{c}(\mathbf{x} - \mathbf{x}^*) &= \mathbf{c_B}(\mathbf{x_B} - \mathbf{x}_\mathbf{B}^*) + \mathbf{c_N}(\mathbf{x_N} - \mathbf{x}_\mathbf{N}^*) \\
&= \mathbf{c_B}\left[\mathbf{B}^{-1}(\mathbf{b} - \mathbf{Nx_N}) - \mathbf{B}^{-1}\mathbf{b}\right] + \mathbf{c_N}(\mathbf{x_N} - \mathbf{0}) \\
&= \mathbf{c_B}\left[\mathbf{B}^{-1}\mathbf{b} - \mathbf{B}^{-1}\mathbf{Nx_N} - \mathbf{B}^{-1}\mathbf{b}\right] + \mathbf{c_N}\mathbf{x_N} \\
&= -\mathbf{c_B}\mathbf{B}^{-1}\mathbf{Nx_N} + \mathbf{c_N}\mathbf{x_N} \\
&= (\mathbf{c_N} - \mathbf{c_B}\mathbf{B}^{-1}\mathbf{N})\mathbf{x_N}
\end{aligned}$$

By hypothesis, $\mathbf{c_N} - \mathbf{c_B}\mathbf{B}^{-1}\mathbf{N} \geq \mathbf{0}$, and, since $\mathbf{x}$ is feasible, $\mathbf{x_N} \geq \mathbf{0}$, so it must be that $\mathbf{c}(\mathbf{x}-\mathbf{x}^*) \geq 0$. Hence $\mathbf{cx} \geq \mathbf{cx}^*$, and therefore $\mathbf{x}^*$ is optimal for the LP. □

When performing the test for optimality on a given bfs $\mathbf{x}=(\mathbf{x_B},\mathbf{x_N})=(\mathbf{B}^{-1}\mathbf{b},\mathbf{0}) \geq \mathbf{0}$, you should first write the m vector $\mathbf{c_B}$ and the $n-m$ vector $\mathbf{c_N}$. It is then best to proceed by computing

1. the m vector $\mathbf{c_B B}^{-1}$ (observe that $\mathbf{c_B}$ is a $1 \times m$ row vector),
2. the $n-m$ vector $(\mathbf{c_B B}^{-1})\mathbf{N}$, and then
3. the $n-m$ vector $\mathbf{c_N}-\mathbf{c_B B}^{-1}\mathbf{N}$.

The vector obtained in step 3 is often referred to as the *reduced costs* of the bfs.

If the reduced costs are all ≥ 0, then Proposition 5.4 states that the bfs is optimal for the LP. To illustrate the test for optimality in Example 5.1, consider the bfs $\mathbf{x}=(0,1,2,0)$, in which x_2 and x_3 are basic variables and x_1 and x_4 are the nonbasic variables. The corresponding matrices are

$$\mathbf{B}=\begin{bmatrix} 0 & 1 \\ 1 & 1 \end{bmatrix}, \quad \mathbf{N}=\begin{bmatrix} 1 & 2 \\ 0 & 2 \end{bmatrix}, \quad \mathbf{B}^{-1}=\begin{bmatrix} -1 & 1 \\ 1 & 0 \end{bmatrix}$$

To perform the test for optimality, note that $\mathbf{c_B}=(c_2,c_3)=(1,-1)$ and $\mathbf{c_N}=(c_1,c_4)=(-3,2)$. Thus one would compute

1. $\mathbf{c_B B}^{-1}=(1,-1)\begin{bmatrix} -1 & 1 \\ 1 & 0 \end{bmatrix}=(-2,1)$,
2. $(\mathbf{c_B B}^{-1})\mathbf{N}=(-2,1)\begin{bmatrix} 1 & 2 \\ 0 & 2 \end{bmatrix}=(-2,-2)$, and
3. $\mathbf{c_N}-\mathbf{c_B B}^{-1}\mathbf{N}=(-3,2)-(-2,-2)=(-1,4)$.

Since these reduced cost are not greater than or equal to 0, this bfs does not pass the test for optimality.

On the other hand, consider the bfs $\mathbf{x}=(2,3,0,0)$ in which x_1 and x_2 are the basic variables and x_3 and x_4 are the nonbasic variables. The corresponding matrices are

$$\mathbf{B}=\begin{bmatrix} 1 & 0 \\ 0 & 1 \end{bmatrix}, \quad \mathbf{N}=\begin{bmatrix} 1 & 2 \\ 1 & 2 \end{bmatrix}, \quad \mathbf{B}^{-1}=\begin{bmatrix} 1 & 0 \\ 0 & 1 \end{bmatrix}$$

To perform the test for optimality note that $\mathbf{c_B}=(c_1,c_2)=(-3,1)$ and $\mathbf{c_N}=(c_3,c_4)=(-1,2)$, so

1. $\mathbf{c_B B}^{-1}=(-3,1)\begin{bmatrix} 1 & 0 \\ 0 & 1 \end{bmatrix}=(-3,1)$,
2. $(\mathbf{c_B B}^{-1})\mathbf{N}=(-3,1)\begin{bmatrix} 1 & 2 \\ 1 & 2 \end{bmatrix}=(-2,-4)$, and
3. $\mathbf{c_N}-\mathbf{c_B B}^{-1}\mathbf{N}=(-1,2)-(-2,-4)=(1,6)$.

Since the reduced costs are nonnegative, the bfs $\mathbf{x}=(2,3,0,0)$ does pass the test for optimality (see Table 5.2 of the previous section).

Another equally valid way to perform the test for optimality on a given bfs $\mathbf{x} = (\mathbf{x_B}, \mathbf{x_N}) = (\mathbf{B}^{-1}\mathbf{b}, \mathbf{0}) \geq \mathbf{0}$ is to check whether the n vector $\mathbf{c} - (\mathbf{c_B}\mathbf{B}^{-1})\mathbf{A} \geq \mathbf{0}$. The reason is that

$$\begin{aligned}\mathbf{c} - (\mathbf{c_B}\mathbf{B}^{-1})\mathbf{A} &= (\mathbf{c_B}, \mathbf{c_N}) - (\mathbf{c_B}\mathbf{B}^{-1})[\mathbf{B}, \mathbf{N}] \\ &= (\mathbf{c_B}, \mathbf{c_N}) - (\mathbf{c_B}\mathbf{I}, \mathbf{c_B}\mathbf{B}^{-1}\mathbf{N}) \\ &= (\mathbf{c_B} - \mathbf{c_B}, \mathbf{c_N} - \mathbf{c_B}\mathbf{B}^{-1}\mathbf{N}) \\ &= (\mathbf{0}, \mathbf{c_N} - \mathbf{c_B}\mathbf{B}^{-1}\mathbf{N})\end{aligned}$$

In other words, the $\mathbf{B}$ components of $\mathbf{c} - (\mathbf{c_B}\mathbf{B}^{-1})\mathbf{A}$ are always 0, and the $\mathbf{N}$ components are the reduced costs. The only advantage of using $\mathbf{c} - (\mathbf{c_B}\mathbf{B}^{-1})\mathbf{A}$ as the test for optimality is its conceptual simplicity. For instance, in Example 5.1, consider the bfs $\mathbf{x} = (0, 1, 0, 1)$, in which x_2 and x_4 are the basic variables and x_1 and x_3 are the nonbasic variables. The corresponding $\mathbf{B}$ and $\mathbf{B}^{-1}$ matrices are

$$\mathbf{B} = \begin{bmatrix} 0 & 2 \\ 1 & 2 \end{bmatrix}, \qquad \mathbf{B}^{-1} = \begin{bmatrix} -1 & 1 \\ \frac{1}{2} & 0 \end{bmatrix}$$

To perform the test for optimality one would compute

$$\mathbf{c_B}\mathbf{B}^{-1} = (1, 2)\begin{bmatrix} -1 & 1 \\ \frac{1}{2} & 0 \end{bmatrix} = (0, 1)$$

and then

$$\begin{aligned}\mathbf{c} - (\mathbf{c_B}\mathbf{B}^{-1}\mathbf{A}) &= (-3, 1, -1, 2) - (0, 1)\begin{bmatrix} 1 & 0 & 1 & 2 \\ 0 & 1 & 1 & 2 \end{bmatrix} \\ &= (-3, 1, -1, 2) - (0, 1, 1, 2) \\ &= (-3, 0, -2, 0)\end{aligned}$$

As expected, the basic components are 0. Since this vector is not greater than or equal to $\mathbf{0}$, $\mathbf{x} = (0, 1, 0, 1)$ does not pass the test for optimality.

This section has described how to test a bfs $\mathbf{x} = (\mathbf{x_B}, \mathbf{x_N}) = (\mathbf{B}^{-1}\mathbf{b}, \mathbf{0}) \geq \mathbf{0}$ for optimality by checking whether the reduced costs are nonnegative. The next section tells you what to do when $\mathbf{x}$ does not pass the test.

EXERCISES

5.2.1. For the LP in Exercise 5.1.1, perform the test for optimality at each bfs by computing **(a)** $\mathbf{c_N} - \mathbf{c_B}\mathbf{B}^{-1}\mathbf{N}$, and **(b)** $\mathbf{c} - \mathbf{c_B}\mathbf{B}^{-1}\mathbf{A}$. Which bfs is optimal?

5.2.2. Explain why it is better to perform the test for optimality by computing $\mathbf{c_B}\mathbf{B}^{-1}$, then $(\mathbf{c_B}\mathbf{B}^{-1})\mathbf{N}$, and finally $\mathbf{c_N} - (\mathbf{c_B}\mathbf{B}^{-1})\mathbf{N}$, instead of computing $\mathbf{B}^{-1}\mathbf{N}$, then $\mathbf{c_B}(\mathbf{B}^{-1}\mathbf{N})$, and finally $\mathbf{c_N} - \mathbf{c_B}(\mathbf{B}^{-1}\mathbf{N})$.

5.2.3. What would be the test for optimality if the original LP had the form of maximizing $\mathbf{cx}$? Prove that your answer is correct.

5.3. DETERMINING A DIRECTION OF MOVEMENT

For a given bfs $\mathbf{x}$ and the corresponding $\mathbf{B}$ and $\mathbf{N}$ matrices, Proposition 5.4 provides a relatively simple algebraic test to determine whether the bfs has the smallest possible objective value. If it does, then the search is over, but if it does not, then a method must somehow be found for moving from the current bfs to a new one. Moreover, to avoid cycling, the new bfs should be chosen to have a strictly smaller objective value. All of this will be accomplished by constructing an n vector $\mathbf{d}$ and then moving from the current point $\mathbf{x}$ by an amount $t^* \geq 0$ in the direction $\mathbf{d}$, ending up at $\mathbf{x} + t^*\mathbf{d}$ (see Figure 5.1.). This section shows how to construct the direction $\mathbf{d}$, and the next section addresses the issue of determining t^*.

In creating the direction $\mathbf{d}$ a primary concern is that the objective value at the new point be less than that at $\mathbf{x}$. The question therefore arises as to what property $\mathbf{d}$ must have to ensure that this will happen. To answer the question, note that the objective value at $\mathbf{x}$ is $\mathbf{cx}$ and that at $\mathbf{x} + t^*\mathbf{d}$ it is $\mathbf{c}(\mathbf{x} + t^*\mathbf{d})$. Thus one would like to have $\mathbf{c}(\mathbf{x} + t^*\mathbf{d}) = \mathbf{cx} + t^*\mathbf{cd} < \mathbf{cx}$ or, equivalently, $t^*\mathbf{cd} < 0$. Although the value for t^* is not yet known, it will not be negative. In the unfortunate situation that $t^* = 0$, it is impossible that $t^*\mathbf{cd} < 0$. On the other hand, if $t^* > 0$, then $t^*\mathbf{cd} < 0$ provided that $\mathbf{cd} < 0$. In other words, if t^* is assumed to be positive, it would be desirable to find a direction $\mathbf{d}$ with the property that $\mathbf{cd} < 0$, for then, the objective value at $\mathbf{x} + t^*\mathbf{d}$ would be strictly smaller than at $\mathbf{x}$. This fact is formally established in the next proposition.

FIGURE 5.1. Moving from $\mathbf{x}$ in the direction $\mathbf{d}$.

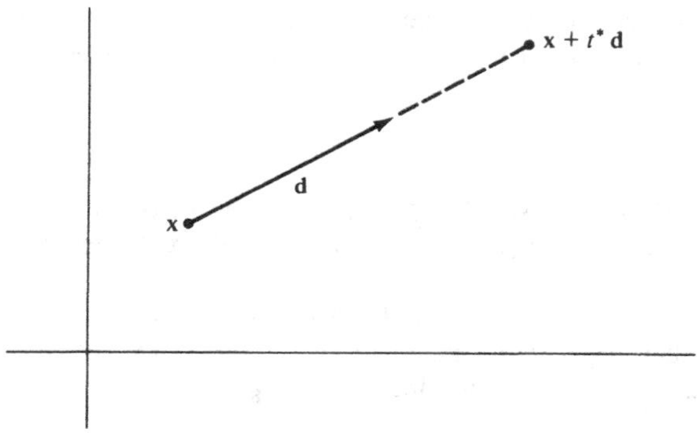

Proposition 5.5. *If* $\mathbf{x}$ *and* $\mathbf{d}$ *are n vectors and* $\mathbf{d}$ *satisfies the property that* $\mathbf{cd} < 0$, *then, for all* $t > 0$, $\mathbf{c}(\mathbf{x} + t\mathbf{d}) < \mathbf{cx}$.

OUTLINE OF PROOF. Recognizing the quantifier "for all' in the conclusion, the choose method is invoked to select a $t > 0$. By working forward from the hypothesis, it can be shown that $\mathbf{c}(\mathbf{x} + t\mathbf{d}) < \mathbf{cx}$. Specifically, since $t > 0$ and $\mathbf{cd} < 0$, it follows that $t\mathbf{cd} < 0$, and, on adding $\mathbf{cx}$ to both sides, one obtains $\mathbf{cx} + t\mathbf{cd} < \mathbf{cx}$ or, equivalently, $\mathbf{c}(\mathbf{x} + t\mathbf{d}) < \mathbf{cx}$.

PROOF OF PROPOSITION 5.5. Let $t > 0$. Since $\mathbf{cd} < 0$, it must be that $t\mathbf{cd} < 0$. On adding $\mathbf{cx}$ to both sides, one has $\mathbf{c}(\mathbf{x} + t\mathbf{d}) = \mathbf{cx} + t\mathbf{cd} < \mathbf{cx}$, as desired. □

Proposition 5.5 establishes that if the direction $\mathbf{d}$ should satisfy the property that $\mathbf{cd} < 0$, then any strictly positive movement t from $\mathbf{x}$ in the direction $\mathbf{d}$ will lead to a strict decrease in the objective value. How can such a direction be found? What is surprising is that, when a bfs fails the test for optimality, that fact itself indicates how to construct a direction $\mathbf{d}$ with $\mathbf{cd} < 0$. Note that, when a bfs $\mathbf{x} = (\mathbf{x_B}, \mathbf{x_N}) = (\mathbf{B}^{-1}\mathbf{b}, \mathbf{0}) \geq \mathbf{0}$ fails the test for optimality [i.e., it is not the case that for all j with $1 \leq j \leq n - m$, $(\mathbf{c_N} - \mathbf{c_B}\mathbf{B}^{-1}\mathbf{N})_j \geq 0$], then there is a j with $1 \leq j \leq n - m$, such that $(\mathbf{c_N} - \mathbf{c_B}\mathbf{B}^{-1}\mathbf{N})_j < 0$. Of course there can be more than one such j, and the next proposition shows that any one can be used to construct the desired direction.

Proposition 5.6. *If* $\mathbf{x} = (\mathbf{x_B}, \mathbf{x_N}) = (\mathbf{B}^{-1}\mathbf{b}, \mathbf{0}) \geq \mathbf{0}$ *is a bfs that does not pass the test for optimality, then for any* j *with* $1 \leq j \leq n - m$ *and* $(\mathbf{c_N} - \mathbf{c_B}\mathbf{B}^{-1}\mathbf{N})_j < 0$, *the direction defined by* $\mathbf{d} = (\mathbf{d_B}, \mathbf{d_N}) = (-\mathbf{B}^{-1}\mathbf{N}_{.j}, \mathbf{I}_{.j})$ *satisfies* $\mathbf{cd} < 0$. [*Here* $\mathbf{I}$ *is the* $(n - m) \times (n - m)$ *identity matrix.*]

OUTLINE OF PROOF. Recognizing the quantifier "for all" that appears in the conclusion, the choose method is invoked to select a j with $1 \leq j \leq n - m$ and $(\mathbf{c_N} - \mathbf{c_B}\mathbf{B}^{-1}\mathbf{N})_j < 0$. We can now show that $\mathbf{cd} < 0$. Specifically, working forward by means of algebraic manipulations from the fact that $\mathbf{d} = (\mathbf{d_B}, \mathbf{d_N}) = (-\mathbf{B}^{-1}\mathbf{N}_{.j}, \mathbf{I}_{.j})$, one can show that $\mathbf{cd} = (\mathbf{c_N} - \mathbf{c_B}\mathbf{B}^{-1}\mathbf{N})_j$ and therefore that $\mathbf{cd} < 0$. Once again, the technique of working with the $\mathbf{B}$ and $\mathbf{N}$ components will be used:

$$\begin{aligned}\mathbf{cd} &= \mathbf{c_B}\mathbf{d_B} + \mathbf{c_N}\mathbf{d_N} = -\mathbf{c_B}\mathbf{B}^{-1}\mathbf{N}_{.j} + \mathbf{c_N}\mathbf{I}_{.j} \\ &= \left(\mathbf{c_N} - \mathbf{c_B}\mathbf{B}^{-1}\mathbf{N}\right)_j\end{aligned}$$

and the proof is complete.

PROOF OF PROPOSITION 5.6. To reach the desired conclusion, let $1 \leq j \leq n - m$ with $(\mathbf{c_N} - \mathbf{c_B}\mathbf{B}^{-1}\mathbf{N})_j < 0$. Then, since $\mathbf{d} = (\mathbf{d_B}, \mathbf{d_N}) = (-\mathbf{B}^{-1}\mathbf{N}_{.j}, \mathbf{I}_{.j})$, it follows that

$$\begin{aligned}\mathbf{cd} &= \mathbf{c_B d_B} + \mathbf{c_N d_N} = -\mathbf{c_B B}^{-1}\mathbf{N}_{.j} + \mathbf{c_N I}_{.j} \\ &= \left(\mathbf{c_N} - \mathbf{c_B B}^{-1}\mathbf{N}\right)_j < 0\end{aligned}$$

as desired. □

In other words, when $\mathbf{x}$ is a bfs that fails the test for optimality, that fact will produce at least one value of j with $1 \le j \le n-m$, for which $(\mathbf{c_N} - \mathbf{c_B B}^{-1}\mathbf{N})_j < 0$. Using any such j it is then possible to compute the direction $\mathbf{d} = (\mathbf{d_B}, \mathbf{d_N}) = (-\mathbf{B}^{-1}\mathbf{N}_{.j}, \mathbf{I}_{.j})$. These computations will now be illustrated for Example 5.1.

Recall the bfs $\mathbf{x} = (0,1,2,0)$, in which $\mathbf{x_B} = (x_2, x_3)$, $\mathbf{x_N} = (x_1, x_4)$, and

$$\mathbf{B} = \begin{bmatrix} 0 & 1 \\ 1 & 1 \end{bmatrix}, \quad \mathbf{B}^{-1} = \begin{bmatrix} -1 & 1 \\ 1 & 0 \end{bmatrix}, \quad \mathbf{N} = \begin{bmatrix} 1 & 2 \\ 0 & 2 \end{bmatrix}$$

This bfs does not pass the test for optimality because $\mathbf{c_N} - \mathbf{c_B B}^{-1}\mathbf{N} = (-1, 4)$ is not greater than or equal to zero. In this case, the value of j for which $(\mathbf{c_N} - \mathbf{c_B B}^{-1}\mathbf{N})_j < 0$ is $j = 1$. Therefore, the vector $\mathbf{d}$ can be computed as follows:

$$\mathbf{d_B} = -\mathbf{B}^{-1}\mathbf{N}_{.1} = -\begin{bmatrix} -1 & 1 \\ 1 & 0 \end{bmatrix}\begin{bmatrix} 1 \\ 0 \end{bmatrix} = \begin{bmatrix} 1 \\ -1 \end{bmatrix} = \begin{bmatrix} d_2 \\ d_3 \end{bmatrix}$$

$$\mathbf{d_N} = \mathbf{I}_{.1} = \begin{bmatrix} 1 \\ 0 \end{bmatrix} = \begin{bmatrix} d_1 \\ d_4 \end{bmatrix}$$

Thus

$$\mathbf{d} = (d_1, d_2, d_3, d_4) = (1, 1, -1, 0)$$

To demonstrate the process once more, consider the bfs $\mathbf{x} = (0,1,0,1)$ for Example 5.1. Here, $\mathbf{x_B} = (x_2, x_4)$, $\mathbf{x_N} = (x_1, x_3)$, and

$$\mathbf{B} = \begin{bmatrix} 0 & 2 \\ 1 & 2 \end{bmatrix}, \quad \mathbf{B}^{-1} = \begin{bmatrix} -1 & 1 \\ \frac{1}{2} & 0 \end{bmatrix}, \quad \mathbf{N} = \begin{bmatrix} 1 & 1 \\ 0 & 1 \end{bmatrix}$$

Once again, the bfs does not pass the test for optimality because $\mathbf{c_N} - \mathbf{c_B B}^{-1}\mathbf{N} = (-3, -2)$ which is not greater than or equal to the zero vector. This time, there are two values of j for which $(\mathbf{c_N} - \mathbf{c_B B}^{-1}\mathbf{N})_j < 0$, namely, $j = 1$ and $j = 2$. Consequently, there are two possible directions that could be constructed. Corresponding to $j = 1$ is the direction $\mathbf{d} =$

$(\mathbf{d_B}, \mathbf{d_N})$ in which

$$\mathbf{d_B} = -\mathbf{B}^{-1}\mathbf{N}_{\cdot 1} = -\begin{bmatrix} -1 & 1 \\ \frac{1}{2} & 0 \end{bmatrix}\begin{bmatrix} 1 \\ 0 \end{bmatrix} = \begin{bmatrix} 1 \\ -\frac{1}{2} \end{bmatrix} = \begin{bmatrix} d_2 \\ d_4 \end{bmatrix}$$

$$\mathbf{d_N} = \mathbf{I}_{\cdot 1} = \begin{bmatrix} 1 \\ 0 \end{bmatrix} = \begin{bmatrix} d_1 \\ d_3 \end{bmatrix}$$

Thus

$$\mathbf{d} = (d_1, d_2, d_3, d_4) = (1, 1, 0, -\tfrac{1}{2})$$

Similarly, corresponding to $j = 2$ is the direction $\mathbf{d} = (\mathbf{d_B}, \mathbf{d_N})$, in which

$$\mathbf{d_B} = -\mathbf{B}^{-1}\mathbf{N}_{\cdot 2} = -\begin{bmatrix} -1 & 1 \\ \frac{1}{2} & 0 \end{bmatrix}\begin{bmatrix} 1 \\ 1 \end{bmatrix} = \begin{bmatrix} 0 \\ -\frac{1}{2} \end{bmatrix} = \begin{bmatrix} d_2 \\ d_4 \end{bmatrix}$$

$$\mathbf{d_N} = \mathbf{I}_{\cdot 2} = \begin{bmatrix} 0 \\ 1 \end{bmatrix} = \begin{bmatrix} d_1 \\ d_3 \end{bmatrix}$$

Thus

$$\mathbf{d} = (d_1, d_2, d_3, d_4) = (0, 0, 1, -\tfrac{1}{2})$$

For computational purposes it is necessary to have a specific rule for selecting a value for j in the event that various choices are possible. There are many conceivable rules; here are three:

Rule 1. First come, first served: Pick the first j for which $(\mathbf{c_N} - \mathbf{c_B}\mathbf{B}^{-1}\mathbf{N})_j < 0$.

Rule 2. Bland's rule: Pick the j with $(\mathbf{c_N} - \mathbf{c_B}\mathbf{B}^{-1}\mathbf{N})_j < 0$ and for which $\mathbf{N}_{\cdot j}$ is the leftmost column of the $\mathbf{A}$ matrix.

Rule 3. Steepest descent: Pick a j that provides the most negative component of $\mathbf{c_N} - \mathbf{c_B}\mathbf{B}^{-1}\mathbf{N}$.

In practice, the actual choice is based on computational considerations, as will be discussed in detail in Chapter 7. Whatever rule is ultimately chosen, the result will be a specific value for j, say, j^*. For the remainder of this chapter, j will refer to any subscript for which $(\mathbf{c_N} - \mathbf{c_B}\mathbf{B}^{-1}\mathbf{N})_j < 0$, and j^* will represent a specific choice for j.

This section has shown how to use the fact that a bfs fails the test for optimality to create a direction $\mathbf{d}$ for which $\mathbf{cd} < 0$. The next section deals with computing the amount $t^* \geq 0$ to move in the direction $\mathbf{d}$.

EXERCISES

5.3.1. For each of the bfs that failed the test for optimality in Exercise 5.2.1, compute all possible directions of movement.

5.3.2. For the LP

$$\begin{aligned} \text{Minimize}\quad & 2x_1 - 3x_2 - x_3 + 4x_4 + x_5 \\ \text{subject to}\quad & 3x_1 + x_2 - x_3 + 2x_4 + x_5 = 5 \\ & x_1 + 2x_2 + 3x_3 - x_4 + 3x_5 = 10 \\ & 2x_1 - 3x_2 + 3x_3 + x_4 - x_5 = 3 \\ & x_1,\ x_2,\ x_3,\ x_4,\ x_5 \geq 0 \end{aligned}$$

if one were to move from the point $\mathbf{x} = (1,2,2,1,0)$ in the direction $\mathbf{d} = (0,1,3,1,2)$, would the objective function improve? Why or why not?

5.3.3. Consider the following LP:

$$\begin{aligned} \text{Minimize}\quad & 2x_1 - x_2 - 4x_3 + x_4 + x_5 \\ \text{subject to}\quad & 2x_1 + 3x_2 + 3x_3 + x_4 = 23 \\ & 3x_1 + 4x_2 - x_3 + x_5 = 31 \\ & x_1,\ x_2,\ x_3,\ x_4,\ x_5 \geq 0 \end{aligned}$$

For the bfs in which the nonbasic variables are x_3, x_2, and x_1 (in that order), determine the direction of movement that would be obtained by

(a) the first come, first served rule,

(b) Bland's rule, and

(c) the rule of steepest descent.

5.4. DETERMINING THE AMOUNT OF MOVEMENT

The previous section has shown how to take a bfs $\mathbf{x} = (\mathbf{x_B}, \mathbf{x_N}) = (\mathbf{B}^{-1}\mathbf{b}, \mathbf{0}) \geq \mathbf{0}$ that fails the test for optimality and use any j with $1 \leq j \leq n - m$ and for which $(\mathbf{c_N} - \mathbf{c_B}\mathbf{B}^{-1}\mathbf{N})_j < 0$ to construct a direction of the form $\mathbf{d} = (\mathbf{d_B}, \mathbf{d_N}) = (-\mathbf{B}^{-1}\mathbf{N}_{.j}, \mathbf{I}_{.j})$. In addition, $\mathbf{d}$ has the property that $\mathbf{cd} < 0$, and, provided that a positive step is taken from $\mathbf{x}$ in the direction $\mathbf{d}$, the objective function must decrease. There are still two issues that must be resolved:

1. how to determine the actual amount t^* to move in the direction $\mathbf{d}$, and
2. how to be sure that the new vector $\mathbf{x} + t^*\mathbf{d}$ is a basic feasible solution.

This section deals with the first question, and the next section addresses the second one. Throughout the remainder of this section, $\mathbf{x}$ and $\mathbf{d}$ will have the above forms.

To determine a value for t^*, consider again what happens to the value of the objective function in moving from $\mathbf{x}$ to $\mathbf{x} + t^*\mathbf{d}$. At $\mathbf{x}$ it is $\mathbf{cx}$, and at $\mathbf{x} + t^*\mathbf{d}$ it is $\mathbf{c}(\mathbf{x} + t^*\mathbf{d}) = \mathbf{cx} + t^*\mathbf{cd}$. Consequently, the amount of decrease in going from $\mathbf{x}$ to $\mathbf{x} + t^*\mathbf{d}$ is $\mathbf{cx} - (\mathbf{cx} + t^*\mathbf{cd}) = -t^*\mathbf{cd}$. It is therefore desirable to make t^* as large as possible.

The question naturally arises as to what prevents t^* from going to infinity. The answer lies in the constraints of the LP, for it can happen that

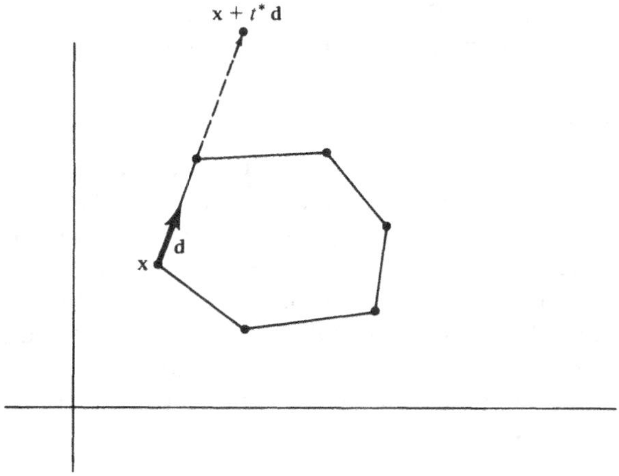

FIGURE 5.2. Infeasibility caused by t^* being too large.

if t^* is chosen too large, then the point $\mathbf{x} + t^*\mathbf{d}$ is outside of the feasible region (see Figure 5.2). What is needed is an algebraic method for determining just how large t can be made before $\mathbf{x} + t\mathbf{d}$ leaves the feasible region. This largest value is t^*.

When determining the value of t^*, three possible cases arise.

Case 1: $t^* = \infty$

In this case the constraints remain feasible no matter how far one moves in the direction **d**. Since the objective value continues to decrease as one moves in this direction, it should come as no surprise that the LP is unbounded, and in fact **d** is the direction of unboundedness.

Case 2: $t^* = 0$

In this case $\mathbf{x} + t^*\mathbf{d} = \mathbf{x} + (0)\mathbf{d} = \mathbf{x}$, and no movement in the direction **d** takes place. Serious problems can now arise because the objective function does not decrease. Section 5.7 will show how to treat this difficulty. The current section only shows the conditions under which $t^* = 0$.

Case 3: $0 < t^* < \infty$

In this normal case a movement from **x** to $\mathbf{x} + t^*\mathbf{d}$ takes place, and a strict decrease in the objective function occurs since $t^* > 0$. It must then be established that $\mathbf{x} + t^*\mathbf{d}$ is indeed a new bfs. Proving that it is a bfs involves producing the new **B** and **N** matrices (say, **B**′ and **N**′) corresponding to

$\mathbf{x} + t^*\mathbf{d}$ and showing that they have all of the necessary properties. Equivalently stated, the new basic and nonbasic variables must be specified for $\mathbf{x} + t^*\mathbf{d}$.

In order to determine which of these three cases actually occurs, what is needed is an algebraic method for computing how large t can be made before $\mathbf{x} + t\mathbf{d}$ leaves the feasible region. For an LP in standard form, there are two possible ways for the point $\mathbf{x} + t\mathbf{d}$ to be outside of the feasible region. Either an equality constraint or else a nonnegativity constraint can become violated. The next proposition shows that, for a direction of the form $\mathbf{d} = (\mathbf{d}_\mathbf{B}, \mathbf{d}_\mathbf{N}) = (-\mathbf{B}^{-1}\mathbf{N}_{.j}, \mathbf{I}_{.j})$, the equality constraints always hold, and so it must be the nonnegativity constraints (if anything) that prevent t from going to infinity.

Proposition 5.7. *If* $\mathbf{x} = (\mathbf{x}_\mathbf{B}, \mathbf{x}_\mathbf{N}) = (\mathbf{B}^{-1}\mathbf{b}, \mathbf{0}) \geq \mathbf{0}$ *is a bfs and* $\mathbf{d} = (\mathbf{d}_\mathbf{B}, \mathbf{d}_\mathbf{N}) = (-\mathbf{B}^{-1}\mathbf{N}_{.j}, \mathbf{I}_{.j})$, *then, for all* $t \geq 0$, $\mathbf{A}(\mathbf{x} + t\mathbf{d}) = \mathbf{b}$.

OUTLINE OF PROOF. From the appearance of the quantifier "for all" in the conclusion, the choose method is appropriate to select a $t \geq 0$, for which it will be shown that $\mathbf{A}(\mathbf{x} + t\mathbf{d}) = \mathbf{b}$. Since $\mathbf{A}(\mathbf{x} + t\mathbf{d}) = \mathbf{A}\mathbf{x} + t\mathbf{A}\mathbf{d}$, the result will follow once it is shown that $\mathbf{A}\mathbf{x} = \mathbf{b}$ and $\mathbf{A}\mathbf{d} = \mathbf{0}$. Both of these facts can be established by working forward. Specifically, since $\mathbf{x}$ is a bfs, it is feasible, and hence $\mathbf{A}\mathbf{x} = \mathbf{b}$. Also, from the hypothesis that $\mathbf{d} = (\mathbf{d}_\mathbf{B}, \mathbf{d}_\mathbf{N}) = (-\mathbf{B}^{-1}\mathbf{N}_{.j}, \mathbf{I}_{.j})$ it follows that

$$\begin{aligned}\mathbf{A}\mathbf{d} &= [\mathbf{B}, \mathbf{N}]\begin{bmatrix}\mathbf{d}_\mathbf{B} \\ \mathbf{d}_\mathbf{N}\end{bmatrix} = \mathbf{B}\mathbf{d}_\mathbf{B} + \mathbf{N}\mathbf{d}_\mathbf{N} \\ &= -\mathbf{B}\mathbf{B}^{-1}\mathbf{N}_{.j} + \mathbf{N}\mathbf{I}_{.j} = -\mathbf{I}\mathbf{N}_{.j} + \mathbf{N}_{.j} \\ &= -\mathbf{N}_{.j} + \mathbf{N}_{.j} = \mathbf{0}\end{aligned}$$

PROOF OF PROPOSITION 5.7. Let $t \geq 0$. It must be shown that $\mathbf{A}(\mathbf{x} + t\mathbf{d}) = \mathbf{A}\mathbf{x} + t\mathbf{A}\mathbf{d} = \mathbf{b}$. By hypothesis, $\mathbf{x}$ is a bfs and hence feasible, so $\mathbf{A}\mathbf{x} = \mathbf{b}$. Since $\mathbf{d} = (\mathbf{d}_\mathbf{B}, \mathbf{d}_\mathbf{N}) = (-\mathbf{B}^{-1}\mathbf{N}_{.j}, \mathbf{I}_{.j})$ it follows that

$$\begin{aligned}\mathbf{A}\mathbf{d} &= \mathbf{B}\mathbf{d}_\mathbf{B} + \mathbf{N}\mathbf{d}_\mathbf{N} = -\mathbf{B}\mathbf{B}^{-1}\mathbf{N}_{.j} + \mathbf{N}\mathbf{I}_{.j} \\ &= -\mathbf{I}\mathbf{N}_{.j} + \mathbf{N}_{.j} = \mathbf{0}\end{aligned}$$

Consequently, $\mathbf{A}(\mathbf{x} + t\mathbf{d}) = \mathbf{A}\mathbf{x} + t\mathbf{A}\mathbf{d} = \mathbf{b} + t(\mathbf{0}) = \mathbf{b}$. □

As a result of Proposition 5.7, only the nonnegativity constraints can prevent t from going to infinity. It therefore becomes necessary to figure out how large t can be made before $\mathbf{x} + t\mathbf{d}$ loses its nonnegativity. The largest such value is t^*.

To see how to compute t^*, it is helpful to consider what happens for a particular choice of **x** and **d**. For example, suppose that $\mathbf{x} = (x_1, x_2, x_3) = (0,4,5)$ and $\mathbf{d} = (d_1, d_2, d_3) = (2, -1, -2)$. Then for $t = 1$, $\mathbf{x} + t\mathbf{d} = (0,4,5) + 1(2, -1, -2) = (2,3,3)$. Since this vector is still greater than or equal to $\mathbf{0}$, t can be made larger; indeed, for $t = 2$, $\mathbf{x} + t\mathbf{d} = (0,4,5) + 2(2, -1, -2) = (4,2,1)$. Once again t can be increased further; however, if it is now chosen to be 3, then $\mathbf{x} + t\mathbf{d} = (0,4,5) + 3(2, -1, -2) = (6,1,-1)$, and all of a sudden the last component of $\mathbf{x} + t\mathbf{d}$ has become negative, indicating that a value of $t = 3$ is too large.

A little effort will shortly convince you that the largest t can be made is $\frac{5}{2}$. The question is how can you use the vectors **x** and **d** to determine this value for t algebraically. In answering this question it is important to note from the example above, that the nonnegative components of **d** play no role in stopping t from going to infinity because those same components of $\mathbf{x} + t\mathbf{d}$ always remain greater than or equal to 0. Consequently, the computation of t^* must be determined by the components of **d** that are negative. If there are none, then nothing stops t from going to infinity. For each component d_i of **d** that is negative, it is possible to compute the value of t for which $x_i + td_i = 0$ by solving for t to obtain $t = -x_i/d_i$ (note that $d_i \neq 0$ since $d_i < 0$). t^* is then chosen as the smallest of all these values:

$$t^* = \min\{-x_i/d_i : 1 \leq i \leq n \text{ and } d_i < 0\}$$

For the example in which $\mathbf{x} = (0,4,5)$ and $\mathbf{d} = (2, -1, -2)$, only components 2 and 3 of **d** are negative, so $t^* = \min\{-4/-1, -5/-2\} = \min\{4, \frac{5}{2}\} = \frac{5}{2}$. In other words, t^* is determined by considering each negative component d_i of **d**, computing the ratio $-x_i/d_i$, and then taking the smallest such value. The next proposition formally establishes that, when t^* is defined in this way, the point $\mathbf{x} + t^*\mathbf{d}$ is feasible for the LP. Ultimately, $\mathbf{x} + t^*\mathbf{d}$ must be shown to be a bfs, but this task will be postponed to the next section.

Proposition 5.8. *If* **x** *is a bfs and* $\mathbf{d} = (\mathbf{d_B}, \mathbf{d_N}) = (-\mathbf{B}^{-1}\mathbf{N}_{\cdot j}, \mathbf{I}_{\cdot j})$ *is a vector that has a negative component, then* $t^* = \min\{-x_i/d_i : 1 \leq i \leq n \text{ and } d_i < 0\}$ *satisfies the property that* $\mathbf{x} + t^*\mathbf{d}$ *is feasible for the LP.*

OUTLINE OF PROOF. The forward–backward method gives rise to the abstraction question "How can I show that a vector (namely, $\mathbf{x} + t^*\mathbf{d}$) is feasible for an LP?" You can answer this by Definition 4.14: It must be shown that $\mathbf{A}(\mathbf{x} + t^*\mathbf{d}) = \mathbf{b}$ and $\mathbf{x} + t^*\mathbf{d} \geq \mathbf{0}$. To see that $\mathbf{A}(\mathbf{x} + t^*\mathbf{d}) = \mathbf{b}$, note that the hypothesis—and hence the conclusion—of Proposition 5.7 is true. On specializing that conclusion to $t = t^*$, it follows that $\mathbf{A}(\mathbf{x} + t^*\mathbf{d}) = \mathbf{b}$.

It remains to establish that $\mathbf{x} + t^*\mathbf{d} \geq \mathbf{0}$. Working backward, you must answer the abstraction question "How can I show that a vector (namely, $\mathbf{x} + t^*\mathbf{d}) \geq \mathbf{0}$?" which can be answered by showing that, for all i with

$1 \le i \le n$, $(\mathbf{x} + t^*\mathbf{d})_i \ge 0$. Hence the choose method is invoked to choose an i with $1 \le i \le n$. There are two possibilities.

Case 1: $d_i \ge 0$

In this case, since $t^* \ge 0$ and $x_i \ge 0$, it must be that $(\mathbf{x} + t^*\mathbf{d})_i = x_i + t^*d_i \ge 0$.

Case 2: $d_i < 0$

In this case, since t^* came from the min ratio formula, it follows that $t^* \le -x_i/d_i$ (because t^* was chosen as the smallest of these numbers). In other words, using the max/min method, $t^* = \min\{-x_i/d_i : d_i < 0\}$ can be converted into the equivalent quantified statement: For all integers k with $1 \le k \le n$ and $d_k < 0$, $t^* \le -x_k/d_k$. Now specialize this statement to the integer i (noting that $d_i < 0$) to claim that $t^* \le -x_i/d_i$.

On multiplying both sides of the inequality $t^* \le -x_i/d_i$ by the negative number d_i (thus reversing the inequality) and adding x_i to both sides, one obtains $(\mathbf{x} + t^*\mathbf{d})_i = x_i + t^*d_i \ge 0$.

PROOF OF PROPOSITION 5.8. To reach the conclusion that $\mathbf{x} + t^*\mathbf{d}$ is feasible, it will be shown that $\mathbf{A}(\mathbf{x} + t^*\mathbf{d}) = \mathbf{b}$ and $\mathbf{x} + t^*\mathbf{d} \ge \mathbf{0}$. By Proposition 5.7, it follows that $\mathbf{A}(\mathbf{x} + t^*\mathbf{d}) = \mathbf{b}$, so all that remains is to show that $\mathbf{x} + t^*\mathbf{d} \ge \mathbf{0}$. To that end, let i with $1 \le i \le n$. If $d_i \ge 0$, then, since $x_i \ge 0$ and $t^* \ge 0$, it follows that $(\mathbf{x} + t^*\mathbf{d})_i = x_i + t^*d_i \ge 0$. On the other hand, if $d_i < 0$, then, since t^* was computed by the min ratio formula, $t^* \le -x_i/d_i$. On multiplying both sides of the inequality by the negative number d_i and then adding x_i to both sides, it again follows that $(\mathbf{x} + t^*\mathbf{d})_i = x_i + t^*d_i \ge 0$, and thus $\mathbf{x} + t^*\mathbf{d}$ is feasible. □

Recall that, for the simplex algorithm, the direction $\mathbf{d}$ has the special form $\mathbf{d} = (\mathbf{d_B}, \mathbf{d_N}) = (-\mathbf{B}^{-1}\mathbf{N}_{.j}, \mathbf{I}_{.j})$. Therefore, when searching for negative components of $\mathbf{d}$, the nonbasic components need not be considered because $\mathbf{d_N} = \mathbf{I}_{.j}$, and $\mathbf{I}_{.j}$ consists exclusively of zeros and ones. Consequently, only the basic components $\mathbf{d_B}$ of $\mathbf{d}$ have to be checked. If $\mathbf{d_B} \ge \mathbf{0}$, then nothing prevents t from going to infinity, and the LP would be unbounded. Otherwise, if $\mathbf{d_B}$ has components that are negative, then t^* can be computed by the *min ratio test*:

$$t^* = \min\{-(\mathbf{x_B})_k/(\mathbf{d_B})_k : 1 \le k \le m \text{ and } (\mathbf{d_B})_k < 0\},$$

as will now be demonstrated on Example 5.1.

Recall the bfs $\mathbf{x} = (x_1, x_2, x_3, x_4) = (0, 1, 2, 0)$ in which the direction of movement was $\mathbf{d} = (d_1, d_2, d_3, d_4) = (1, 1, -1, 0)$. Here, $\mathbf{x_B} = (x_2, x_3) =$

(1, 2) and $\mathbf{d_B} = (d_2, d_3) = (1, -1)$. Since $(\mathbf{d_B})_2 < 0$, the min ratio test yields

$$t^* = \min\{-(\mathbf{x_B})_2/(\mathbf{d_B})_2\} = \min\{-2/-1\} = \min\{2\} = 2$$

Just as the test for optimality can be thought of as producing a value for j^* with $1 \le j^* \le n-m$, so too the min ratio test can be thought of as producing a value for k^* with $1 \le k^* \le m$; i.e., k^* is any integer for which $t^* = -(\mathbf{x_B})_{k^*}/(\mathbf{d_B})_{k^*}$. For the bfs $\mathbf{x} = (0, 1, 2, 0)$ above, k^* is 2. It is possible that there can be various choices for k^*. For computational purposes, it is necessary to have a specific rule for selecting a value of k^* in the event of ties, such as the following:

Rule 1. First come, first served: Pick the first k for which $t^* = -(\mathbf{x_B})_k/(\mathbf{d_B})_k$.

Rule 2. Bland's rule: In the event that there is more than one k for which $t^* = -(\mathbf{x_B})_k/(\mathbf{d_B})_k$, pick the one for which $\mathbf{B}_{.k}$ is the leftmost column of the original $\mathbf{A}$ matrix.

Whatever rule is used, the result will be a unique value for k, say, k^*.

The next proposition formally establishes that if $\mathbf{d_B} \ge \mathbf{0}$ then the LP is unbounded.

Proposition 5.9. *If* $\mathbf{x} = (\mathbf{x_B}, \mathbf{x_N}) = (\mathbf{B}^{-1}\mathbf{b}, \mathbf{0}) \ge \mathbf{0}$ *is a bfs and* $\mathbf{d} = (\mathbf{d_B}, \mathbf{d_N}) = (-\mathbf{B}^{-1}\mathbf{N}_{.j}, \mathbf{I}_{.j})$ *has the property that* $\mathbf{d_B} \ge \mathbf{0}$, *then the LP is unbounded.*

OUTLINE OF PROOF. The forward–backward method gives rise to the abstraction question "How can I show that an LP is unbounded?" This question is answered by Definition 4.19, whereby it must be shown that (1) the LP is feasible and (2) there is a vector $\mathbf{d} \ge \mathbf{0}$ such that $\mathbf{Ad} = \mathbf{0}$ and $\mathbf{cd} < 0$. Working forward from the hypothesis immediately establishes (1), because $\mathbf{x}$ is a bfs and hence it is feasible for the LP. Also, the desired direction for (2) is $\mathbf{d} = (\mathbf{d_B}, \mathbf{d_N}) = (-\mathbf{B}^{-1}\mathbf{N}_{.j}, \mathbf{I}_{.j})$. To see that $\mathbf{d}$ satisfies all of the desired properties, note that, in Proposition 5.6, it was established that $\mathbf{cd} < 0$ and in the proof of Proposition 5.7 it was shown that $\mathbf{Ad} = \mathbf{0}$. The remaining property that $\mathbf{d} \ge \mathbf{0}$ is obtained by showing that $\mathbf{d_B} \ge \mathbf{0}$ and $\mathbf{d_N} \ge \mathbf{0}$. The former is true by the hypothesis. For the latter, note that $\mathbf{d_N} = \mathbf{I}_{.j}$ and hence also $\mathbf{d_N} \ge \mathbf{0}$.

PROOF OF PROPOSITION 5.9. Since $\mathbf{x}$ is a bfs, the LP is feasible. To show that the LP is unbounded, the direction $\mathbf{d} = (\mathbf{d_B}, \mathbf{d_N}) = (-\mathbf{B}^{-1}\mathbf{N}_{.j}, \mathbf{I}_{.j})$ must be shown to satisfy (1) $\mathbf{d} \ge \mathbf{0}$, (2) $\mathbf{Ad} = \mathbf{0}$, and (3) $\mathbf{cd} < 0$. To see (1), note that $\mathbf{d_B} \ge \mathbf{0}$ by hypothesis, and $\mathbf{d_N} = \mathbf{I}_{.j} \ge \mathbf{0}$, and so, $\mathbf{d} = (\mathbf{d_B}, \mathbf{d_N}) \ge \mathbf{0}$. The fact that $\mathbf{Ad} = \mathbf{0}$ was established in the proof of Proposition 5.7. Finally, Proposition 5.6 shows that $\mathbf{cd} < 0$. □

The result of Proposition 5.9 is that if $\mathbf{d_B} \geq \mathbf{0}$, then the condition of unboundedness has been detected and the simplex algorithm will stop. On the other hand, if $\mathbf{d_B}$ is not, then t^* can be computed by the min ratio test

$$t^* = \min\{-(\mathbf{x_B})_k/(\mathbf{d_B})_k : 1 \leq k \leq m \text{ and } (\mathbf{d_B})_k < 0\}$$

Now that t^* can be computed by the min ratio test, perhaps it is time to move from $\mathbf{x}$ to $\mathbf{x} + t^*\mathbf{d}$. Unfortunately, there is still one serious problem that can arise: What happens if $t^* = 0$? In this case, $\mathbf{x} + t^*\mathbf{d} = \mathbf{x} + (0)\mathbf{d} = \mathbf{x}$, and no movement is taking place! Consequently, the objective function is not decreased, thus destroying the entire finiteness argument. Ultimately, this *degenerate* situation must be addressed properly, and that will be done in Section 5.7. For now, observe from the min ratio test that, for t^* to be 0, there must be a k with $1 \leq k \leq m$, such that both $(\mathbf{x_B})_k = 0$ and $(\mathbf{d_B})_k < 0$. Thus one way to avoid t^* from being 0 is to assume that the values of the basic variables are always positive. Since this nondegeneracy assumption is not always realistic, one must eventually deal with the case when $t^* = 0$. Nonetheless, for a preliminary understanding of the simplex algorithm, it is quite acceptable to pretend that t^* is never 0. All of the issues relating to degeneracy are dealt with in Section 5.7, so, for now, suppose that $0 < t^* < \infty$.

It is a simple matter to move from $\mathbf{x}$ to the new point $\mathbf{x}' = \mathbf{x} + t^*\mathbf{d}$, but how do you know that $\mathbf{x}'$ is going to be a bfs? Indeed, it is, but to prove it formally, the appropriate matrices $\mathbf{B}'$ and $\mathbf{N}'$ corresponding to $\mathbf{x}'$ must be found. Conveniently, $\mathbf{B}'$ is obtained from $\mathbf{B}$ by replacing column k^* of $\mathbf{B}$ with column j^* of $\mathbf{N}$. This switching of columns is referred to as a pivot operation and will be described in detail in the next section. The purpose of this section has been to describe how to compute a value for t^* and what to do if $t^* = \infty$, $t^* = 0$, or $0 < t^* < \infty$. The information is summarized in Table 5.3.

TABLE 5.3. The Possible Outcomes for t^*

Value for t^*	Cause	Action
$t^* = \infty$	$\mathbf{d_B} \geq \mathbf{0}$	Stop, the LP is unbounded
$0 < t^* < \infty$	$(\mathbf{x_B})_{k^*} > 0$, $(\mathbf{d_B})_{k^*} < 0$	Pivot (see Section 5.5)
$t^* = 0$	$(\mathbf{x_B})_{k^*} = 0$, $(\mathbf{d_B})_{k^*} < 0$	Handle degeneracy (see Section 5.7)

EXERCISES

5.4.1. For each of the directions computed in Exercise 5.3.1 determine t^* and k^*.

5.4.2. Let $\mathbf{x}$ be a feasible point for the following LP:

$$\begin{array}{ll} \text{Minimize} & \mathbf{cx} \\ \text{subject to} & \mathbf{Ax} \leq \mathbf{b} \\ & \mathbf{x} \geq \mathbf{0} \end{array}$$

and let $\mathbf{d}$ be a given direction of movement. Derive, in algebraic terms, the min ratio test for computing t^*. When will the LP be unbounded?

5.4.3. For a bfs $\mathbf{x}$ in which $x_3 = 1$ and $x_2 = 2$ are basic, in that order, and for the direction $\mathbf{d} = (d_1, d_2, d_3, d_4) = (1, -\frac{2}{3}, -\frac{1}{3}, 0)$, apply the min ratio test to determine the value for k^* by using
(a) Bland's rule and
(b) the first come, first served rule.

5.4.4. Consider the following rule for determining the value of j^* (not k^*).

Maximum Improvement Rule: Among all possible values of j for which $(\mathbf{c_N} - \mathbf{c_B B^{-1} N})_j < 0$, choose the one such that the resulting vector $\mathbf{x} + t^*\mathbf{d}$ has the smallest objective value. Describe in algebraic terms how to compute the value for j^*. What advantages and disadvantages might this rule have?

5.4.5. Consider the following LP:

$$\begin{array}{lrrrrrl} \text{Minimize} & -7x_1 & +6x_2 & +4x_3 & & & \\ \text{subject to} & -7x_1 & +2x_2 & + x_3 & +x_4 & & = 25 \\ & -8x_1 & +9x_2 & +6x_3 & & +x_5 & = 35 \\ & x_1, & x_2, & x_3, & x_4, & x_5 & \geq 0 \end{array}$$

For the bfs in which x_4 and x_5 are the basic variables, find the direction of movement by using the rule of steepest descent. Find t^* for this direction.

5.5. MOVING: THE PIVOT OPERATION

The previous section presented an algebraic method for determining a value for t^*—the amount of movement to be taken from the current bfs $\mathbf{x} = (\mathbf{x_B}, \mathbf{x_N}) = (\mathbf{B}^{-1}\mathbf{b}, \mathbf{0}) \geq \mathbf{0}$ in the direction $\mathbf{d} = (\mathbf{d_B}, \mathbf{d_N}) = (-\mathbf{B}^{-1}\mathbf{N}_{.j^*}, \mathbf{I}_{.j^*})$. In the event that $\mathbf{d_B} \geq \mathbf{0}$, $t^* = \infty$ and the LP is unbounded. If $\mathbf{d_B} \geq \mathbf{0}$ fails to hold, then the min ratio test can be used to compute

$$t^* = \min\{-(\mathbf{x_B})_k/(\mathbf{d_B})_k : 1 \leq k \leq m \text{ and } (\mathbf{d_B})_k < 0\}$$

as well as an integer $1 \leq k^* \leq m$ for which $t^* = -(\mathbf{x_B})_{k^*}/(\mathbf{d_B})_{k^*}$. If $t^* = 0$, then the techniques of Section 5.7 should be used, so, for the moment,

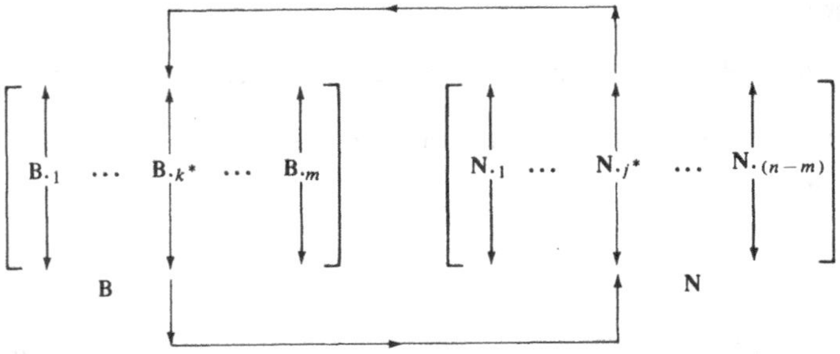

FIGURE 5.3. Pivot operation.

assume that $0 < t^* < \infty$. Before moving from the current point $\mathbf{x}$ to the new point $\mathbf{x}' = \mathbf{x} + t^*\mathbf{d}$, it would be nice to know that $\mathbf{x}'$ is going to be a bfs. Indeed it is, as will be established in this section.

In order to show that $\mathbf{x}'$ is a bfs, the new basis matrix $\mathbf{B}'$ (and corresponding matrix $\mathbf{N}'$) must be found. Simply stated, $\mathbf{B}'$ is obtained by replacing column k^* of $\mathbf{B}$ with column j^* from $\mathbf{N}$. In other words, when the current bfs $\mathbf{x}$ fails the test for optimality, a value for j^* is produced that represents the column of $\mathbf{N}$ that is about to enter the basis matrix $\mathbf{B}$. Similarly, if $\mathbf{d_B} \geq \mathbf{0}$ does not hold, then the min ratio test produces a value for k^* that is the column of $\mathbf{B}$ to be replaced by $\mathbf{N}_{.j^*}$. This switching of columns is referred to as a *pivot operation* (see Figure 5.3). Of course, it is necessary to establish formally that the new matrices $\mathbf{B}'$ and $\mathbf{N}'$ have all of the desired properties for $\mathbf{x}'$ to be a bfs; $\mathbf{B}'$ is nonsingular, $\mathbf{x}'_{\mathbf{B}'} = \mathbf{B}'^{-1}\mathbf{b}$, $\mathbf{x}'_{\mathbf{B}'} \geq \mathbf{0}$, and $\mathbf{x}'_{\mathbf{N}'} = \mathbf{0}$. Before doing so, however, it is instructive to see why $\mathbf{B}'$ and $\mathbf{N}'$ are constructed in this particular manner.

Recall from Section 5.1 that one way to find the basis matrix for $\mathbf{x}' = \mathbf{x} + t^*\mathbf{d}$ is to find those components of $\mathbf{x}'$ that are positive and to put the corresponding columns of $\mathbf{A}$ into $\mathbf{B}'$. The remaining components of $\mathbf{x}'$ should have the value 0, and those columns of $\mathbf{A}$ should constitute $\mathbf{N}'$.

In order to determine which components of $\mathbf{x}'$ are positive, it is best to start with the original bfs $\mathbf{x} = (\mathbf{x_B}, \mathbf{x_N}) = (\mathbf{B}^{-1}\mathbf{b}, \mathbf{0}) \geq \mathbf{0}$ and consider what happens when moving an amount $t^* > 0$ in the direction $\mathbf{d} = (\mathbf{d_B}, \mathbf{d_N}) = (-\mathbf{B}^{-1}\mathbf{N}_{.j^*}, \mathbf{I}_{.j^*})$. Picture the basic variables together with their values of $\mathbf{B}^{-1}\mathbf{b}$ on one side, and the nonbasic variables together with their values of 0 on the other side (see Figure 5.4). When moving an amount $t^* > 0$ in the

FIGURE 5.4. **(a)** Basic and **(b)** nonbasic variables.

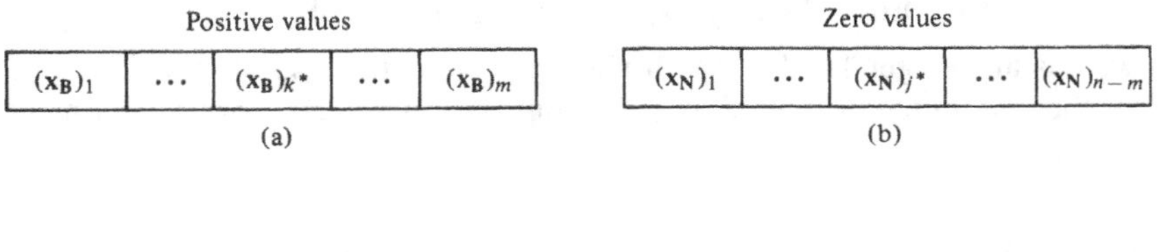

direction $\mathbf{d} = (\mathbf{d_B}, \mathbf{d_N})$, the values of the basic variables change from $\mathbf{x_B} = \mathbf{B}^{-1}\mathbf{b}$ to $\mathbf{x_B} + t^*\mathbf{d_B}$ and the values of the nonbasic variable change from $\mathbf{x_N} = \mathbf{0}$ to $\mathbf{x_N} + t^*\mathbf{d_N}$.

Consider first what happens to the nonbasic components

$$\mathbf{x_N} + t^*\mathbf{d_N} = \mathbf{x_N} + t^*\mathbf{I}_{.j^*} = \mathbf{0} + t^*\mathbf{I}_{.j^*} = t^*\begin{bmatrix} 0 \\ \vdots \\ 1 \\ \vdots \\ 0 \end{bmatrix} \leftarrow j^* = \begin{bmatrix} 0 \\ \vdots \\ t^* \\ \vdots \\ 0 \end{bmatrix} \leftarrow j^*$$

In other words, all of the nonbasic variables remain at the value 0 except that $(\mathbf{x_N} + t^*\mathbf{d_N})_{j^*} = t^*$. If $t^* > 0$, then this variable must now become basic since its value is positive, but all the rest of the nonbasic variables can remain nonbasic since their values stay at 0.

In an analogous fashion, consider what happens to the values of the basic variables as one moves an amount t^* in the direction $\mathbf{d_B}$. They change from $\mathbf{x_B} = \mathbf{B}^{-1}\mathbf{b}$ to $\mathbf{x_B} + t^*\mathbf{d_B}$. Some of these variables decrease in value [when $(\mathbf{d_B})_k < 0$], and the rest increase or remain the same. The min ratio test was designed to make sure that none of the basic variables become negative in value. In fact, if $t^* = -(\mathbf{x_B})_{k^*}/(\mathbf{d_B})_{k^*}$, then

$$\begin{aligned}
(\mathbf{x_B} + t^*\mathbf{d_B})_{k^*} &= (\mathbf{x_B})_{k^*} + t^*(\mathbf{d_B})_{k^*} \\
&= (\mathbf{x_B})_{k^*} - [(\mathbf{x_B})_{k^*}/(\mathbf{d_B})_{k^*}](\mathbf{d_B})_{k^*} \\
&= (\mathbf{x_B})_{k^*} - (\mathbf{x_B})_{k^*} \\
&= 0
\end{aligned}$$

In other words, the value of this basic variable has decreased to 0, and thus can now become nonbasic together with its associated column $\mathbf{B}_{.k^*}$, as depicted in Figure 5.5.

FIGURE 5.5. The new basic and nonbasic variables.

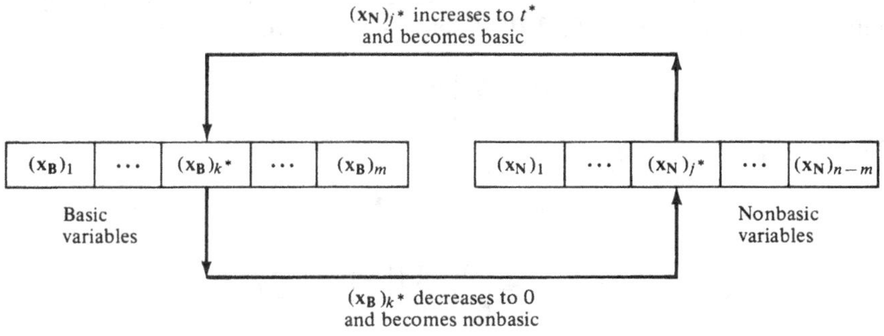

To summarize the pivot operation, in moving from $\mathbf{x}$ to $\mathbf{x} + t^*\mathbf{d}$, the value of $(\mathbf{x_N})_{j^*}$ increases from 0 to t^* and thus becomes a basic variable. Its associated column $\mathbf{N}_{.j^*}$ enters the basis matrix while all other nonbasic variables remain at the value 0. Simultaneously, some of the basic variables decrease while others do not. The first basic variable (indicated by k^*) whose value decreases to 0 is removed along with the corresponding column $\mathbf{B}_{.k^*}$ from the basis. The pivot operation is now complete and will be demonstrated on Example 5.1.

Recall the bfs $\mathbf{x} = (x_1, x_2, x_3, x_4) = (0,1,2,0)$, in which $\mathbf{x_B} = (x_2, x_3) = (1,2)$, $\mathbf{x_N} = (x_1, x_4) = (0,0)$, and

$$\mathbf{B} = \begin{bmatrix} 0 & 1 \\ 1 & 1 \end{bmatrix}, \qquad \mathbf{N} = \begin{bmatrix} 1 & 2 \\ 0 & 2 \end{bmatrix}$$

The vector $\mathbf{x}$ fails the test for optimality because $\mathbf{c_N} - \mathbf{c_B}\mathbf{B}^{-1}\mathbf{N} = (-1,4)$, and so $j^* = 1$. The corresponding direction is computed by

$$\mathbf{d_B} = \begin{bmatrix} d_2 \\ d_3 \end{bmatrix} = -\mathbf{B}^{-1}\mathbf{N}_{.1} = \begin{bmatrix} 1 \\ -1 \end{bmatrix}$$

$$\mathbf{d_N} = \begin{bmatrix} d_1 \\ d_4 \end{bmatrix} = \mathbf{I}_{.1} = \begin{bmatrix} 1 \\ 0 \end{bmatrix}$$

Thus

$$\mathbf{d} = (d_1, d_2, d_3, d_4) = (1, 1, -1, 0)$$

Since $\mathbf{d_B} \geq \mathbf{0}$ fails to hold, the min ratio test yields $t^* = \min\{-(\mathbf{x_B})_2/(\mathbf{d_B})_2\} = \min\{-2/-1\} = 2$, and so $k^* = 2$. Thus $\mathbf{x}' = \mathbf{x} + t^*\mathbf{d} = (0,1,2,0) + 2(1,1,-1,0) = (2,3,0,0)$. To obtain the new basis matrix $\mathbf{B}'$ and the matrix $\mathbf{N}'$, simply replace column $k^* = 2$ of $\mathbf{B}$ with column $j^* = 1$ of $\mathbf{N}$, so

$$\mathbf{B}' = \begin{bmatrix} 0 & 1 \\ 1 & 0 \end{bmatrix}, \qquad \mathbf{N}' = \begin{bmatrix} 1 & 2 \\ 1 & 2 \end{bmatrix}$$

Look carefully at the matrix $\mathbf{B}'$. Observe that the new basic variables are $\mathbf{x}'_{\mathbf{B}'} = (x_2, x_1)$, not (x_1, x_2).

It is interesting to note that there are two ways to compute the values of the new basic variables. One way is to look at the $\mathbf{B}'$ components of the vector $\mathbf{x} + t^*\mathbf{d} = (2,3,0,0)$. Thus $\mathbf{x}'_{\mathbf{B}'} = (\mathbf{x}_2, \mathbf{x}_1) = (3,2)$. The second way of computing the values is to find $\mathbf{B}'^{-1}$ first, which in this case is

$$\mathbf{B}'^{-1} = \begin{bmatrix} 0 & 1 \\ 1 & 0 \end{bmatrix}$$

for then

$$\mathbf{x}'_{\mathbf{B}'} = \begin{bmatrix} x_2 \\ x_1 \end{bmatrix} = \mathbf{B}'^{-1}\mathbf{b}$$

$$= \begin{bmatrix} 0 & 1 \\ 1 & 0 \end{bmatrix}\begin{bmatrix} 2 \\ 3 \end{bmatrix} = \begin{bmatrix} 3 \\ 2 \end{bmatrix}$$

The remainder of this section is devoted to establishing formally that the new vector $\mathbf{x}' = \mathbf{x} + t^*\mathbf{d}$ is a bfs, i.e., that the matrix $\mathbf{B}'$ satisfies the conditions that (1) $\mathbf{B}'$ is nonsingular, (2) $\mathbf{x}'_{\mathbf{B}'} = \mathbf{B}'^{-1}\mathbf{b}$, (3) $\mathbf{x}'_{\mathbf{B}'} \geq \mathbf{0}$, and (4) $\mathbf{x}'_{\mathbf{N}'} = \mathbf{0}$. These four statements are established in the next four lemmas, respectively.

Lemma 5.1. *If* $\mathbf{x} = (\mathbf{x}_{\mathbf{B}}, \mathbf{x}_{\mathbf{N}}) = (\mathbf{B}^{-1}\mathbf{b}, \mathbf{0}) \geq \mathbf{0}$ *is a bfs,* $\mathbf{d} = (\mathbf{d}_{\mathbf{B}}, \mathbf{d}_{\mathbf{N}}) = (-\mathbf{B}^{-1}\mathbf{N}_{.j^*}, \mathbf{I}_{.j^*})$ *and* $\mathbf{t}^* = -(\mathbf{x}_{\mathbf{B}})_{k^*}/(\mathbf{d}_{\mathbf{B}})_{k^*} = \min\{-(\mathbf{x}_{\mathbf{B}})_k/(\mathbf{d}_{\mathbf{B}})_k : 1 \leq k \leq m \text{ and } (\mathbf{d}_{\mathbf{B}})_k < 0\}$, *then the matrix* $\mathbf{B}'$ *obtained by replacing column* k^* *of* $\mathbf{B}$ *with column* j^* *of* $\mathbf{N}$ *is nonsingular.*

OUTLINE OF PROOF. The forward–backward method gives rise to the abstraction question "How can I show that a matrix (namely, $\mathbf{B}'$) is nonsingular?" The question is answered by means of Proposition 4.3, whereby it will be shown that $\mathbf{B}'$ can be written as the product of two nonsingular matrices, namely, $\mathbf{B}$ (which is nonsingular because $\mathbf{x}$ is a bfs) and a new matrix $\mathbf{E}$. Consequently, the remainder of the proof is devoted to constructing the matrix $\mathbf{E}$ and showing that (1) $\mathbf{B}' = \mathbf{BE}$ and (2) $\mathbf{E}$ is nonsingular.

The direction $\mathbf{d}_{\mathbf{B}}$ is the key to constructing $\mathbf{E}$. Specifically, $\mathbf{E}$ is obtained by replacing column k^* of the $m \times m$ identity matrix $\mathbf{I}$ with the vector $-\mathbf{d}_{\mathbf{B}}$.

To see that $\mathbf{B}' = \mathbf{BE}$, it will be shown that each column of $\mathbf{B}'$ equals the corresponding column of $\mathbf{BE}$, i.e., that for all k with $1 \leq k \leq m$, $\mathbf{B}'_{.k} = (\mathbf{BE})_{.k}$. The appearance of the quantifier "for all" in the backward process suggests using the choose method, whereby one selects a k with $1 \leq k \leq m$, for which it must be shown that $\mathbf{B}'_{.k} = (\mathbf{BE})_{.k}$. There are two cases to consider:

Case 1: $k \neq k^*$

In this case, since $\mathbf{E}_{.k} = \mathbf{I}_{.k}$ and $\mathbf{B}'_{.k} = \mathbf{B}_{.k}$, it follows that $(\mathbf{BE})_{.k} = \mathbf{B}(\mathbf{E}_{.k}) = \mathbf{B}(\mathbf{I}_{.k}) = \mathbf{B}_{.k} = \mathbf{B}'_{.k}$.

Case 2: $k = k^*$

In this case, since $\mathbf{E}_{.k} = -\mathbf{d}_{\mathbf{B}}$ and $\mathbf{d}_{\mathbf{B}} = -\mathbf{B}^{-1}\mathbf{N}_{.j^*}$, it follows that $(\mathbf{BE})_{.k} = \mathbf{B}(\mathbf{E}_{.k}) = -\mathbf{B}\mathbf{d}_{\mathbf{B}} = \mathbf{B}(\mathbf{B}^{-1}\mathbf{N}_{.j^*}) = \mathbf{I}\mathbf{N}_{.j^*} = \mathbf{N}_{.j^*} = \mathbf{B}'_{.k^*}$.

All that remains is to show that $\mathbf{E}$ is nonsingular, and this is accomplished by writing the inverse matrix $\mathbf{E}^{-1}$. Specifically, since $\mathbf{E}$ differs from the identity matrix in only one column and since $(\mathbf{d}_{\mathbf{B}})_{k^*} \neq 0$, $\mathbf{E}^{-1}$ can be obtained by replacing column k^* of the identity matrix with the vector $((\mathbf{d}_{\mathbf{B}})_1/(\mathbf{d}_{\mathbf{B}})_{k^*}, \ldots, -1/(\mathbf{d}_{\mathbf{B}})_{k^*}, \ldots, (\mathbf{d}_{\mathbf{B}})_m/(\mathbf{d}_{\mathbf{B}})_{k^*})$ (see Section 4.4 on eta matrices).

PROOF OF LEMMA 5.1. To show that $\mathbf{B}'$ is nonsingular, a nonsingular matrix $\mathbf{E}$ will be constructed such that $\mathbf{B}' = \mathbf{BE}$, for then, since $\mathbf{B}$ is also nonsingular, Proposition 4.3 ensures that the product $\mathbf{BE}$ is nonsingular too. The matrix $\mathbf{E}$ is obtained by replacing column k^* of the $m \times m$ identity matrix with the vector $-\mathbf{d_B}$. To see that $\mathbf{B}' = \mathbf{BE}$, let k with $1 \le k \le m$. If $k \ne k^*$, then $(\mathbf{BE})_{.k} = \mathbf{BE}_{.k} = \mathbf{BI}_{.k} = \mathbf{B}_{.k} = \mathbf{B}'_{.k}$; if $k = k^*$, then $(\mathbf{BE})_{.k} = \mathbf{BE}_{.k^*} = \mathbf{B}(-\mathbf{d_B}) = \mathbf{B}(\mathbf{B}^{-1}\mathbf{N}_{.j^*}) = \mathbf{N}_{.j^*} = \mathbf{B}'_{.k^*}$. In either case, column k of $\mathbf{B}'$ equals column k of $\mathbf{BE}$, and so $\mathbf{B}' = \mathbf{BE}$.

It remains to show that $\mathbf{E}$ is nonsingular. Since $\mathbf{E}$ differs from the identity matrix only in column k^* and since $(\mathbf{d_B})_{k^*} \ne 0$, Section 4.4 ensures that $\mathbf{E}^{-1}$ can be obtained by replacing column k^* of the identity matrix with the vector $((\mathbf{d_B})_1/(\mathbf{d_B})_{k^*}, \ldots, -1/(\mathbf{d_B})_{k^*}, \ldots, (\mathbf{d_B})_m/(\mathbf{d_B})_{k^*})$, and the proof is complete. □

Lemma 5.2. *If* $\mathbf{x} = (\mathbf{x_B}, \mathbf{x_N}) = (\mathbf{B}^{-1}\mathbf{b}, \mathbf{0}) \ge \mathbf{0}$ *is a bfs,* $\mathbf{d} = (\mathbf{d_B}, \mathbf{d_N}) = (-\mathbf{B}^{-1}\mathbf{N}_{.j^*}, \mathbf{I}_{.j^*})$, $t^* = -(\mathbf{x_B})_{k^*}/(\mathbf{d_B})_{k^*} = \min\{-(\mathbf{x_B})_k/(\mathbf{d_B})_k : 1 \le k \le$ *m and* $(\mathbf{d_B})_k < 0\}$, $\mathbf{x}' = \mathbf{x} + t^*\mathbf{d}$, *and* $\mathbf{B}'$ *and* $\mathbf{N}'$ *are defined as above, then* $\mathbf{x}'_{\mathbf{N}'} = 0$.

OUTLINE OF PROOF. The forward–backward method gives rise to the abstraction question "How can I show that a vector (namely, $\mathbf{x}'_{\mathbf{N}'}$) $= \mathbf{0}$?" One answer is to show that for all j with $1 \le j \le n - m$, $(\mathbf{x}'_{\mathbf{N}'})_j = 0$. The choose method is then invoked to select a j with $1 \le j \le n - m$, for which it will be shown that $(\mathbf{x}'_{\mathbf{N}'})_j = 0$. There are two cases.

Case 1: $j \ne j^*$

In this case, $(\mathbf{x}'_{\mathbf{N}'})_j = (\mathbf{x_N})_j$, and since by hypothesis $\mathbf{x}$ is a bfs, $(\mathbf{x_N})_j = 0$.

Case 2: $j = j^*$

In this case

$$\begin{aligned}(\mathbf{x}'_{\mathbf{N}'})_j &= (\mathbf{x}'_{\mathbf{N}'})_{j^*} = (\mathbf{x_B} + t^*\mathbf{d_B})_{k^*} \\ &= (\mathbf{x_B})_{k^*} + t^*(\mathbf{d_B})_{k^*} = (\mathbf{x_B})_{k^*} - [(\mathbf{x_B})_{k^*}/(\mathbf{d_B})_{k^*}](\mathbf{d_B})_{k^*} \\ &= 0\end{aligned}$$

PROOF OF LEMMA 5.2. To show that $\mathbf{x}'_{\mathbf{N}'} = \mathbf{0}$, let j with $1 \le j \le n - m$. If $j \ne j^*$, then $(\mathbf{x}'_{\mathbf{N}'})_j = (\mathbf{x_N})_j$, and since $\mathbf{x}$ is a bfs, $(\mathbf{x_N})_j = 0$. On the other hand, if $j = j^*$, then

$$\begin{aligned}(\mathbf{x}'_{\mathbf{N}'})_j &= (\mathbf{x}'_{\mathbf{N}'})_{j^*} = (\mathbf{x_B} + t^*\mathbf{d_B})_{k^*} \\ &= (\mathbf{x_B})_{k^*} + t^*(\mathbf{d_B})_{k^*} = (\mathbf{x_B})_{k^*} - [(\mathbf{x_B})_{k^*}/(\mathbf{d_B})_{k^*}](\mathbf{d_B})_{k^*} \\ &= 0\end{aligned}$$

□

Lemma 5.3. *If* $\mathbf{x}$, $\mathbf{d}$, $\mathbf{x}'$, $\mathbf{B}'$, $\mathbf{N}'$ *are as in Lemma* 5.2, *then* $\mathbf{x}'_{\mathbf{B}'} = \mathbf{B}'^{-1}\mathbf{b}$.

OUTLINE OF PROOF. To see that $\mathbf{x}'_{\mathbf{B}'} = \mathbf{B}'^{-1}\mathbf{b}$, it will actually be shown that $\mathbf{B}'\mathbf{x}'_{\mathbf{B}'} = \mathbf{b}$, for then, since $\mathbf{B}'$ is nonsingular (see Lemma 5.1), it follows that $\mathbf{x}'_{\mathbf{B}'} = \mathbf{B}'^{-1}\mathbf{b}$.

The forward process is used to establish that $\mathbf{B}'\mathbf{x}'_{\mathbf{B}'} = \mathbf{b}$. Specifically, since $\mathbf{x}'$ is feasible for the LP (see Proposition 5.8), $\mathbf{b} = \mathbf{A}\mathbf{x}' = \mathbf{B}'\mathbf{x}'_{\mathbf{B}'} + \mathbf{N}'\mathbf{x}'_{\mathbf{N}'}$ and, from Lemma 5.2, $\mathbf{x}'_{\mathbf{N}'} = \mathbf{0}$, so $\mathbf{b} = \mathbf{B}'\mathbf{x}'_{\mathbf{B}'}$.

PROOF OF LEMMA 5.3. From Proposition 5.8, $\mathbf{x}'$ is feasible for the LP, so $\mathbf{b} = \mathbf{A}\mathbf{x}' = \mathbf{B}'\mathbf{x}'_{\mathbf{B}'} + \mathbf{N}'\mathbf{x}'_{\mathbf{N}'}$. By Lemma 5.2, $\mathbf{x}'_{\mathbf{N}'} = \mathbf{0}$ so $\mathbf{b} = \mathbf{B}'\mathbf{x}'_{\mathbf{B}'}$. Finally, since $\mathbf{B}'$ is nonsingular by Lemma 5.1, it must be that $\mathbf{x}'_{\mathbf{B}'} = \mathbf{B}'^{-1}\mathbf{b}$. □

Lemma 5.4. *If* $\mathbf{x}$, $\mathbf{d}$, t^*, $\mathbf{x}'$, $\mathbf{B}'$, $\mathbf{N}'$ *are as in Lemma* 5.2, *then* $\mathbf{x}'_{\mathbf{B}'} \geq \mathbf{0}$.

OUTLINE OF PROOF. The forward–backward method gives rise to the abstraction question "How can I show that a vector (namely, $\mathbf{x}'_{\mathbf{B}'}$) is $\geq \mathbf{0}$?" This question is answered by showing that the entire vector $\mathbf{x}' \geq \mathbf{0}$ for then, $\mathbf{x}'_{\mathbf{B}'} \geq \mathbf{0}$. To see that $\mathbf{x}' \geq \mathbf{0}$, recall that Proposition 5.8 established that $\mathbf{x}'$ is feasible for the LP, and so it must be that $\mathbf{x}' \geq \mathbf{0}$.

PROOF OF LEMMA 5.4. By Proposition 5.8, $\mathbf{x}'$ is feasible for the LP, thus $\mathbf{x}' \geq 0$, and so $\mathbf{x}'_{\mathbf{B}'} \geq \mathbf{0}$. □

These four lemmas can be combined to yield the next theorem.

Theorem 5.1. *If* $\mathbf{x}$ *is an initial bfs for the LP, then the simplex algorithm will terminate in a finite number of iterations with an optimal solution, or else detect unboundedness, provided that, at each bfs, the value of* t^* *is positive.*

PROOF OF THEOREM 5.1. The previous four lemmas have established that the simplex algorithm moves from one bfs to another at each iteration. Each time a bfs fails the test for optimality, it is possible to move to another bfs, unless unboundedness is detected. As long as $t^* > 0$, the objective value must strictly decrease, and thus no bfs can be repeated. Since there are only a finite number of bfs, either an optimal one will be found, or else the condition of unboundedness will be detected. □

This section has described the pivot operation for the case in which $0 < t^* < \infty$. It is important to note that, even when $t^* = 0$, the pivot operation can still be performed in the sense that it is possible to bring $\mathbf{N}_{.j^*}$ into the basis matrix while removing $\mathbf{B}_{.k^*}$. One could then repeat the steps of the simplex algorithm with the new matrices, hoping this time that the amount of movement will be positive. Unfortunately, it might not be, and that can cause serious problems in the form of cycling. Extensive computational experience has shown that, from a practical point of view, cycling does not occur, even though the objective function might not strictly decrease at each iteration. Nonetheless, it is important to resolve this problem, at least theoretically, as will be done in Section 5.7. The next section provides a summary of the steps of the simplex algorithm and demonstrates its application with several examples.

EXERCISES

5.5.1. Consider the following LP:

$$\begin{array}{lrrrrrrl}
\text{Minimize} & x_1 & -2x_2 & +x_3 & & -4x_5 & +x_6 & \\
\text{subject to} & x_1 & +x_2 & -2x_3 & -x_4 & & +x_6 & =2 \\
& 2x_1 & & +x_3 & +2x_4 & +x_5 & -x_6 & =3 \\
& x_1, & x_2, & x_3, & x_4, & x_5, & x_6 & \geq 0
\end{array}$$

After performing the test for optimality on the bfs $\mathbf{x} = (0, 2, 0, 0, 3, 0)$, the following data have been obtained: $j^* = 4$, $\mathbf{d} = (0, -1, 0, 0, 1, 1)$, $t^* = 2$, $k^* = 1$.

(a) What are the new basic and nonbasic variables?
(b) What are the new **B** and **N** matrices?
(c) Compute the new inverse matrix.
(d) Compute the values of the new basic variables by using the new inverse matrix.
(e) Compute the values of the new basic variables by using **x**, **d**, and t^*.
(f) Test the new bfs for optimality.

5.5.2. Consider the following LP:

$$\begin{array}{lrrrrrrl}
\text{Minimize} & x_1 & +2x_2 & -3x_3 & -4x_4 & +x_5 & -x_6 & \\
\text{subject to} & x_1 & & -4x_3 & +5x_4 & & +x_6 & =2 \\
& & x_2 & +2x_3 & -2x_4 & & +2x_6 & =4 \\
& & & x_3 & +x_4 & +x_5 & -x_6 & =3 \\
& x_1, & x_2, & x_3, & x_4, & x_5, & x_6 & \geq 0
\end{array}$$

After performing the test for optimality on the bfs $\mathbf{x} = (2, 4, 0, 0, 3, 0)$, the following data have been obtained: $j^* = 2$, $\mathbf{d} = (-5, 2, 0, 1, -1, 0)$, $t^* = \frac{2}{5}$, $k^* = 1$.

(a) What are the new basic and nonbasic variables?
(b) What are the new **B** and **N** matrices?
(c) Compute the values of the new basic variables by using **x**, **d**, and t^*.

5.5.3. Consider the following LP:

$$\begin{array}{lrrrrrl}
\text{Minimize} & x_1 & -2x_2 & -3x_3 & +x_4 & -x_5 & \\
\text{subject to} & 2x_1 & -x_2 & +x_3 & & -x_5 & =0 \\
& & 2x_2 & +x_3 & +3x_4 & +2x_5 & =9 \\
& x_1, & x_2, & x_3, & x_4, & x_5 & \geq 0
\end{array}$$

After testing the bfs in which x_1 and x_4 are the basic variables, the following data have been obtained: $j^* = 2$, $\mathbf{d} = (-\frac{1}{2}, 0, 1, -\frac{1}{3}, 0)$, $t^* = 0$, $k^* = 1$.

(a) What are the new basic and nonbasic variables?
(b) What are the new **B** and **N** matrices?
(c) Compute the new inverse matrix.
(d) Compute the values of the new basic variables by using the new inverse matrix.
(e) Compute the values of the new basic variables by using **x**, **d**, and t^*.
(f) How does the new bfs compare with the original one?

5.6. A SUMMARY OF THE SIMPLEX ALGORITHM

The simplex algorithm is now complete. This section will provide a summary and flow chart of the steps of the algorithm, and several examples will be solved in complete detail. Specifically, given an initial bfs $\mathbf{x} = (\mathbf{x_B}, \mathbf{x_N}) = (\mathbf{B}^{-1}\mathbf{b}, \mathbf{0}) \geq \mathbf{0}$ together with the matrix $\mathbf{B}^{-1}$, these are the steps of the simplex algorithm:

Step 1. Testing for optimality: Compute the reduced costs $\mathbf{c_N} - \mathbf{c_B}\mathbf{B}^{-1}\mathbf{N}$. If this vector is greater than or equal to $\mathbf{0}$, then the current bfs is optimal; otherwise select any j with $1 \leq j \leq n - m$ and for which $(\mathbf{c_N} - \mathbf{c_B}\mathbf{B}^{-1}\mathbf{N})_j < 0$ and call it j^*. Go to step 2.

Step 2. Computing the direction: Using the value of j^* obtained in step 1, compute the direction $\mathbf{d_B} = -\mathbf{B}^{-1}\mathbf{N}_{.j^*}$ (note that $\mathbf{d_N} = \mathbf{I}_{.j^*}$ is not really needed). Go to step 3.

Step 3. Computing t^*: If the vector $\mathbf{d_B}$ (obtained in step 2) $\geq \mathbf{0}$, then the LP is unbounded; otherwise, select any k^* with $1 \leq k^* \leq m$ and for which $t^* = \min\{-(\mathbf{x_B})_k/(\mathbf{d_B})_k : 1 \leq k \leq m$ and $(\mathbf{d_B})_k < 0\} = -(\mathbf{x_B})_{k^*}/(\mathbf{d_B})_{k^*}$. Go to step 4.

Step 4. Pivoting: Using the value of j^* and k^* obtained in steps 2 and 3, respectively, create the new basis matrix $\mathbf{B}'$ by replacing column k^* of $\mathbf{B}$ with column j^* of $\mathbf{N}$. Before returning to step 1, compute the new inverse matrix $\mathbf{B}'^{-1}$ and the values of the new basic variables, namely, $\mathbf{B}'^{-1}\mathbf{b}$.

These steps will now be demonstrated with several examples using the rule of steepest descent.

EXAMPLE 5.2:

$$\begin{array}{lrrrrrl}
\text{Minimize} & -2x_1 & -3x_2 & -\ x_3 & & & \\
\text{subject to} & x_1 & -\ x_2 & +2x_3 & +x_4 & & = 1 \\
 & 4x_1 & +2x_2 & -\ x_3 & & +x_5 & = 2 \\
 & x_1, & x_2, & x_3, & x_4, & x_5 & \geq 0
\end{array}$$

Thus

$$\mathbf{c} = (-2, -3, -1, 0, 0)$$
$$\mathbf{A} = \begin{bmatrix} 1 & -1 & 2 & 1 & 0 \\ 4 & 2 & -1 & 0 & 1 \end{bmatrix}$$
$$\mathbf{b} = \begin{bmatrix} 1 \\ 2 \end{bmatrix}$$

Iteration 1

Step 0. The bfs for this iteration:

$$\mathbf{B} = [\mathbf{A}_{.4}, \mathbf{A}_{.5}] = \begin{bmatrix} 1 & 0 \\ 0 & 1 \end{bmatrix}$$
$$\mathbf{N} = [\mathbf{A}_{.1}, \mathbf{A}_{.2}, \mathbf{A}_{.3}] = \begin{bmatrix} 1 & -1 & 2 \\ 4 & 2 & -1 \end{bmatrix}$$
$$\mathbf{x_B} = \begin{bmatrix} x_4 \\ x_5 \end{bmatrix} = \mathbf{B}^{-1}\mathbf{b} = \begin{bmatrix} 1 & 0 \\ 0 & 1 \end{bmatrix}\begin{bmatrix} 1 \\ 2 \end{bmatrix} = \begin{bmatrix} 1 \\ 2 \end{bmatrix}$$
$$\mathbf{x_N} = (x_1, x_2, x_3) = (0, 0, 0)$$
$$\mathbf{x} = (0, 0, 0, 1, 2)$$

Step 1. Testing for optimality:

$$\begin{aligned} \mathbf{c_N} - \mathbf{c_B}\mathbf{B}^{-1}\mathbf{N} &= (-2, -3, -1) - (0, 0)\begin{bmatrix} 1 & 0 \\ 0 & 1 \end{bmatrix}\begin{bmatrix} 1 & -1 & 2 \\ 4 & 2 & -1 \end{bmatrix} \\ &= (-2, -3, -1) - (0, 0)\begin{bmatrix} 1 & -1 & 2 \\ 4 & 2 & -1 \end{bmatrix} \\ &= (-2, -3, -1) - (0, 0, 0) \\ &= (-2, -3, -1) \qquad \text{so} \quad j^* = 2 \end{aligned}$$

Step 2. Computing the direction:

$$\mathbf{d_B} = \begin{bmatrix} d_4 \\ d_5 \end{bmatrix} = -\mathbf{B}^{-1}\mathbf{N}_{.j^*} = -\begin{bmatrix} 1 & 0 \\ 0 & 1 \end{bmatrix}\begin{bmatrix} -1 \\ 2 \end{bmatrix} = \begin{bmatrix} 1 \\ -2 \end{bmatrix}$$
$$\mathbf{d_N} = (\mathbf{d}_1, \mathbf{d}_2, \mathbf{d}_3) = \mathbf{I}_{.j^*} = \mathbf{I}_{.2} = (0, 1, 0)$$
$$\mathbf{d} = (0, 1, 0, 1, -2)$$

Step 3. Computing t^*:

$$\begin{aligned} t^* &= \min\{-(\mathbf{x_B})_k/(\mathbf{d_B})_k : (\mathbf{d_B})_k < 0\} = \min\{-(\mathbf{x_B})_2/(\mathbf{d_B})_2\} \\ &= \min\{-2/(-2)\} \\ &= 1 \qquad \text{so} \quad k^* = 2 \end{aligned}$$

Step 4. Pivoting: Column $k^* = 2$ of $\mathbf{B}$ is replaced with column $j^* = 2$ of $\mathbf{N}$, so

$$\mathbf{B}' = \begin{bmatrix} 1 & -1 \\ 0 & 2 \end{bmatrix}$$

$$\mathbf{N}' = \begin{bmatrix} 1 & 0 & 2 \\ 4 & 1 & -1 \end{bmatrix}$$

$$\mathbf{x}' = \mathbf{x} + t^*\mathbf{d} = (0,0,0,1,2) + 1(0,1,0,1,-2)$$
$$= (0,1,0,2,0)$$

Iteration 2

Step 0. The bfs for this iteration:

$$\mathbf{B} = [\mathbf{A}_{\cdot 4}, \mathbf{A}_{\cdot 2}] = \begin{bmatrix} 1 & -1 \\ 0 & 2 \end{bmatrix}$$

$$\mathbf{N} = [\mathbf{A}_{\cdot 1}, \mathbf{A}_{\cdot 5}, \mathbf{A}_{\cdot 3}] = \begin{bmatrix} 1 & 0 & 2 \\ 4 & 1 & -1 \end{bmatrix}$$

$$\mathbf{x_B} = \begin{bmatrix} x_4 \\ x_2 \end{bmatrix} = \mathbf{B}^{-1}\mathbf{b} = \begin{bmatrix} 1 & \frac{1}{2} \\ 0 & \frac{1}{2} \end{bmatrix}\begin{bmatrix} 1 \\ 2 \end{bmatrix} = \begin{bmatrix} 2 \\ 1 \end{bmatrix}$$

$$\mathbf{x_N} = (x_1, x_5, x_3) = (0,0,0)$$

$$\mathbf{x} = (0,1,0,2,0)$$

Step 1. Testing for optimality:

$$\mathbf{c_N} - \mathbf{c_B}\mathbf{B}^{-1}\mathbf{N} = (-2,0,-1) - (0,-3)\begin{bmatrix} 1 & \frac{1}{2} \\ 0 & \frac{1}{2} \end{bmatrix}\begin{bmatrix} 1 & 0 & 2 \\ 4 & 1 & -1 \end{bmatrix}$$

$$= (-2,0,-1) - (0,-\tfrac{3}{2})\begin{bmatrix} 1 & 0 & 2 \\ 4 & 1 & -1 \end{bmatrix}$$

$$= (-2,0,-1) - (-6,-\tfrac{3}{2},\tfrac{3}{2})$$

$$= (4,\tfrac{3}{2},-\tfrac{5}{2}) \qquad \text{so} \quad j^* = 3$$

Step 2. Computing the direction:

$$\mathbf{d_B} = \begin{bmatrix} d_4 \\ d_2 \end{bmatrix} = -\mathbf{B}^{-1}\mathbf{N}_{\cdot j^*} = -\begin{bmatrix} 1 & \frac{1}{2} \\ 0 & \frac{1}{2} \end{bmatrix}\begin{bmatrix} 2 \\ -1 \end{bmatrix} = \begin{bmatrix} -\frac{3}{2} \\ \frac{1}{2} \end{bmatrix}$$

$$\mathbf{d_N} = (d_1, d_5, d_3) = \mathbf{I}_{\cdot j^*} = (0,0,1)$$

$$\mathbf{d} = (0,\tfrac{1}{2},1,-\tfrac{3}{2},0)$$

Step 3. Computing t^*:

$$\begin{aligned} t^* &= \min\{-(\mathbf{x_B})_k/(\mathbf{d_B})_k : (\mathbf{d_B})_k < 0\} = \min\{-(\mathbf{x_B})_1/(\mathbf{d_B})_1\} \\ &= \min\{2/\tfrac{3}{2}\} \\ &= \tfrac{4}{3} \qquad \text{so} \quad k^* = 1 \end{aligned}$$

Step 4. Pivoting: Column $k^* = 1$ of $\mathbf{B}$ is replaced with column $j^* = 3$ of $\mathbf{N}$, so

$$\mathbf{B}' = \begin{bmatrix} 2 & -1 \\ -1 & 2 \end{bmatrix}$$

$$\mathbf{N}' = \begin{bmatrix} 1 & 0 & 1 \\ 4 & 1 & 0 \end{bmatrix}$$

$$\begin{aligned} \mathbf{x}' &= \mathbf{x} + t^*\mathbf{d} \\ &= (0,1,0,2,0) + \tfrac{4}{3}(0,\tfrac{1}{2},1,-\tfrac{3}{2},0) \\ &= (0,\tfrac{5}{3},\tfrac{4}{3},0,0) \end{aligned}$$

Iteration 3

Step 0. The bfs for this iteration:

$$\mathbf{B} = [\mathbf{A}_{\cdot 3}, \mathbf{A}_{\cdot 2}] = \begin{bmatrix} 2 & -1 \\ -1 & 2 \end{bmatrix}$$

$$\mathbf{N} = [\mathbf{A}_{\cdot 1}, \mathbf{A}_{\cdot 5}, \mathbf{A}_{\cdot 4}] = \begin{bmatrix} 1 & 0 & 1 \\ 4 & 1 & 0 \end{bmatrix}$$

$$\mathbf{x_B} = \begin{bmatrix} x_3 \\ x_2 \end{bmatrix} = \mathbf{B}^{-1}\mathbf{b} = \begin{bmatrix} \frac{2}{3} & \frac{1}{3} \\ \frac{1}{3} & \frac{2}{3} \end{bmatrix}\begin{bmatrix} 1 \\ 2 \end{bmatrix} = \begin{bmatrix} \frac{4}{3} \\ \frac{5}{3} \end{bmatrix}$$

$$\mathbf{x_N} = (x_1, x_5, x_4) = (0,0,0)$$

$$\mathbf{x} = (0,\tfrac{5}{3},\tfrac{4}{3},0,0)$$

Step 1. Testing for optimality:

$$\begin{aligned} \mathbf{c_N} - \mathbf{c_B}\mathbf{B}^{-1}\mathbf{N} &= (-2,0,0) - (-1,-3)\begin{bmatrix} \frac{2}{3} & \frac{1}{3} \\ \frac{1}{3} & \frac{2}{3} \end{bmatrix}\begin{bmatrix} 1 & 0 & 1 \\ 4 & 1 & 0 \end{bmatrix} \\ &= (-2,0,0) - (-\tfrac{5}{3},-\tfrac{7}{3})\begin{bmatrix} 1 & 0 & 1 \\ 4 & 1 & 0 \end{bmatrix} \\ &= (-2,0,0) - (-11,-\tfrac{7}{3},-\tfrac{5}{3}) \\ &= (9,\tfrac{7}{3},\tfrac{5}{3}) \end{aligned}$$

Since $\mathbf{c_N} - \mathbf{c_B}\mathbf{B}^{-1}\mathbf{N} \geq \mathbf{0}$, the current solution $\mathbf{x} = (0, \frac{5}{3}, \frac{4}{3}, 0, 0)$ is optimal. The objective value is $\mathbf{cx} = -\frac{19}{3}$.

EXAMPLE 5.3:

$$\begin{array}{lrrrrrl}
\text{Minimize} & -3x_1 & -x_2 & -4x_3 & & & \\
\text{subject to} & x_1 & -x_2 & +2x_3 & +x_4 & & =2 \\
 & 3x_1 & -x_2 & +2x_3 & & +x_5 & =1 \\
 & x_1, & x_2, & x_3, & x_4, & x_5 & \geq 0
\end{array}$$

Thus

$$\mathbf{c} = (-3, -1, -4, 0, 0)$$

$$\mathbf{A} = \begin{bmatrix} 1 & -1 & 2 & 1 & 0 \\ 3 & -1 & 2 & 0 & 1 \end{bmatrix}$$

$$\mathbf{b} = \begin{bmatrix} 2 \\ 1 \end{bmatrix}$$

Iteration 1

Step 0. The bfs for this iteration:

$$\mathbf{B} = [\mathbf{A}_{.4}, \mathbf{A}_{.5}] = \begin{bmatrix} 1 & 0 \\ 0 & 1 \end{bmatrix}$$

$$\mathbf{N} = [\mathbf{A}_{.1}, \mathbf{A}_{.2}, \mathbf{A}_{.3}] = \begin{bmatrix} 1 & -1 & 2 \\ 3 & -1 & 2 \end{bmatrix}$$

$$\mathbf{x_B} = \begin{bmatrix} x_4 \\ x_5 \end{bmatrix} = \mathbf{B}^{-1}\mathbf{b} = \begin{bmatrix} 1 & 0 \\ 0 & 1 \end{bmatrix}\begin{bmatrix} 2 \\ 1 \end{bmatrix} = \begin{bmatrix} 2 \\ 1 \end{bmatrix}$$

$$\mathbf{x_N} = (x_1, x_2, x_3) = (0, 0, 0)$$

$$\mathbf{x} = (0, 0, 0, 2, 1)$$

Step 1. Testing for optimality:

$$\mathbf{c_N} - \mathbf{c_B}\mathbf{B}^{-1}\mathbf{N} = (-3, -1, -4) - (0, 0)\begin{bmatrix} 1 & 0 \\ 0 & 1 \end{bmatrix}\begin{bmatrix} 1 & -1 & 2 \\ 3 & -1 & 2 \end{bmatrix}$$

$$= (-3, -1, -4) \qquad \text{so} \quad j^* = 3$$

Step 2. Computing the direction:

$$\mathbf{d_B} = \begin{bmatrix} d_4 \\ d_5 \end{bmatrix} = -\mathbf{B}^{-1}\mathbf{N}_{.j^*} = -\begin{bmatrix} 1 & 0 \\ 0 & 1 \end{bmatrix}\begin{bmatrix} 2 \\ 2 \end{bmatrix} = \begin{bmatrix} -2 \\ -2 \end{bmatrix}$$

$$\mathbf{d_N} = (d_1, d_2, d_3) = \mathbf{I}_{.j^*} = (0, 0, 1)$$

$$\mathbf{d} = (0, 0, 1, -2, -2)$$

Step 3. Computing t^*:

$$\begin{aligned} t^* &= \min\{-(\mathbf{x_B})_k/(\mathbf{d_B})_k : (\mathbf{d_B})_k < 0\} \\ &= \min\{-(\mathbf{x_B})_1/(\mathbf{d_B})_1, -(\mathbf{x_B})_2/(\mathbf{d_B})_2\} \\ &= \min\{-2/(-2), -1/(-2)\} \\ &= \tfrac{1}{2} \qquad \text{so} \quad k^* = 2 \end{aligned}$$

Step 4. Pivoting: Column $k^* = 2$ of $\mathbf{B}$ is replaced with column $j^* = 3$ of $\mathbf{N}$, so

$$\mathbf{B}' = \begin{bmatrix} 1 & 2 \\ 0 & 2 \end{bmatrix}$$

$$\mathbf{N}' = \begin{bmatrix} 1 & -1 & 0 \\ 3 & -1 & 1 \end{bmatrix}$$

$$\begin{aligned} \mathbf{x}' &= \mathbf{x} + t^*\mathbf{d} \\ &= (0,0,0,2,1) + (\tfrac{1}{2})(0,0,1,-2,-2) \\ &= (0,0,\tfrac{1}{2},1,0) \end{aligned}$$

Iteration 2

Step 0. The bfs for this iteration:

$$\mathbf{B} = [\mathbf{A}_{.4}, \mathbf{A}_{.3}] = \begin{bmatrix} 1 & 2 \\ 0 & 2 \end{bmatrix}$$

$$\mathbf{N} = [\mathbf{A}_{.1}, \mathbf{A}_{.2}, \mathbf{A}_{.5}] = \begin{bmatrix} 1 & -1 & 0 \\ 3 & -1 & 1 \end{bmatrix}$$

$$\mathbf{x_B} = \begin{bmatrix} x_4 \\ x_3 \end{bmatrix} = \mathbf{B}^{-1}\mathbf{b} = \begin{bmatrix} 1 & -1 \\ 0 & \tfrac{1}{2} \end{bmatrix}\begin{bmatrix} 2 \\ 1 \end{bmatrix} = \begin{bmatrix} 1 \\ \tfrac{1}{2} \end{bmatrix}$$

$$\mathbf{x_N} = (x_1, x_2, x_5) = (0,0,0)$$

$$\mathbf{x} = (0,0,\tfrac{1}{2},1,0)$$

Step 1. Testing for optimality:

$$\begin{aligned} \mathbf{c_N} - \mathbf{c_B}\mathbf{B}^{-1}\mathbf{N} &= (-3,-1,0) - (0,-4)\begin{bmatrix} 1 & -1 \\ 0 & \tfrac{1}{2} \end{bmatrix}\begin{bmatrix} 1 & -1 & 0 \\ 3 & -1 & 1 \end{bmatrix} \\ &= (-3,-1,0) - (0,-2)\begin{bmatrix} 1 & -1 & 0 \\ 3 & -1 & 1 \end{bmatrix} \\ &= (-3,-1,0) - (-6,2,-2) \\ &= (3,-3,2) \qquad \text{so} \quad j^* = 2 \end{aligned}$$

Step 2. Computing the direction:

$$\mathbf{d_B} = \begin{bmatrix} d_4 \\ d_3 \end{bmatrix} = -\mathbf{B}^{-1}\mathbf{N}_{.j*} = -\begin{bmatrix} 1 & -1 \\ 0 & \frac{1}{2} \end{bmatrix}\begin{bmatrix} -1 \\ -1 \end{bmatrix} = \begin{bmatrix} 0 \\ \frac{1}{2} \end{bmatrix}$$

$$\mathbf{d_N} = (d_1, d_2, d_5) = \mathbf{I}_{.j*} = (0,1,0)$$

Since $\mathbf{d_B} \geq \mathbf{0}$, the LP is unbounded, and the direction of unboundedness is $\mathbf{d} = (0, 1, \frac{1}{2}, 0, 0)$.

EXAMPLE 5.4:

$$\begin{array}{lrrrrrrl} \text{Minimize} & -2x_1 & -3x_2 & -4x_3 & & & & \\ \text{subject to} & x_1 & +\ x_2 & +2x_3 & +x_4 & & & =2 \\ & x_1 & +4x_2 & -\ x_3 & & +x_5 & & =1 \\ & x_1 & +2x_2 & -4x_3 & & & +x_6 & =1 \\ & x_1, & x_2, & x_3, & x_4, & x_5, & x_6 & \geq 0 \end{array}$$

Thus

$$\mathbf{c} = (-2, -3, -4, 0, 0, 0)$$

$$\mathbf{A} = \begin{bmatrix} 1 & 1 & 2 & 1 & 0 & 0 \\ 1 & 4 & -1 & 0 & 1 & 0 \\ 1 & 2 & -4 & 0 & 0 & 1 \end{bmatrix}$$

$$\mathbf{b} = \begin{bmatrix} 2 \\ 1 \\ 1 \end{bmatrix}$$

Iteration 1

Step 0. The bfs for this iteration:

$$\mathbf{B} = [\mathbf{A}_{.4}, \mathbf{A}_{.5}, \mathbf{A}_{.6}] = \begin{bmatrix} 1 & 0 & 0 \\ 0 & 1 & 0 \\ 0 & 0 & 1 \end{bmatrix}$$

$$\mathbf{N} = [\mathbf{A}_{.1}, \mathbf{A}_{.2}, \mathbf{A}_{.3}] = \begin{bmatrix} 1 & 1 & 2 \\ 1 & 4 & -1 \\ 1 & 2 & -4 \end{bmatrix}$$

$$\mathbf{x_B} = \begin{bmatrix} x_4 \\ x_5 \\ x_6 \end{bmatrix} = \mathbf{B}^{-1}\mathbf{b} = \begin{bmatrix} 1 & 0 & 0 \\ 0 & 1 & 0 \\ 0 & 0 & 1 \end{bmatrix}\begin{bmatrix} 2 \\ 1 \\ 1 \end{bmatrix} = \begin{bmatrix} 2 \\ 1 \\ 1 \end{bmatrix}$$

$$\mathbf{x_N} = (x_1, x_2, x_3) = (0,0,0)$$

$$\mathbf{x} = (0,0,0,2,1,1)$$

Step 1. Testing for optimality:

$$\mathbf{c_N} - \mathbf{c_B}\mathbf{B}^{-1}\mathbf{N} = (-2, -3, -4) - (0,0,0)\begin{bmatrix} 1 & 0 & 0 \\ 0 & 1 & 0 \\ 0 & 0 & 1 \end{bmatrix}\begin{bmatrix} 1 & 1 & 2 \\ 1 & 4 & -1 \\ 1 & 2 & -4 \end{bmatrix}$$

$$= (-2, -3, -4) - (0,0,0)\begin{bmatrix} 1 & 1 & 2 \\ 1 & 4 & -1 \\ 1 & 2 & -4 \end{bmatrix}$$

$$= (-2, -3, -4) - (0,0,0)$$

$$= (-2, -3, -4) \qquad \text{so} \quad j^* = 3$$

Step 2. Computing the direction:

$$\mathbf{d_B} = \begin{bmatrix} d_4 \\ d_5 \\ d_6 \end{bmatrix} = -\mathbf{B}^{-1}\mathbf{N}_{.j^*} = -\begin{bmatrix} 1 & 0 & 0 \\ 0 & 1 & 0 \\ 0 & 0 & 1 \end{bmatrix}\begin{bmatrix} 2 \\ -1 \\ -4 \end{bmatrix} = \begin{bmatrix} -2 \\ 1 \\ 4 \end{bmatrix}$$

$$\mathbf{d_N} = (d_1, d_2, d_3) = \mathbf{I}_{.j^*} = \mathbf{I}_{.3} = (0,0,1)$$

$$\mathbf{d} = (0,0,1,-2,1,4)$$

Step 3. Computing t^*:

$$t^* = \min\{-(\mathbf{x_B})_k/(\mathbf{d_B})_k : (\mathbf{d_B})_k < 0\}$$

$$= \min\{-(\mathbf{x_B})_1/(\mathbf{d_B})_1\} = \min\{2/2\}$$

$$= 1 \qquad \text{so} \quad k^* = 1$$

Step 4. Pivoting: Column $k^* = 1$ of **B** is replaced with column $j^* = 3$ of **N**, so

$$\mathbf{B}' = \begin{bmatrix} 2 & 0 & 0 \\ -1 & 1 & 0 \\ -4 & 0 & 1 \end{bmatrix}$$

$$\mathbf{N}' = \begin{bmatrix} 1 & 1 & 1 \\ 1 & 4 & 0 \\ 1 & 2 & 0 \end{bmatrix}$$

$$\mathbf{x}' = \mathbf{x} + t^*\mathbf{d} = (0,0,0,2,1,1) + 1(0,0,1,-2,1,4)$$

$$= (0,0,1,0,2,5)$$

Iteration 2

Step 0. The bfs for this iteration:

$$\mathbf{B}=[\mathbf{A}_{\cdot 3},\mathbf{A}_{\cdot 5},\mathbf{A}_{\cdot 6}]=\begin{bmatrix} 2 & 0 & 0 \\ -1 & 1 & 0 \\ -4 & 0 & 1 \end{bmatrix}$$

$$\mathbf{N}=[\mathbf{A}_{\cdot 1},\mathbf{A}_{\cdot 2},\mathbf{A}_{\cdot 4}]=\begin{bmatrix} 1 & 1 & 1 \\ 1 & 4 & 0 \\ 1 & 2 & 0 \end{bmatrix}$$

$$\mathbf{x_B}=\begin{bmatrix} x_3 \\ x_5 \\ x_6 \end{bmatrix}=\mathbf{B}^{-1}\mathbf{b}=\begin{bmatrix} \frac{1}{2} & 0 & 0 \\ \frac{1}{2} & 1 & 0 \\ 2 & 0 & 1 \end{bmatrix}\begin{bmatrix} 2 \\ 1 \\ 1 \end{bmatrix}=\begin{bmatrix} 1 \\ 2 \\ 5 \end{bmatrix}$$

$$\mathbf{x_N}=(x_1,x_2,x_4)=(0,0,0)$$

$$\mathbf{x}=(0,0,1,0,2,5)$$

Step 1. Testing for optimality:

$$\mathbf{c_N}-\mathbf{c_B}\mathbf{B}^{-1}\mathbf{N}=(-2,-3,0)-(-4,0,0)\begin{bmatrix} \frac{1}{2} & 0 & 0 \\ \frac{1}{2} & 1 & 0 \\ 2 & 0 & 1 \end{bmatrix}\begin{bmatrix} 1 & 1 & 1 \\ 1 & 4 & 0 \\ 1 & 2 & 0 \end{bmatrix}$$

$$=(-2,-3,0)-(-2,0,0)\begin{bmatrix} 1 & 1 & 1 \\ 1 & 4 & 0 \\ 1 & 2 & 0 \end{bmatrix}$$

$$=(-2,-3,0)-(-2,-2,-2)$$

$$=(0,-1,2) \qquad \text{so} \quad j^*=2$$

Step 2. Computing the direction:

$$\mathbf{d_B}=\begin{bmatrix} d_3 \\ d_5 \\ d_6 \end{bmatrix}=-\mathbf{B}^{-1}\mathbf{N}_{\cdot j^*}=-\begin{bmatrix} \frac{1}{2} & 0 & 0 \\ \frac{1}{2} & 1 & 0 \\ 2 & 0 & 1 \end{bmatrix}\begin{bmatrix} 1 \\ 4 \\ 2 \end{bmatrix}=\begin{bmatrix} -\frac{1}{2} \\ -\frac{9}{2} \\ -4 \end{bmatrix}$$

$$\mathbf{d_N}=(d_1,d_2,d_4)=\mathbf{I}_{\cdot j^*}=(0,1,0)$$

$$\mathbf{d}=(0,1,-\tfrac{1}{2},0,-\tfrac{9}{2},-4)$$

Step 3. Computing t^*:

$$t^*=\min\{-(\mathbf{x_B})_k/(\mathbf{d_B})_k : (\mathbf{d_B})_k<0\}$$

$$=\min\{-(\mathbf{x_B})_1/(\mathbf{d_B})_1,-(\mathbf{x_B})_2/(\mathbf{d_B})_2,-(\mathbf{x_B})_3/(\mathbf{d_B})_3\}$$

$$=\min\{1/\tfrac{1}{2},2/\tfrac{9}{2},\tfrac{5}{4}\}$$

$$=\tfrac{4}{9} \qquad \text{so} \quad k^*=2$$

Step 4. Pivoting: Column $k^* = 2$ of $\mathbf{B}$ is replaced with column $j^* = 2$ of $\mathbf{N}$, so

$$\mathbf{B}' = \begin{bmatrix} 2 & 1 & 0 \\ -1 & 4 & 0 \\ -4 & 2 & 1 \end{bmatrix}$$

$$\mathbf{N}' = \begin{bmatrix} 1 & 0 & 1 \\ 1 & 1 & 0 \\ 1 & 0 & 0 \end{bmatrix}$$

$$\begin{aligned} \mathbf{x}' &= \mathbf{x} + t^*\mathbf{d} \\ &= (0,0,1,0,2,5) + \tfrac{4}{9}(0,1,-\tfrac{1}{2},0,-\tfrac{9}{2},-4) \\ &= (0,\tfrac{4}{9},\tfrac{7}{9},0,0,\tfrac{29}{9}) \end{aligned}$$

Iteration 3

Step 0. The bfs for this iteration:

$$\mathbf{B} = [\mathbf{A}_{\cdot 3}, \mathbf{A}_{\cdot 2}, \mathbf{A}_{\cdot 6}] = \begin{bmatrix} 2 & 1 & 0 \\ -1 & 4 & 0 \\ -4 & 2 & 1 \end{bmatrix}$$

$$\mathbf{N} = [\mathbf{A}_{\cdot 1}, \mathbf{A}_{\cdot 5}, \mathbf{A}_{\cdot 4}] = \begin{bmatrix} 1 & 0 & 1 \\ 1 & 1 & 0 \\ 1 & 0 & 0 \end{bmatrix}$$

$$\mathbf{x_B} = \begin{bmatrix} x_3 \\ x_2 \\ x_6 \end{bmatrix} = \mathbf{B}^{-1}\mathbf{b} = \begin{bmatrix} \frac{4}{9} & -\frac{1}{9} & 0 \\ \frac{1}{9} & \frac{2}{9} & 0 \\ \frac{14}{9} & -\frac{8}{9} & 1 \end{bmatrix} \begin{bmatrix} 2 \\ 1 \\ 1 \end{bmatrix} = \begin{bmatrix} \frac{7}{9} \\ \frac{4}{9} \\ \frac{29}{9} \end{bmatrix}$$

$$\mathbf{x_N} = (x_1, x_5, x_4) = (0,0,0)$$

$$\mathbf{x} = (0,\tfrac{4}{9},\tfrac{7}{9},0,0,\tfrac{29}{9})$$

Step 1. Testing for optimality:

$$\mathbf{c_N} - \mathbf{c_B}^{-1}\mathbf{N} = (-2,0,0) - (-4,-3,0) \begin{bmatrix} \frac{4}{9} & -\frac{1}{9} & 0 \\ \frac{1}{9} & \frac{2}{9} & 0 \\ \frac{14}{9} & -\frac{8}{9} & 1 \end{bmatrix} \begin{bmatrix} 1 & 0 & 1 \\ 1 & 1 & 0 \\ 1 & 0 & 0 \end{bmatrix}$$

$$= (-2,0,0) - (-\tfrac{19}{9}, -\tfrac{2}{9}, 0) \begin{bmatrix} 1 & 0 & 1 \\ 1 & 1 & 0 \\ 1 & 0 & 0 \end{bmatrix}$$

$$= (-2,0,0) - (-\tfrac{21}{9}, -\tfrac{2}{9}, -\tfrac{19}{9})$$

$$= (\tfrac{1}{3}, \tfrac{2}{9}, \tfrac{19}{9})$$

Since $\mathbf{c_N} - \mathbf{c_B B^{-1} N} \geq \mathbf{0}$, the current solution $\mathbf{x} = (0, \frac{4}{9}, \frac{7}{9}, 0, 0, \frac{29}{9})$ is optimal, and the objective value is $\mathbf{cx} = -\frac{40}{9}$.

This section has reviewed and illustrated the steps of the simplex algorithm. A flow chart summary is given in Figure 5.6. The final section of this chapter will address the issue of what to do when $t^* = 0$.

FIGURE 5.6. Flow chart of the simplex algorithm.

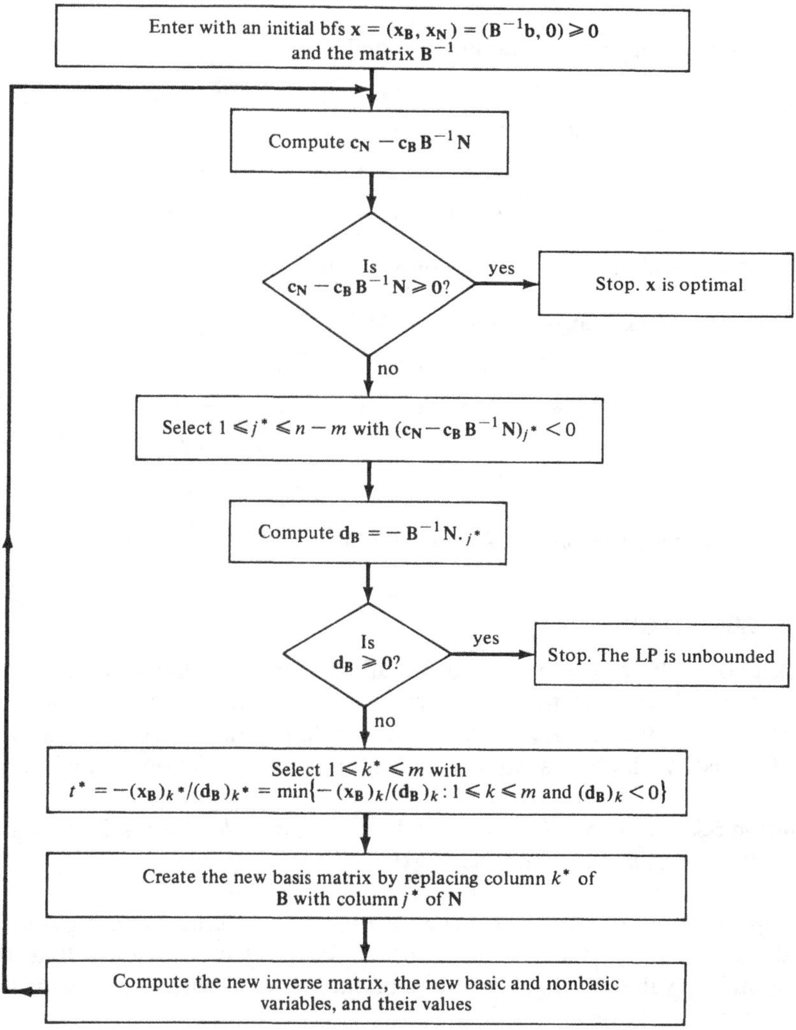

EXERCISES

Note: In these exercises, use the rule of steepest descent.

5.6.1. Use the simplex algorithm to solve the LP

$$\begin{array}{rrrrrl} \text{Minimize} & x_1 & -2x_2 & -3x_3 & +x_4 & \\ \text{subject to} & x_1 & +2x_2 & -3x_3 & & =3 \\ & & -3x_2 & +5x_3 & +x_4 & =2 \\ & x_1, & x_2, & x_3, & x_4 & \geq 0 \end{array}$$

starting with the initial bfs $\mathbf{x} = (3,0,0,2)$.

5.6.2. Use the simplex algorithm to solve the LP

$$\begin{array}{rrrrrrl} \text{Minimize} & x_1 & +x_2 & -4x_3 & +x_4 & -5x_5 & \\ \text{subject to} & 2x_1 & +x_2 & -\ x_3 & & +4x_5 & =8 \\ & x_1 & & +2x_3 & +x_4 & -\ x_5 & =2 \\ & x_1, & x_2, & x_3, & x_4, & x_5 & \geq 0 \end{array}$$

starting with the initial bfs $\mathbf{x} = (0,8,0,2,0)$.

5.6.3. Use the simplex algorithm to solve the LP

$$\begin{array}{rrrrrl} \text{Minimize} & x_1 & -\ x_2 & -8x_3 & +x_4 & \\ \text{subject to} & x_1 & -\ x_2 & -\ x_3 & & =3 \\ & & -3x_2 & +2x_3 & +x_4 & =2 \\ & x_1, & x_2, & x_3, & x_4 & \geq 0 \end{array}$$

starting with the initial bfs $\mathbf{x} = (3,0,0,2)$.

5.7. DEGENERACY

This section will resolve once and for all the problem that can be caused when $t^* = 0$. Recall, from the min ratio test of Section 5.4, that the only way t^* can be 0 is if $(\mathbf{x_B})_{k^*} = 0$ and $(\mathbf{d_B})_{k^*} < 0$. In other words, t^* can be 0 only if a basic variable has value 0, thus motivating the following definition.

Definition 5.2. A bfs $\mathbf{x} = (\mathbf{x_B}, \mathbf{x_N}) = (\mathbf{B}^{-1}\mathbf{b}, \mathbf{0}) \geq \mathbf{0}$ is *degenerate* if there is a k with $1 \leq k \leq m$ such that $(\mathbf{x_B})_k = 0$.

It is important to note that, just because a bfs is degenerate (say, $(\mathbf{x_B})_k = 0$), this does not mean that t^* is going to be 0. The reason is that the corresponding value of $(\mathbf{d_B})_k$ might be nonnegative, in which case $(\mathbf{x_B})_k$ will not be involved in the min ratio test.

In two dimensions, the geometry of degeneracy is very revealing. Consider the feasible region depicted in Figure 5.7 and imagine that you are standing at the extreme point **x**. If the simplex algorithm happened to produce the direction **d** pointing straight up, as is shown in Figure 5.7, then it should come as no surprise that in this event $t^* = 0$ because the largest t that can be made before $\mathbf{x} + t\mathbf{d}$ leaves the feasible region is 0.

Figure 5.7 shows how degeneracy can arise; it also reveals why the difficulty can potentially be resolved. If one could somehow choose the direction indicated in Figure 5.8 instead of the one in Figure 5.7, then everything would work out fine, and t^* would be positive. The question is how to go about obtaining that direction. Such a method is the ultimate goal of this section, but first it is important to understand precisely what difficulties can arise in the simplex algorithm if $t^* = 0$.

Imagine that $\mathbf{x} = (\mathbf{x_B}, \mathbf{x_N}) = (\mathbf{B}^{-1}\mathbf{b}, \mathbf{0}) \geq \mathbf{0}$ is a bfs that has failed the test for optimality and that, as a result, a direction $\mathbf{d} = (\mathbf{d_B}, \mathbf{d_N}) = (-\mathbf{B}^{-1}\mathbf{N}_{.j^*}, \mathbf{I}_{.j^*})$ has been computed. In going from **x** to $\mathbf{x} + t^*\mathbf{d}$, no movement is taking place if $t^* = 0$. Thus you will end up at the same bfs **x** with the same objective value when, in fact, you wanted to move to a new bfs with a strictly smaller objective value. Although it is true that geometrically you are standing still when $t^* = 0$, algebraically you are not. Recall from Section 5.6, that, even though $t^* = 0$, it is still possible to perform the actual pivot operation by entering $\mathbf{N}_{.j^*}$ into the basis matrix **B** and simultaneously removing column k^*. Exchanging these columns would produce new matrices $\mathbf{B}'$ and $\mathbf{N}'$, although geometrically you would remain at the same point **x**. Both **B** and $\mathbf{B}'$ can be basis matrices for **x**, because the definition of a bfs never required that a basis matrix had to be unique—in fact, there could be still other basis matrices corresponding to **x**.

Nevertheless, it is reasonable to ask what happens if one simply proceeds with the simplex algorithm starting with the new matrices $[\mathbf{B}', \mathbf{N}']$, having

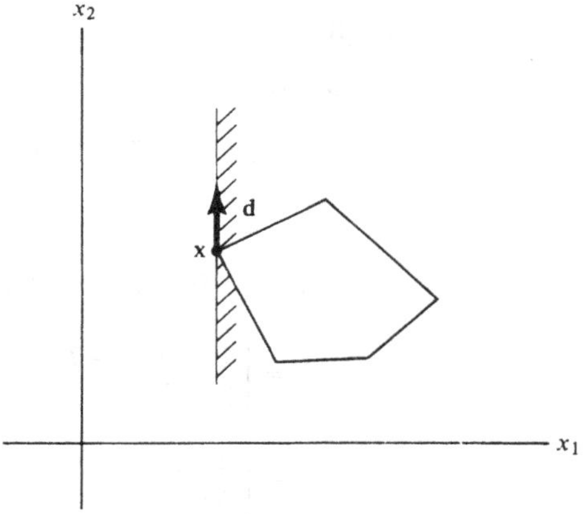

FIGURE 5.7. A direction causing t^* to be 0.

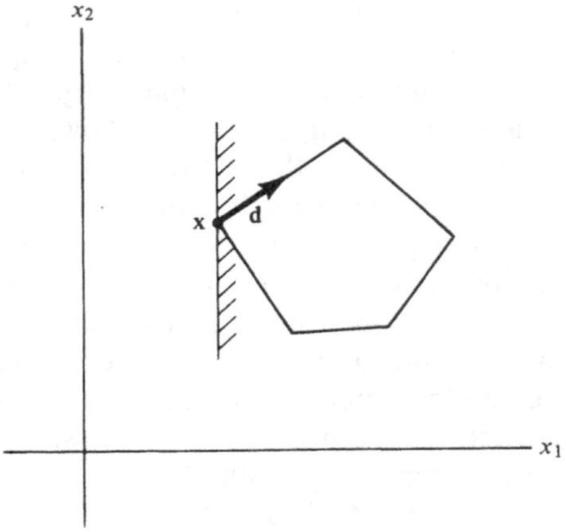

FIGURE 5.8. A direction that avoids $t^* = 0$.

just moved algebraically from $[\mathbf{B}, \mathbf{N}]$. Although it is technically possible to continue with the simplex algorithm, a serious problem can arise. Suppose that from $[\mathbf{B}', \mathbf{N}']$ the simplex algorithm leads to yet a third basis matrix $\mathbf{B}''$ for the same bfs $\mathbf{x}$. Once again it is possible to continue with the algorithm, but disaster can be imminent. What would happen if, from $[\mathbf{B}'', \mathbf{N}'']$, you were led back to the first basis matrix $\mathbf{B}$? In this case you would have traveled algebraically from $\mathbf{B}$ to $\mathbf{B}'$ to $\mathbf{B}''$ and come full circle to $\mathbf{B}$. Nothing says that this is going to happen, but, if it should, then you would forever *cycle* from $\mathbf{B}$ to $\mathbf{B}'$ to $\mathbf{B}''$ and back to $\mathbf{B}$. The situation of cycling destroys the entire finiteness of the simplex algorithm and must be avoided at all costs. The next example serves to demonstrate this undesirable sequence of events.

EXAMPLE 5.5:

$$
\begin{array}{llllllll}
\text{Minimize} & & & -2x_3 & -2x_4 & +8x_5 & +2x_6 & \\
\text{subject to} & x_1 & & -7x_3 & -3x_4 & +7x_5 & +2x_6 & = 0 \\
 & & x_2 & +2x_3 & +\ x_4 & -3x_5 & -\ x_6 & = 0 \\
 & x_1, & x_2, & x_3, & x_4, & x_5, & x_6 & \geq 0
\end{array}
$$

$$\mathbf{c} = (0, 0, -2, -2, 8, 2)$$

$$\mathbf{A} = \begin{bmatrix} 1 & 0 & -7 & -3 & 7 & 2 \\ 0 & 1 & 2 & 1 & -3 & -1 \end{bmatrix}$$

$$\mathbf{b} = \begin{bmatrix} 0 \\ 0 \end{bmatrix}$$

Iteration 1

$$\mathbf{B} = [\mathbf{A}_{\cdot 1}, \mathbf{A}_{\cdot 2}] = \begin{bmatrix} 1 & 0 \\ 0 & 1 \end{bmatrix}$$

$$\mathbf{N} = [\mathbf{A}_{\cdot 3}, \mathbf{A}_{\cdot 4}, \mathbf{A}_{\cdot 5}, \mathbf{A}_{\cdot 6}] = \begin{bmatrix} -7 & -3 & 7 & 2 \\ 2 & 1 & -3 & -1 \end{bmatrix}$$

$$\mathbf{x} = (0,0,0,0,0,0)$$

$$\mathbf{c_N} - \mathbf{c_B}\mathbf{B}^{-1}\mathbf{N} = (-2,-2,8,2) - (0,0)\begin{bmatrix} 1 & 0 \\ 0 & 1 \end{bmatrix}\begin{bmatrix} -7 & -3 & 7 & 2 \\ 2 & 1 & -3 & -1 \end{bmatrix}$$

$$= (-2,-2,8,2) \qquad \text{so choose} \quad j^* = 1$$

$$\mathbf{d_B} = -\mathbf{B}^{-1}\mathbf{N}_{\cdot j^*} = -\begin{bmatrix} 1 & 0 \\ 0 & 1 \end{bmatrix}\begin{bmatrix} -7 \\ 2 \end{bmatrix} = \begin{bmatrix} 7 \\ -2 \end{bmatrix}$$

Thus $k^* = 2$ [because only $(\mathbf{d_B})_2 < 0$].

Iteration 2

$$\mathbf{B} = [\mathbf{A}_{\cdot 1}, \mathbf{A}_{\cdot 3}] = \begin{bmatrix} 1 & -7 \\ 0 & 2 \end{bmatrix}$$

$$\mathbf{N} = [\mathbf{A}_{\cdot 2}, \mathbf{A}_{\cdot 4}, \mathbf{A}_{\cdot 5}, \mathbf{A}_{\cdot 6}] = \begin{bmatrix} 0 & -3 & 7 & 2 \\ 1 & 1 & -3 & -1 \end{bmatrix}$$

$$\mathbf{c_N} - \mathbf{c_B}\mathbf{B}^{-1}\mathbf{N} = (0,-2,8,2) - (0,-2)\begin{bmatrix} 1 & \frac{7}{2} \\ 0 & \frac{1}{2} \end{bmatrix}\begin{bmatrix} 0 & -3 & 7 & 2 \\ 1 & 1 & -3 & -1 \end{bmatrix}$$

$$= (1,-1,5,1) \qquad \text{so} \quad j^* = 2$$

$$\mathbf{d_B} = -\mathbf{B}^{-1}\mathbf{N}_{\cdot j^*} = -\begin{bmatrix} 1 & \frac{7}{2} \\ 0 & \frac{1}{2} \end{bmatrix}\begin{bmatrix} -3 \\ 1 \end{bmatrix} = \begin{bmatrix} -\frac{1}{2} \\ -\frac{1}{2} \end{bmatrix}$$

Choose $k^* = 1$ (since $\mathbf{x} = \mathbf{0}$).

Iteration 3

$$\mathbf{B} = [\mathbf{A}_{\cdot 4}, \mathbf{A}_{\cdot 3}] = \begin{bmatrix} -3 & -7 \\ 1 & 2 \end{bmatrix}$$

$$\mathbf{N} = [\mathbf{A}_{\cdot 2}, \mathbf{A}_{\cdot 1}, \mathbf{A}_{\cdot 5}, \mathbf{A}_{\cdot 6}] = \begin{bmatrix} 0 & 1 & 7 & 2 \\ 1 & 0 & -3 & -1 \end{bmatrix}$$

$$\mathbf{c_N} - \mathbf{c_B}\mathbf{B}^{-1}\mathbf{N} = (0,0,8,2) - (-2,-2)\begin{bmatrix} 2 & 7 \\ -1 & -3 \end{bmatrix}\begin{bmatrix} 0 & 1 & 7 & 2 \\ 1 & 0 & -3 & -1 \end{bmatrix}$$

$$= (8,2,-2,-2) \qquad \text{so choose} \quad j^* = 3$$

$$\mathbf{d_B} = -\mathbf{B}^{-1}\mathbf{N}_{\cdot j^*} = -\begin{bmatrix} 2 & 7 \\ -1 & -3 \end{bmatrix}\begin{bmatrix} 7 \\ -3 \end{bmatrix} = \begin{bmatrix} 7 \\ -2 \end{bmatrix}$$

Thus $k^* = 2$ [because only $(\mathbf{d_B})_2 < 0$].

Iteration 4

$$\mathbf{B} = [\mathbf{A}_{\cdot 4}, \mathbf{A}_{\cdot 5}] = \begin{bmatrix} -3 & 7 \\ 1 & -3 \end{bmatrix}$$

$$\mathbf{N} = [\mathbf{A}_{\cdot 2}, \mathbf{A}_{\cdot 1}, \mathbf{A}_{\cdot 3}, \mathbf{A}_{\cdot 6}] = \begin{bmatrix} 0 & 1 & -7 & 2 \\ 1 & 0 & 2 & -1 \end{bmatrix}$$

$$\mathbf{c_N} - \mathbf{c_B}\mathbf{B}^{-1}\mathbf{N} = (0,0,-2,2) - (-2,8)\begin{bmatrix} -\frac{3}{2} & -\frac{7}{2} \\ -\frac{1}{2} & -\frac{3}{2} \end{bmatrix}\begin{bmatrix} 0 & 1 & -7 & 2 \\ 1 & 0 & 2 & -1 \end{bmatrix}$$

$$= (5,1,1,-1) \qquad \text{so} \quad j^* = 4$$

$$\mathbf{d_B} = -\mathbf{B}^{-1}\mathbf{N}_{\cdot j^*} = -\begin{bmatrix} -\frac{3}{2} & -\frac{7}{2} \\ -\frac{1}{2} & -\frac{3}{2} \end{bmatrix}\begin{bmatrix} 2 \\ -1 \end{bmatrix} = \begin{bmatrix} -\frac{1}{2} \\ -\frac{1}{2} \end{bmatrix}$$

Choose $k^* = 1$ (since $\mathbf{x} = \mathbf{0}$).

Iteration 5

$$\mathbf{B} = [\mathbf{A}_{\cdot 6}, \mathbf{A}_{\cdot 5}] = \begin{bmatrix} 2 & 7 \\ -1 & -3 \end{bmatrix}$$

$$\mathbf{N} = [\mathbf{A}_{\cdot 2}, \mathbf{A}_{\cdot 1}, \mathbf{A}_{\cdot 3}, \mathbf{A}_{\cdot 4}] = \begin{bmatrix} 0 & 1 & -7 & -3 \\ 1 & 0 & 2 & 1 \end{bmatrix}$$

$$\mathbf{c_N} - \mathbf{c_B}\mathbf{B}^{-1}\mathbf{N} = (0,0,-2,-2) - (2,8)\begin{bmatrix} -3 & -7 \\ 1 & 2 \end{bmatrix}\begin{bmatrix} 0 & 1 & -7 & -3 \\ 1 & 0 & 2 & 1 \end{bmatrix}$$

$$= (-2,-2,8,2) \qquad \text{so choose} \quad j^* = 1$$

$$\mathbf{d_B} = -\mathbf{B}^{-1}\mathbf{N}_{\cdot j^*} = -\begin{bmatrix} -3 & -7 \\ 1 & 2 \end{bmatrix}\begin{bmatrix} 0 \\ 1 \end{bmatrix} = \begin{bmatrix} 7 \\ -2 \end{bmatrix}$$

Thus $k^* = 2$ [because only $(\mathbf{d_B})_2 < 0$].

Iteration 6

$$\mathbf{B} = [\mathbf{A}_{\cdot 6}, \mathbf{A}_{\cdot 2}] = \begin{bmatrix} 2 & 0 \\ -1 & 1 \end{bmatrix}$$

$$\mathbf{N} = [\mathbf{A}_{\cdot 5}, \mathbf{A}_{\cdot 1}, \mathbf{A}_{\cdot 3}, \mathbf{A}_{\cdot 4}] = \begin{bmatrix} 7 & 1 & -7 & -3 \\ -3 & 0 & 2 & 1 \end{bmatrix}$$

$$\mathbf{c_N} - \mathbf{c_B}\mathbf{B}^{-1}\mathbf{N} = (8,0,-2,-2) - (2,0)\begin{bmatrix} \frac{1}{2} & 0 \\ \frac{1}{2} & 1 \end{bmatrix}\begin{bmatrix} 7 & 1 & -7 & -3 \\ -3 & 0 & 2 & 1 \end{bmatrix}$$

$$= (1,-1,5,1) \qquad \text{so} \quad j^* = 2$$

$$\mathbf{d_B} = -\mathbf{B}^{-1}\mathbf{N}_{\cdot j^*} = -\begin{bmatrix} \frac{1}{2} & 0 \\ \frac{1}{2} & 1 \end{bmatrix}\begin{bmatrix} 1 \\ 0 \end{bmatrix} = \begin{bmatrix} -\frac{1}{2} \\ -\frac{1}{2} \end{bmatrix}$$

Choose $k^* = 1$ (since $\mathbf{x} = \mathbf{0}$).

Iteration 7

$$\mathbf{B} = [\mathbf{A}_{.1}, \mathbf{A}_{.2}],$$
$$\mathbf{N} = [\mathbf{A}_{.5}, \mathbf{A}_{.6}, \mathbf{A}_{.3}, \mathbf{A}_{.4}]$$

The initial basis matrix is obtained, and the cycle is complete.

From Example 5.5 it is important to observe that the issue of cycling is highly dependent on the specific rule that is used for choosing the values of j^* and k^*. Indeed, in the first iteration of Example 5.5, if j^* were chosen to be 2 instead of 1 then, at the next iteration, optimality would be detected, and hence cycling would not occur. Thus, for each rule, it is reasonable to ask if there is an example of cycling. For instance, recall the rule of steepest descent in which j^* is chosen as the most negative component of $\mathbf{c_N} - \mathbf{c_B}\mathbf{B}^{-1}\mathbf{N}$ (see Section 5.3) and k^* is chosen arbitrarily. The steps in Example 5.5 are precisely those obtained by following the rule of steepest descent (with the added stipulation that, in the event of ties for the most negative component of the reduced costs, the first such component be chosen).

Clearly, what is needed is a rule for moving algebraically from one basis matrix of a bfs $\mathbf{x}$ to another in a guaranteed noncycling fashion. Unfortunately, the objective function is not useful here. The reason is that the objective value associated with each basis matrix of a bfs $\mathbf{x}$ is the same, namely, $\mathbf{cx}$. Thus a strict decrease does not occur, and so another noncycling mechanism must be found. Many such rules have been found, but none is easier to state and to work with than Bland's rule.

Bland's rule. If $\mathbf{x} = (\mathbf{x_B}, \mathbf{x_N}) = (\mathbf{B}^{-1}\mathbf{b}, \mathbf{0}) \geq \mathbf{0}$ is a bfs that fails the test for optimality, then among all possible values of j with $1 \leq j \leq n - m$ and $(\mathbf{c_N} - \mathbf{c_B}\mathbf{B}^{-1}\mathbf{N})_j < 0$, choose the one for which $\mathbf{N}_{.j}$ corresponds to the lowest numbered column in the original $\mathbf{A}$ matrix. Similarly, among all values of k with $1 \leq k \leq m$ and $t^* = -(\mathbf{x_B})_k/(\mathbf{d_B})_k$, choose the one for which $\mathbf{B}_{.k}$ corresponds to the lowest numbered column in the original $\mathbf{A}$ matrix.

The next theorem shows that, starting from an initial bfs $\mathbf{x}$ and using Bland's rule, the simplex algorithm cannot cycle.

Theorem 5.2. *If* $\mathbf{x} = (\mathbf{x_B}, \mathbf{x_N}) = (\mathbf{B}^{-1}\mathbf{b}, \mathbf{0}) \geq \mathbf{0}$ *is an initial bfs and Bland's rule is used, then the simplex algorithm cannot repeat any basis matrix.*

OUTLINE OF PROOF. The appearance of the word "cannot" in the conclusion suggests using the contradiction method, whereby it is assumed that, using Bland's rule, some basis matrix is repeated. In other words, it will be assumed that, starting with the matrices $[\mathbf{B}^1, \mathbf{N}^1]$ and using Bland's rule, the

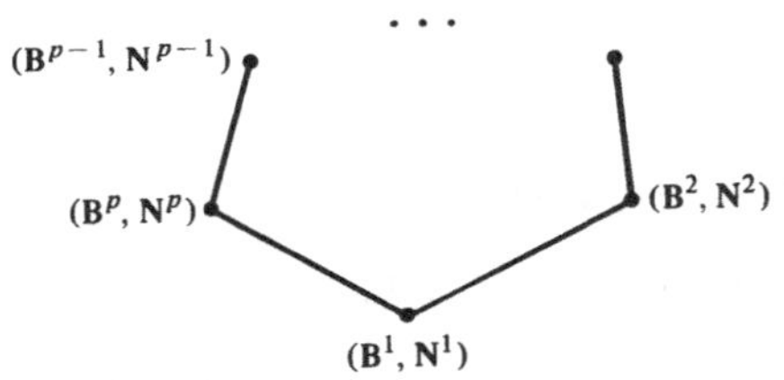

FIGURE 5.9. A cycle of basic matrices.

simplex algorithm algebraically visits $[\mathbf{B}^2, \mathbf{N}^2], \ldots, [\mathbf{B}^P, \mathbf{N}^P]$ and then returns to $[\mathbf{B}^1, \mathbf{N}^1]$ (see Figure 5.9). A remarkable contradiction will be reached by producing one of the basis matrices from the cycle together with a particular basic variable that fails to leave that basis even though Bland's rule says it must!

One has to be very clever in choosing the proper variable. To see how, consider all the variables that enter and leave a basis at some time during the cycle. Let $I' = \{1 \le i \le n$: the variable x_i enters and leaves a basis sometime during the cycle$\}$. The actual variable that produces the contradiction is x_{i^*}, where $i^* = \max\{i: i$ is in $I'\}$. Suppose that Bland's rule dictates that the variable x_{i^*} enters the basis matrix $\mathbf{B}$ from $\mathbf{N}$ and, at some point later in the cycle, leaves the basis $\mathbf{B}'$ into $\mathbf{N}'$. By carefully considering the matrices $[\mathbf{B}', \mathbf{N}']$, a contradiction can be reached by producing a variable x_i (different from x_{i^*}) that is basic in $\mathbf{B}'$ and showing that it is the one to leave $\mathbf{B}'$ instead of x_{i^*}.

Once the value for i has been chosen, it must then be shown that, by Bland's rule, x_i and not x_{i^*} leaves the basis $\mathbf{B}'$. To do this it is necessary to establish that (1) x_i is a basic variable in $\mathbf{B}'$; (2) x_i is a candidate to leave the basis in the sense that $x_i = 0$; and (3) $i < i^*$ so that, by Bland's rule, x_i leaves instead of x_{i^*}. All three will be accomplished by working forward from the fact that (a) x_{i^*} enters the basis $\mathbf{B}$ from $\mathbf{N}$, (b) x_{i^*} leaves the basis $\mathbf{B}'$ into $\mathbf{N}'$, and (c) i^* is chosen in a special way. In the proof that follows, you will have to be able to think about the steps of the simplex algorithm when performed with the matrices $[\mathbf{B}, \mathbf{N}]$ as well as $[\mathbf{B}', \mathbf{N}']$.

PROOF OF THEOREM 5.2. Suppose, contrary to the conclusion, that the simplex algorithm does repeat a basis matrix when using Bland's rule, and thus a cycle is formed as in Figure 5.9. Let $I' = \{1 \le i \le n$: the variable x_i enters and leaves a basis some time during the cycle$\}$, and let $i^* = \max\{i: i$ is in $I'\}$. Since i^* is in I', x_{i^*} must enter and leave a basis. Suppose that x_{i^*} enters a basis $\mathbf{B}$ from $\mathbf{N}$ and leaves at some point later in the cycle from $\mathbf{B}'$ into $\mathbf{N}'$. A contradiction is reached by producing an $i \ne i^*$ for which x_i is basic in $[\mathbf{B}', \mathbf{N}']$ and yet, by Bland's rule, x_i instead of x_{i^*} leaves $\mathbf{B}'$.

To produce the value for i, consider first what happens when x_{i^*} is about to enter $\mathbf{B}$ from $\mathbf{N}$. Let $\mathbf{u} = \mathbf{c_B B}^{-1}$ and note that the reason x_{i^*} is a candidate to enter the basis is because $(\mathbf{c} - \mathbf{uA})_{i^*} < 0$. Now consider what happens when x_{i^*} is about to leave $\mathbf{B}'$ and is being replaced by some nonbasic variable, say, x_p in $\mathbf{N}'$. Let $\mathbf{d}'$ be the direction of movement for $[\mathbf{B}', \mathbf{N}']$, so $\mathbf{A}\mathbf{d}' = \mathbf{0}$, $\mathbf{c}\mathbf{d}' < 0$, and, since x_{i^*} is about to leave, $d'_{i^*} < 0$. It then follows that $(\mathbf{c} - \mathbf{u}\mathbf{A})\mathbf{d}' = \mathbf{c}\mathbf{d}' - \mathbf{u}(\mathbf{A}\mathbf{d}') = \mathbf{c}\mathbf{d}' < 0$, i.e., $\Sigma_{j=1}^{n}(\mathbf{c} - \mathbf{u}\mathbf{A})_j d'_j < 0$, and, since $(\mathbf{c} - \mathbf{u}\mathbf{A})_{i^*} < 0$ and $d'_{i^*} < 0$, there must be an i with $1 \le i \le n$ such that $(\mathbf{c} - \mathbf{u}\mathbf{A})_i d'_i < 0$. Clearly, $i \neq i^*$ because $(\mathbf{c} - \mathbf{u}\mathbf{A})_{i^*} d'_{i^*} > 0$ and $(\mathbf{c} - \mathbf{u}\mathbf{A})_i d'_i < 0$. Also, note that $(\mathbf{c} - \mathbf{u}\mathbf{A})_i \neq 0$ and $d'_i \neq 0$.

It must now be shown that, in $[\mathbf{B}', \mathbf{N}']$, the variable x_i is to leave the basis by Bland's rule. To this end it must be shown that (1) x_i is a basic variable in $\mathbf{B}'$, (2) x_i is a candidate to leave the basis $\mathbf{B}'$ (in the sense that $x_i = 0$ and hence its t^* value is also 0), and (3) $i < i^*$ so that Bland's rule would select x_i instead of x_{i^*} to leave $\mathbf{B}'$. All of these statements can be proved best by first showing that i is in I', i.e., that x_i enters and leaves a basis at some time during the cycle. This, in turn, is accomplished by noting first that x_i is nonbasic in $[\mathbf{B}, \mathbf{N}]$ because if it were basic, $(\mathbf{c} - \mathbf{u}\mathbf{A})_i$ would have to be 0, but it is already known that $(\mathbf{c} - \mathbf{u}\mathbf{A})_i \neq 0$. Thus x_i is nonbasic in $[\mathbf{B}, \mathbf{N}]$. To firmly establish that i is in I' it will be shown that either x_i is basic in $\mathbf{B}'$ or just about to become basic (i.e., $i = p$). The reason is that $d'_i \neq 0$ [since $(\mathbf{c} - \mathbf{u}\mathbf{A})_i d'_i < 0$] and the only nonzero components of $\mathbf{d}'$ are the basic ones and the nonbasic one that is about to enter. Thus it has been shown that x_i is nonbasic in $[\mathbf{B}, \mathbf{N}]$ and either basic in $[\mathbf{B}', \mathbf{N}']$ or about to become basic.

To complete the proof, it must still be shown that x_i is basic in $[\mathbf{B}', \mathbf{N}']$, that it is a candidate to leave the basis, and that $i < i^*$ so that, by Bland's rule, it leaves instead of x_{i^*}.

It is easy to see that $i < i^*$. The reason is that since i is in I' and i^* is the largest element of I', it must be that $i < i^*$.

To establish that x_i is actually basic in $[\mathbf{B}', \mathbf{N}']$, it will be shown that $d'_i < 0$, for only basic components of $\mathbf{d}'$ can be negative. Since $(\mathbf{c} - \mathbf{u}\mathbf{A})_i d'_i < 0$ the desired conclusion that $d'_i < 0$ will follow once it is shown that $(\mathbf{c} - \mathbf{u}\mathbf{A})_i > 0$. Recall that x_{i^*} was chosen to enter $[\mathbf{B}, \mathbf{N}]$ by Bland's rule; thus, since $i < i^*$, it had better be the case that $(\mathbf{c} - \mathbf{u}\mathbf{A})_i \ge 0$; otherwise, x_i would have been chosen to enter. Since it was already shown that $(\mathbf{c} - \mathbf{u}\mathbf{A})_i \neq 0$, the only alternative is that $(\mathbf{c} - \mathbf{u}\mathbf{A})_i > 0$, whereby $d'_i < 0$, and so x_i must be a basic variable in $\mathbf{B}'$.

To complete the proof, it must be shown that x_i is a candidate to leave the basis $\mathbf{B}'$, and this is done by showing that $x_i = 0$. Recall Figure 5.9, which shows the cycle of basis matrices. The only way that the simplex algorithm can form such a cycle is if a strict decrease in the objective function does not occur, i.e., if at each iteration $t^* = 0$, and this can only happen if the variables that enter and leave the basis during the cycle each have value 0. Since x_i is one such variable, it must be that $x_i = 0$. □

If Bland's rule is applied to Example 5.5, then the first four iterations remain the same; however, in the fifth iteration, $j^* = 2$ instead of 1 since

$\mathbf{N}_{.2} = \mathbf{A}_{.1}$ is the lowest numbered column in the **A** matrix. Then, at the next iteration, optimality will be detected, as shown in Exercises 5.7.4.

Although the proof of Theorem 5.2 is complicated, computationally Bland's rule is very easy to implement, and yet no professional code uses it. In fact, professional codes competely disregard the possibility of cycling. The reason is that years of computational tests on thousands of real world problems have established that there is virtually no danger of cycling. Since there is no practical danger of cycling, the rule for choosing the values of j^* and k^* can be based on other considerations, such as computational efficiency. That is to say, the actual rule that is used by professional simplex codes is to select values of j^* and k^* so as to try to reduce the total number of basic feasible solutions that must be examined by the algorithm before an optimal one is found. More will be said about such a rule in Chapter 7.

This section has shown that the potential problem of degeneracy (i.e., a basic variable having the value 0) is really no problem at all. Just because a bfs is degenerate does not mean that $t^* = 0$. Moreover, even if $t^* = 0$, the pivot operation of the simplex algorithm can still be performed. If this is done, then there is a theoretical possibility of cycling; however, computational tests indicate that it simply will not happen. Finally, Bland's rule was shown to resolve the issue once and for all, even though it is not needed computationally. The result of this chapter is that if an initial bfs can be found, then, in a finite number of iterations, the simplex algorithm will either produce an optimal solution or else detect the condition of unboundedness. It would seem that the last step is to find an initial bfs (or determine that there is none). This is the topic of the next chapter.

EXERCISES

5.7.1. Consider the following LP:

$$\begin{array}{lrcl}
\text{Minimize} & -x_1 - 3x_2 & & \\
\text{subject to} & x_1 + x_2 + x_3 & & = 5 \\
& x_1 + 2x_2 & + x_4 & = 10 \\
& x_1, \quad x_2, \quad x_3, & x_4 & \geq 0
\end{array}$$

(a) Show that the bfs $\mathbf{x} = (0,5,0,0)$ in which x_1 and x_2 are the basic variables fails the test for optimality.

(b) Using your results from part (a), perform one iteration of the simplex algorithm to show that the next bfs [which is still $\mathbf{x} = (0,5,0,0)$] does pass the test for optimality.

(c) How it is possible that the optimal bfs of part (a) failed the test for optimality?

5.7.2. For the LP

$$\begin{array}{llllllll} \text{Minimize} & -2x_1 & +3x_2 & - \; x_3 & +3x_4 & -5x_5 & -3x_6 & \\ \text{subject to} & 4x_1 & + \; x_2 & + \; x_3 & +2x_4 & +8x_5 & & =4 \\ & 2x_1 & +2x_2 & +3x_3 & + \; x_4 & +4x_5 & + \; x_6 & =2 \\ & x_1, & x_2, & x_3, & x_4, & x_5, & x_6 & \geq 0 \end{array}$$

and for each of the given degenerate bfs, perform two iterations of the simplex algorithm using the rule of steepest descent. Clearly indicate the values of t^* at each iteration.

(a) x_4 and x_6 are the basic variables.
(b) x_3 and x_5 are the basic variables.
(c) x_1 and x_2 are the basic variables.

5.7.3. Solve the following LP using Bland's rule starting with the bfs in which x_1 and x_2 are the basic variables:

$$\begin{array}{llllllll} \text{Minimize} & & & -2x_3 & +x_4 & - \; x_5 & \\ \text{subject to} & x_1 & & +3x_3 & +x_4 & - \; x_5 & =0 \\ & & x_2 & + \; x_3 & -x_4 & +3x_5 & =0 \\ & x_1, & x_2, & x_3, & x_4, & x_5 & \geq 0 \end{array}$$

5.7.4. Show that if Bland's rule is used in the fifth iteration of Example 5.2, then optimality is detected at the next iteration.

DISCUSSION

Steps of the Simplex Algorithm

As presented in this chapter, each time the pivot operation is performed the columns of the **B** and **N** matrices are being rearranged, thus requiring care to keep subscripts and components in order. There is another method that enables one to perform the steps of the simplex algorithm while keeping all of the columns of the **A** matrix in their original order. This approach is referred to as the tableau method, and it has been the most common method for teaching the steps of the simplex algorithm. This approach is presented in Appendix A, and each step of the tableau method is related to the corresponding step that was described in this chapter. Since most of the textbooks on LP-related subjects use the tableau method, it will be beneficial to learn that method in addition.

Degeneracy

Prior to Bland's rule [1], another method, often referred to as epsilon perturbation, was developed by Charnes [2] for resolving the issue of degeneracy. Essentially, by changing the values of the **b** vector slightly in a particular way, it was possible to obtain a new problem that had no degenerate bfs and whose solution could be made arbitrarily close to the solution of the original problem. Geometrically, perturbing the **b** vector would eliminate the situation shown in Figure 5.7. Independently, another method for handling degeneracy was developed by Dantzig, Orden, and Wolfe [3], which resulted in what is called the lexicographic pivoting rule. Although these two methods were obtained from different points of view, their computational implementation is identical. It is interesting to note that the perturbation approach of Charnes is designed to eliminate degenerate bfs altogether, whereas the approaches of Bland and Dantzig, Orden, and Wolfe are designed to avoid repeating a basis matrix (even though there can be degenerate bfs).

REFERENCES

1. R. G. Bland, "New Finite Pivoting Rules for the Simplex Method," *Mathematics of Operations Research* **2**(2), 103–107 (1977).
2. A. Charnes, "Optimality and Degeneracy in Linear Programming," *Econometrica* **20**(2), 160–170 (1952).
3. G. B. Dantzig, A. Orden, and P. Wolfe, "The Generalized Simplex Method for Minimizing a Linear Form Under Linear Inequality Restraints," *Pacific Journal of Mathematics* **5**(2), 183–195 (1955).

Chapter 6

Phase 1

The result of Chapter 5 is an algorithm that, for any LP in standard form, terminates in a finite number of steps in either an unbounded or optimal condition, provided only that an initial bfs can be found. The purpose of this chapter is to describe a method for finding an initial bfs or determining that none exists. One way to proceed is by the trial-and-error approach of Section 5.2, whereby one guesses a basis matrix **B** and then checks whether (1) **B** is nonsingular and (2) $\mathbf{B}^{-1}\mathbf{b} \geq \mathbf{0}$. However, a more systematic method has been developed, known as *phase 1*.

6.1. THE PHASE 1 PROBLEM

The goal of finding an initial bfs or determining that none exists will be accomplished in two steps. The first step is designed to check if the original LP is feasible or not, for in the latter case it will be impossible to produce an initial bfs. On the other hand, if the LP is determined to be feasible, then there is hope of finding an initial bfs, and the second step of phase 1 is to produce such a vector. Once an initial bfs has been constructed, the simplex algorithm of Chapter 5 can then be applied to solve the original LP in the process of what is known as *phase 2*.

This section will address the first step of phase 1, namely, determining if the original LP is feasible or not. To that end, a new LP (called the *phase 1 LP*) will be created from the original problem. Then it will be shown that the simplex algorithm of Chapter 5 is always able to produce an optimal solution to the phase 1 LP and, furthermore, that a solution thus obtained will indicate whether the original LP is feasible or not.

In order to be able to apply the simplex algorithm to the phase 1 LP, an initial bfs is needed. Fortunately, the phase 1 LP has an obvious bfs from which to get started. The results in Chapter 5 then ensure that the simplex algorithm must terminate in a finite number of steps in either an optimal or unbounded condition. Conveniently, unbounded termination is impossible for the phase 1 LP. Thus the simplex algorithm must produce an optimal solution to the phase 1 LP. By looking at the solution thus obtained, it can then be determined whether the original LP is feasible.

To construct the phase 1 LP, consider the original LP:

$$\begin{aligned} \text{Minimize} \quad & \mathbf{cx} \\ \text{subject to} \quad & \mathbf{Ax} = \mathbf{b} \\ & \mathbf{x} \geq \mathbf{0} \end{aligned}$$

The phase 1 LP is obtained by adding m new variables $\mathbf{y} = (y_1, \ldots, y_m)$ to the original LP, one for each of the constraints. The new variables are called *artificial variables*, and they will be restricted to be nonnegative. Observe that the original objective function $\mathbf{cx}$ has nothing to do with determining whether the original LP is feasible, and hence it can be discarded temporarily. To replace it, the objective of phase 1 is to force all the artificial variables to have the value 0, for then it can be shown that the original LP is feasible; otherwise, it is not. The problem is thus to devise an objective function that has the effect of driving the artificial variables to 0. One solution is to minimize the sum of the artificial variables. The phase 1 LP can be written in matrix/vector notation:

$$\begin{aligned} \text{Minimize} \quad & \sum_{i=1}^{m} y_i = \mathbf{ey} \\ \text{subject to} \quad & \mathbf{Ax} + \mathbf{Iy} = \mathbf{b} \\ & \mathbf{x}, \quad \mathbf{y} \geq \mathbf{0} \end{aligned}$$

where, as before, the vector $\mathbf{e} = (1, \ldots, 1)$, and $\mathbf{I}$ is the $m \times m$ identity matrix. The next examples serve to demonstrate the construction of the phase 1 LP.

EXAMPLE 6.1: *Original LP.*

$$\begin{aligned} \text{Minimize} \quad & 2x_1 + 3x_2 + x_3 + 2x_4 \\ \text{subject to} \quad & x_1 + x_2 + x_3 + 2x_4 - x_5 = 5 \\ & 2x_1 + x_2 + 3x_3 - x_4 - x_6 = 3 \\ & x_1, \quad x_2, \quad x_3, \quad x_4, \quad x_5, \quad x_6 \geq 0 \end{aligned}$$

Phase 1 LP.

$$\begin{array}{llllllllll} \text{Minimize} & & & & & & & y_1 & +y_2 & \\ \text{subject to} & x_1 & +x_2 & +x_3 & +2x_4 & -x_5 & & +y_1 & & =5 \\ & 2x_1 & +x_2 & +3x_3 & -\ x_4 & & -x_6 & & +y_2 & =3 \\ & x_1, & x_2, & x_3, & x_4, & x_5, & x_6, & y_1, & y_2 & \geq 0 \end{array}$$

EXAMPLE 6.2: *Original LP.*

$$\begin{array}{llllllll} \text{Minimize} & 3x_1 & +\ x_2 & +2x_3 & -\ x_4 & & & \\ \text{subject to} & 2x_1 & +3x_2 & +\ x_3 & +\ x_4 & -x_5 & & =7 \\ & x_1 & & +\ x_3 & +2x_4 & & -x_6 & =5 \\ & x_1 & +\ x_2 & +\ x_3 & +5x_4 & & & =9 \\ & x_1, & x_2, & x_3, & x_4, & x_5, & x_6 & \geq 0 \end{array}$$

Phase 1 LP.

$$\begin{array}{lllllllllll} \text{Minimize} & & & & & & & y_1 & +y_2 & +y_3 & \\ \text{subject to} & 2x_1 & +3x_2 & +x_3 & +\ x_4 & -x_5 & & +y_1 & & & =7 \\ & x_1 & & +x_3 & +2x_4 & & -x_6 & & +y_2 & & =5 \\ & x_1 & +\ x_2 & +x_3 & +5x_4 & & & & & +y_3 & =9 \\ & x_1, & x_2, & x_3, & x_4, & x_5, & x_6, & y_1, & y_2, & y_3 & \geq 0 \end{array}$$

The next two propositions establish that an optimal solution to the phase 1 LP can be used to determine whether the original LP is feasible.

Proposition 6.1. *If* $(\mathbf{x}^*, \mathbf{y}^*)$ *is an optimal solution to the phase* 1 *LP and* $\mathbf{y}^* = \mathbf{0}$, *then* $\mathbf{x}^*$ *is feasible for the original LP.*

OUTLINE OF PROOF. The forward–backward method gives rise to the abstraction question "How can I show that a vector (namely, $\mathbf{x}^*$) is feasible for an LP?" One answer is to use the definition, whereby it must be shown that (1) $\mathbf{Ax}^* = \mathbf{b}$ and (2) $\mathbf{x}^* \geq \mathbf{0}$. Both of these statements are established by working forward from the hypothesis that $(\mathbf{x}^*, \mathbf{y}^*)$ solves the phase 1 LP and that $\mathbf{y}^* = \mathbf{0}$. Specifically, since $(\mathbf{x}^*, \mathbf{y}^*)$ is an optimal solution, $(\mathbf{x}^*, \mathbf{y}^*)$ is feasible for the phase 1 LP, and so $\mathbf{Ax}^* + \mathbf{Iy}^* = \mathbf{b}$, $\mathbf{x}^* \geq \mathbf{0}$, and $\mathbf{y}^* \geq \mathbf{0}$. This immediately establishes (2). To see (1), note that the hypothesis also states that $\mathbf{y}^* = \mathbf{0}$, and so $\mathbf{b} = \mathbf{Ax}^* + \mathbf{Iy}^* = \mathbf{Ax}^* + \mathbf{I(0)} = \mathbf{Ax}^*$.

PROOF OF PROPOSITION 6.1. Since $(\mathbf{x}^*, \mathbf{y}^*)$ is optimal for the phase 1 LP, it must be feasible, so $\mathbf{Ax}^* + \mathbf{Iy}^* = \mathbf{b}$, $\mathbf{x}^* \geq \mathbf{0}$, and $\mathbf{y}^* \geq \mathbf{0}$. Also, by the hypothesis, $\mathbf{y}^* = \mathbf{0}$, and thus $\mathbf{b} = \mathbf{Ax}^* + \mathbf{Iy}^* = \mathbf{Ax}^* + \mathbf{I(0)} = \mathbf{Ax}^*$. In other words, $\mathbf{Ax}^* = \mathbf{b}$ and $\mathbf{x}^* \geq \mathbf{0}$, so $\mathbf{x}^*$ is feasible for the original LP. □

The result of Proposition 6.1 is that if an optimal solution $(\mathbf{x}^*, \mathbf{y}^*)$ to the phase 1 LP can be produced, and if the artificial variables all have value 0 (i.e., $\mathbf{y}^* = \mathbf{0}$), then the original LP is feasible—$\mathbf{x}^*$ being a feasible solution. The next proposition shows that if $(\mathbf{x}^*, \mathbf{y}^*)$ is an optimal solution for the phase 1 LP and not all of the artificial variables are 0, then the original LP is infeasible.

Proposition 6.2. *If $(\mathbf{x}^*, \mathbf{y}^*)$ is an optimal solution for the phase 1 LP and $\mathbf{y}^* \neq \mathbf{0}$, then the original LP is not feasible.*

OUTLINE OF PROOF. Because of the appearance of the word "not" in the conclusion, the contradiction method will be used. Thus it can be assumed that $(\mathbf{x}^*, \mathbf{y}^*)$ is optimal for the phase 1 LP, $\mathbf{y}^* \neq \mathbf{0}$ and, in addition, that the original LP is feasible, so suppose that $\mathbf{x}$ is such a feasible solution. With the vector $\mathbf{x}$, a contradiction to the hypothesis is reached upon showing that $(\mathbf{x}^*, \mathbf{y}^*)$ is not an optimal solution to the phase 1 problem. Specifically, it will be shown that $(\mathbf{x}, \mathbf{0})$ is a feasible vector for the phase 1 LP and yet has a strictly smaller objective value than does $(\mathbf{x}^*, \mathbf{y}^*)$. This, of course, cannot happen if $(\mathbf{x}^*, \mathbf{y}^*)$ is an optimal solution to the phase 1 problem.

PROOF OF PROPOSITION 6.2. Assume, contrary to the conclusion, that the original LP is feasible, i.e., there is an $\mathbf{x} \geq \mathbf{0}$ such that $\mathbf{Ax} = \mathbf{b}$. A contradiction is established by showing that the vector $(\mathbf{x}, \mathbf{0})$ is feasible for the phase 1 LP and has a strictly smaller objective value than $(\mathbf{x}^*, \mathbf{y}^*)$, which cannot happen if $(\mathbf{x}^*, \mathbf{y}^*)$ is an optimal solution.

To see that $(\mathbf{x}, \mathbf{0})$ is feasible for the phase 1 LP, note that $\mathbf{Ax} + \mathbf{I}(\mathbf{0}) = \mathbf{Ax} = \mathbf{b}$ and $(\mathbf{x}, \mathbf{0}) \geq \mathbf{0}$. Furthermore, the phase 1 objective value at $(\mathbf{x}, \mathbf{0})$ is $\mathbf{e}(\mathbf{0}) = 0$, while that at $(\mathbf{x}^*, \mathbf{y}^*)$ is $\mathbf{ey}^*$. But $\mathbf{ey}^* > 0$ because $\mathbf{y}^* \geq \mathbf{0}$ and $\mathbf{y}^* \neq \mathbf{0}$, and this contradiction establishes the claim. □

This section has described how to construct the phase 1 LP from the original problem. In addition, it was shown how an optimal solution to the phase 1 LP can easily be used to determine whether the original LP is feasible. The next section shows that the simplex algorithm of Chapter 5 can always obtain an optimal solution to the phase 1 problem.

EXERCISES

6.1.1. For each of the following LPs, write the associated phase 1 LP.

(a)

$$\begin{array}{lrrrrrrl}
\text{Minimize} & 4x_1 & +3x_2 & +4x_3 & & & & \\
\text{subject to} & x_1 & & +3x_3 & -x_4 & & & = 5 \\
& 2x_1 & + x_2 & & & -x_5 & & = 7 \\
& & x_2 & + x_3 & & & -x_6 & = 3 \\
& x_1, & x_2, & x_3, & x_4, & x_5, & x_6 & \geq 0
\end{array}$$

(b) Maximize $5x_1 - 2x_2 + x_3$

subject to

$$\begin{aligned} x_1 - x_2 \quad &\geq 1 \\ x_2 + 3x_3 &\leq -5 \\ x_1, \quad x_2, \quad x_3 &\geq 0 \end{aligned}$$

6.1.2. For each of the following LPs, write the associated phase 1 LP in matrix/vector notation. (Assume that $\mathbf{b} \geq \mathbf{0}$.)

(a) Minimize $\mathbf{cx}$

subject to $\mathbf{Ax} \geq \mathbf{b}$

$\mathbf{x} \geq \mathbf{0}$

(b) Maximize $\mathbf{cx}$

subject to $\mathbf{Ax} = \mathbf{b}$

$\mathbf{x} \leq \mathbf{0}$

6.1.3. Consider the following phase 1 LP:

Minimize $\mathbf{ey}$

subject to $\mathbf{Ax} + \mathbf{Iy} = \mathbf{b}$

$\mathbf{x}, \quad \mathbf{y} \geq \mathbf{0}$

Write the conditions for a direction $\mathbf{d} = (\mathbf{d}_x, \mathbf{d}_y)$ to be a direction of unboundedness.

6.1.4. Let $\mathbf{c}'$ be an m vector in which each component is greater than 0, and consider the following "phase 1" LP:

Minimize $\mathbf{c}'\mathbf{y}$

subject to $\mathbf{Ax} + \mathbf{Iy} = \mathbf{b}$

$\mathbf{x}, \quad \mathbf{y} \geq \mathbf{0}$

State and prove propositions analogous to Propositions 6.1 and 6.2.

6.2. SOLVING THE PHASE 1 PROBLEM

The phase 1 LP of the previous section has two remarkable properties that enable the simplex algorithm of Chapter 5 to compute an optimal solution to the problem. First, the phase 1 LP has an "obvious" initial bfs from which to get started, and thus the simplex algorithm must terminate in either an unbounded or optimal condition. Second, the phase 1 LP cannot be unbounded, and hence the simplex algorithm will always produce an optimal solution that can then be used to determine if the original LP is feasible or not (see Propositions 6.1 and 6.2). The remainder of this section is devoted to establishing the two properties of the phase 1 LP.

The first job is to find an initial bfs for the phase 1 LP:

$$\begin{array}{ll} \text{Minimize} & \mathbf{ey} \\ \text{subject to} & \mathbf{Ax} + \mathbf{Iy} = \mathbf{b} \\ & \mathbf{x}, \quad \mathbf{y} \geq \mathbf{0} \end{array}$$

That is,

$$\begin{array}{ll} \text{Minimize} & (\mathbf{0}, \mathbf{e})\begin{bmatrix} \mathbf{x} \\ \mathbf{y} \end{bmatrix} \\ \text{subject to} & [\mathbf{A}, \mathbf{I}]\begin{bmatrix} \mathbf{x} \\ \mathbf{y} \end{bmatrix} = \mathbf{b} \\ & \mathbf{x}, \mathbf{y} \geq \mathbf{0} \end{array}$$

Constructing the bfs requires locating (within the matrix $[\mathbf{A}, \mathbf{I}]$) a basis matrix $\mathbf{B}$ and verifying that $\mathbf{B}$ is nonsingular and $\mathbf{B}^{-1}\mathbf{b} \geq \mathbf{0}$. The $\mathbf{B}$ matrix turns out to be the $m \times m$ identity matrix. Hence the artificial variables (namely, $\mathbf{y}$) become the basic ones, and the $\mathbf{x}$ variables are nonbasic and have the value 0.

To verify that the identity matrix has the desired properties, it is easy to see that $\mathbf{I}$ is nonsingular (since $\mathbf{I}^{-1} = \mathbf{I}$), but how do you know whether $\mathbf{B}^{-1}\mathbf{b} = \mathbf{I}^{-1}\mathbf{b} = \mathbf{Ib} = \mathbf{b} \geq \mathbf{0}$? Indeed, $\mathbf{b}$ is nonnegative because, in standard form, the $\mathbf{b}$ vector is required to be so (see Section 4.5). The next proposition formally establishes that the vector $(\mathbf{x}, \mathbf{y}) = (\mathbf{0}, \mathbf{b})$ is an initial bfs for the phase 1 LP.

Proposition 6.3. *The vector* $(\mathbf{x}, \mathbf{y}) = (\mathbf{0}, \mathbf{b})$ *is an initial bfs for the phase* 1 *LP.*

OUTLINE OF PROOF. The forward–backward method gives rise to the abstraction question "How can I show that a vector (namely, $(\mathbf{0}, \mathbf{b})$) is a bfs?" One answer comes from the definition, whereby the appropriate $\mathbf{B}$ matrix is identified and all of the properties are established. The $\mathbf{B}$ matrix will be the $m \times m$ identity matrix $\mathbf{I}$. Thus the $\mathbf{y}$ variables are the basic ones, and their values are $\mathbf{b}$. The $\mathbf{x}$ variables are nonbasic, and their values are $\mathbf{0}$. Clearly, the identity matrix is nonsingular (since $\mathbf{I}^{-1} = \mathbf{I}$), and the basic variables $\mathbf{y}$ satisfy $\mathbf{y} = \mathbf{I}^{-1}\mathbf{b} = \mathbf{Ib} = \mathbf{b}$. The last property to be verified is that $\mathbf{I}^{-1}\mathbf{b} \geq \mathbf{0}$, but this is true because, for an LP in standard form, $\mathbf{b} \geq \mathbf{0}$.

PROOF OF PROPOSITION 6.3. To show that $(\mathbf{x}, \mathbf{y}) = (\mathbf{0}, \mathbf{b})$ is a bfs for the phase 1 LP, let the $m \times m$ identity matrix $\mathbf{I}$ be the basis matrix. Thus the $\mathbf{y}$ variables are basic and have value $\mathbf{b}$, and the $\mathbf{x}$ variables are nonbasic and have value $\mathbf{0}$. Clearly, $\mathbf{I}$ is nonsingular, $\mathbf{y} = \mathbf{I}^{-1}\mathbf{b} = \mathbf{b}$, $\mathbf{I}^{-1}\mathbf{b} = \mathbf{b} \geq \mathbf{0}$, and $\mathbf{x} = \mathbf{0}$. Hence the vector $(\mathbf{x}, \mathbf{y}) = (\mathbf{0}, \mathbf{b})$ is a bfs. □

As a result of Proposition 6.3, the simplex algorithm of Chapter 5 can be applied to the phase 1 LP by starting with the initial bfs $(\mathbf{x},\mathbf{y}) = (\mathbf{0},\mathbf{b})$. Hence, in a finite number of iterations the algorithm must terminate in an optimal or unbounded condition. The next proposition shows that, for the phase 1 LP, the simplex algorithm cannot end in an unbounded condition.

Proposition 6.4. *The phase* 1 *LP is not unbounded.*

OUTLINE OF PROOF. The appearance of the word "not" suggests using the contradiction method, whereby it is assumed that the phase 1 LP is unbounded and, as a result, that there is a direction $\mathbf{d}$ of unboundedness (see Definition 4.19). Using the properties of the direction $\mathbf{d}$, we can reach a contradiction by showing that, on one hand, all components of $\mathbf{d}$ are greater than or equal to 0 and, on the other hand, one component of $\mathbf{d}$ is less than zero.

Working forward from the assumption that the phase 1 LP is unbounded, there is a direction $\mathbf{d}$ satisfying the properties of Definition 4.19. Let $\mathbf{d}_{\mathbf{x}}$ and $\mathbf{d}_{\mathbf{y}}$ be those components of $\mathbf{d}$ corresponding to the $\mathbf{x}$ and $\mathbf{y}$ variables, respectively. From Definition 4.19 it must be that

$$\mathbf{d} = (\mathbf{d}_{\mathbf{x}}, \mathbf{d}_{\mathbf{y}}) \geq \mathbf{0} \tag{1}$$

$$(\mathbf{0},\mathbf{e})\begin{bmatrix} \mathbf{d}_{\mathbf{x}} \\ \mathbf{d}_{\mathbf{y}} \end{bmatrix} = \mathbf{e}\mathbf{d}_{\mathbf{y}} < 0 \tag{2}$$

From (1) it follows that $\mathbf{d} \geq \mathbf{0}$. It remains to construct a component of $\mathbf{d}$ that is less than 0, but from (2), since $\mathbf{e}\mathbf{d}_{\mathbf{y}} = \sum_{i=1}^{m}(\mathbf{d}_{\mathbf{y}})_i < 0$, at least one component of $\mathbf{d}_{\mathbf{y}}$ must be strictly negative.

PROOF OF PROPOSITION 6.4. Assume, contrary to the conclusion, that the phase 1 LP is unbounded. Hence there is a direction $\mathbf{d}$ of unboundedness satisfying the properties of Definition 4.19. Let $\mathbf{d}_{\mathbf{x}}$ and $\mathbf{d}_{\mathbf{y}}$ be the components of $\mathbf{d}$ corresponding to the $\mathbf{x}$ and $\mathbf{y}$ variables, respectively. According to Definition 4.19, it must be that $\mathbf{d} = (\mathbf{d}_{\mathbf{x}}, \mathbf{d}_{\mathbf{y}}) \geq \mathbf{0}$, and also,

$$(\mathbf{0},\mathbf{e})\begin{bmatrix} \mathbf{d}_{\mathbf{x}} \\ \mathbf{d}_{\mathbf{y}} \end{bmatrix} = \mathbf{e}\mathbf{d}_{\mathbf{y}} = \sum_{i=1}^{m} (\mathbf{d}_{\mathbf{y}})_i < 0$$

Since $\sum_{i=1}^{m}(\mathbf{d}_{\mathbf{y}})_i < 0$, there must be a component of $\mathbf{d}_{\mathbf{y}}$ that is less than 0, but this contradicts the fact that $\mathbf{d}_{\mathbf{y}} \geq \mathbf{0}$. □

Proposition 6.4 has established that the phase 1 LP is not unbounded. Together with Proposition 6.3, this means that the simplex algorithm must produce an optimal solution to the phase 1 LP in a finite number of iterations. The optimal solution thus obtained serves to determine if the original LP is feasible or not, as illustrated in the next examples.

EXAMPLE 6.3: For the phase 1 LP in Example 6.1, one has

$$\mathbf{c} = (0,0,0,0,0,0,1,1)$$

$$\mathbf{A} = \begin{bmatrix} 1 & 1 & 1 & 2 & -1 & 0 & 1 & 0 \\ 2 & 1 & 3 & -1 & 0 & -1 & 0 & 1 \end{bmatrix}$$

$$\mathbf{b} = \begin{bmatrix} 5 \\ 3 \end{bmatrix}$$

Iteration 1

$$\mathbf{B} = [\mathbf{A}_{\cdot 7}, \mathbf{A}_{\cdot 8}] = \begin{bmatrix} 1 & 0 \\ 0 & 1 \end{bmatrix}$$

$$\mathbf{N} = [\mathbf{A}_{\cdot 1}, \mathbf{A}_{\cdot 2}, \mathbf{A}_{\cdot 3}, \mathbf{A}_{\cdot 4}, \mathbf{A}_{\cdot 5}, \mathbf{A}_{\cdot 6}] = \begin{bmatrix} 1 & 1 & 1 & 2 & -1 & 0 \\ 2 & 1 & 3 & -1 & 0 & -1 \end{bmatrix}$$

$$\mathbf{x_B} = \begin{bmatrix} y_1 \\ y_2 \end{bmatrix} = \mathbf{B}^{-1}\mathbf{b} = \begin{bmatrix} 1 & 0 \\ 0 & 1 \end{bmatrix}\begin{bmatrix} 5 \\ 3 \end{bmatrix} = \begin{bmatrix} 5 \\ 3 \end{bmatrix}$$

$$\begin{aligned} \mathbf{c_N} - \mathbf{c_B}\mathbf{B}^{-1}\mathbf{N} &= (0,0,0,0,0,0) \\ &\quad -(1,1)\begin{bmatrix} 1 & 0 \\ 0 & 1 \end{bmatrix}\begin{bmatrix} 1 & 1 & 1 & 2 & -1 & 0 \\ 2 & 1 & 3 & -1 & 0 & -1 \end{bmatrix} \\ &= (-3, -2, -4, -1, 1, 1) \qquad \text{so} \quad j^* = 3 \end{aligned}$$

$$\mathbf{d_B} = -\mathbf{B}^{-1}\mathbf{N}_{\cdot j^*} = -\begin{bmatrix} 1 & 0 \\ 0 & 1 \end{bmatrix}\begin{bmatrix} 1 \\ 3 \end{bmatrix} = \begin{bmatrix} -1 \\ -3 \end{bmatrix}$$

$$\begin{aligned} t^* &= \min\{-(\mathbf{x_B})_1/(\mathbf{d_B})_1, -(\mathbf{x_B})_2/(\mathbf{d_B})_2\} = \min\{5,1\} \\ &= 1 \qquad \text{so} \quad k^* = 2 \end{aligned}$$

Column $k^* = 2$ of $\mathbf{B}$ is replaced with column $j^* = 3$ of $\mathbf{N}$.

Iteration 2

$$\mathbf{B} = [\mathbf{A}_{\cdot 7}, \mathbf{A}_{\cdot 3}] = \begin{bmatrix} 1 & 1 \\ 0 & 3 \end{bmatrix}$$

$$\begin{aligned} \mathbf{N} &= [\mathbf{A}_{\cdot 1}, \mathbf{A}_{\cdot 2}, \mathbf{A}_{\cdot 8}, \mathbf{A}_{\cdot 4}, \mathbf{A}_{\cdot 5}, \mathbf{A}_{\cdot 6}] \\ &= \begin{bmatrix} 1 & 1 & 0 & 2 & -1 & 0 \\ 2 & 1 & 1 & -1 & 0 & -1 \end{bmatrix} \end{aligned}$$

$$\mathbf{x_B} = \begin{bmatrix} y_1 \\ x_3 \end{bmatrix} = \mathbf{B}^{-1}\mathbf{b} = \begin{bmatrix} 1 & -\frac{1}{3} \\ 0 & \frac{1}{3} \end{bmatrix}\begin{bmatrix} 5 \\ 3 \end{bmatrix} = \begin{bmatrix} 4 \\ 1 \end{bmatrix}$$

$$\mathbf{c_N} - \mathbf{c_B B^{-1} N} = (0,0,1,0,0,0) - (1,0)\begin{bmatrix} 1 & -\frac{1}{3} \\ 0 & \frac{1}{3} \end{bmatrix}$$

$$\times \begin{bmatrix} 1 & 1 & 0 & 2 & -1 & 0 \\ 2 & 1 & 1 & -1 & 0 & -1 \end{bmatrix}$$

$$= (-\tfrac{1}{3}, -\tfrac{2}{3}, \tfrac{4}{3}, -\tfrac{7}{3}, 1, -\tfrac{1}{3}) \qquad \text{so} \quad j^* = 4$$

$$\mathbf{d_B} = -\mathbf{B}^{-1}\mathbf{N}_{\cdot j^*} = -\begin{bmatrix} 1 & -\frac{1}{3} \\ 0 & \frac{1}{3} \end{bmatrix}\begin{bmatrix} 2 \\ -1 \end{bmatrix} = \begin{bmatrix} -\frac{7}{3} \\ \frac{1}{3} \end{bmatrix}$$

Since only $(\mathbf{d_B})_1 < 0$, $k^* = 1$, and so column $k^* = 1$ of $\mathbf{B}$ is replaced by column $j^* = 4$ of $\mathbf{N}$.

Iteration 3

$$\mathbf{B} = [\mathbf{A}_{\cdot 4}, \mathbf{A}_{\cdot 3}] = \begin{bmatrix} 2 & 1 \\ -1 & 3 \end{bmatrix}$$

$$\mathbf{N} = [\mathbf{A}_{\cdot 1}, \mathbf{A}_{\cdot 2}, \mathbf{A}_{\cdot 8}, \mathbf{A}_{\cdot 7}, \mathbf{A}_{\cdot 5}, \mathbf{A}_{\cdot 6}]$$

$$= \begin{bmatrix} 1 & 1 & 0 & 1 & -1 & 0 \\ 2 & 1 & 1 & 0 & 0 & -1 \end{bmatrix}$$

$$\mathbf{x_B} = \begin{bmatrix} x_4 \\ x_3 \end{bmatrix} = \mathbf{B}^{-1}\mathbf{b} = \begin{bmatrix} \frac{3}{7} & -\frac{1}{7} \\ \frac{1}{7} & \frac{2}{7} \end{bmatrix}\begin{bmatrix} 5 \\ 3 \end{bmatrix} = \begin{bmatrix} \frac{12}{7} \\ \frac{11}{7} \end{bmatrix}$$

$$\mathbf{c_N} - \mathbf{c_B B^{-1} N} = (0,0,1,1,0,0) - (0,0)\begin{bmatrix} \frac{3}{7} & -\frac{1}{7} \\ \frac{1}{7} & \frac{2}{7} \end{bmatrix}$$

$$\times \begin{bmatrix} 1 & 1 & 0 & 1 & -1 & 0 \\ 2 & 1 & 1 & 0 & 0 & -1 \end{bmatrix}$$

$$= (0,0,1,1,0,0)$$

The test for optimality has been passed. Since y_1 and y_2 are nonbasic, their values are 0. Hence the vector $\mathbf{x} = (0, 0, \frac{11}{7}, \frac{12}{7}, 0, 0)$ is feasible for the original LP.

EXAMPLE 6.4: *Original LP.*

$$\begin{array}{lrcl}
\text{Minimize} & 3x_1 + 6x_2 + 7x_3 + x_4 & & \\
\text{subject to} & x_1 + x_2 + x_3 + x_4 & = & 1 \\
& x_1 + 3x_2 + x_3 + 4x_4 - x_5 & = & 100 \\
& x_1, \quad x_2, \quad x_3, \quad x_4, \quad x_5 & \geq & 0
\end{array}$$

Phase 1 LP.

$$\begin{array}{lllllllll} \text{Minimize} & & & & & & y_1 & +y_2 & \\ \text{subject to} & x_1 + & x_2 & +x_3 + & x_4 & & +y_1 & & = 1 \\ & x_1 & +3x_2 & +x_3 & +4x_4 & -x_5 & & +y_2 & = 100 \\ & x_1, & x_2, & x_3, & x_4, & x_5, & y_1, & y_2 \geq & 0 \end{array}$$

Iteration 1

$$\mathbf{B} = [\mathbf{A}_{.6}, \mathbf{A}_{.7}] = \begin{bmatrix} 1 & 0 \\ 0 & 1 \end{bmatrix}$$

$$\mathbf{N} = [\mathbf{A}_{.1}, \mathbf{A}_{.2}, \mathbf{A}_{.3}, \mathbf{A}_{.4}, \mathbf{A}_{.5}] = \begin{bmatrix} 1 & 1 & 1 & 1 & 0 \\ 1 & 3 & 1 & 4 & -1 \end{bmatrix}$$

$$\mathbf{x_B} = \begin{bmatrix} y_1 \\ y_2 \end{bmatrix} = \mathbf{B}^{-1}\mathbf{b} = \begin{bmatrix} 1 & 0 \\ 0 & 1 \end{bmatrix} \begin{bmatrix} 1 \\ 100 \end{bmatrix} = \begin{bmatrix} 1 \\ 100 \end{bmatrix}$$

$$\mathbf{c_N} - \mathbf{c_B}\mathbf{B}^{-1}\mathbf{N} = (0,0,0,0,0,0) - (1,1) \begin{bmatrix} 1 & 0 \\ 0 & 1 \end{bmatrix} \begin{bmatrix} 1 & 1 & 1 & 1 & 0 \\ 1 & 3 & 1 & 4 & -1 \end{bmatrix}$$

$$= (-2, -4, -2, -5, 1) \qquad \text{so} \quad j^* = 4$$

$$\mathbf{d_B} = -\mathbf{B}^{-1}\mathbf{N}_{.j^*} = -\begin{bmatrix} 1 & 0 \\ 0 & 1 \end{bmatrix} \begin{bmatrix} 1 \\ 4 \end{bmatrix} = \begin{bmatrix} -1 \\ -4 \end{bmatrix}$$

$$t^* = \min\{-(\mathbf{x_B})_1/(\mathbf{d_B})_1, -(\mathbf{x_B})_2/(\mathbf{d_B})_2\} = \min\{1, 25\}$$

$$= 1 \qquad \text{so} \quad k^* = 1$$

Column $k^* = 1$ of $\mathbf{B}$ is replaced with column $j^* = 4$ of $\mathbf{N}$.

Iteration 2

$$\mathbf{B} = [\mathbf{A}_{.4}, \mathbf{A}_{.7}] = \begin{bmatrix} 1 & 0 \\ 4 & 1 \end{bmatrix}$$

$$\mathbf{N} = [\mathbf{A}_{.1}, \mathbf{A}_{.2}, \mathbf{A}_{.3}, \mathbf{A}_{.6}, \mathbf{A}_{.5}] = \begin{bmatrix} 1 & 1 & 1 & 1 & 0 \\ 1 & 3 & 1 & 0 & -1 \end{bmatrix}$$

$$\mathbf{x_B} = \begin{bmatrix} x_4 \\ y_2 \end{bmatrix} = \mathbf{B}^{-1}\mathbf{b} = \begin{bmatrix} 1 & 0 \\ -4 & 1 \end{bmatrix} \begin{bmatrix} 1 \\ 100 \end{bmatrix} = \begin{bmatrix} 1 \\ 96 \end{bmatrix}$$

$$\mathbf{c_N} - \mathbf{c_B}\mathbf{B}^{-1}\mathbf{N} = (0,0,0,1,0) - (0,1) \begin{bmatrix} 1 & 0 \\ -4 & 1 \end{bmatrix} \begin{bmatrix} 1 & 1 & 1 & 1 & 0 \\ 1 & 3 & 1 & 0 & -1 \end{bmatrix}$$

$$= (3, 1, 3, 5, 1)$$

The test for optimality has been passed. Since y_2 is basic at value 96, the original LP is infeasible.

This section has shown that the simplex algorithm of Chapter 5 always produces an optimal solution to the phase 1 LP in a finite number of iterations. The reason is that the phase 1 LP (a) has an initial bfs from which to get started and (b) cannot be unbounded. The solution thus obtained can easily be used to determine if the original LP is feasible or not. In the latter case, there is no hope of finding an initial bfs for the original LP. The next section explains how to find an initial bfs for the original LP in the former case.

EXERCISES

6.2.1. Use the simplex algorithm to solve each of the following phase 1 problems. Using the solution thus obtained, indicate whether the original LP is feasible or not. For those that are feasible, is the resulting feasible solution a bfs for the original LP? Why or why not? Explain. (For future reference, the original minimization objective functions are also given.)

(a) Minimize $y_1 + y_2$

subject to

$$\begin{aligned} x_1 + x_2 + y_1 \quad &= 3 \\ 2x_1 + 6x_2 \quad + y_2 &= 7 \\ x_1, \; x_2, \; y_1, \; y_2 &\geq 0 \end{aligned}$$

The original objective function is $x_1 + x_2$.

(b) Minimize $y_1 + y_2$

subject to

$$\begin{aligned} x_1 + x_2 - x_3 + y_1 \quad &= 1 \\ x_1 + \tfrac{1}{2}x_2 - x_3 \quad + y_2 &= 6 \\ x_1, \; x_2, \; x_3, \; y_1, \; y_2 &\geq 0 \end{aligned}$$

The original objective function is $x_1 - 2x_2 + 3x_3$.

(c) Minimize $y_1 + y_2$

subject to

$$\begin{aligned} x_1 - 2x_2 + 3x_3 + y_1 \quad &= 1 \\ 2x_1 - 4x_2 + 6x_3 \quad + y_2 &= 2 \\ x_1, \; x_2, \; x_3, \; y_1, \; y_2 &\geq 0 \end{aligned}$$

The original objective function is $3x_1 - x_2 - 2x_3$.

6.3. INITIATING PHASE 2

Recall that the ultimate goal of phase 1 is to find an initial bfs for the original LP (or determine that none exists). Suppose that the simplex algorithm has produced an optimal solution $(\mathbf{x}^*, \mathbf{y}^*)$ to the phase 1 LP. In the event that $\mathbf{y}^* \neq \mathbf{0}$, the original LP is infeasible, and hence no initial bfs can be found (see Proposition 6.2). On the other hand, if $\mathbf{y}^* = \mathbf{0}$, then $\mathbf{x}^*$ is feasible for the original LP (see Proposition 6.1). In this event, it would be convenient if, in addition, $\mathbf{x}^*$ were a bfs from which to initiate the simplex algorithm on the original LP. Unfortunately, it is not always the case that the feasible vector $\mathbf{x}^*$ obtained at the end of phase 1 is a bfs for the original LP. Nonetheless, $\mathbf{x}^*$ can always be used to produce an initial bfs from which to proceed. The process of solving the original LP after phase 1 is completed is referred to as *phase 2*. This section will describe how to initiate phase 2.

The first step in this direction is to determine under what conditions one would be fortunate enough to be able to initiate phase 2 directly from $\mathbf{x}^*$. This is the purpose of the next proposition.

Proposition 6.5. *If* $(\mathbf{x}^*, \mathbf{y}^*)$ *is a bfs for the phase* 1 *LP and all of the* $\mathbf{y}$ *variables are nonbasic, then* $\mathbf{x}^*$ *is a bfs for the original LP.*

OUTLINE OF PROOF. The forward–backward method gives rise to the abstraction question "How can I show that a vector (namely, $\mathbf{x}^*$) is a bfs for the original LP?" One answer is to use the definition, whereby the appropriate submatrix $\mathbf{B}$ of $\mathbf{A}$ is produced and all of the desired properties are shown to hold. The $\mathbf{B}$ matrix will be constructed by working forward from the knowledge that $(\mathbf{x}^*, \mathbf{y}^*)$ is a bfs for the phase 1 LP and all the $\mathbf{y}^*$ variables are nonbasic. Specifically, at the optimal solution $(\mathbf{x}^*, \mathbf{y}^*)$, since the $\mathbf{y}^*$ variables are nonbasic, m of the $\mathbf{x}^*$ variables must be basic. These same m variables also serve to show that $\mathbf{x}^*$ is a bfs for the original LP, and the associated columns of the $\mathbf{A}$ matrix will make up the desired $\mathbf{B}$ matrix. The properties that $\mathbf{B}$ is nonsingular, $\mathbf{x}^*_\mathbf{B} = \mathbf{B}^{-1}\mathbf{b} \geq \mathbf{0}$, and $\mathbf{x}^*_\mathbf{N} = \mathbf{0}$ are obtained by working forward from the hypothesis that $(\mathbf{x}^*, \mathbf{y}^*)$ is a bfs for the phase 1 LP.

PROOF OF PROPOSITION 6.5. Since $(\mathbf{x}^*, \mathbf{y}^*)$ is a bfs for the phase 1 LP, and none of the $\mathbf{y}^*$ variables are basic, m of the $\mathbf{x}$ variables must be basic. These variables together with their associated columns of the $\mathbf{A}$ matrix will make up the basis matrix for the original LP. The $\mathbf{B}$ matrix thus obtained satisfies all of the desired properties because $(\mathbf{x}^*, \mathbf{y}^*)$ is a bfs for the phase 1 LP. Specifically, $\mathbf{B}$ is nonsingular, and, since $\mathbf{y}^* = \mathbf{0}$,

$$\begin{aligned}\mathbf{A}\mathbf{x}^* + \mathbf{I}\mathbf{y}^* &= \mathbf{B}\mathbf{x}^*_\mathbf{B} + \mathbf{N}\mathbf{x}^*_\mathbf{N} + \mathbf{I}\mathbf{y}^* = \mathbf{B}\mathbf{x}^*_\mathbf{B} + \mathbf{N}(\mathbf{0}) + \mathbf{I}(\mathbf{0})\\ &= \mathbf{B}\mathbf{x}^*_\mathbf{b} = \mathbf{b}\end{aligned}$$

It then follows that

1. $\mathbf{x}^*_\mathbf{B} = \mathbf{B}^{-1}\mathbf{b}$,
2. $\mathbf{B}^{-1}\mathbf{b} \geq \mathbf{0}$, and
3. $\mathbf{x}^*_\mathbf{N} = \mathbf{0}$.

Hence $\mathbf{x}$ is a bfs for the original LP, as desired. □

In Example 6.3, when phase 1 was completed both y_1 and y_2 were nonbasic variables, and x_3 and x_4 were basic at the values $\frac{11}{7}$ and $\frac{12}{7}$, respectively. Since y_1 and y_2 are nonbasic, the vector $\mathbf{x} = (0, 0, \frac{11}{7}, \frac{12}{7}, 0, 0)$ is an initial bfs for the original LP of Example 6.1. Indeed,

$$\mathbf{B} = [\mathbf{A}_{\cdot 3}, \mathbf{A}_{\cdot 4}] = \begin{bmatrix} 1 & 2 \\ 3 & -1 \end{bmatrix}$$

$$\mathbf{N} = [\mathbf{A}_{\cdot 1}, \mathbf{A}_{\cdot 2}, \mathbf{A}_{\cdot 5}, \mathbf{A}_{\cdot 6}] = \begin{bmatrix} 1 & 1 & -1 & 0 \\ 2 & 1 & 0 & -1 \end{bmatrix}$$

and, on performing the test for optimality using the original cost vector $\mathbf{c} = (2, 3, 1, 2, 0, 0)$, one has

$$\begin{aligned} \mathbf{c_N} - \mathbf{c_B}\mathbf{B}^{-1}\mathbf{N} &= (2,3,0,0) - (1,2)\begin{bmatrix} \frac{1}{7} & \frac{2}{7} \\ \frac{3}{7} & -\frac{1}{7} \end{bmatrix}\begin{bmatrix} 1 & 1 & -1 & 0 \\ 2 & 1 & 0 & -1 \end{bmatrix} \\ &= (1, 2, 1, 0) \end{aligned}$$

Since the reduced costs are nonnegative, the bfs $\mathbf{x} = (0, 0, \frac{11}{7}, \frac{12}{7}, 0, 0)$ is optimal for the original LP of Example 6.1.

It is indeed unfortunate that there are times when $\mathbf{x}^*$ cannot be used as a bfs for the original LP even though $\mathbf{y}^* = \mathbf{0}$. The reason is that some of the $\mathbf{y}$ variables may still be in the basis when phase 1 is finished, as demonstrated in the next example.

EXAMPLE 6.5: *Original LP.*

$$\begin{array}{ll} \text{Minimize} & 2x_1 + x_2 + 3x_3 - x_4 \\ \text{subject to} & 2x_1 + 3x_2 - x_3 + x_4 - x_5 = 2 \\ & 4x_1 + 6x_2 - 2x_3 + 2x_4 = 4 \\ & x_1, \quad x_2, \quad x_3, \quad x_4, \quad x_5 \geq 0 \end{array}$$

Phase 1 LP.

$$\begin{array}{ll} \text{Minimize} & y_1 + y_2 \\ \text{subject to} & 2x_1 + 3x_2 - x_3 + x_4 - x_5 + y_1 = 2 \\ & 4x_1 + 6x_2 - 2x_3 + 2x_4 + y_2 = 4 \\ & x_1, \quad x_2, \quad x_3, \quad x_4, \quad x_5, \quad y_1, \quad y_2 \geq 0 \end{array}$$

For the phase 1 LP

$$\mathbf{c} = (0, 0, 0, 0, 0, 1, 1)$$

$$\mathbf{A} = \begin{bmatrix} 2 & 3 & -1 & 1 & -1 & 1 & 0 \\ 4 & 6 & -2 & 2 & 0 & 0 & 1 \end{bmatrix}$$

$$\mathbf{b} = \begin{bmatrix} 2 \\ 4 \end{bmatrix}$$

Iteration 1

$$\mathbf{B} = [\mathbf{A}_{.6}, \mathbf{A}_{.7}] = \begin{bmatrix} 1 & 0 \\ 0 & 1 \end{bmatrix}$$

$$\mathbf{N} = [\mathbf{A}_{.1}, \mathbf{A}_{.2}, \mathbf{A}_{.3}, \mathbf{A}_{.4}, \mathbf{A}_{.5}] = \begin{bmatrix} 2 & 3 & -1 & 1 & -1 \\ 4 & 6 & -2 & 2 & 0 \end{bmatrix}$$

$$\mathbf{x_B} = \begin{bmatrix} y_1 \\ y_2 \end{bmatrix} = \mathbf{B}^{-1}\mathbf{b} = \begin{bmatrix} 1 & 0 \\ 0 & 1 \end{bmatrix}\begin{bmatrix} 2 \\ 4 \end{bmatrix} = \begin{bmatrix} 2 \\ 4 \end{bmatrix}$$

$$\mathbf{c_N} - \mathbf{c_B}\mathbf{B}^{-1}\mathbf{N} = (0,0,0,0,0) - (1,1)\begin{bmatrix} 1 & 0 \\ 0 & 1 \end{bmatrix}\begin{bmatrix} 2 & 3 & -1 & 1 & -1 \\ 4 & 6 & -2 & 2 & 0 \end{bmatrix}$$

$$= (-6, -9, 3, -3, 1) \qquad \text{so} \quad j^* = 2$$

$$\mathbf{d_B} = -\mathbf{B}^{-1}\mathbf{N}_{.j^*} = -\begin{bmatrix} 1 & 0 \\ 0 & 1 \end{bmatrix}\begin{bmatrix} 3 \\ 6 \end{bmatrix} = \begin{bmatrix} -3 \\ -6 \end{bmatrix}$$

$$t^* = \min\{-(\mathbf{x_B})_1/(\mathbf{d_B})_1, -(\mathbf{x_B})_2/(\mathbf{d_B})_2\} = \min\{\tfrac{2}{3}, \tfrac{4}{6}\}$$

$$= \tfrac{2}{3} \qquad \text{so} \quad k^* = 1 \text{ or } 2$$

Choose $k^* = 2$. Column $k^* = 2$ of $\mathbf{B}$ is replaced with column $j^* = 2$ of $\mathbf{N}$.

Iteration 2

$$\mathbf{B} = [\mathbf{A}_{.6}, \mathbf{A}_{.2}] = \begin{bmatrix} 1 & 3 \\ 0 & 6 \end{bmatrix}$$

$$\mathbf{N} = [\mathbf{A}_{.1}, \mathbf{A}_{.7}, \mathbf{A}_{.3}, \mathbf{A}_{.4}, \mathbf{A}_{.5}] = \begin{bmatrix} 2 & 0 & -1 & 1 & -1 \\ 4 & 1 & -2 & 2 & 0 \end{bmatrix}$$

$$\mathbf{x_B} = \begin{bmatrix} y_1 \\ x_2 \end{bmatrix} = \mathbf{B}^{-1}\mathbf{b} = \begin{bmatrix} 1 & -\frac{1}{2} \\ 0 & \frac{1}{6} \end{bmatrix}\begin{bmatrix} 2 \\ 4 \end{bmatrix} = \begin{bmatrix} 0 \\ \frac{2}{3} \end{bmatrix}$$

$$\mathbf{c_N} - \mathbf{c_B}\mathbf{B}^{-1}\mathbf{N} = (0,1,0,0,0) - (1,0)\begin{bmatrix} 1 & -\frac{1}{2} \\ 0 & \frac{1}{6} \end{bmatrix}\begin{bmatrix} 2 & 0 & -1 & 1 & -1 \\ 4 & 1 & -2 & 2 & 0 \end{bmatrix}$$

$$= (0, \tfrac{3}{2}, 0, 0, 1)$$

Since the reduced costs are nonnegative, optimality has been detected. Furthermore, y_2 is nonbasic and hence has the value 0. Also, y_1 is basic at the value 0. Thus $\mathbf{x} = (0, \frac{2}{3}, 0, 0, 0)$ is feasible for the original LP. Unfortunately, since y_1 is basic, $\mathbf{x}$ cannot be used as an initial bfs for phase 2.

One possibility for dealing with such a problem is to attempt to replace all the basic $\mathbf{y}$ variables with some of the nonbasic $\mathbf{x}$ variables. Such an approach has been developed, but it will not be described here. Instead, a new *auxiliary* LP will be created from the phase 1 LP. The advantage of the new LP is that the solution $(\mathbf{x}^*, \mathbf{y}^*)$ obtained at the end of phase 1 can always be used to construct an initial bfs for the auxiliary LP. The simplex algorithm of Chapter 5 can then be applied to the auxiliary LP and will terminate in a finite number of steps in an unbounded or optimal condition. Whichever occurs, the same condition can be shown to apply to the original LP.

To construct the auxiliary LP, begin with the phase 1 LP. Then

1. replace the objective function **ey** with the original objective function **cx**, and
2. add one new constraint of the form $\mathbf{ey} = 0$ so as to force all of the **y** variables to remain at 0 value.

For reasons that will soon become clear, a single new variable z is also added. Thus, in matrix/vector notation, the auxiliary LP is to

$$\begin{array}{llll} \text{minimize} & \mathbf{cx} & & \\ \text{subject to} & \mathbf{Ax} & +\mathbf{Iy} & = \mathbf{b} \\ & & \mathbf{ey} \ +z & = 0 \\ & \mathbf{x}, & \mathbf{y}, & z \geq \mathbf{0} \end{array}$$

or, equivalently, to

$$\begin{array}{ll} \text{minimize} & (\mathbf{c}, \mathbf{0}, 0)\begin{bmatrix} \mathbf{x} \\ \mathbf{y} \\ z \end{bmatrix} \\ \text{subject to} & \begin{bmatrix} \mathbf{A} & \mathbf{I} & \mathbf{0} \\ \mathbf{0} & \mathbf{e} & 1 \end{bmatrix}\begin{bmatrix} \mathbf{x} \\ \mathbf{y} \\ z \end{bmatrix} = \begin{bmatrix} \mathbf{b} \\ 0 \end{bmatrix} \\ & \mathbf{x}, \mathbf{y}, z \geq \mathbf{0} \end{array}$$

The constraint $\mathbf{ey} + z = 0$ together with $\mathbf{y}, z \geq \mathbf{0}$ guarantees that an optimal solution to the auxiliary LP produces an optimal solution to the original LP, as shown in the next proposition.

Proposition 6.6. *If* $(\mathbf{x}', \mathbf{y}', z')$ *is an optimal solution to the auxiliary LP, then* $\mathbf{x}'$ *is optimal for the original LP.*

OUTLINE OF PROOF. The forward–backward method gives rise to the abstraction question "How can I show that a vector (namely, $\mathbf{x}'$) is optimal for the original LP?" One answer is to use the definition, whereby it must be shown that

1. $\mathbf{Ax}' = \mathbf{b}$ and $\mathbf{x}' \geq \mathbf{0}$ and,
2. for all $\mathbf{x}$ with $\mathbf{Ax} = \mathbf{b}$ and $\mathbf{x} \geq \mathbf{0}$, $\mathbf{cx}' \leq \mathbf{cx}$.

Both of these statements are obtained by working forward from the optimality of $(\mathbf{x}', \mathbf{y}', z')$ for the auxiliary LP. Specifically, $(\mathbf{x}', \mathbf{y}', z')$ must be feasible for the auxiliary LP, so $\mathbf{Ax}' + \mathbf{Iy}' = \mathbf{b}$, $\mathbf{ey}' + z' = 0$, and $\mathbf{x}', \mathbf{y}', z' \geq \mathbf{0}$. Thus $\mathbf{x}' \geq \mathbf{0}$, but it must still be shown that $\mathbf{Ax}' = \mathbf{b}$, which will be accomplished by showing that $\mathbf{y}' = \mathbf{0}$, for then $\mathbf{Ax}' + \mathbf{Iy}' = \mathbf{Ax}' + \mathbf{I}(\mathbf{0}) = \mathbf{Ax}' = \mathbf{b}$. To see that $\mathbf{y}' = \mathbf{0}$, note that $\mathbf{ey}' + z' = 0$ and $\mathbf{y}', z' \geq \mathbf{0}$, i.e., $\sum_{i=1}^{m} y_i' + z' = 0$ and $\mathbf{y}', z' \geq \mathbf{0}$. The only way that the nonnegative numbers $y_1, y_2, \ldots, y_m$, and z can add up to 0 is if each and every one of them is 0, for if one of them were strictly positive, another one would have to be strictly negative so that the sum would be 0. Thus $\mathbf{y}' = \mathbf{0}$ and $z' = 0$, and hence $\mathbf{x}'$ is feasible for the original LP.

It remains to show that, for all $\mathbf{x}$ with $\mathbf{Ax} = \mathbf{b}$ and $\mathbf{x} \geq \mathbf{0}$, $\mathbf{cx}' \leq \mathbf{cx}$. The appearance of the quantifier "for all" in the backward process suggests using the choose method, whereby an $\mathbf{x}$ is chosen with the property that $\mathbf{Ax} = \mathbf{b}$ and $\mathbf{x} \geq \mathbf{0}$ and it must be shown that $\mathbf{cx}' \leq \mathbf{cx}$. Now it is time to use the fact that $(\mathbf{x}', \mathbf{y}', z')$ is optimal for the auxiliary LP.

Since $\mathbf{Ax} = \mathbf{b}$ and $\mathbf{x} \geq \mathbf{0}$, the vector $(\mathbf{x}, \mathbf{0}, 0)$ is feasible for the auxiliary LP because $\mathbf{Ax} + \mathbf{I}(\mathbf{0}) = \mathbf{Ax} = \mathbf{b}$, $\mathbf{e}(\mathbf{0}) + 0 = 0$, and $(\mathbf{x}, \mathbf{0}, 0) \geq \mathbf{0}$. Hence the objective value of the auxiliary LP at the optimal solution $(\mathbf{x}', \mathbf{y}', z')$ must be better than at the feasible solution $(\mathbf{x}, \mathbf{0}, 0)$; i.e., $\mathbf{cx}' \leq \mathbf{cx}$.

PROOF OF PROPOSITION 6.6. First it will be shown that $\mathbf{x}'$ is feasible for the original LP. Since $(\mathbf{x}', \mathbf{y}', z')$ is feasible for the auxiliary LP, $\mathbf{Ax}' + \mathbf{Iy}' = \mathbf{b}$, $\mathbf{ey}' + z' = 0$, and $\mathbf{x}', \mathbf{y}', z' \geq 0$. Thus $\mathbf{x}' \geq \mathbf{0}$, and, since $\mathbf{ey}' + z' = \sum_{i=1}^{m} y_i' + z' = 0$ and $\mathbf{y}', z' \geq 0$, it must be the case that $\mathbf{y}' = \mathbf{0}$ and $z' = 0$. Therefore $\mathbf{Ax}' + \mathbf{Iy}' = \mathbf{Ax}' + \mathbf{I}(\mathbf{0}) = \mathbf{Ax}' = \mathbf{b}$. In other words, $\mathbf{Ax}' = \mathbf{b}$ and $\mathbf{x}' \geq \mathbf{0}$, whereby $\mathbf{x}'$ is feasible for the original LP.

It remains to show that $\mathbf{x}'$ is optimal for the original LP. To that end, let $\mathbf{x}$ be any feasible solution to the original LP, so $\mathbf{Ax} = \mathbf{b}$ and $\mathbf{x} \geq \mathbf{0}$. Hence $(\mathbf{x}, \mathbf{0}, 0)$ is feasible for the auxiliary LP. Thus it must be that $\mathbf{cx}' \leq \mathbf{cx}$, and so $\mathbf{x}'$ is optimal for the original LP. □

The previous proposition established that if an optimal solution $(\mathbf{x}', \mathbf{y}', z')$ to the auxiliary LP can be found, then $\mathbf{x}'$ is an optimal solution to the original LP. The next proposition shows that if the auxiliary LP is unbounded, then so is the original one.

Proposition 6.7. *If the auxiliary LP is unbounded, then the original LP is unbounded.*

OUTLINE OF PROOF. The forward–backward method gives rise to the abstraction question "How can I show that an LP (namely, the original one) is unbounded?" One answer is by the definition, whereby it must be established that

1. the original LP is feasible and
2. there is a direction of unboundedness.

Both of these statements can be obtained by working forward from the hypothesis that the auxiliary LP is unbounded. Hence, the auxiliary LP is (a) feasible and (b) has a direction $\mathbf{d}$ of unboundedness. Suppose that $(\mathbf{x}, \mathbf{y}, z)$ is feasible for the auxiliary LP. As in the proof of Proposition 6.6, it can be shown that $\mathbf{y} = \mathbf{0}$ and $z = 0$, and hence $\mathbf{x}$ is feasible for the original LP.

All that remains is to use the direction of unboundedness $\mathbf{d}$ for the auxiliary LP to construct the corresponding direction of unboundedness for the original LP. To this end, let $\mathbf{d}_\mathbf{x}$, $\mathbf{d}_\mathbf{y}$, and d_z be the components of $\mathbf{d}$ corresponding to the $\mathbf{x}$, $\mathbf{y}$, and z variables, respectively. Hence $\mathbf{A}\mathbf{d}_\mathbf{x} + \mathbf{I}\mathbf{d}_\mathbf{y} = \mathbf{0}$, $\mathbf{e}\mathbf{d}_\mathbf{y} + d_z = 0$, $(\mathbf{d}_\mathbf{x}, \mathbf{d}_\mathbf{y}, d_z) \geq \mathbf{0}$, and $\mathbf{c}\mathbf{d}_\mathbf{x} < 0$. Using this information, it can be shown that $\mathbf{d}_\mathbf{x}$ is the desired direction of unboundedness for the original LP:

1. $\mathbf{A}\mathbf{d}_\mathbf{x} = \mathbf{0}$.
2. $\mathbf{d}_\mathbf{x} \geq \mathbf{0}$.
3. $\mathbf{c}\mathbf{d}_\mathbf{x} < 0$.

Clearly, conditions 2 and 3 are true because $\mathbf{d}$ is a direction of unboundedness for the auxiliary LP. It remains to work forward from $\mathbf{A}\mathbf{d}_\mathbf{x} + \mathbf{I}\mathbf{d}_\mathbf{y} = \mathbf{0}$ and $\mathbf{e}\mathbf{d}_\mathbf{y} + d_z = 0$ to show that $\mathbf{A}\mathbf{d}_\mathbf{x} = \mathbf{0}$. Since $\mathbf{A}\mathbf{d}_\mathbf{x} = -\mathbf{I}\mathbf{d}_\mathbf{y}$, the result will follow if $\mathbf{d}_\mathbf{y}$ can be shown to be $\mathbf{0}$. But $\mathbf{d}_\mathbf{y}$, $d_z \geq \mathbf{0}$ and $\mathbf{e}\mathbf{d}_\mathbf{y} + d_z = 0$. As in the proof of Proposition 6.6, the only way that the sum of the nonnegative numbers $(\mathbf{d}_\mathbf{y})_1, \ldots, (\mathbf{d}_\mathbf{y})_m$, and d_z can be 0 is for all of the numbers to be 0, so $\mathbf{d}_\mathbf{y} = \mathbf{0}$ and $\mathbf{A}\mathbf{d}_\mathbf{x} = -\mathbf{I}\mathbf{d}_\mathbf{y} = \mathbf{0}$.

PROOF OF PROPOSITION 6.7. Since the auxiliary LP is unbounded, then by definition,

1. there is a vector $(\mathbf{x}, \mathbf{y}, z)$ such that $\mathbf{A}\mathbf{x} + \mathbf{I}\mathbf{y} = \mathbf{b}$, $\mathbf{e}\mathbf{y} + z = 0$, and $(\mathbf{x}, \mathbf{y}, z) \geq 0$, and
2. there is a direction $\mathbf{d}$ of unboundedness.

Let $\mathbf{d}_\mathbf{x}$, $\mathbf{d}_\mathbf{y}$, and d_z be the components of the vector $\mathbf{d}$ corresponding to the $\mathbf{x}$, $\mathbf{y}$, and z variables, respectively, Thus $\mathbf{A}\mathbf{d}_\mathbf{x} + \mathbf{I}\mathbf{d}_\mathbf{y} = \mathbf{0}$, $\mathbf{e}\mathbf{d}_\mathbf{y} + d_z = 0$, $(\mathbf{d}_\mathbf{x}, \mathbf{d}_\mathbf{y}, d_z) \geq \mathbf{0}$, and $\mathbf{c}\mathbf{d}_\mathbf{x} < 0$.

To show that the original LP is unbounded, it will be established that $\mathbf{x}$ is feasible for the original LP and that $\mathbf{d}_\mathbf{x}$ is the direction of unboundedness. To see that $\mathbf{x}$ is feasible, note that $\mathbf{x} \geq \mathbf{0}$ and $\mathbf{A}\mathbf{x} + \mathbf{I}\mathbf{y} = \mathbf{b}$. As in the proof of Proposition 6.6, it can be shown that $\mathbf{y} = \mathbf{0}$, and so $\mathbf{A}\mathbf{x} = \mathbf{b}$.

The direction $\mathbf{d}_\mathbf{x}$ must be shown to satisfy $\mathbf{A}\mathbf{d}_\mathbf{x} = \mathbf{0}$, $\mathbf{d}_\mathbf{x} \geq \mathbf{0}$, and $\mathbf{c}\mathbf{d}_\mathbf{x} < 0$. Since $\mathbf{d} = (\mathbf{d}_\mathbf{x}, \mathbf{d}_\mathbf{y}, d_z)$ is a direction of unboundedness for the auxiliary LP, it immediately follows that $\mathbf{d}_\mathbf{x} \geq \mathbf{0}$ and $\mathbf{c}\mathbf{d}_\mathbf{x} < 0$. To see that $\mathbf{A}\mathbf{d}_\mathbf{x} = \mathbf{0}$, observe that $\mathbf{A}\mathbf{d}_\mathbf{x} = -\mathbf{I}\mathbf{d}_\mathbf{y}$, and thus it will be established that $\mathbf{d}_\mathbf{y} = \mathbf{0}$. But $\mathbf{e}\mathbf{d}_\mathbf{y} + d_z = 0$ and $\mathbf{d}_\mathbf{y}$, $d_z \geq \mathbf{0}$. The only way that the sum of the nonnegative numbers $(\mathbf{d}_\mathbf{y})_1, \ldots, (\mathbf{d}_\mathbf{y})_m$, and d_z can be 0 is for all of the numbers to be 0. Thus $\mathbf{d}_\mathbf{y} = \mathbf{0}$, and so $\mathbf{A}\mathbf{d}_\mathbf{x} = -\mathbf{I}\mathbf{d}_\mathbf{y} = \mathbf{0}$. □

The previous two propositions have established the equivalence of the auxiliary LP and the original one in the sense that a solution to the auxiliary LP provides one for the original LP, and the unboundedness of the auxiliary LP indicates that the original LP is unbounded too. The last task is to show that, for the auxiliary LP, an initial bfs can be constructed from an optimal bfs $(\mathbf{x}^*, \mathbf{y}^*)$ obtained from the phase 1 problem. Observe that a bfs for the auxiliary LP would require $m+1$ variables because there are $m+1$ constraints. Since phase 1 can produce only m basic variables, one more must be found. That is the purpose of the z variable, as shown in the next proposition.

Proposition 6.8. *If* $(\mathbf{x}^*, \mathbf{y}^*)$ *is an optimal bfs for the phase* 1 *LP and* $\mathbf{y}^* = \mathbf{0}$, *then the vector* $(\mathbf{x}, \mathbf{y}, z) = (\mathbf{x}^*, \mathbf{y}^*, 0)$ *is a bfs for the auxiliary LP.*

OUTLINE OF PROOF. The forward–backward method gives rise to the abstraction question "How can I show that a vector (namely, $(\mathbf{x}^*, \mathbf{y}^*, 0)$) is a bfs?" One answer is the definition, whereby the appropriate $(m+1) \times (m+1)$ basis matrix must be produced along with the $m+1$ basic variables, for which it must be shown that all of the desired properties hold. To do so requires working forward from the hypothesis that $(\mathbf{x}^*, \mathbf{y}^*)$ is an optimal bfs for the phase 1 LP in which $\mathbf{y}^* = \mathbf{0}$. Specifically, there is an $m \times m$ nonsingular basis matrix $\mathbf{B}$ of $[\mathbf{A}, \mathbf{I}]$ such that

1. $\mathbf{B}^{-1}\mathbf{b} = (\mathbf{x}^*, \mathbf{y}^*)_{\mathbf{B}}$,
2. $\mathbf{B}^{-1}\mathbf{b} \geq \mathbf{0}$, and
3. $(\mathbf{x}^*, \mathbf{y}^*)_{\mathbf{N}} = \mathbf{0}$.

To construct the $(m+1) \times (m+1)$ basis matrix for the auxiliary LP, recall that $\mathbf{B}$ consists of m columns of $[\mathbf{A}, \mathbf{I}]$. The corresponding columns of

$$\begin{bmatrix} \mathbf{A} & \mathbf{I} \\ \mathbf{0} & \mathbf{e} \end{bmatrix}$$

together with the column

$$\begin{bmatrix} \mathbf{0} \\ 1 \end{bmatrix}$$

(for the z variable) make up the desired basis matrix. To write it, let $\mathbf{f}$ be the vector consisting of the appropriate components of the vector $(\mathbf{0}, \mathbf{e})$; that is, $\mathbf{f}$ is the m vector $(\mathbf{0}, \mathbf{e})_{\mathbf{B}}$. Then, the $(m+1) \times (m+1)$ basis matrix for the auxiliary LP is

$$\begin{bmatrix} \mathbf{B} & \mathbf{0} \\ \mathbf{f} & 1 \end{bmatrix}$$

To verify that this matrix has all of the desired properties, the corresponding properties of the $\mathbf{B}$ matrix will be helpful. For instance, to see that

$$\begin{bmatrix} \mathbf{B} & \mathbf{0} \\ \mathbf{f} & 1 \end{bmatrix}$$

is nonsingular, the inverse can be written using $\mathbf{B}^{-1}$. Specifically,

$$\begin{bmatrix} \mathbf{B} & \mathbf{0} \\ \mathbf{f} & 1 \end{bmatrix}^{-1} = \begin{bmatrix} \mathbf{B}^{-1} & \mathbf{0} \\ -\mathbf{f}\mathbf{B}^{-1} & 1 \end{bmatrix}$$

because

$$\begin{aligned} \begin{bmatrix} \mathbf{B} & \mathbf{0} \\ \mathbf{f} & 1 \end{bmatrix}\begin{bmatrix} \mathbf{B} & \mathbf{0} \\ \mathbf{f} & 1 \end{bmatrix}^{-1} &= \begin{bmatrix} \mathbf{B} & \mathbf{0} \\ \mathbf{f} & 1 \end{bmatrix}\begin{bmatrix} \mathbf{B}^{-1} & \mathbf{0} \\ -\mathbf{f}\mathbf{B}^{-1} & 1 \end{bmatrix} \\ &= \begin{bmatrix} \mathbf{B}\mathbf{B}^{-1} - (\mathbf{0})\mathbf{f}\mathbf{B}^{-1} & \mathbf{B}(\mathbf{0}) + \mathbf{0}(1) \\ \mathbf{f}\mathbf{B}^{-1} - (1)\mathbf{f}\mathbf{B}^{-1} & \mathbf{f}(\mathbf{0}) + 1(1) \end{bmatrix} \\ &= \begin{bmatrix} \mathbf{I} & \mathbf{0} \\ \mathbf{0} & 1 \end{bmatrix} \end{aligned}$$

the $(m+1)\times(m+1)$ identity matrix.

The next property to be verified is that the values of the basic variables for the auxiliary LP can be computed properly using the inverse. In so doing, it is helpful to make two observations:

1. $(\mathbf{x}^*,\mathbf{y}^*)_\mathbf{B} = \mathbf{B}^{-1}\mathbf{b}$ [because $(\mathbf{x}^*,\mathbf{y}^*)$ is a bfs for the phase 1 LP].
2. $\mathbf{f}(\mathbf{B}^{-1}\mathbf{b}) = 0$ [because $\mathbf{f}(\mathbf{B}^{-1}\mathbf{b}) = \Sigma_{i=1}^m f_i(\mathbf{B}^{-1}\mathbf{b})_i$ and because each component of $\mathbf{B}^{-1}\mathbf{b}$ that is not 0 corresponds to an $\mathbf{x}$ variable, since $\mathbf{y}^* = \mathbf{0}$, and so the corresponding component of $\mathbf{f} = (\mathbf{0},\mathbf{e})_\mathbf{B}$ is 0].

Thus

$$\begin{aligned} \begin{bmatrix} \mathbf{B}^{-1} & \mathbf{0} \\ -\mathbf{f}\mathbf{B}^{-1} & 1 \end{bmatrix}\begin{bmatrix} \mathbf{b} \\ 0 \end{bmatrix} &= \begin{bmatrix} \mathbf{B}^{-1}\mathbf{b} + \mathbf{0}(0) \\ -\mathbf{f}\mathbf{B}^{-1}\mathbf{b} + 1(0) \end{bmatrix} = \begin{bmatrix} \mathbf{B}^{-1}\mathbf{b} \\ 0 \end{bmatrix} \\ &= \begin{bmatrix} (\mathbf{x}^*,\mathbf{y}^*)_\mathbf{B} \\ z \end{bmatrix} \end{aligned}$$

The third property to check is that the values of the basic variables are nonnegative. Again, work forward from the fact that $(\mathbf{x}^*,\mathbf{y}^*)$ is a bfs for the phase 1 LP. Now, $\mathbf{B}^{-1}\mathbf{b} \geq \mathbf{0}$, so

$$\begin{bmatrix} (\mathbf{x}^*,\mathbf{y}^*)_\mathbf{B} \\ z \end{bmatrix} = \begin{bmatrix} \mathbf{B}^{-1}\mathbf{b} \\ 0 \end{bmatrix} \geq \mathbf{0}$$

The final property to be verified is that the nonbasic variables of the auxiliary LP have value 0. But the nonbasic variables of the auxiliary LP are precisely those of the phase 1 LP, so again, since $(\mathbf{x}^*,\mathbf{y}^*)$ is a bfs for the phase 1 LP, $(\mathbf{x}^*,\mathbf{y}^*)_\mathbf{N} = \mathbf{0}$.

PROOF OF PROPOSITION 6.8. Since $(\mathbf{x}^*,\mathbf{y}^*)$ is a bfs for the phase 1 LP, there is a nonsingular $m \times m$ submatrix $\mathbf{B}$ of $[\mathbf{A},\mathbf{I}]$ such that

1. $(\mathbf{x}^*,\mathbf{y}^*)_\mathbf{B} = \mathbf{B}^{-1}\mathbf{b}$,
2. $\mathbf{B}^{-1}\mathbf{b} \geq \mathbf{0}$, and
3. $(\mathbf{x}^*,\mathbf{y}^*)_\mathbf{N} = \mathbf{0}$.

To construct the $(m+1)\times(m+1)$ basis matrix for the bfs $(\mathbf{x}^*, \mathbf{y}^*, 0)$, recall that $\mathbf{B}$ consists of m columns from $[\mathbf{A}, \mathbf{I}]$. The corresponding columns of

$$\begin{bmatrix} \mathbf{A} & \mathbf{I} \\ \mathbf{0} & \mathbf{e} \end{bmatrix}$$

together with the column

$$\begin{bmatrix} \mathbf{0} \\ 1 \end{bmatrix}$$

(for the z variable) make up the desired matrix. To write this matrix, let $\mathbf{f}$ be the vector consisting of the appropriate components of the vector $(\mathbf{0}, \mathbf{e})$; that is, $\mathbf{f}$ is the m vector $(\mathbf{0}, \mathbf{e})_{\mathbf{B}}$. Then the basis matrix for the auxiliary LP is

$$\begin{bmatrix} \mathbf{B} & \mathbf{0} \\ \mathbf{f} & 1 \end{bmatrix}$$

To verify that this matrix is nonsingular, one writes the inverse

$$\begin{bmatrix} \mathbf{B}^{-1} & \mathbf{0} \\ -\mathbf{f}\mathbf{B}^{-1} & 1 \end{bmatrix}$$

because

$$\begin{bmatrix} \mathbf{B} & \mathbf{0} \\ \mathbf{f} & 1 \end{bmatrix}\begin{bmatrix} \mathbf{B}^{-1} & \mathbf{0} \\ -\mathbf{f}\mathbf{B}^{-1} & 1 \end{bmatrix} = \begin{bmatrix} \mathbf{I} & \mathbf{0} \\ \mathbf{0} & 1 \end{bmatrix}$$

The values of the basic variables for the auxiliary LP can be obtained by using the inverse matrix. In so doing, recall that $(\mathbf{x}^*, \mathbf{y}^*)_{\mathbf{B}} = \mathbf{B}^{-1}\mathbf{b}$. Also note that $\mathbf{f}(\mathbf{B}^{-1}\mathbf{b}) = \sum_{i=1}^{m} f_i(\mathbf{B}^{-1}\mathbf{b})_i = 0$ because each component of $\mathbf{B}^{-1}\mathbf{b}$ that is not 0 corresponds to an $\mathbf{x}$ variable (since $\mathbf{y}^* = \mathbf{0}$), and so the corresponding component of $\mathbf{f}$ is 0. Thus

$$\begin{bmatrix} \mathbf{B}^{-1} & \mathbf{0} \\ -\mathbf{f}\mathbf{B}^{-1} & 1 \end{bmatrix}\begin{bmatrix} \mathbf{b} \\ 0 \end{bmatrix} = \begin{bmatrix} \mathbf{B}^{-1}\mathbf{b} \\ -\mathbf{f}\mathbf{B}^{-1}\mathbf{b} \end{bmatrix} = \begin{bmatrix} \mathbf{B}^{-1}\mathbf{b} \\ 0 \end{bmatrix}$$
$$= \begin{bmatrix} (\mathbf{x}^*, \mathbf{y}^*)_{\mathbf{B}} \\ z \end{bmatrix}$$

Since $\mathbf{B}^{-1}\mathbf{b} \geq \mathbf{0}$,

$$\begin{bmatrix} \mathbf{B}^{-1}\mathbf{b} \\ 0 \end{bmatrix} \geq \mathbf{0}$$

Thus the basic variables of the auxiliary LP are nonnegative. Finally, the values of the nonbasic variables of the auxiliary LP are $(\mathbf{x}^*, \mathbf{y}^*)_{\mathbf{N}}$ and are 0. □

Proposition 6.8 shows how to use an optimal bfs to the phase 1 LP to produce an initial bfs for the auxiliary LP. The proof also shows how to create the new basis inverse from the old one. The process is illustrated for Example 6.5.

EXAMPLE 6.6: The auxiliary LP associated with Example 6.5 is to

$$\begin{array}{llllllllll}
\text{minimize} & 2x_1 & + x_2 & +3x_3 & - x_4 & & & & & \\
\text{subject to} & 2x_1 & +3x_2 & - x_3 & + x_4 & -x_5 & +y_1 & & & =2 \\
& 4x_1 & +6x_2 & -2x_3 & +2x_4 & & & +y_2 & & =4 \\
& & & & & & y_1 & +y_2 & +z & =0 \\
& x_1, & x_2, & x_3, & x_4, & x_5, & y_1, & y_2, & z & \geq 0
\end{array}$$

Iteration 1

$$\mathbf{c} = (2,1,3,-1,0,0,0,0)$$

$$\mathbf{A} = \begin{bmatrix} 2 & 3 & -1 & 1 & -1 & 1 & 0 & 0 \\ 4 & 6 & -2 & 2 & 0 & 0 & 1 & 0 \\ 0 & 0 & 0 & 0 & 0 & 1 & 1 & 1 \end{bmatrix}$$

$$\mathbf{b} = \begin{bmatrix} 2 \\ 4 \\ 0 \end{bmatrix}$$

From the last iteration of phase 1 in Example 6.5, the initial bfs for the auxiliary LP is

$$\mathbf{B} = [\mathbf{A}_{.6}, \mathbf{A}_{.2}, \mathbf{A}_{.8}] = \begin{bmatrix} 1 & 3 & 0 \\ 0 & 6 & 0 \\ 1 & 0 & 1 \end{bmatrix}$$

$$\mathbf{N} = [\mathbf{A}_{.1}, \mathbf{A}_{.7}, \mathbf{A}_{.3}, \mathbf{A}_{.4}, \mathbf{A}_{.5}] = \begin{bmatrix} 2 & 0 & -1 & 1 & -1 \\ 4 & 1 & -2 & 2 & 0 \\ 0 & 1 & 0 & 0 & 0 \end{bmatrix}$$

$$\mathbf{x_B} = \begin{bmatrix} y_1 \\ x_2 \\ z \end{bmatrix} = \mathbf{B}^{-1}\mathbf{b} = \begin{bmatrix} 1 & -\frac{1}{2} & 0 \\ 0 & \frac{1}{6} & 0 \\ -1 & \frac{1}{2} & 1 \end{bmatrix} \begin{bmatrix} 2 \\ 4 \\ 0 \end{bmatrix} = \begin{bmatrix} 0 \\ \frac{2}{3} \\ 0 \end{bmatrix}$$

$$\mathbf{c_N} - \mathbf{c_B}\mathbf{B}^{-1}\mathbf{N} = (2,0,3,-1,0) - (0,1,0) \begin{bmatrix} 1 & -\frac{1}{2} & 0 \\ 0 & \frac{1}{6} & 0 \\ -1 & \frac{1}{2} & 1 \end{bmatrix}$$

$$\times \begin{bmatrix} 2 & 0 & -1 & 1 & -1 \\ 4 & 1 & -2 & 2 & 0 \\ 0 & 1 & 0 & 0 & 0 \end{bmatrix}$$

$$= \left(\tfrac{4}{3}, -\tfrac{1}{6}, \tfrac{10}{3}, -\tfrac{4}{3}, 0\right) \qquad \text{so} \quad j^* = 4$$

$$\mathbf{d_B} = -\mathbf{B}^{-1}\mathbf{N}_{\cdot j^*} = -\begin{bmatrix} 1 & -\frac{1}{2} & 0 \\ 0 & \frac{1}{6} & 0 \\ -1 & \frac{1}{2} & 1 \end{bmatrix}\begin{bmatrix} 1 \\ 2 \\ 0 \end{bmatrix} = \begin{bmatrix} 0 \\ -\frac{1}{3} \\ 0 \end{bmatrix} \text{ so } \quad k^* = 2$$

Column $k^* = 2$ of $\mathbf{B}$ is replaced with column $j^* = 4$ of $\mathbf{N}$.

Iteration 2

$$\mathbf{B} = [\mathbf{A}_{\cdot 6}, \mathbf{A}_{\cdot 4}, \mathbf{A}_{\cdot 8}] = \begin{bmatrix} 1 & 1 & 0 \\ 0 & 2 & 0 \\ 1 & 0 & 1 \end{bmatrix}$$

$$\mathbf{N} = [\mathbf{A}_{\cdot 1}, \mathbf{A}_{\cdot 7}, \mathbf{A}_{\cdot 3}, \mathbf{A}_{\cdot 2}, \mathbf{A}_{\cdot 5}] = \begin{bmatrix} 2 & 0 & -1 & 3 & -1 \\ 4 & 1 & -2 & 6 & 0 \\ 0 & 1 & 0 & 0 & 0 \end{bmatrix}$$

$$\mathbf{x_B} = \begin{bmatrix} y_1 \\ x_4 \\ z \end{bmatrix} = \mathbf{B}^{-1}\mathbf{b} = \begin{bmatrix} 1 & -\frac{1}{2} & 0 \\ 0 & \frac{1}{2} & 0 \\ -1 & \frac{1}{2} & 1 \end{bmatrix}\begin{bmatrix} 2 \\ 4 \\ 0 \end{bmatrix} = \begin{bmatrix} 0 \\ 2 \\ 0 \end{bmatrix}$$

$$\mathbf{c_N} - \mathbf{c_B}\mathbf{B}^{-1}\mathbf{N} = (2,0,3,1,0) - (0,-1,0)\begin{bmatrix} 1 & -\frac{1}{2} & 0 \\ 0 & \frac{1}{2} & 0 \\ -1 & \frac{1}{2} & 1 \end{bmatrix}$$

$$\times \begin{bmatrix} 2 & 0 & -1 & 3 & -1 \\ 4 & 1 & -2 & 6 & 0 \\ 0 & 1 & 0 & 0 & 0 \end{bmatrix}$$

$$= (4, \tfrac{1}{2}, 2, 4, 0)$$

Since the reduced costs are nonnegative, optimality has been detected. The optimal solution is $(x_1, x_2, x_3, x_4, x_5, y_1, y_2, z) = (0,0,0,2,0,0,0,0)$. The solution for the original LP is $\mathbf{x} = (0,0,0,2,0)$.

With the work of this chapter, it is now possible to state that every LP (in standard form or otherwise) is either (1) infeasible, (2) unbounded, or (3) has an optimal solution. The reason is that any LP can first be put into standard form (see Section 4.6). Then, the simplex algorithm can be applied to the phase 1 LP and, according to Section 6.2, an optimal solution $(\mathbf{x}^*, \mathbf{y}^*)$ can be produced in a finite number of iterations. If $\mathbf{y}^* \neq \mathbf{0}$, then the original LP is infeasible and case 1 pertains. On the other hand, if $\mathbf{y}^* = \mathbf{0}$, then the vector $(\mathbf{x}^*, \mathbf{y}^*)$ can be used to create an initial bfs for the auxiliary LP. Therefore the simplex algorithm can be applied to the auxiliary LP. Once again, the algorithm must terminate in a finite number of steps in either an unbounded or optimal condition. Whichever occurs, the same condition applies to the original LP. These observations are summarized in the next theorem.

Theorem 6.1. *Any LP is either infeasible, unbounded, or optimal.*

Theorem 6.1 provides a useful technique for concluding that a given LP has an optimal solution. To do so, one only has to show that the particular LP of interest is neither infeasible nor unbounded, for then the only alternative is for it to have an optimal solution.

This chapter has described phase 1, a finite procedure for attempting to find an initial bfs for the original LP by applying the simplex algorithm to the phase 1 problem. The result is a solution $(\mathbf{x}^*, \mathbf{y}^*)$ for the phase 1 LP. If $\mathbf{y}^* \neq \mathbf{0}$, then the original LP is infeasible, and there is no hope of finding an initial bfs. On the other hand, if $\mathbf{y}^* = \mathbf{0}$, under fortunate circumstances, $\mathbf{x}^*$ is an initial bfs from which to initiate phase 2. Otherwise, $\mathbf{x}^*$ can be used to construct an initial bfs for the auxiliary LP. In any event, the simplex algorithm determines, in a finite number of iterations, if an LP is infeasible, optimal, or unbounded. If it is optimal, an optimal solution will be computed, but if unboundedness is detected, a direction of unboundedness is produced. Now that a method exists for solving an LP, it is natural to seek the most efficient computational method. That is the topic of the next chapter.

EXERCISES

6.3.1. For each of the phase 1 LPs in Exercise 6.2.1 that terminated with an initial bfs for the original LP, use the simplex algorithm to solve the original LP.

6.3.2. For each of the phase 1 LPs in Exercise 6.2.1 that terminated with a feasible solution for the original LP (but not an initial bfs), formulate the auxiliary LP.

6.3.3. The optimal bfs to the following phase 1 LP has y_1, x_5, and y_3 as basic variables with $y_1 = y_3 = 0$:

$$\begin{array}{llrrrrrrrrl}
\text{Minimize} & & & & & & & y_1 & +y_2 & +y_3 & \\
\text{subject to} & 2x_1 & & -3x_3 & -3x_4 & +4x_5 & +y_1 & & & & = 8 \\
 & -x_1 & +x_2 & +2x_3 & +4x_4 & +2x_5 & & +y_2 & & & = 4 \\
 & -4x_1 & +x_2 & +x_3 & -3x_4 & +x_5 & & & +y_3 & & = 2 \\
 & x_1, & x_2, & x_3, & x_4, & x_5, & y_1, & y_2, & y_3 & \geq 0 &
\end{array}$$

Formulate the auxiliary LP using the original objective function: Minimize $3x_1 - x_3 + 2x_5$.

6.3.4. Prove that if an LP in standard form is feasible and if the phase 1 LP does not have a degenerate bfs, then the phase 1 LP can be used to produce an initial bfs for the original LP; i.e., the auxiliary LP is not needed.

DISCUSSION

More on Phase 1 Procedures

There are many other approaches and improvements to finding an initial basic feasible solution. Some of them will be presented in Chapter 7. As mentioned in this chapter, when artificial variables appear in the final phase 1 basis at the value 0, it is possible to develop a procedure for attempting to interchange those variables with the nonbasic original variables. Such a procedure can be used to identify redundant equations, i.e., those equations whose removal would not change the geometry of the feasible region. Also, the procedure will determine when, in a given **A** matrix, it is not possible to find a nonsingular $m \times m$ **B** matrix. More specifically, the procedure can be used to determine the rank of the **A** matrix, i.e., the maximum number of linearly independent columns in the **A** matrix. These and related topics are discussed by Bazaraa and Jarvis [1].

REFERENCES

1. M. S. Bazaraa and J. J. Jarvis, *Linear Programming and Network Flows*, Wiley, New York, 1977.

Chapter 7

Computational Implementation

The time has come to discuss the implementation of the steps of the simplex algorithm so that phase 1 and phase 2 can be carried out on a computer. In designing a computer code, one should strive for

1. computational efficiency, so as to reduce the amount of computer time and storage needed to solve an LP;
2. numerical accuracy, so that the solutions obtained are precise and reliable;
3. ease of use for the problem solver.

Each of these topics will be addressed in this chapter.

7.1. COMPUTATIONAL CONSIDERATIONS

The result of Chapters 5 and 6 is an algorithm that, in a finite number of steps, will determine if a given LP is infeasible, optimal, or unbounded. Now that a method exists, it is natural to seek the best method. The question therefore arises as to what constitutes a good method. In addressing that issue, it is helpful to consider an *iteration* of the simplex algorithm that consists of starting with a bfs $\mathbf{x} = (\mathbf{x}_\mathbf{B}, \mathbf{x}_\mathbf{N})$ together with the matrix $\mathbf{B}^{-1}$ and

1. performing the test for optimality by computing the vector $\mathbf{c}_\mathbf{N} - \mathbf{c}_\mathbf{B}\mathbf{B}^{-1}\mathbf{N}$;
2. computing the direction of movement $\mathbf{d} = (\mathbf{d}_\mathbf{B}, \mathbf{d}_\mathbf{N})$;
3. performing the min ratio test to obtain t^*;

4. moving to the new bfs $\mathbf{x} + t^*\mathbf{d}$ and performing the pivot operation by exchanging the appropriate columns of **B** and **N**, and then computing the new inverse matrix.

With the concept of an iteration, it seems reasonable to define the amount of effort that is needed to solve an LP as the total number of iterations required to obtain a solution times the amount of work (additions, multiplications, etc.) that is done in each iteration. Consequently, there are at least two ways to improve the efficiency of the simplex algorithm: (1) reduce the total number of iterations needed to solve an LP and (2) reduce the amount of work that is done in each iteration.

It is often (but not always) the case that a strategy designed to reduce the total number of iterations does so at the expense of increasing the amount of work done per iteration, and vice versa. Thus the ultimate value of a particular strategy usually rests on its actual performance in computational tests. Nonetheless, it is advisable to seek such strategies.

The total number of iterations required to solve an LP is the sum of the iterations in phase 1 plus those in phase 2. Therefore several ways to reduce the number of iterations in each of the two phases will be described in this chapter.

The amount of work done in each iteration can take various forms. For example, when performing the test for optimality on the current bfs $\mathbf{x} = (\mathbf{x_B}, \mathbf{x_N}) = (\mathbf{B}^{-1}\mathbf{b}, \mathbf{0}) \geq \mathbf{0}$, it is necessary to compute $\mathbf{c_N} - \mathbf{c_B}\mathbf{B}^{-1}\mathbf{N}$, and this operation requires a certain number of additions, multiplications, and so on. Since multiplication requires substantially more computational effort than addition, it is reasonable to ask how the computation of the reduced costs can be accomplished so as to require the fewest multiplications. For instance, if one were to compute $\mathbf{B}^{-1}\mathbf{N}$ first, then $\mathbf{c_B}(\mathbf{B}^{-1}\mathbf{N})$, and finally $\mathbf{c_N} - \mathbf{c_B}(\mathbf{B}^{-1}\mathbf{N})$, it would require a total of $m^2(n-m) + m(n-m)$ multiplications. Compare that approach with the more efficient method of first computing $\mathbf{c_B}\mathbf{B}^{-1}$, then $(\mathbf{c_B}\mathbf{B}^{-1})\mathbf{N}$, and finally $\mathbf{c_N} - (\mathbf{c_B}\mathbf{B}^{-1})\mathbf{N}$, which requires a total of only $m^2 + m(n-m)$ multiplications. For $n = 501$ and $m = 100$, the difference in these two approaches amounts to a savings of four million multiplications each time the test for optimality is performed! It clearly pays to take a careful look at the various operations that are performed during an iteration of the simplex algorithm so as to find efficient ways of executing them.

Sometimes the concept of work per iteration is a bit more subtle than this. Consider, for example, the logistics of storing and accessing the **A** matrix in the computer. In general, the data can be stored either in *primary* memory (which has a relatively limited capacity but rapid access time), or else in *secondary* memory such as magnetic drums or disks (which have a vast capacity but slow retrieval time). In large problems it will not be possible to store the entire **A** matrix in primary memory, and so secondary memory will have to be used. Consequently, a large fraction of the time needed to perform an iteration of the simplex algorithm can be spent in retrieving the data. Several techniques will be presented to help in this area.

It is often the case that computational efficiency can be improved by exploiting special structure that appears in the constraints of the LP. For instance, consider a problem in which $m = 500$ and $n = 2000$. The **A** matrix for such a problem would contain $500 \times 2000 = 1{,}000{,}000$ entries. It would be a formidable task for someone to collect that much information and to type it into a computer. Fortunately, no one has to because problems of this size usually exhibit the special structure known as *sparsity*. Sparsity means that the vast majority (perhaps 95% or more) of the entries in the **A** matrix are 0. One way to take advantage of sparsity is to store only those entries that are not zero. For example, the matrix

$$\mathbf{A} = \begin{bmatrix} 0 & 1 & 0 & 0 \\ 1 & 4 & 0 & 0 \\ 0 & 0 & 3 & 0 \\ 0 & 0 & 0 & 5 \end{bmatrix}$$

would be stored in the computer as follows:

1. Row 1, column 2 contains 1.
2. Row 2, column 1 contains 1.
3. Row 2, column 2 contains 4.
4. Row 3, column 3 contains 3.
5. Row 4, column 4 contains 5.

Sparsity can also be exploited in another way. Suppose, for instance, that **c** and **x** are two sparse n vectors. Normally, it would require n multiplications to compute $\mathbf{cx} = \sum_{i=1}^{n} c_i x_i$. The number **cx** can be obtained with fewer multiplications by noting that only those components of **c** and **x** that are both nonzero need to be multiplied. More will be said about sparsity later. In addition to sparsity, there are many other forms of special structure that can, and will, be exploited. The conclusion, however, is to look whenever there is special structure for ways of enhancing the performance of the simplex algorithm based on that structure.

In addition to efficiency, there is another concern regarding the proper computational implementation of the simplex algorithm, and that is to maintain numerical accuracy. To see how complications can arise, recall that, on a computer, real numbers such as $\frac{1}{3} = 0.333\ldots$ can only be represented with a finite number of digits, the actual number being dependent on the particular computer system that is used. Currently, most computers *truncate* numbers that are too long; i.e., the extra digits are discarded. For example, the number 0.8179 would become 0.817 when truncated to three digits of accuracy. Thus, after many iterations, truncation will reduce the accuracy of numbers. Truncation is "biased" in the sense that numbers always get smaller. An alternative to truncation is known as *rounding*, wherein the number is rounded up or down depending on which of the two results is closer to the original number. For example, 0.8179 would be rounded up to 0.818, but 0.8174 would be rounded down to 0.817. The strategy of rounding requires more effort than that of truncation, and therefore it is not used in most computers. Other approaches exist for dealing with this problem, but they are beyond the scope of this text.

The loss of digits gives rise to what is known as *round-off error*. Another source of round-off error occurs when very large numbers are added to very small numbers. In any event, round-off error can appear inconsequential at first glance; however, after a large number of arithmetic operations the amount of accumulated round-off error can grow to such proportions as to make the LP solution completely useless. It has been observed that one primary source of round-off error occurs when relatively large numbers are divided by small ones. A remarkably accurate implementation of the simplex algorithm will be described in Section 7.3, and its success is attributable to avoiding such divisions whenever possible.

Another problem that results from round-off error is the inability to recognize the number 0. The need to do so arises in two steps of the simplex algorithm. The first such place is in the test for optimality, where it is necessary to check whether each component of the reduced costs $\mathbf{c_N} - \mathbf{c_B B}^{-1}\mathbf{N}$ is nonnegative. Suppose that, because of round-off error, there is a j for which $(\mathbf{c_N} - \mathbf{c_B B}^{-1}\mathbf{N})_j = -0.000001$ when it should have been 0. If one does not take precautions, the computer will take this variable as a candidate to become basic when in reality it is not.

A similar situation occurs in the min ratio test, where it is necessary to see whether $(\mathbf{d_B})_k < 0$. Again because of round-off error, it could turn out that $(\mathbf{d_B})_k = -0.000001$ when in reality it should have been $+0.000001$. It is thus possible that one might select the wrong variable to leave the basis. To help avoid these types of problems, professional codes ask the user to supply *zero tolerances* in the form of two real numbers $z_1, z_2 > 0$. The computer code will allow a variable $(\mathbf{x_N})_j$ to enter the basis only if $(\mathbf{c_N} - \mathbf{c_B B}^{-1}\mathbf{N})_j < -z_1 < 0$; $(\mathbf{x_B})_k$ will be considered as a leaving variable only if $(\mathbf{d_B})_k < -z_2 < 0$. Choosing the correct values for z_1 and z_2 is more an art than a science. Most professional codes automatically supply values for z_1 and z_2 but allow the user to change them if desired.

Speaking of the user, a good computer program for the simplex algorithm will facilitate data entry. The user must supply the data in the form of the nonzero values of the n vector $\mathbf{c}$, the m vector $\mathbf{b}$, and the $m \times n$ matrix $\mathbf{A}$. Even when $\mathbf{A}$ is sparse, there can be a lot of data that have to be read into the computer. Separate computer programs have been written for the sole purpose of helping the user enter the data. One such program is called MAGEN, which stands for "matrix generator." MAGEN is particularly useful when the entries of the $\mathbf{A}$ matrix can be computed by a formula of some sort. In this case MAGEN allows the user to give instructions as to how to compute these values, rather than requiring that each value be input separately. In addition, MAGEN allows the user to create more descriptive names for the variables, such as "coal," "gas," "nuclear," and "solar" instead of x_1, x_2, x_3, and x_4. Also, after the simplex algorithm is finished, professional codes will display the solution in an easily understandable format. These input and output programs will not be described further in this text.

This section has discussed some of the considerations that are involved in producing an efficient and numerically accurate computer program that can be used easily to solve linear programming problems. The next section begins a more detailed study of computational efficiency.

7.2. THE REVISED SIMPLEX ALGORITHM

This section will begin the development of an efficient computational implementation of the simplex algorithm. It is natural to begin by looking at the most costly operations to be performed during an iteration. One of them is in the pivot step, where it is necessary to compute the inverse of the new basis matrix $\mathbf{B}'$. When $\mathbf{B}'$ is a 2×2 matrix, its inverse can be computed quickly and easily; when the size of the matrix is larger, however, finding its inverse can be quite time-consuming. Since $\mathbf{B}'$ differs from $\mathbf{B}$ in only one column, it is reasonable to conjecture that $\mathbf{B}'^{-1}$ can somehow be computed from $\mathbf{B}^{-1}$, which is already available. In fact, there are several such procedures. The process of obtaining $\mathbf{B}'^{-1}$ from $\mathbf{B}^{-1}$ is referred to as *updating*.

The next proposition provides one way of updating $\mathbf{B}^{-1}$. For the rest of this chapter, $\mathbf{B}$ will be the current basis matrix, $\mathbf{B}'$ will be the new basis matrix obtained by replacing column k^* of $\mathbf{B}$ with column j^* of $\mathbf{N}$, and, of course, $\mathbf{d_B} = -\mathbf{B}^{-1}\mathbf{N}_{\cdot j^*}$. It is important to note that, since k^* came from the min ratio test, $(\mathbf{d_B})_{k^*} < 0$ and can therefore be used as a denominator.

Proposition 7.1. *The rows of the matrix* $\mathbf{B}'^{-1}$ *can be computed from those of* $\mathbf{B}^{-1}$ *according to the following formula*:

$$(\mathbf{B}'^{-1})_{i\cdot} = \begin{cases} (\mathbf{B}^{-1})_{i\cdot} - [(\mathbf{d_B})_i/(\mathbf{d_B})_{k^*}](\mathbf{B}^{-1})_{k^*\cdot} & \text{if} \quad i \neq k^* \\ -[1/(\mathbf{d_B})_{k^*}](\mathbf{B}^{-1})_{k^*\cdot} & \text{if} \quad i = k^* \end{cases}$$

The proof of Proposition 7.1 is deferred until Proposition 7.2 is introduced.

The next example demonstrates the use of the formula in Proposition 7.1.

EXAMPLE 7.1: Suppose that the current inverse matrix is

$$\mathbf{B}^{-1}=\begin{bmatrix} 1 & 1 & 1\\ -4 & -9 & -5\\ 1 & 2 & 1 \end{bmatrix}$$

and that

$$\mathbf{d_B}=\begin{bmatrix} -3\\ 18\\ -4 \end{bmatrix}$$

If $k^*=3$, then the rows of $\mathbf{B}'^{-1}$ are

$$\begin{aligned}(\mathbf{B}'^{-1})_{1\cdot} &= (\mathbf{B}^{-1})_{1\cdot}-\left[(\mathbf{d_B})_1/(\mathbf{d_B})_3\right](\mathbf{B}^{-1})_{3\cdot}=(1,1,1)-(-3/-4)(1,2,1)\\ &=(\tfrac{1}{4},-\tfrac{1}{2},\tfrac{1}{4})\\ (\mathbf{B}'^{-1})_{2\cdot} &= (\mathbf{B}^{-1})_{2\cdot}-\left[(\mathbf{d_B})_2/(\mathbf{d_B})_3\right](\mathbf{B}^{-1})_{3\cdot}=(-4,-9,-5)\\ &\qquad -(18/-4)(1,2,1)=(\tfrac{1}{2},0,-\tfrac{1}{2})\\ (\mathbf{B}'^{-1})_{3\cdot} &= -\left[1/(\mathbf{d_B})_3\right](\mathbf{B}^{-1})_{3\cdot}=(-1/-4)(1,2,1)\\ &=(\tfrac{1}{4},\tfrac{1}{2},\tfrac{1}{4})\end{aligned}$$

so

$$\mathbf{B}'^{-1}=\begin{bmatrix} \frac{1}{4} & -\frac{1}{2} & \frac{1}{4}\\ \frac{1}{2} & 0 & -\frac{1}{2}\\ \frac{1}{4} & \frac{1}{2} & \frac{1}{4} \end{bmatrix}$$

Some of the best methods for updating are based on the observation that the inverse of the basis matrix is not really needed in the simplex algorithm. To see why, consider carefully the two steps in which $\mathbf{B}^{-1}$ is used:

1. computing the vector $\mathbf{u}=\mathbf{c_B}\mathbf{B}^{-1}$, so that the test for optimality can be performed, and
2. computing $\mathbf{d_B}=-\mathbf{B}^{-1}\mathbf{N}_{\cdot j^*}$.

Careful restatement leads to these steps:

1. Use the **B** matrix to solve the linear system of (row) equations $\mathbf{uB}=\mathbf{c_B}$ for **u**.
2. Use the **B** matrix to solve the linear system of (column) equations $\mathbf{B}\mathbf{d_B}=-\mathbf{N}_{\cdot j^*}$ for $\mathbf{d_B}$.

Thus the matrix $\mathbf{B}^{-1}$ can be replaced with any "device" that is capable of using the **B** matrix to solve linear row and column equations. Indeed, very efficient methods already exist for solving such equations. Whatever procedure is used in the simplex algorithm, it must be updated when a column enters and leaves the basis matrix. The next section discusses these advanced basis handling techniques.

The next area in which computational savings can be achieved is the rule that is used to determine the value for j^*. Recall, from Section 5.2, that several such rules were given. The first come, first served rule suggests taking the first value of j for which $(\mathbf{c_N} - \mathbf{c_B B}^{-1}\mathbf{N})_j < 0$. The rule of steepest descent requires finding the most negative component of $\mathbf{c_N} - \mathbf{c_B B}^{-1}\mathbf{N}$. The former rule has the advantage of requiring less work per iteration than the latter one because, in general, the entire vector $\mathbf{c_N} - \mathbf{c_B B}^{-1}\mathbf{N}$ need not be computed. On the other hand, computational experiments have shown that the first come, first served rule also requires many more iterations to solve the LP than does the rule of steepest descent. The actual approach that is used in most professional codes is known as *partial pricing*, and it attempts to combine the best of both rules. (The term "pricing" refers to the operation of computing $\mathbf{c_N} - \mathbf{c_B B}^{-1}\mathbf{N}$ or any part of it.)

With partial pricing, only p of the $n - m$ components of $\mathbf{c_N} - \mathbf{c_B B}^{-1}\mathbf{N}$ are computed. From these p components, the most negative one is selected to enter the basis. If there are no negative components in this portion of $\mathbf{c_N} - \mathbf{c_B B}^{-1}\mathbf{N}$, then the next p components are computed, and so on. Observe that if the $\mathbf{N}$ matrix has to be kept in secondary memory, then partial pricing can greatly reduce the retrieval costs by accessing only p columns at a time.

Most professional codes combine the idea of partial pricing with that of *multiple pricing*. To understand multiple pricing, look again at the rule of steepest descent, whereby the most negative component of $\mathbf{c_N} - \mathbf{c_B B}^{-1}\mathbf{N}$ is used to determine the entering variable. Now consider the second most negative component of $\mathbf{c_N} - \mathbf{c_B B}^{-1}\mathbf{N}$ (if there is one). Although that corresponding variable was not chosen to enter the basis, it certainly could have been. There is thus good reason to suspect that, on the next iteration of the simplex algorithm, this variable will again be a candidate to enter the basis (provided, of course, that its new reduced cost is negative). Therefore, the idea of multiple pricing is to select the q most negative components of $\mathbf{c_N} - \mathbf{c_B B}^{-1}\mathbf{N}$ (q is usually 5 or 10 in practice) and then to pretend that the remaining nonbasic components of the problem do not exist. To be specific, if $\mathbf{N}'$ is the $m \times q$ submatrix of $\mathbf{N}$ consisting of those columns corresponding to the q most negative components of $\mathbf{c_N} - \mathbf{c_B B}^{-1}\mathbf{N}$, then the LP obtained by throwing away the remaining nonbasic variables is to

$$\begin{aligned}
\text{minimize} \quad & \mathbf{c_B x_B} + \mathbf{c_{N'} x_{N'}} \\
\text{subject to} \quad & \mathbf{B x_B} + \mathbf{N' x_{N'}} = \mathbf{b} \\
& \mathbf{x_B}, \quad \mathbf{x_{N'}} \geq \mathbf{0}
\end{aligned}$$

Once this *restricted problem* has been formed, it can then be solved efficiently to completion by using the rule of steepest descent since there are only q nonbasic columns to work with. Moreover, because q is small, the restricted problem can easily fit into the primary memory for fast processing. After the restricted problem is solved, the pricing operation is then repeated on the full set of nonbasic columns from the original problem. Note that if the restricted problem is unbounded, then so is the original one (see Exercise 7.2.4).

The *revised simplex algorithm* consists of updating the basis inverse together with the ideas of partial and multiple pricing. The steps are as follows:

Step 0. Initialization: Begin with an initial bfs $\mathbf{x} = (\mathbf{x}_\mathbf{B}, \mathbf{x}_\mathbf{N})$ and the matrix $\mathbf{B}^{-1}$. Go to step 1.

Step 1. Partial pricing: Select p of the columns of $\mathbf{N}$ and compute the corresponding components of $\mathbf{c}_\mathbf{N} - \mathbf{c}_\mathbf{B}\mathbf{B}^{-1}\mathbf{N}$. If they are all nonnegative, then select the next p columns, and so on. If all components of $\mathbf{c}_\mathbf{N} - \mathbf{c}_\mathbf{B}\mathbf{B}^{-1}\mathbf{N}$ are nonnegative, then the current bfs is optimal; otherwise, go to step 2.

Step 2. Multiple pricing: Of the p components of $\mathbf{c}_\mathbf{N} - \mathbf{c}_\mathbf{B}\mathbf{B}^{-1}\mathbf{N}$, select the q most negative ones and form the restricted problem. Go to step 3.

Step 3. Solving the restricted problem: Solve the restricted problem to completion by (1) using the rule of steepest descent to choose the incoming column, (2) determining the direction of movement in the usual manner, (3) performing the min ratio test and stopping if unboundedness is detected, and (4) selecting the leaving variable by any rule, and updating the basis inverse according to Proposition 7.1. Return to step 1.

These steps are demonstrated on the next example with $p = 5$ and $q = 2$.

EXAMPLE 7.2: Consider the following LP:

$$\begin{aligned} \text{Minimize} \quad & \mathbf{cx} \\ \text{subject to} \quad & \mathbf{Ax} = \mathbf{b} \\ & \mathbf{x} \geq \mathbf{0} \end{aligned}$$

in which

$$\mathbf{c} = (1, -1, 2, 4, -2, -1, 1, -4, 2, 4, 0, 0)$$

$$\mathbf{A} = \begin{bmatrix} 2 & 1 & 4 & 2 & 0 & 2 & 0 & 1 & 1 & 1 & 1 & 0 \\ 1 & 2 & 2 & 2 & 4 & 0 & 2 & 1 & 0 & 1 & 0 & 1 \end{bmatrix}$$

$$\mathbf{b} = \begin{bmatrix} 1 \\ 1 \end{bmatrix}$$

Iteration 1

Step 0. The bfs for this iteration:

$$\mathbf{B} = [\mathbf{A}_{\cdot 11}, \mathbf{A}_{\cdot 12}] = \begin{bmatrix} 1 & 0 \\ 0 & 1 \end{bmatrix}$$

$$N = [\mathbf{A}_{\cdot 1}, \mathbf{A}_{\cdot 2}, \mathbf{A}_{\cdot 3}, \mathbf{A}_{\cdot 4}, \mathbf{A}_{\cdot 5}, \mathbf{A}_{\cdot 6}, \mathbf{A}_{\cdot 7}, \mathbf{A}_{\cdot 8}, \mathbf{A}_{\cdot 9}, \mathbf{A}_{\cdot 10}]$$

$$= \begin{bmatrix} 2 & 1 & 4 & 2 & 0 & 2 & 0 & 1 & 1 & 1 \\ 1 & 2 & 2 & 2 & 4 & 0 & 2 & 1 & 0 & 1 \end{bmatrix}$$

$$\mathbf{x_B} = \begin{bmatrix} x_{11} \\ x_{12} \end{bmatrix} = \mathbf{B}^{-1}\mathbf{b} = \begin{bmatrix} 1 & 0 \\ 0 & 1 \end{bmatrix} \begin{bmatrix} 1 \\ 1 \end{bmatrix} = \begin{bmatrix} 1 \\ 1 \end{bmatrix}$$

$$\mathbf{x_N} = (x_1, x_2, x_3, x_4, x_5, x_6, x_7, x_8, x_9, x_{10}) = (0,0,0,0,0,0,0,0,0,0)$$

$$\mathbf{x} = (0,0,0,0,0,0,0,0,0,0,1,1)$$

Step 1. Partial pricing: The reduced costs are computed using the first five columns of **N**. Thus

$$\mathbf{c_N} - \mathbf{c_B}\mathbf{B}^{-1}\mathbf{N} = (1,-1,2,4,-2) - (0,0)\begin{bmatrix} 1 & 0 \\ 0 & 1 \end{bmatrix}\begin{bmatrix} 2 & 1 & 4 & 2 & 0 \\ 1 & 2 & 2 & 2 & 4 \end{bmatrix}$$

$$= (1,-1,2,4,-2)$$

Step 2. Multiple pricing: The basic variables together with the two most negative components of the reduced costs are used to form the restricted LP:

$$\begin{array}{llllll} \text{Minimize} & -x_2 & -2x_5 & & & \\ \text{subject to} & x_2 & & +x_{11} & & = 1 \\ & 2x_2 & +4x_5 & & +x_{12} & = 1 \\ & x_2, & x_5, & x_{11}, & x_{12} & \geq 0 \end{array}$$

Starting with x_{11} and x_{12} as basic variables, the simplex algorithm (using the rule of steepest descent) will obtain the following optimal solution to the restricted problem in two iterations:

$$\mathbf{B} = [\mathbf{A}_{\cdot 11}, \mathbf{A}_{\cdot 5}] = \begin{bmatrix} 1 & 0 \\ 0 & 4 \end{bmatrix}$$

$$\mathbf{x_B} = \begin{bmatrix} x_{11} \\ x_5 \end{bmatrix} = \mathbf{B}^{-1}\mathbf{b} = \begin{bmatrix} 1 & 0 \\ 0 & \frac{1}{4} \end{bmatrix} \begin{bmatrix} 1 \\ 1 \end{bmatrix} = \begin{bmatrix} 1 \\ \frac{1}{4} \end{bmatrix}$$

Iteration 2

Step 0. The bfs for this iteration:

$$\mathbf{B} = [\mathbf{A}_{.11}, \mathbf{A}_{.5}] = \begin{bmatrix} 1 & 0 \\ 0 & 4 \end{bmatrix}$$

$$\mathbf{N} = [\mathbf{A}_{.1}, \mathbf{A}_{.2}, \mathbf{A}_{.3}, \mathbf{A}_{.4}, \mathbf{A}_{.12}, \mathbf{A}_{.6}, \mathbf{A}_{.7}, \mathbf{A}_{.8}, \mathbf{A}_{.9}, \mathbf{A}_{.10}]$$

$$= \begin{bmatrix} 2 & 1 & 4 & 2 & 0 & 2 & 0 & 1 & 1 & 1 \\ 1 & 2 & 2 & 2 & 1 & 0 & 2 & 1 & 0 & 1 \end{bmatrix}$$

$$\mathbf{x_B} = \begin{bmatrix} x_{11} \\ x_5 \end{bmatrix} = \mathbf{B}^{-1}\mathbf{b} = \begin{bmatrix} 1 & 0 \\ 0 & \frac{1}{4} \end{bmatrix} \begin{bmatrix} 1 \\ 1 \end{bmatrix} = \begin{bmatrix} 1 \\ \frac{1}{4} \end{bmatrix}$$

$$\mathbf{x_N} = (x_1, x_2, x_3, x_4, x_{12}, x_6, x_7, x_8, x_9, x_{10}) = (0,0,0,0,0,0,0,0,0,0)$$

Step 1. Partial pricing: The reduced costs are computed using the last five columns of **N**. Thus

$$\mathbf{c_N} - \mathbf{c_B}\mathbf{B}^{-1}\mathbf{N} = (-1,1,-4,2,4) - (0,-2)\begin{bmatrix} 1 & 0 \\ 0 & \frac{1}{4} \end{bmatrix}\begin{bmatrix} 2 & 0 & 1 & 1 & 1 \\ 0 & 2 & 1 & 0 & 1 \end{bmatrix}$$

$$= (-1, 2, -\tfrac{7}{2}, 2, \tfrac{9}{2})$$

Step 2. Multiple pricing: The basic variables together with the two most negative components of the reduced costs are used to form the restricted LP:

$$\begin{aligned} \text{Minimize} \quad & -2x_5 - x_6 - 4x_8 \\ \text{subject to} \quad & 2x_6 + x_8 + x_{11} = 1 \\ & 4x_5 + x_8 = 1 \\ & x_5, \ x_6, \ x_8, \ x_{11} \geq 0 \end{aligned}$$

Starting with x_{11} and x_5 as basic variables, the simplex algorithm (using the rule of steepest descent) will obtain the following optimal solution to the restricted problem in two iterations:

$$\mathbf{B} = [\mathbf{A}_{.8}, \mathbf{A}_{.5}] = \begin{bmatrix} 1 & 0 \\ 1 & 4 \end{bmatrix}$$

$$\mathbf{x_B} = \begin{bmatrix} x_8 \\ x_5 \end{bmatrix} = \mathbf{B}^{-1}\mathbf{b} = \begin{bmatrix} 1 & 0 \\ -\frac{1}{4} & \frac{1}{4} \end{bmatrix} \begin{bmatrix} 1 \\ 1 \end{bmatrix} = \begin{bmatrix} 1 \\ 0 \end{bmatrix}$$

Iteration 3

Step 0. The bfs for this iteration:

$$\mathbf{B} = [\mathbf{A}_{\cdot 8}, \mathbf{A}_{\cdot 5}] = \begin{bmatrix} 1 & 0 \\ 1 & 4 \end{bmatrix}$$

$$\mathbf{N} = [\mathbf{A}_{\cdot 1}, \mathbf{A}_{\cdot 2}, \mathbf{A}_{\cdot 3}, \mathbf{A}_{\cdot 4}, \mathbf{A}_{\cdot 12}, \mathbf{A}_{\cdot 6}, \mathbf{A}_{\cdot 7}, \mathbf{A}_{\cdot 11}, \mathbf{A}_{\cdot 9}, \mathbf{A}_{\cdot 10}]$$
$$= \begin{bmatrix} 2 & 1 & 4 & 2 & 0 & 2 & 0 & 1 & 1 & 1 \\ 1 & 2 & 2 & 2 & 1 & 0 & 2 & 0 & 0 & 1 \end{bmatrix}$$

$$\mathbf{x_B} = \begin{bmatrix} x_8 \\ x_5 \end{bmatrix} = \mathbf{B}^{-1}\mathbf{b} = \begin{bmatrix} 1 & 0 \\ -\frac{1}{4} & \frac{1}{4} \end{bmatrix}\begin{bmatrix} 1 \\ 1 \end{bmatrix} = \begin{bmatrix} 1 \\ 0 \end{bmatrix}$$

$$\mathbf{x_N} = (x_1, x_2, x_3, x_4, x_{12}, x_6, x_7, x_{11}, x_9, x_{10}) = (0,0,0,0,0,0,0,0,0,0)$$

Step 1. Partial pricing: The reduced costs are computed using the first five columns of **N**. Thus

$$\mathbf{c_N} - \mathbf{c_B}\mathbf{B}^{-1}\mathbf{N} = (1, -1, 2, 4, 0) - (-4, -2)\begin{bmatrix} 1 & 0 \\ -\frac{1}{4} & \frac{1}{4} \end{bmatrix}\begin{bmatrix} 2 & 1 & 4 & 2 & 0 \\ 1 & 2 & 2 & 2 & 1 \end{bmatrix}$$
$$= \left(\tfrac{17}{2}, \tfrac{7}{2}, 17, 12, \tfrac{1}{2}\right)$$

Since all of these reduced costs are nonnegative, the last five columns of **N** are used, and so

$$\mathbf{c_N} - \mathbf{c_B}\mathbf{B}^{-1}\mathbf{N} = (-1, 1, 0, 2, 4) - (-4, -2)\begin{bmatrix} 1 & 0 \\ -\frac{1}{4} & \frac{1}{4} \end{bmatrix}\begin{bmatrix} 2 & 0 & 1 & 1 & 1 \\ 0 & 2 & 0 & 0 & 1 \end{bmatrix}$$
$$= \left(6, 2, \tfrac{7}{2}, \tfrac{11}{2}, 8\right)$$

Since all of these reduced costs are nonnegative and there are no other columns of **N**, the current bfs $\mathbf{x} = (0,0,0,0,0,0,0,1,0,0,0,0)$ is optimal.

This section has described a revised simplex algorithm that not only updates the basis inverse rather than computes it from scratch, but also uses partial and multiple pricing to determine the column to enter the basis. The next section describes even better methods for handling the basis matrix.

EXERCISES

7.2.1. With the following data, use Proposition 7.1 to compute $\mathbf{B}'^{-1}$.

(a)

$$\mathbf{B}^{-1} = \begin{bmatrix} 1 & 9 & 0 \\ 4 & 1 & 6 \\ 3 & -4 & -2 \end{bmatrix}$$

$$\mathbf{d_B} = \begin{bmatrix} -2 \\ -1 \\ 3 \end{bmatrix}, \qquad k^* = 1$$

(b)

$$\mathbf{B}^{-1}=\begin{bmatrix} 24 & 20 & 0 & 8 \\ -20 & 0 & 16 & 12 \\ 0 & -36 & 0 & 0 \\ -4 & 4 & 0 & 32 \end{bmatrix}$$

$$\mathbf{d_B}=\begin{bmatrix} -2 \\ -4 \\ -3 \\ 5 \end{bmatrix}, \qquad k^*=1$$

7.2.2. For each of the following LPs and given **B** matrices and values of $\mathbf{N}_{.j*}$, write the two linear systems of equations that must be solved when performing the simplex algorithm.

(a) Minimize $-6x_1+3x_2+2x_4-x_5+5x_6-3x_7$

subject to

$$\begin{aligned} x_1+3x_2+x_4-3x_5+x_6+x_7&=2\\ -x_3-x_4+x_7&=5\\ 2x_1-x_2+x_3+2x_5-x_6+x_7&=9\\ x_1,\ x_2,\ x_3,\ x_4,\ x_5,\ x_6,\ x_7&\geq 0 \end{aligned}$$

Given: $\mathbf{B}=[\mathbf{A}_{.1},\mathbf{A}_{.4},\mathbf{A}_{.7}],\mathbf{N}_{.J*}=\mathbf{A}_{.3}$.

(b) Minimize $5x_1+10x_2-5x_3+x_4-3x_5+2x_6+3x_7-x_8$

subject to

$$\begin{aligned} x_2+x_3+5x_4-x_7+x_8&=9\\ x_1-x_3+x_5-x_6+2x_7&=3\\ 2x_1+x_2-x_4+3x_5+5x_6-x_7+x_8&=18\\ -3x_2+x_3+x_4-3x_6&=0\\ x_1,\ x_2,\ x_3,\ x_4,\ x_5,\ x_6,\ x_7,\ x_8&\geq 0 \end{aligned}$$

Given: $\mathbf{B}=[\mathbf{A}_{.5},\mathbf{A}_{.6},\mathbf{A}_{.7},\mathbf{A}_{.8}],\mathbf{N}_{.j*}=\mathbf{A}_{.1}$.

7.2.3. Consider an LP in which

$$\mathbf{c}=(0,0,-1,-2,3,-3,-1,-4,-2,-3)$$

$$\mathbf{A}=\begin{bmatrix} 1 & 0 & -2 & -1 & 0 & -3 & 4 & 0 & -2 & 1 \\ 0 & 1 & -1 & 1 & 0 & -1 & 5 & 1 & -1 & 2 \end{bmatrix}$$

$$\mathbf{b}=\begin{bmatrix} 2 \\ 1 \end{bmatrix}$$

Using $p=4$ and $q=2$, formulate the two restricted LPs that would be obtained when $\mathbf{B}=[\mathbf{A}_{.1},\mathbf{A}_{.2}]$.

7.2.4. Prove that if the restricted LP is unbounded, then so is the original LP.

7.2.5. Using the update formula in Proposition 7.1, solve the following LP by starting with the initial bfs in which the basic variables are x_4, x_5, and x_6.

$$\begin{array}{lrrrrrrl}
\text{Maximize} & 5x_1 & +3x_2 & +2x_3 & & & & \\
\text{subject to} & 4x_1 & +3x_2 & +2x_3 & +x_4 & & & = 150 \\
& 5x_1 & +3x_2 & +2x_3 & & +x_5 & & = 125 \\
& 2x_1 & +\ x_2 & -2x_3 & & & +x_6 & = 200 \\
& x_1, & x_2, & x_3, & x_4, & x_5, & x_6 & \geq\ 0
\end{array}$$

7.3. ADVANCED BASIS HANDLING TECHNIQUES

This section will describe some of the best-known methods for using the basis matrix $\mathbf{B}$ to solve the two linear systems of equations $\mathbf{uB} = \mathbf{c_B}$ and $\mathbf{Bd_B} = -\mathbf{N}_{.j^*}$. These advanced methods are needed not only for reasons of computational efficiency, but also to achieve numerical accuracy. The round-off errors caused by the computer's limited precision can accumulate over time and have devastating results, as the next example shows.

EXAMPLE 7.3: Consider the following LP:

$$\begin{array}{lrrrrl}
\text{Minimize} & \frac{25}{12}x_1 & +\frac{77}{60}x_2 & +\frac{57}{60}x_3 & +\frac{319}{420}x_4 & \\
\text{subject to} & x_1 & +\ \frac{1}{2}x_2 & +\ \frac{1}{3}x_3 & +\ \frac{1}{4}x_4 & \leq \frac{25}{12} \\
& \frac{1}{2}x_1 & +\ \frac{1}{3}x_2 & +\ \frac{1}{4}x_3 & +\ \frac{1}{5}x_4 & \leq \frac{77}{60} \\
& \frac{1}{3}x_1 & +\ \frac{1}{4}x_2 & +\ \frac{1}{5}x_3 & +\ \frac{1}{6}x_4 & \leq \frac{57}{60} \\
& \frac{1}{4}x_1 & +\ \frac{1}{5}x_2 & +\ \frac{1}{6}x_3 & +\ \frac{1}{7}x_4 & \leq \frac{319}{420} \\
& x_1, & x_2, & x_3, & x_4 & \geq\ 0
\end{array}$$

This LP has the optimal solution $\mathbf{x}^* = (15, 1, 1, 1)$. Allowing five digits of accuracy, one particular computer code produced the solution $\mathbf{x}^* = (15.001, 0.98393, 1.0385, 0.97502)$. Allowing three digits of accuracy, the same code produced the solution $\mathbf{x}^* = (15.0, 0.287, 2.60, 0.00)$.

The advanced basis handling techniques consist of storing just enough information to be able to solve the two linear systems $\mathbf{uB} = \mathbf{c_B}$ and $\mathbf{Bd_B} = -\mathbf{N}_{.j^*}$ efficiently and accurately. Each time the pivot step of the simplex algorithm is performed, the stored information must be updated to reflect the fact that $\mathbf{N}_{.j^*}$ enters the basis matrix while $\mathbf{B}_{.k^*}$ leaves. The updating is often accomplished by storing additional information rather than by actually changing the existing information, as was done in the previous sections. The additional information can then be used to solve the new systems $\mathbf{uB}' = \mathbf{c_{B'}}$ and $\mathbf{B'd_{B'}} = -\mathbf{N}'_{.j^*}$.

Product Form of the Inverse

The first method to be described is known as the *product form of the inverse.* Instead of computing $\mathbf{B}'^{-1}$ explicitly from $\mathbf{B}^{-1}$ as was done in Proposition 7.1, a mechanism is provided that enables one to compute $\mathbf{B}'^{-1}$ whenever it is needed. That mechanism manifests itself in the form of a nonsingular $m \times m$ matrix $\mathbf{E}$ from which $\mathbf{B}'^{-1}$ can be computed as $\mathbf{E}^{-1}\mathbf{B}^{-1}$.

The desired $\mathbf{E}$ matrix is obtained by replacing column k^* of the $m \times m$ identity matrix with $-\mathbf{d_B}$. Thus

$$\mathbf{E} = \begin{bmatrix} 1 & & \overset{k^*}{\uparrow} & & 0 \\ 0 & & & & \vdots \\ \vdots & \cdots & -\mathbf{d_B} & \cdots & \\ & & & & 0 \\ 0 & & \downarrow & & 1 \end{bmatrix}$$

Recall from Section 4.4 that $\mathbf{E}$ is called an eta matrix, meaning that it differs from the identity matrix in only one column. To compute $\mathbf{B}'^{-1}$, what is really needed is $\mathbf{E}^{-1}$, not $\mathbf{E}$. Because of the simple structure of $\mathbf{E}$ and because $(\mathbf{d_B})_{k^*} \neq 0$, the matrix $\mathbf{E}^{-1}$ can be found easily. From Section 4.4, $\mathbf{E}^{-1}$ is also an eta matrix and is obtained by replacing column k^* of the $m \times m$ identity matrix by the m vector $\mathbf{w}$ defined by

$$w_i = \begin{cases} -(\mathbf{d_B})_i/(\mathbf{d_B})_{k^*} & \text{if} \quad i \neq k^* \\ -1/(\mathbf{d_B})_{k^*} & \text{if} \quad i = k^* \end{cases}$$

so

$$\mathbf{E}^{-1} = \begin{bmatrix} 1 & & \overset{k^*}{\uparrow} & & 0 \\ 0 & & | & & \vdots \\ \vdots & \cdots & w & \cdots & 0 \\ 0 & & \downarrow & & 1 \end{bmatrix}$$

The next proposition formally establishes that $\mathbf{B}'^{-1} = \mathbf{E}^{-1}\mathbf{B}^{-1}$.

Proposition 7.2. *If* $\mathbf{E}$ *is the eta matrix obtained by replacing column* k^* *of* $\mathbf{I}$ *with* $-\mathbf{d_B}$, *then* $\mathbf{E}$ *is nonsingular and* $\mathbf{B}'^{-1} = \mathbf{E}^{-1}\mathbf{B}^{-1}$.

OUTLINE OF PROOF. The forward–backward method gives rise to the abstraction question "How can I show that a matrix (namely, $\mathbf{E}$) is nonsingular?" One answer is the definition, whereby the inverse matrix must be produced. Specifically, $\mathbf{E}^{-1}$ is the eta matrix described in the paragraph preceding this proposition, and it is not hard to verify that $\mathbf{EE}^{-1} = \mathbf{E}^{-1}\mathbf{E} = \mathbf{I}$.

It remains to show that $\mathbf{B}'^{-1} = \mathbf{E}^{-1}\mathbf{B}^{-1}$. The forward–backward method now gives rise to the rather interesting abstraction question "How can I show that the inverse of a matrix (namely, $\mathbf{B}'$) is the product of the inverse of two matrices (namely, $\mathbf{E}$ and $\mathbf{B}$)?" One answer is found in the conclusion of Proposition 4.3, whereby one need only show that $\mathbf{B}$ and $\mathbf{E}$ are nonsingular and $\mathbf{B}' = \mathbf{BE}$.

Clearly, $\mathbf{B}$ is nonsingular because $\mathbf{x}$ is a bfs, and the first part of this proof established that $\mathbf{E}$ is nonsingular. All that remains is to show that $\mathbf{B}' = \mathbf{BE}$, and an associated abstraction question is "How can I show that two matrices (namely, $\mathbf{B}'$ and $\mathbf{BE}$) are equal?" One answer is to show that their corresponding columns are equal, i.e., that for all i with $1 \le i \le m$, $\mathbf{B}'_{\cdot i} = (\mathbf{BE})_{\cdot i}$. The appearance of the quantifier "for all" suggests using the choose method to select an i with $1 \le i \le m$, for which it must be shown that $\mathbf{B}'_{\cdot i} = (\mathbf{BE})_{\cdot i}$. There seem to be two cases.

Case 1: $i \ne k^*$

In this case, $\mathbf{E}_{\cdot i} = \mathbf{I}_{\cdot i}$ and $\mathbf{B}'_{\cdot i} = \mathbf{B}_{\cdot i}$, so $(\mathbf{BE})_{\cdot i} = \mathbf{BE}_{\cdot i} = \mathbf{BI}_{\cdot i} = \mathbf{B}_{\cdot i} = \mathbf{B}'_{\cdot i}$, as desired.

Case 2: $i = k^*$

In this case, $\mathbf{E}_{\cdot i} = \mathbf{E}_{\cdot k^*} = -\mathbf{d}_{\mathbf{B}}$ and $\mathbf{B}'_{\cdot i} = \mathbf{B}'_{\cdot k^*} = \mathbf{N}_{\cdot j^*}$, so $(\mathbf{BE})_{\cdot i} = \mathbf{BE}_{\cdot i} = -\mathbf{Bd}_{\mathbf{B}} = \mathbf{N}_{\cdot j^*} = \mathbf{B}'_{\cdot k^*} = \mathbf{B}'_{\cdot i}$, as desired.

PROOF OF PROPOSITION 7.2. The inverse of the matrix $\mathbf{E}$ has just been given, and thus $\mathbf{E}$ is nonsingular.

It remains to show that $\mathbf{B}'^{-1} = \mathbf{E}^{-1}\mathbf{B}^{-1}$. By Proposition 4.3, one need only establish that $\mathbf{B}' = \mathbf{BE}$, for then $\mathbf{B}^{-1} = (\mathbf{BE})^{-1} = \mathbf{E}^{-1}\mathbf{B}^{-1}$. To see that $\mathbf{B}' = \mathbf{BE}$, let i with $1 \le i \le m$. If $i \ne k^*$, then $(\mathbf{BE})_{\cdot i} = \mathbf{BE}_{\cdot i} = \mathbf{BI}_{\cdot i} = \mathbf{B}_{\cdot i} = \mathbf{B}'_{\cdot i}$; if $i = k^*$, then $(\mathbf{BE})_{\cdot i} = \mathbf{BE}_{\cdot i} = \mathbf{BE}_{\cdot k^*} = -\mathbf{Bd}_{\mathbf{B}} = \mathbf{N}_{\cdot j^*} = \mathbf{B}'_{\cdot k^*} = \mathbf{B}'_{\cdot i}$. □

It is interesting to note that computing $\mathbf{B}'^{-1}$ as $\mathbf{E}^{-1}\mathbf{B}^{-1}$ is the same as computing $\mathbf{B}^{-1}$ by the formula of Proposition 7.1. It is reasonable to wonder why the $\mathbf{E}$ matrix should be used at all. Furthermore, each time the pivot step is performed, another $\mathbf{E}^{-1}$ matrix must be found and stored, so after k iterations, one has the original $\mathbf{B}^{-1}$ matrix together with the *eta file* consisting of the k matrices $\mathbf{E}_1^{-1}, \mathbf{E}_2^{-1}, \ldots, \mathbf{E}_k^{-1}$. With the eta file, it is then possible to compute the current basis inverse by $(\mathbf{E}_k^{-1} \cdots \mathbf{E}_1^{-1})\mathbf{B}^{-1}$.

At first glance it may appear that the use of the eta file for solving the two linear equations is more inefficient than updating the basis at each step. Not only does one have to store the eta matrices $\mathbf{E}_1^{-1}, \ldots, \mathbf{E}_k^{-1}$, but also the length of the computation needed to find the current inverse matrix grows as k gets larger. Why, then, is the product form of the inverse so successful?

To answer the question, let us begin by looking at the problem of storing the k matrices in the eta file. Since each $\mathbf{E}^{-1}$ matrix is an eta matrix, only the column that differs from that of the identity matrix needs to be stored. Moreover, only the nonzero elements of that column need to be saved. Consequently, as long as k is not permitted to get too large (however big that is), storage will not cause much of a problem.

Turning to the issue of computational efficiency, it is perhaps surprising to discover that the current linear systems can be solved more efficiently with the eta file than with the explicit form of the inverse, provided that k is not allowed to get too large. The reason is that the eta file tends to have many more zero elements than the explicit form of the inverse (at least when k is small). It is this fact that makes the solution of the two linear systems so efficient. Because of the large number of zeros in the eta file, each time the systems are solved, it requires fewer multiplications using the product form of the inverse than the explicit form. As noted earlier, this statement is only true when k is not too large—but what is "too large" and what happens when k takes on that value?

Unfortunately, there is no definite answer. In practice, k is usually not allowed to exceed 50 or 60. When k reaches this value, the current basis inverse can be computed explicitly as $\hat{\mathbf{B}}^{-1} = \mathbf{E}_k^{-1} \cdots \mathbf{E}_1^{-1}\mathbf{B}^{-1}$. However, in practice, the $\hat{\mathbf{B}}$ matrix is formed using the appropriate columns of the original $\mathbf{A}$ matrix, and then $\hat{\mathbf{B}}$ is inverted from scratch by a special state-of-the-art matrix inversion routine to be discussed in the next section. The main advantage of performing the reinversion from scratch is that, by so doing, the accumulated round-off errors are greatly reduced. Once this is accomplished, the new inverse matrix serves as $\mathbf{B}^{-1}$ and the eta file is started all over. The next example demonstrates the use of the product form of the inverse.

EXAMPLE 7.4:

$$\begin{array}{llcccccl}
\text{Minimize} & -x_1 - \frac{3}{2}x_2 - 2x_3 \\
\text{subject to} & x_1 - x_2 + x_4 = 1 \\
& -2x_1 + \frac{17}{8}x_2 + 2x_3 + x_5 = 1 \\
& x_1, \quad x_2, \quad x_3, \quad x_4, \quad x_5 \geq 0
\end{array}$$

so that

$$\mathbf{c} = \left(-1, -\tfrac{3}{2}, -2, 0, 0\right)$$

$$\mathbf{A} = \begin{bmatrix} 1 & -1 & 0 & 1 & 0 \\ -2 & \frac{17}{8} & 2 & 0 & 1 \end{bmatrix}$$

$$\mathbf{b} = \begin{bmatrix} 1 \\ 1 \end{bmatrix}$$

Iteration 1

Step 0. The bfs for this iteration:

$$\mathbf{B} = [\mathbf{A}_{.4}, \mathbf{A}_{.5}] = \begin{bmatrix} 1 & 0 \\ 0 & 1 \end{bmatrix}$$

$$\mathbf{N} = [\mathbf{A}_{.1}, \mathbf{A}_{.2}, \mathbf{A}_{.3}] = \begin{bmatrix} 1 & -1 & 0 \\ -2 & \frac{17}{8} & 2 \end{bmatrix}$$

$$\mathbf{x_B} = \begin{bmatrix} x_4 \\ x_5 \end{bmatrix} = \mathbf{B}^{-1}\mathbf{b} = \begin{bmatrix} 1 & 0 \\ 0 & 1 \end{bmatrix}\begin{bmatrix} 1 \\ 1 \end{bmatrix} = \begin{bmatrix} 1 \\ 1 \end{bmatrix}$$

Step 1. Testing for optimality:

$$\mathbf{c_N} - \mathbf{c_B}\mathbf{B}^{-1}\mathbf{N} = (-1, -\tfrac{3}{2}, -2) - (0,0)\begin{bmatrix} 1 & 0 \\ 0 & 1 \end{bmatrix}\begin{bmatrix} 1 & -1 & 0 \\ -2 & \frac{17}{8} & 2 \end{bmatrix}$$

$$= (-1, -\tfrac{3}{2}, -2) \qquad \text{so} \quad j^* = 3$$

Step 2. Computing the direction of movement:

$$\mathbf{d_B} = -\mathbf{B}^{-1}\mathbf{N}_{.j^*} = -\begin{bmatrix} 1 & 0 \\ 0 & 1 \end{bmatrix}\begin{bmatrix} 0 \\ 2 \end{bmatrix} = \begin{bmatrix} 0 \\ -2 \end{bmatrix}$$

Step 3. Computing t^*:

$$t^* = \min\{-(\mathbf{x_B})_2/(\mathbf{d_B})_2\}$$

$$= \tfrac{1}{2} \qquad \text{so} \quad k^* = 2$$

Step 4. Pivoting: $\mathbf{E}_1$ is obtained by replacing column $k^* = 2$ of $\mathbf{I}$ with $-\mathbf{d_B}$, so

$$\mathbf{E}_1 = \begin{bmatrix} 1 & 0 \\ 0 & 2 \end{bmatrix}$$

$$\mathbf{E}_1^{-1} = \begin{bmatrix} 1 & 0 \\ 0 & \frac{1}{2} \end{bmatrix}$$

$$\mathbf{x_B} = \begin{bmatrix} x_4 \\ x_3 \end{bmatrix} = \mathbf{E}^{-1}(\mathbf{B}^{-1}\mathbf{b}) = \begin{bmatrix} 1 & 0 \\ 0 & \frac{1}{2} \end{bmatrix}\begin{bmatrix} 1 \\ 1 \end{bmatrix} = \begin{bmatrix} 1 \\ \frac{1}{2} \end{bmatrix}$$

$$\mathbf{x_N} = (x_1, x_2, x_5) = (0,0,0)$$

Iteration 2

Step 1. Testing for optimality:

$$\mathbf{u} = \mathbf{c_B}(\mathbf{E}_1^{-1}\mathbf{B}^{-1}) = (0, -2)\begin{bmatrix} 1 & 0 \\ 0 & \frac{1}{2} \end{bmatrix}\begin{bmatrix} 1 & 0 \\ 0 & 1 \end{bmatrix} = (0, -1)$$

$$\mathbf{c_N} - \mathbf{uN} = (-1, -\tfrac{3}{2}, 0) - (0, -1)\begin{bmatrix} 1 & -1 & 0 \\ -2 & \frac{17}{8} & 1 \end{bmatrix}$$

$$= (-3, \tfrac{5}{8}, 1) \qquad \text{so} \quad j^* = 1$$

Step 2. Computing the direction of movement:

$$\mathbf{d_B} = -\mathbf{E}_1^{-1}\left(\mathbf{B}^{-1}\mathbf{N}_{\cdot j^*}\right) = -\begin{bmatrix} 1 & 0 \\ 0 & \frac{1}{2} \end{bmatrix}\begin{bmatrix} 1 & 0 \\ 0 & 1 \end{bmatrix}\begin{bmatrix} 1 \\ -2 \end{bmatrix} = \begin{bmatrix} -1 \\ 1 \end{bmatrix}$$

Step 3. Computing t^*:

$$t^* = \min\{-(\mathbf{x_B})_1/(\mathbf{d_B})_1\}$$
$$= 1 \qquad \text{so} \quad k^* = 1$$

Step 4. Pivoting: $\mathbf{E}_2$ is obtained by replacing column $k^* = 1$ of $\mathbf{I}$ with $-\mathbf{d_B}$, so

$$\mathbf{E}_2 = \begin{bmatrix} 1 & 0 \\ -1 & 1 \end{bmatrix}$$

$$\mathbf{E}_2^{-1} = \begin{bmatrix} 1 & 0 \\ 1 & 1 \end{bmatrix}$$

$$\mathbf{x_B} = \begin{bmatrix} x_1 \\ x_3 \end{bmatrix} = \mathbf{E}_2^{-1}\left(\mathbf{E}_1^{-1}\mathbf{B}^{-1}\mathbf{b}\right)$$
$$= \begin{bmatrix} 1 & 0 \\ 1 & 1 \end{bmatrix}\begin{bmatrix} 1 \\ \frac{1}{2} \end{bmatrix}$$
$$= \begin{bmatrix} 1 \\ \frac{3}{2} \end{bmatrix}$$

$$\mathbf{x_N} = (x_4, x_2, x_5) = (0,0,0)$$

Iteration 3

Step 1. Testing for optimality:

$$\mathbf{u} = \mathbf{c_B}\left(\mathbf{E}_2^{-1}\mathbf{E}_1^{-1}\mathbf{B}^{-1}\right) = (-1,-2)\begin{bmatrix} 1 & 0 \\ 1 & 1 \end{bmatrix}\begin{bmatrix} 1 & 0 \\ 0 & \frac{1}{2} \end{bmatrix}\begin{bmatrix} 1 & 0 \\ 0 & 1 \end{bmatrix} = (-3,-1)$$

$$\mathbf{c_N} - \mathbf{uN} = \left(0, -\tfrac{3}{2}, 0\right) - (-3,-1)\begin{bmatrix} 1 & -1 & 0 \\ 0 & \frac{17}{8} & 1 \end{bmatrix}$$
$$= \left(3, -\tfrac{19}{8}, 1\right) \qquad \text{so} \quad j^* = 2$$

Step 2. Computing the direction of movement:

$$\mathbf{d_B} = -\mathbf{E}_2^{-1}\left(\mathbf{E}_1^{-1}\mathbf{B}^{-1}\mathbf{N}_{\cdot j^*}\right) = -\begin{bmatrix} 1 & 0 \\ 1 & 1 \end{bmatrix}\begin{bmatrix} 1 & 0 \\ 0 & \frac{1}{2} \end{bmatrix}\begin{bmatrix} 1 & 0 \\ 0 & 1 \end{bmatrix}\begin{bmatrix} -1 \\ \frac{17}{8} \end{bmatrix}$$
$$= \begin{bmatrix} 1 \\ -\frac{1}{16} \end{bmatrix}$$

Step 3. Computing t^*:

$$t^* = \min\{-(\mathbf{x_B})_2/(\mathbf{d_B})_2\}$$
$$= 24 \qquad \text{so} \quad k^* = 2$$

Step 4. Pivoting: $\mathbf{E}_3$ is obtained by replacing column $k^* = 2$ of $\mathbf{I}$ with $-\mathbf{d}_{\mathbf{B}}$, so

$$\mathbf{E}_3 = \begin{bmatrix} 1 & -1 \\ 0 & \frac{1}{16} \end{bmatrix}$$

$$\mathbf{E}_3^{-1} = \begin{bmatrix} 1 & 16 \\ 0 & 16 \end{bmatrix}$$

$$\mathbf{xb} = \begin{bmatrix} x_1 \\ x_2 \end{bmatrix} = \mathbf{E}_3^{-1}(\mathbf{E}_2^{-1}\mathbf{E}_1^{-1}\mathbf{B}^{-1}\mathbf{b})$$

$$= \begin{bmatrix} 1 & 16 \\ 0 & 16 \end{bmatrix}\begin{bmatrix} 1 \\ \frac{3}{2} \end{bmatrix}$$

$$= \begin{bmatrix} 25 \\ 24 \end{bmatrix}$$

$$\mathbf{x_N} = (x_4, x_3, x_5) = (0,0,0)$$

Iteration 4

Step 1. Testing for optimality:

$$\mathbf{u} = \mathbf{c_B}(\mathbf{E}_3^{-1}\mathbf{E}_2^{-1}\mathbf{E}_1^{-1}\mathbf{B}^{-1})$$

$$= (-1, -\tfrac{3}{2})\begin{bmatrix} 1 & 16 \\ 0 & 16 \end{bmatrix}\begin{bmatrix} 1 & 0 \\ 1 & 1 \end{bmatrix}\begin{bmatrix} 1 & 0 \\ 0 & \frac{1}{2} \end{bmatrix}$$

$$= (-41, -20)$$

$$\mathbf{c_N} - \mathbf{uN} = (0, -2, 0) - (-41, -20)\begin{bmatrix} 1 & 0 & 0 \\ 0 & 2 & 1 \end{bmatrix}$$

$$= (41, 38, 20)$$

Since the reduced costs are nonnegative, the current bfs $\mathbf{x} = (25, 24, 0, 0, 0)$ is optimal.

LU Decomposition

Now that a product form of the inverse exists, it is natural to seek the best product form. Here, "best" is meant in the sense of (1) computational efficiency in solving the current linear systems and (2) maintaining numerical accuracy.

One of the best product forms arises from yet another approach to solving the two systems of equations $\mathbf{uB} = \mathbf{c_B}$ and $\mathbf{Bd_B} = -\mathbf{N}_{.j^*}$. It came from the recognition that if $\mathbf{B}$ is upper triangular, then the column system $\mathbf{Bd_B} = -\mathbf{N}_{.j^*}$ can be solved very quickly by the process known as back substitution (see Section 4.4).

The problem is that, in general, $\mathbf{B}$ is not upper triangular. A clever method has been found for "making" the $\mathbf{B}$ matrix upper triangular. That method consists of attempting to find a nonsingular lower triangular matrix $\mathbf{L}$ (see Section 4.4) for which $\mathbf{LB}$ is upper triangular. Then, to solve the system $\mathbf{Bd_B} = -\mathbf{N}_{.j*}$, one might just as well solve the system $(\mathbf{LB})\mathbf{d_B} = -\mathbf{LN}_{.j*}$, for the latter system has the desirable property that $\mathbf{LB}$ is upper triangular. Ways for finding the desired $\mathbf{L}$ matrix are discussed in the next section; for now, assume that such a matrix has been found. The upper triangular matrix $\mathbf{LB}$ will be denoted $\mathbf{U}$.

The next question is how to use the $\mathbf{L}$ matrix to solve the row system $\mathbf{uB} = \mathbf{c_B}$. The answer is to solve the row system $\mathbf{w}(\mathbf{LB}) = \mathbf{c_B}$ instead of $\mathbf{uB} = \mathbf{c_B}$. Once $\mathbf{w}$ has been found, $\mathbf{u}$ can be computed as $\mathbf{wL}$, for then $\mathbf{uB} = (\mathbf{wL})\mathbf{B} = \mathbf{w}(\mathbf{LB}) = \mathbf{c_B}$. The advantage of finding $\mathbf{w}$ first is that, in solving the equations $\mathbf{w}(\mathbf{LB}) = \mathbf{c_B}$, the matrix $\mathbf{LB}$ is upper triangular, thus making the computation of $\mathbf{w}$ by forward substitution very efficient. The use of the $\mathbf{L}$ matrix for solving the two systems is demonstrated in the next example.

EXAMPLE 7.5: Suppose that $\mathbf{B}$, $\mathbf{N}_{.j*}$, and $\mathbf{c_B}$ are given by

$$\mathbf{B} = \begin{bmatrix} -1 & 2 & 3 \\ 1 & 3 & -4 \\ -1 & -8 & 9 \end{bmatrix}, \qquad \mathbf{N}_{.j*} = \begin{bmatrix} -1 \\ 1 \\ 0 \end{bmatrix}, \qquad \mathbf{c_B} = (1,3,0)$$

If the lower triangular matrix is

$$\mathbf{L} = \begin{bmatrix} 1 & 0 & 0 \\ 1 & 1 & 0 \\ 1 & 2 & 1 \end{bmatrix}$$

then

$$\mathbf{LB} = \mathbf{U} = \begin{bmatrix} 1 & 0 & 0 \\ 1 & 1 & 0 \\ 1 & 2 & 1 \end{bmatrix} \begin{bmatrix} -1 & 2 & 3 \\ 1 & 3 & -4 \\ -1 & -8 & 9 \end{bmatrix} = \begin{bmatrix} -1 & 2 & 3 \\ 0 & 5 & -1 \\ 0 & 0 & 4 \end{bmatrix}$$

and

$$-\mathbf{LN}_{.j*} = -\begin{bmatrix} 1 & 0 & 0 \\ 1 & 1 & 0 \\ 1 & 2 & 1 \end{bmatrix} \begin{bmatrix} -1 \\ 1 \\ 0 \end{bmatrix} = \begin{bmatrix} 1 \\ 0 \\ -1 \end{bmatrix}$$

Now $\mathbf{d_B}$ can be obtained by solving $(\mathbf{LB})\mathbf{d_B} = -\mathbf{LN}_{.j*}$:

$$\begin{bmatrix} -1 & 2 & 3 \\ 0 & 5 & -1 \\ 0 & 0 & 4 \end{bmatrix} \begin{bmatrix} (\mathbf{d_B})_1 \\ (\mathbf{d_B})_2 \\ (\mathbf{d_B})_3 \end{bmatrix} = \begin{bmatrix} 1 \\ 0 \\ -1 \end{bmatrix}$$

By back substitution,

$$(\mathbf{d_B})_3 = -\tfrac{1}{4}$$
$$(\mathbf{d_B})_2 = \tfrac{1}{5}(\mathbf{d_B})_3 = -\tfrac{1}{20}$$
$$(\mathbf{d_B})_1 = -1 + 2(\mathbf{d_B})_2 + 3(\mathbf{d_B})_3 = -\tfrac{37}{20}$$

To solve the system $\mathbf{uB} = \mathbf{c_B}$ for $\mathbf{u}$, one first solves the system $\mathbf{w(LB)} = \mathbf{c_B}$ for $\mathbf{w}$;

$$(w_1, w_2, w_3)\begin{bmatrix} -1 & 2 & 3 \\ 0 & 5 & -1 \\ 0 & 0 & 4 \end{bmatrix} = (1,3,0)$$

By forward substitution

$$w_1 = -1$$
$$w_2 = \tfrac{1}{5}(3 - 2w_1) = 1$$
$$w_3 = \tfrac{1}{4}(-3w_1 + w_2) = 1$$

Then $\mathbf{u} = \mathbf{wL}$, so

$$\mathbf{u} = (-1,1,1)\begin{bmatrix} 1 & 0 & 0 \\ 1 & 1 & 0 \\ 1 & 2 & 1 \end{bmatrix} = (1,3,1)$$

To repeat, an **LU** *decomposition* of **B** consists of finding a lower triangular matrix **L** for which $\mathbf{LB} = \mathbf{U}$ is upper triangular. As with every other basis handling technique, the issue of updating **L** must be addressed. The update involves producing a new matrix $\mathbf{L}'$ such that $\mathbf{L}'\mathbf{B}'$ is again upper triangular. As in the product form of the inverse, instead of changing **L** into $\mathbf{L}'$, an eta matrix **E** is found so that $\mathbf{L}'$ can be computed as **EL**. Each iteration produces another eta matrix, so, after k iterations, the eta file consists of $\mathbf{E}_1, \mathbf{E}_2, \ldots, \mathbf{E}_k$ with the property that $(\mathbf{E}_k \cdots \mathbf{E}_1\mathbf{L})\mathbf{B}$ is upper triangular. The eta file can then be used efficiently to solve the current linear systems. Once again, when k gets too large, the entire process is started over by constructing the current basis matrix using the columns of **A**; then a new **L** matrix is found, and the eta file is erased.

Because an **LU** decomposition is not unique (i.e., there are various ways to choose an **L** matrix so that **LB** is upper triangular), it is possible to select **L** in some "intelligent" way. One such method chooses **L** so as to control round-off error as much as possible. It has been observed that numerical accuracy can often be maintained by judiciously avoiding the division of relatively large numbers by relatively small ones. The slight additional effort that is needed to avoid such divisions is amply rewarded by the numerically accurate solutions that are obtained.

Suppose that **L** is a given lower triangular matrix for which $\mathbf{LB} = \mathbf{U}$ is upper triangular. The issue is how to update **L** when $\mathbf{N}_{\cdot j*}$ enters the basis matrix and $\mathbf{B}_{\cdot k*}$ leaves. Furthermore, the update must be done in such a way that the new matrix $\mathbf{L}'$ has the property that $\mathbf{L}'\mathbf{B}' = \mathbf{U}'$ is upper triangular and, also, so that round-off error is controlled.

$$\left[\begin{array}{ccccc} \uparrow & & \uparrow & & \uparrow \\ \mathbf{B}_{\cdot 1}, & \ldots, & \mathbf{B}_{\cdot k^*}, & \ldots, & \mathbf{B}_{\cdot m} \\ \downarrow & & \downarrow & & \downarrow \end{array}\right], \left[\begin{array}{c} \uparrow \\ \mathbf{N}_{\cdot j^*} \\ \downarrow \end{array}\right] = \left[\begin{array}{cccccc} \uparrow & & \uparrow & \uparrow & & \uparrow \quad \uparrow \\ \mathbf{B}_{\cdot 1}, & \ldots, & \mathbf{B}_{\cdot k^*-1}, & \mathbf{B}_{\cdot k^*+1}, & \ldots, & \mathbf{B}_{\cdot m}, \mathbf{N}_{\cdot j^*} \\ \downarrow & & \downarrow & \downarrow & & \downarrow \quad \downarrow \end{array}\right]$$

FIGURE 7.1. Placing $\mathbf{N}_{\cdot j^*}$ at the end of **B**.

Unlike previous versions of the simplex algorithm in which $\mathbf{N}_{\cdot j^*}$ replaces column k^* of **B**, here column k^* of **B** is removed, and $\mathbf{N}_{\cdot j^*}$ is added on as the last column of **B**, and the remaining columns of **B** are squeezed together to make room for $\mathbf{N}_{\cdot j^*}$ at the end (see Figure 7.1). Since $\mathbf{LB} = \mathbf{U}$ is to be updated into $\mathbf{L}'\mathbf{B}' = \mathbf{U}'$, let us start with $\mathbf{LB}'$ and see what has to be done in order to make it upper triangular. Since **LB** is upper triangular and $\mathbf{B}'$ differs from **B** only slightly, one might expect $\mathbf{LB}'$ to be almost upper triangular. This is indeed the case. The only nonzero elements of $\mathbf{LB}'$ that appear below the diagonal do so immediately below it, as will be seen in the next example. Such a matrix is called *upper Hessenberg*.

EXAMPLE 7.6: If in Example 7.5 $\mathbf{N}_{\cdot j^*}$ enters and column 1 of **B** leaves, then

$$\mathbf{B}' = \begin{bmatrix} 2 & 3 & -1 \\ 3 & -4 & 1 \\ -8 & 9 & 0 \end{bmatrix}$$

and

$$\mathbf{LB}' = \begin{bmatrix} 1 & 0 & 0 \\ 1 & 1 & 0 \\ 1 & 2 & 1 \end{bmatrix} \begin{bmatrix} 2 & 3 & -1 \\ 3 & -4 & 1 \\ -8 & 9 & 0 \end{bmatrix} = \begin{bmatrix} 2 & 3 & -1 \\ 5 & -1 & 0 \\ 0 & 4 & 1 \end{bmatrix}$$

Since $\mathbf{LB}'$ is upper Hessenberg, it is not difficult to update **L** into $\mathbf{L}'$ so that $\mathbf{L}'\mathbf{B}'$ becomes upper triangular. Moreover, the update will be done in such a way as to avoid the division of large numbers by small ones whenever possible. In so doing, it occasionally becomes necessary to interchange the rows of $\mathbf{LB}'$, which can be accomplished by the use of permutation matrices (see Section 4.3).

As in the product form of the inverse, the update is accomplished by storing additional information rather than actually changing **L** into $\mathbf{L}'$. In particular, to eliminate the possible $m-1$ nonzero elements of $\mathbf{LB}'$ appearing below the diagonal, nonsingular lower triangular matrices $\mathbf{L}_1, \ldots, \mathbf{L}_{m-1}$ and permutation matrices $\mathbf{P}_1, \ldots, \mathbf{P}_{m-1}$ will be found with the property that $\mathbf{L}_{m-1}\mathbf{P}_{m-1} \cdots \mathbf{L}_1\mathbf{P}_1(\mathbf{LB}')$ is upper triangular. All of these matrices are extremely sparse, differing from the identity matrix in only one element. Moreover, some of the permutation matrices might be the identity matrix and need not be stored at all. Each pair of matrices $\mathbf{L}_i$ and $\mathbf{P}_i$ is designed to remove a nonzero element in row $i+1$ and column i of $\mathbf{LB}'$, as shown in the next example.

EXAMPLE 7.7: For the matrix $\mathbf{LB}'$ of Example 7.6, to remove the nonzero entry in $(\mathbf{LB}')_{21}$, let

$$\mathbf{L}_1 = \begin{bmatrix} 1 & 0 & 0 \\ -\frac{5}{2} & 1 & 0 \\ 0 & 0 & 1 \end{bmatrix}, \qquad \mathbf{P}_1 = \mathbf{I}$$

then

$$\mathbf{L}_1\mathbf{P}_1(\mathbf{LB}') = \begin{bmatrix} 1 & 0 & 0 \\ -\frac{5}{2} & 1 & 0 \\ 0 & 0 & 1 \end{bmatrix} \begin{bmatrix} 2 & 3 & -1 \\ 5 & -1 & 0 \\ 0 & 4 & 1 \end{bmatrix} = \begin{bmatrix} 2 & 3 & -1 \\ 0 & -\frac{17}{2} & \frac{5}{2} \\ 0 & 4 & 1 \end{bmatrix}$$

To remove the nonzero entry in $(\mathbf{L}_1\mathbf{P}_1\mathbf{LB}')_{32}$, let

$$\mathbf{L}_2 = \begin{bmatrix} 1 & 0 & 0 \\ 0 & 1 & 0 \\ 0 & \frac{8}{17} & 1 \end{bmatrix}, \qquad \mathbf{P}_2 = \mathbf{I}$$

Then

$$\mathbf{L}_2\mathbf{P}_2\mathbf{L}_1\mathbf{P}_1(\mathbf{LB}') = \begin{bmatrix} 1 & 0 & 0 \\ 0 & 1 & 0 \\ 0 & \frac{8}{17} & 1 \end{bmatrix} \begin{bmatrix} 2 & 3 & -1 \\ 0 & -\frac{17}{2} & \frac{5}{2} \\ 0 & 4 & 1 \end{bmatrix} = \begin{bmatrix} 2 & 3 & -1 \\ 0 & -\frac{17}{2} & \frac{5}{2} \\ 0 & 0 & \frac{37}{17} \end{bmatrix}$$

For Example 7.7, $\mathbf{P}_1$ is the identity matrix. In general, $\mathbf{P}_1$ is chosen so as to avoid the division of large numbers by small ones that can occur when the element $(\mathbf{LB}')_{11}$ is small, for it is this number that will be used as a denominator when computing the $\mathbf{L}_1$ matrix. For the matrix $\mathbf{LB}'$ in Example 7.6, note that rows 1 and 2 can be interchanged without affecting the form of the matrix, and it is desirable to do so because $(\mathbf{LB}')_{21}$ can then be used as the denominator instead of $(\mathbf{LB}')_{11}$, and this is better since $(\mathbf{LB}')_{21} > (\mathbf{LB}')_{11}$. In general, at each step, row i and row $i+1$ are interchanged (by using $\mathbf{P}_i$) if the resulting diagonal element has a larger absolute value than the previous diagonal element. Then, when constructing the matrix $\mathbf{L}_i$, round-off errors are reduced. These operations are now demonstrated for the previous example.

EXAMPLE 7.8: For the matrix $\mathbf{LB}'$ in Example 7.6, rows 1 and 2 can be interchanged by using the permutation matrix

$$\mathbf{P}_1 = \begin{bmatrix} 0 & 1 & 0 \\ 1 & 0 & 0 \\ 0 & 0 & 1 \end{bmatrix}$$

Then

$$\mathbf{P}_1\mathbf{LB}' = \begin{bmatrix} 0 & 1 & 0 \\ 1 & 0 & 0 \\ 0 & 0 & 1 \end{bmatrix} \begin{bmatrix} 2 & 3 & -1 \\ 5 & -1 & 0 \\ 0 & 4 & 1 \end{bmatrix} = \begin{bmatrix} 5 & -1 & 0 \\ 2 & 3 & -1 \\ 0 & 4 & 1 \end{bmatrix}$$

Now, to remove the nonzero entry in $(\mathbf{P}_1\mathbf{LB}')_{21}$, let

$$\mathbf{L}_1 = \begin{bmatrix} 1 & 0 & 0 \\ -\frac{2}{5} & 1 & 0 \\ 0 & 0 & 1 \end{bmatrix}$$

Then

$$\mathbf{L}_1\mathbf{P}_1\mathbf{LB}' = \begin{bmatrix} 1 & 0 & 0 \\ -\frac{2}{5} & 1 & 0 \\ 0 & 0 & 1 \end{bmatrix} \begin{bmatrix} 5 & -1 & 0 \\ 2 & 3 & -1 \\ 0 & 4 & -1 \end{bmatrix} = \begin{bmatrix} 5 & -1 & 0 \\ 0 & \frac{17}{5} & -1 \\ 0 & 4 & 1 \end{bmatrix}$$

This section has described several advanced basis handling techniques, including the product form of the inverse, and various **LU** decompositions. All these procedures require periodic reinversion whereby the accumulated eta file is erased and the process is started over. The next section deals with various efficient and accurate reinversion routines.

EXERCISES

7.3.1. For the data in Exercise 7.2.1(a) and 7.2.1(b), find the eta matrices $\mathbf{E}$ and $\mathbf{E}^{-1}$ that would be used for updating the basis.

7.3.2. Given the basis inverse matrix

$$\mathbf{B}^{-1} = \begin{bmatrix} 1 & 0 & 4 & 0 \\ 1 & 3 & 2 & 0 \\ 0 & 0 & 1 & 1 \\ 2 & 1 & 0 & 1 \end{bmatrix}$$

suppose that, after two iterations of the simplex algorithm, the following inverse eta matrices have been stored (in sparse form):

$\mathbf{E}_1^{-1}$, which contains -1 in row 2, column 2, and 3 in row 4, column 2, and $\mathbf{E}_2^{-1}$, which contains 2 in row 2, column 3, and -2 in row 3, column 3.

Use these data to solve the current systems of equations in which $\mathbf{c}_\mathbf{B} = (1,-1,0,3)$ and $\mathbf{N}_{.j*} = (2,-2,-1,1)$. (Note that the current inverse matrix is $\mathbf{E}_2^{-1}\mathbf{E}_1^{-1}\mathbf{B}^{-1}$.)

7.3.3. Given the following basis matrix $\mathbf{B}$ and the lower triangular matrix $\mathbf{L}$ for which $\mathbf{LB}$ is upper triangular, together with $\mathbf{N}_{.j*}$ and $\mathbf{c}_\mathbf{B}$, solve the two linear systems by forward and back substitution.

$$\mathbf{B} = \begin{bmatrix} 1 & 2 & 5 \\ -1 & -1 & -2 \\ 0 & -1 & -4 \end{bmatrix}, \qquad \mathbf{L} = \begin{bmatrix} 1 & 0 & 0 \\ 1 & 1 & 0 \\ 1 & 1 & 1 \end{bmatrix}$$

$$\mathbf{N}_{.j*} = \begin{bmatrix} 1 \\ -2 \\ 1 \end{bmatrix}, \qquad \mathbf{c}_\mathbf{B} = (3,3,4)$$

7.4. MATRIX INVERSION

All of the basis handling methods of this chapter begin with some form of $\mathbf{B}^{-1}$, which is updated at the end of each iteration. Regardless of the particular method used, computational efficiency and numerical accuracy can only be maintained if the current basis inverse is periodically recomputed from scratch by using the appropriate columns from the original $\mathbf{A}$ matrix. This section describes several such methods.

Gaussian Elimination with Partial Pivoting

For each form of the inverse there is a corresponding method for performing the reinversion. For instance, one approach is to compute an explicit form of $\mathbf{B}^{-1}$. One way to find an explicit form of the inverse is to produce m eta matrices, $\mathbf{E}_1, \ldots, \mathbf{E}_m$, for which $\mathbf{E}_m \cdots \mathbf{E}_1\mathbf{B} = \mathbf{I}$, for then $\mathbf{B}^{-1} = \mathbf{E}_m \cdots \mathbf{E}_1$. The matrix $\mathbf{E}_1$ is chosen so that the first column of $\mathbf{E}_1\mathbf{B}$ is the first column of $\mathbf{I}$. Then, starting with $\mathbf{E}_1\mathbf{B}$, $\mathbf{E}_2$ is chosen so that the first two columns of $\mathbf{E}_2(\mathbf{E}_1\mathbf{B})$ are the first two columns of $\mathbf{I}$. In general, starting with $\mathbf{E}_{k-1} \cdots \mathbf{E}_1\mathbf{B}$, $\mathbf{E}_k$ is chosen so that the first k columns of $\mathbf{E}_k \cdots \mathbf{E}_1\mathbf{B}$ are the first k columns of $\mathbf{I}$. Thus, after m iterations, one finds $\mathbf{E}_m \cdots \mathbf{E}_1\mathbf{B}$, the identity matrix. The process of finding the m eta matrices is known as Gaussian elimination and was described in Section 4.4. Recall that during Gaussian elimination it may be necessary to use permutation matrices to avoid dividing by 0. Specifically, if $\mathbf{C} = \mathbf{E}_{k-1} \cdots \mathbf{E}_1\mathbf{B}$, and $\mathbf{C}_{kk} = 0$, then a permutation matrix is needed to interchange some row of $\mathbf{C}$ (say, $r > k$) with row k. There can, of course, be many choices for r. For the purpose of maintaining numerical accuracy, it is desirable to choose a permutation matrix in such a way that, when $\mathbf{E}_k$ is computed, the division of large numbers by small ones is avoided. Doing so requires choosing $\mathbf{P}$ so that $(\mathbf{PC})_{kk}$ has as large an absolute value as possible, for $(\mathbf{PC})_{kk}$ is used as the denominator when computing $\mathbf{E}_k$.

In other words, to determine which row of $\mathbf{C}$ to interchange with row k of $\mathbf{C}$, scan down column k of $\mathbf{C}$ (below row k) and find the nonzero element having the largest absolute value. This row is interchanged with row k and then $\mathbf{E}_k$ is computed. Choosing the row by this method is called *partial pivoting*, and it is illustrated in the next example.

EXAMPLE 7.9: To illustrate Gaussian elimination with partial pivoting, suppose that the current matrix is

$$\mathbf{C} = \begin{bmatrix} 1 & 0 & -1 & 1 & \frac{3}{5} \\ 0 & 1 & 1 & 0 & \frac{1}{5} \\ 0 & 0 & 1 & 2 & \frac{11}{5} \\ 0 & 0 & 0 & -1 & 3 \\ 0 & 0 & 2 & 6 & -2 \end{bmatrix}$$

Before finding an eta matrix $\mathbf{E}_3$ where the first three columns of $\mathbf{E}_3\mathbf{C}$ are those of the identity matrix, scan down column 3 of **C** (below row 3) to find the element with the largest absolute value (in this case, $\mathbf{C}_{53} = 2$). Then row 5 of **C** can be interchanged with row 3 by using the permutation matrix

$$\mathbf{P}_3 = \begin{bmatrix} 1 & 0 & 0 & 0 & 0 \\ 0 & 1 & 0 & 0 & 0 \\ 0 & 0 & 0 & 0 & 1 \\ 0 & 0 & 0 & 1 & 0 \\ 0 & 0 & 1 & 0 & 0 \end{bmatrix}$$

to obtain

$$\mathbf{P}_3\mathbf{C} = \begin{bmatrix} 1 & 0 & -1 & 1 & \frac{3}{5} \\ 0 & 1 & 1 & 0 & \frac{1}{5} \\ 0 & 0 & 2 & 6 & -2 \\ 0 & 0 & 0 & -1 & 3 \\ 0 & 0 & 1 & 2 & \frac{11}{5} \end{bmatrix}$$

Gaussian elimination can now be applied to $\mathbf{P}_3\mathbf{C}$ as described in Section 4.4.

Scaling

Because of the way in which computers store and divide real numbers, partial pivoting is effective only if all nonzero entries in the **B** matrix have approximately the same absolute value. In other words, before performing the inversion, it is necessary to prepare or *scale* the **B** matrix. In practice, the entire **A** matrix is scaled once and for all at the very beginning of the computer program, thus avoiding the need to scale the **B** matrix each time reinversion is to be performed.

Scaling is accomplished by multiplying **A** on the left by a nonsingular $m \times m$ diagonal matrix **S** of the form

$$\begin{bmatrix} s_1 & & 0 \\ 0 & & \vdots \\ \vdots & \ddots & \\ 0 & & s_m \end{bmatrix}$$

and then on the right by a nonsingular $n \times n$ diagonal matrix **T** of the form:

$$\begin{bmatrix} t_1 & & 0 \\ 0 & & \vdots \\ \vdots & \ddots & \\ 0 & & t_n \end{bmatrix}$$

The particular choice of **S** and **T** is made so as to ensure that, in the resulting matrix

$$\begin{bmatrix} s_1\mathbf{A}_{11}t_1 & \cdots & s_1\mathbf{A}_{1n}t_n \\ \vdots & & \vdots \\ s_m\mathbf{A}_{m1}t_1 & \cdots & s_m\mathbf{A}_{mn}t_n \end{bmatrix}$$

first, the largest absolute value of a nonzero element in each row is between 0.1 and 1.0 and, second, the largest absolute value of a nonzero element in each column is between 0.1 and 1.0. (After scaling, however, both conditions need not hold.)

First $s_1, \ldots, s_m$ are chosen so that every nonzero element of **SA** has an absolute value of at most 1.0. Then $t_1, \ldots, t_n$ are selected so as to ensure that the largest absolute value of a nonzero entry in each column of **SAT** is between 0.1 and 1.0.

To compute the correct value for s_i, look at row i of the **A** matrix and find the element having the largest absolute value (say, $\mathbf{A}_{ij}$). If k is an integer for which $10^{k-1} \le |\mathbf{A}_{ij}| \le 10^k$, then $s_i = 10^{-k}$. Repeating the procedure for each row of **A** yields values for $s_1, \ldots, s_m$. It can then be seen that each element of **SA** has an absolute value of at most 1.0.

To compute the value for t_j, look at column j of **SA** and find the nonzero element having the largest absolute value [say, $(\mathbf{SA})_{ij}$]. If k is an integer for which $10^{-k-1} \le |(\mathbf{SA})_{ij}| \le 10^{-k}$, then $t_j = 10^k$.

Once the **A** matrix has been well scaled in this manner, the submatrix **SBT** will also be well scaled. Gaussian elimination with partial pivoting can then be applied to **SBT** to obtain its inverse $\mathbf{E}_m \cdots \mathbf{E}_1$. Then $\mathbf{B}^{-1}$ can be computed as $\mathbf{TE}_m \cdots \mathbf{E}_1\mathbf{S}$. From now on it will be assumed that the **B** matrix, whose inverse is to be computed, has been well scaled.

Sparsity

As indicated in the previous section, the matrix $\mathbf{B}^{-1}$ is not really needed. What is needed is a fast, accurate mechanism for solving the two linear systems $\mathbf{uB} = \mathbf{c_B}$ and $\mathbf{Bd_B} = -\mathbf{N}_{.j*}$. These mechanisms are the various product forms of the inverse. The best product forms are those which are

1. sparse, thus allowing for a fast solution of the linear systems, and
2. numerically accurate because they carefully avoid the division of large numbers by small ones.

Gaussian elimination with partial pivoting can be modified easily to produce an **LU** decomposition of **B**. The idea is to start with **B** and to produce lower triangular matrices $\mathbf{L}_1, \ldots, \mathbf{L}_{m-1}$ and permutation matrices $\mathbf{P}_1, \ldots, \mathbf{P}_{m-1}$ such that the matrix $\mathbf{L}_{m-1}\mathbf{P}_{m-1} \cdots \mathbf{L}_1\mathbf{P}_1\mathbf{B}$ is upper triangular. The details are given in Exercises 7.4.3.

So far all of the ideas of this section have been directed toward producing a numerically stable product form of the inverse. Now it is time to address the issue of computational efficiency. Rather than attempt to reduce the amount of work that it takes to compute a product form of the inverse, it is better to spend additional effort to seek methods for producing a sparse product form. The extra effort is compensated by the fact that a sparse product form will then enable one to solve the two linear systems $\mathbf{uB} = \mathbf{c_B}$ and $\mathbf{Bd_B} = -\mathbf{N}_{.j*}$ more rapidly at each iteration.

Although the **B** matrix can itself be sparse, there is no guarantee that the product form produced by Gaussian elimination with partial pivoting will be sparse. Indeed, sparsity can disappear altogether, as shown in the next example, (in which all zeros have been omitted).

EXAMPLE 7.10:

$$\mathbf{B} = \begin{bmatrix} 5 & 1 & 1 & 1 & 1 \\ 1 & 1 & & & \\ 1 & & 1 & & \\ 1 & & & 1 & \\ 1 & & & & \end{bmatrix} \qquad \mathbf{E}_1 = \begin{bmatrix} \frac{1}{5} & & & & \\ -\frac{1}{5} & 1 & & & \\ -\frac{1}{5} & & 1 & & \\ -\frac{1}{5} & & & 1 & \\ -\frac{1}{5} & & & & 1 \end{bmatrix}$$

$$\mathbf{E}_1\mathbf{B} = \begin{bmatrix} 1 & \frac{1}{5} & \frac{1}{5} & \frac{1}{5} & \frac{1}{5} \\ & \frac{4}{5} & -\frac{1}{5} & -\frac{1}{5} & -\frac{1}{5} \\ & -\frac{1}{5} & \frac{4}{5} & -\frac{1}{5} & -\frac{1}{5} \\ & -\frac{1}{5} & -\frac{1}{5} & \frac{4}{5} & -\frac{1}{5} \\ & -\frac{1}{5} & -\frac{1}{5} & -\frac{1}{5} & -\frac{1}{5} \end{bmatrix}, \qquad \mathbf{E}_2 = \begin{bmatrix} 1 & -\frac{1}{4} & & & \\ & \frac{5}{4} & & & \\ & \frac{1}{4} & 1 & & \\ & \frac{1}{4} & & 1 & \\ & \frac{1}{4} & & & 1 \end{bmatrix}$$

$$\mathbf{E}_2\mathbf{E}_1\mathbf{B} = \begin{bmatrix} 1 & & \frac{1}{4} & \frac{1}{4} & \frac{1}{4} \\ & 1 & -\frac{1}{4} & -\frac{1}{4} & -\frac{1}{4} \\ & & \frac{3}{4} & -\frac{1}{4} & -\frac{1}{4} \\ & & -\frac{1}{4} & \frac{3}{4} & -\frac{1}{4} \\ & & -\frac{1}{4} & -\frac{1}{4} & -\frac{1}{4} \end{bmatrix}, \qquad \mathbf{E}_3 = \begin{bmatrix} 1 & & -\frac{1}{3} & & \\ & 1 & \frac{1}{3} & & \\ & & \frac{4}{3} & & \\ & & \frac{1}{3} & 1 & \\ & & \frac{1}{3} & & 1 \end{bmatrix}$$

$$\mathbf{E}_3\mathbf{E}_2\mathbf{E}_1\mathbf{B} = \begin{bmatrix} 1 & & & \frac{1}{3} & \frac{1}{3} \\ & 1 & & -\frac{1}{3} & -\frac{1}{3} \\ & & 1 & -\frac{1}{3} & -\frac{1}{3} \\ & & & \frac{2}{3} & -\frac{1}{3} \\ & & & -\frac{1}{3} & -\frac{1}{3} \end{bmatrix}, \qquad \mathbf{E}_4 = \begin{bmatrix} 1 & & & -\frac{1}{2} & \\ & 1 & & \frac{1}{2} & \\ & & 1 & \frac{1}{2} & \\ & & & \frac{3}{2} & \\ & & & \frac{1}{2} & 1 \end{bmatrix}$$

$$\mathbf{E}_4\mathbf{E}_3\mathbf{E}_2\mathbf{E}_1\mathbf{B} = \begin{bmatrix} 1 & & & & \frac{1}{2} \\ & 1 & & & -\frac{1}{2} \\ & & 1 & & -\frac{1}{2} \\ & & & 1 & -\frac{1}{2} \\ & & & & -\frac{1}{2} \end{bmatrix}, \qquad \mathbf{E}_5 = \begin{bmatrix} 1 & & & & 1 \\ & 1 & & & -1 \\ & & 1 & & -1 \\ & & & 1 & -1 \\ & & & & -2 \end{bmatrix}$$

$$\mathbf{E}_5\mathbf{E}_4\mathbf{E}_3\mathbf{E}_2\mathbf{E}_1\mathbf{B} = \mathbf{I}$$

After the very first step of Gaussian elimination, all signs of sparsity have disappeared. Losing sparsity, as in Example 7.10, is referred to as *fill-in*. Notice that all of the eta vectors in Example 7.10 are *dense* (i.e., have all nonzero entries), as is the noneliminated portion of the basis matrix at each iteration. It has been observed that if **B** is lower triangular, then no fill-in occurs, provided that all of the permutation matrices in partial pivoting are the identity. With this in mind, the idea is to interchange the rows and columns of **B** somehow so as to make it as lower triangular as possible. For the matrix in Example 7.10, this could be accomplished by interchanging rows 1 and 5. Now when Gaussian elimination is performed, notice the lack of fill-in, as shown in the next example. Specifically, all of the eta vectors (except the first one) are sparse.

EXAMPLE 7.11:

$$\mathbf{B}^* = \begin{bmatrix} 1 & & & & \\ 1 & 1 & & & \\ 1 & & 1 & & \\ 1 & & & 1 & \\ 5 & 1 & 1 & 1 & 1 \end{bmatrix}, \qquad \mathbf{E}_1 = \begin{bmatrix} 1 & & & & \\ -1 & 1 & & & \\ -1 & & 1 & & \\ -1 & & & 1 & \\ -5 & & & & 1 \end{bmatrix}$$

$$\mathbf{E}_1\mathbf{B}^* = \begin{bmatrix} 1 & & & & \\ & 1 & & & \\ & & 1 & & \\ & & & 1 & \\ & 1 & 1 & 1 & 1 \end{bmatrix}, \qquad \mathbf{E}_2 = \begin{bmatrix} 1 & & & & \\ & 1 & & & \\ & & 1 & & \\ & & & 1 & \\ & -1 & & & 1 \end{bmatrix}$$

$$\mathbf{E}_2\mathbf{E}_1\mathbf{B}^* = \begin{bmatrix} 1 & & & & \\ & 1 & & & \\ & & 1 & & \\ & & & 1 & \\ & & 1 & 1 & 1 \end{bmatrix}, \qquad \mathbf{E}_3 = \begin{bmatrix} 1 & & & & \\ & 1 & & & \\ & & 1 & & \\ & & & 1 & \\ & & -1 & & 1 \end{bmatrix}$$

$$\mathbf{E}_3\mathbf{E}_2\mathbf{E}_1\mathbf{B}^* = \begin{bmatrix} 1 & & & & \\ & 1 & & & \\ & & 1 & & \\ & & & 1 & \\ & & & 1 & 1 \end{bmatrix}, \qquad \mathbf{E}_4 = \begin{bmatrix} 1 & & & & \\ & 1 & & & \\ & & 1 & & \\ & & & 1 & \\ & & & -1 & 1 \end{bmatrix}$$

$$\mathbf{E}_4\mathbf{E}_3\mathbf{E}_2\mathbf{E}_1\mathbf{B}^* = \mathbf{I}$$

In the general case, heuristics are used to permute the columns and rows of **B** prior to inversion so as to make **B** as lower triangular as possible. At the end of the procedure, all columns that still have nonzero entries above the diagonal are called *spikes*, and it is these columns that cause fill-in. The heuristics are first applied to the rows of **B** and then to the columns of the resulting matrix. Specifically, the following algorithm attempts to put the **B** matrix into lower triangular form.

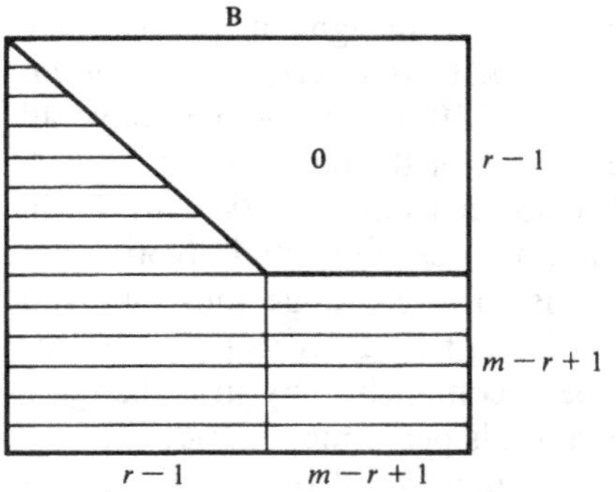

FIGURE 7.2. **B** after permuting the rows.

Step 1. Start with $r = 1$. As long as $r < m$, see if there is a row $i \geq r + 1$ for which there is a unique column $j \geq r$ such that $\mathbf{B}_{ij} \neq 0$. If so, then interchange rows i and r, interchange columns j and r, and increase r by 1; otherwise, go to step 2.

After termination, the **B** matrix will have the form indicated in Figure 7.2. Further improvements can then be made by rearranging the rows and columns of the $(m - r + 1) \times (m - r + 1)$ submatrix in the lower right-hand corner of Figure 7.2. The procedure for doing this is the same as that in step 1 except that it works with the columns of the matrix, starting with the last one:

Step 2. Start with the value of r from step 1 and set $s = m$. As long as $s \geq r + 1$, see if there is a column j with $r \leq j \leq s - 1$, for which there is a unique row i with $r \leq i \leq s$, such that $\mathbf{B}_{ij} \neq 0$. If so, interchange columns s and j, interchange rows s and i, and decrease s by 1; otherwise, stop.

After completion, **B** will appear as in Figure 7.3. The shaded portion above the diagonal is called the *bump*. For large problems, the number of columns in the bump rarely exceeds 10% of the number of columns in the **B**

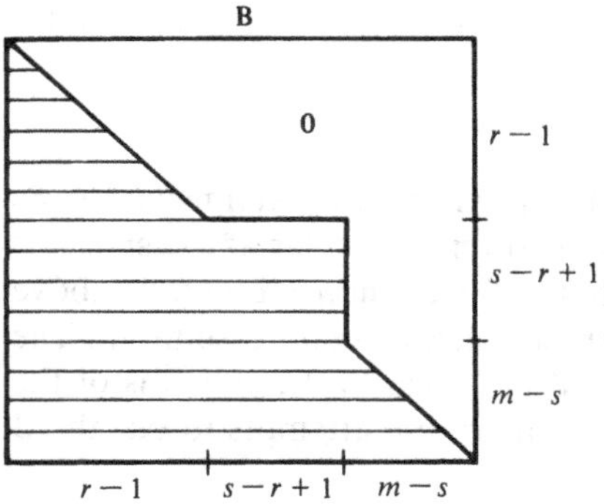

FIGURE 7.3. **B** after permuting the rows and columns.

matrix. Once this final matrix has been obtained, Gaussian elimination with partial pivoting can be used to produce a sparse product form of the inverse. It should be noted that the permutation matrices that arise in partial pivoting can cause fill-in by disturbing the lower triangular structure of **B** that has been created. Similar sparsity techniques have been designed for keeping the **L** and **U** matrices sparse in an **LU** decomposition of **B**.

This section has presented various ways for solving the two linear systems $\mathbf{uB} = \mathbf{c_B}$ and $\mathbf{Bd_B} = -\mathbf{N}_{.j*}$. Particular attention was given to maintaining numerical accuracy as well as computational efficiency. The next section explores various techniques for improving the efficiency of the simplex algorithm by exploiting special structure that often arises in the constraints.

EXERCISES

7.4.1. Find the eta and permutation matrices of the following 4×4 matrix that would be obtained by Gaussian elimination with partial pivoting.

$$\begin{bmatrix} 1 & 1 & \frac{1}{2} & 3 \\ 2 & 2 & 1 & 2 \\ 1 & 0 & 3 & 1 \\ 3 & 3 & 0 & 2 \end{bmatrix}$$

7.4.2. Scale the following matrix by producing the **S** and **T** matrices as described in this section.

$$\begin{bmatrix} 58 & 0.01 & -2 & 12 \\ -3 & 0.07 & 0.1 & 2.5 \\ 500 & 3 & -50 & 200 \end{bmatrix}$$

7.4.3. Develop an algorithm that applies partial pivoting to a nonsingular matrix **B** to produce lower triangular eta matrices $\mathbf{L}_1, \ldots, \mathbf{L}_{m-1}$ and permutation matrices $\mathbf{P}_1, \ldots, \mathbf{P}_{m-1}$ so that $(\mathbf{L}_{m-1}\mathbf{P}_{m-1} \cdots \mathbf{L}_1\mathbf{P}_1)\mathbf{B}$ is upper triangular.

7.4.4. Apply the heuristics in the text to the following 5×5 matrix to produce as lower triangular a matrix as possible.

$$\begin{bmatrix} 0 & 1 & 2 & 3 & -9 \\ 0 & 1 & 0 & 2 & 12 \\ 5 & 0 & 0 & 0 & 1 \\ 0 & 0 & 0 & 0 & 1 \\ 5 & 1 & 0 & 1 & 3 \end{bmatrix}$$

7.5. SPECIAL STRUCTURE

The previous sections of this chapter have dealt with the computational implementation of the revised simplex algorithm in which particular attention was given to the basis handling techniques. Additional improvements can often be achieved by exploiting the *special structure* that arises in the constraints of certain LPs. Several such examples will be presented here, and some of the ideas are developed in much greater detail in Chapters 10 and 11.

One instance of special structure occurs when some of the constraints have a particularly simple form. Recall the bound constraints in which a single variable x_i is restricted to be less than or equal to some given number u_i, called an *upper bound*, as well as being nonnegative or, more generally, greater than or equal to some *lower bound* l_i. It is certainly possible to convert each of these constraints to standard form by adding and subtracting slack variables; however, doing so increases the number of variables as well as the size of the basis matrix. Because of the simplicity of the constraints $l_i \le x_i \le u_i$, a slightly modified simplex algorithm, in which each nonbasic variable is allowed to assume a value of either l_i or u_i (instead of 0), has been devised to avoid the necessity of adding slack variables. The most general procedure for handling bounds is known as *generalized upper bounding* and will be described in Chapter 10. Section 7.6 will show how it is sometimes possible to use lower and upper bounds to actually decrease the number of iterations needed to solve the LP.

Special structure is often encountered in large scale problems containing thousands of variables and hundreds (or thousands) of constraints. One special structure has already been exploited. In Section 7.4 sparsity was exploited by attempting to maintain sparsity in the product form of the basis inverse so that the two linear systems $\mathbf{uB} = \mathbf{c_B}$ and $\mathbf{Bd_B} = -\mathbf{N}_{.j*}$ can be solved quickly.

Additional structure of large-scale LPs can often be found in the very form of the constraints. Several examples are illustrated in Figure 7.4. These types of constraints are particularly prevalent in time-dependent linear programming models, i.e., models in which each block of constraints corresponds to a different time period. Each particular structure gives rise to its own special method for improving the simplex algorithm. One such approach is that of *decomposition*, whereby the original problem is decom-

FIGURE 7.4. Special structure in the constraints.

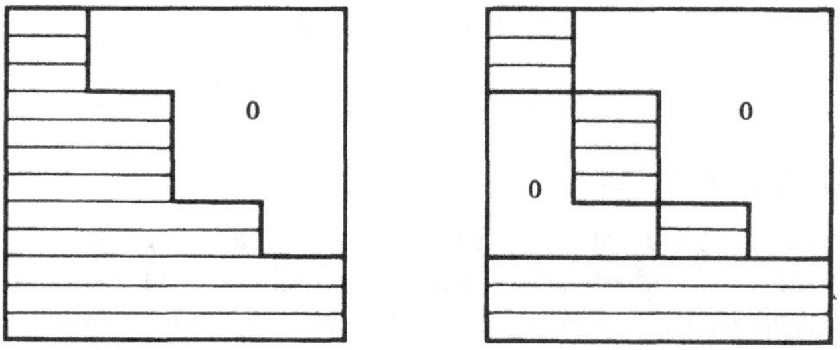

posed into smaller subproblems, each of which can be solved independently of the others. The solutions thus obtained are then used to find the solution to the original problem.

One might think that degeneracy in linear programming is hardly a special structure. Nonetheless, it has been observed that much time is wasted in the simplex algorithm when a degenerate pivot is performed. A large effort is expended in updating the basis, yet the value of the objective function does not improve since $t^* = 0$. Recently, a method has been developed for doing less work than a full pivot when degeneracy occurs. In problems where a large number of degenerate pivots are performed, this idea can lead to significant computational savings.

Finally, there are certain highly structured problems for which it is possible to perform the steps of the simplex algorithm in a greatly simplified manner—so much so that the test for optimality, the min ratio test, and the update operations require no multiplications or divisions. These problems will be analyzed in Chapter 11.

The lesson to be learned from this section is that whenever there is special structure in the LP, look for ways to enhance the performance of the simplex algorithm based on that structure. The next section explores various computational improvements that can be attained when solving the phase 1 LP.

7.6. COMPUTATIONAL IMPROVEMENTS IN PHASE 1

In this section a careful examination of phase 1 will be conducted with the purpose of reducing the amount of work that is done in each iteration. In addition, several methods will be suggested for decreasing the total number of iterations needed to solve the original LP. Since the total number of iterations that are required to solve a given LP is the sum of the iterations in phase 1 and those in phase 2, it is worthwhile seeking ways to reduce the number of iterations required in each of the two phases.

Dropping Artificial Variables

For example, recall the phase 1 LP,

$$\begin{array}{ll} \text{Minimize} & \mathbf{e}\mathbf{y} \\ \text{subject to} & \mathbf{A}\mathbf{x} + \mathbf{I}\mathbf{y} = \mathbf{b} \\ & \mathbf{x},\quad \mathbf{y} \geq \mathbf{0} \end{array}$$

which is solved by starting with the initial bfs $(\mathbf{x}, \mathbf{y}) = (\mathbf{0}, \mathbf{b})$. The $\mathbf{y}$ variables are basic at value $\mathbf{b}$ and the $\mathbf{x}$ variables are nonbasic at value $\mathbf{0}$. During the first iteration, a nonbasic $\mathbf{x}$ variable will become basic and replace one of the basic y variables. It is perhaps surprising that each time an artificial y variable becomes nonbasic, it can be kept nonbasic forever (or better yet, discarded completely). In general, it is not true that a variable that leaves the basis can be kept nonbasic forever, for that variable might just end up in the final basis. For the phase 1 LP, however, once an artificial variable leaves the basis it can be dropped permanently, and one can proceed to solve the reduced LP. The reason is that the objective of the phase 1 LP is to see if all of the artificial variables can be decreased to value 0. Consequently, as each one leaves the basis, it becomes nonbasic at value 0 and is no longer needed. To be a bit more precise, consider the reduced phase 1 LP obtained by discarding the artificial variable as it leaves the basis. If the values of the remaining artificial variables can be driven to 0, then the original LP is feasible (see Proposition 6.1); otherwise, it is not (see Proposition 6.2).

Using Fewer Artificial Variables

Another idea for improving the efficiency of phase 1 involves reducing the number of artificial variables needed to form the phase 1 LP. For example, suppose that, within the original **A** matrix, one could find an $m \times m$ identity matrix. In this favorable case, phase 1 could be eliminated altogether because an initial bfs for the original LP would already exist, as shown in the next example.

EXAMPLE 7.12:

$$\begin{array}{llllll} \text{Minimize} & 3x_1 & -3x_2 & +5x_3 & -4x_4 & +\ x_5 \\ \text{subject to} & 2x_1 & +\ x_2 & & +\ x_4 & +7x_5 = 3 \\ & x_1 & +6x_2 & +\ x_3 & & -4x_5 = 5 \\ & x_1, & x_2, & x_3, & x_4, & x_5 \geq 0 \end{array}$$

so

$$\mathbf{c} = (3, -3, 5, -4, 1)$$

$$\mathbf{A} = \begin{bmatrix} 2 & 1 & 0 & 1 & 7 \\ 1 & 6 & 1 & 0 & -4 \end{bmatrix}$$

$$\mathbf{b} = \begin{bmatrix} 3 \\ 5 \end{bmatrix}$$

Since $\mathbf{B} = [\mathbf{A}_{.4}, \mathbf{A}_{.3}]$ is the identity matrix, an initial bfs for this LP is

$$\mathbf{B} = [\mathbf{A}_{.4}, \mathbf{A}_{.3}] = \begin{bmatrix} 1 & 0 \\ 0 & 1 \end{bmatrix}$$

$$\mathbf{N} = [\mathbf{A}_{.1}, \mathbf{A}_{.2}, \mathbf{A}_{.5}] = \begin{bmatrix} 2 & 1 & 7 \\ 1 & 6 & -4 \end{bmatrix}$$

$$\mathbf{x_B} = \begin{bmatrix} x_4 \\ x_3 \end{bmatrix} = \mathbf{B}^{-1}\mathbf{b} = \begin{bmatrix} 1 & 0 \\ 0 & 1 \end{bmatrix}\begin{bmatrix} 3 \\ 5 \end{bmatrix} = \begin{bmatrix} 3 \\ 5 \end{bmatrix}$$

$$\mathbf{x_N} = (x_1, x_2, x_5) = (0,0,0)$$

One instance in which an $m \times m$ identity matrix will appear in the **A** matrix is an original LP (before being put into standard form) having $\mathbf{b} \geq \mathbf{0}$ and only $\leq$ constraints. Then, when put into standard form, m slack variables will be added to each of the m constraints, thus giving rise to the desired $m \times m$ identity matrix, as demonstrated in the next example.

EXAMPLE 7.13: *Original LP.*

$$\begin{array}{ll} \text{Minimize} & -10x_1 - 11x_2 - 9x_3 \\ \text{subject to} & x_1 + 2x_2 + x_3 \leq 4 \\ & 5x_1 \qquad\quad - 4x_3 \leq 1 \\ & x_1, \quad x_2, \quad x_3 \geq 0 \end{array}$$

Standard form LP.

$$\begin{array}{ll} \text{Minimize} & -10x_1 - 11x_2 - 9x_3 \\ \text{subject to} & x_1 + 2x_2 + x_3 + s_1 \qquad = 4 \\ & 5x_1 \qquad\quad - 4x_3 \qquad + s_2 = 1 \\ & x_1, \quad x_2, \quad x_3, \quad s_1, \quad s_2 \geq 0 \end{array}$$

so

$$\mathbf{c} = (-10, -11, -9, 0, 0)$$

$$\mathbf{A} = \begin{bmatrix} 1 & 2 & 1 & 1 & 0 \\ 5 & 0 & -4 & 0 & 1 \end{bmatrix}$$

$$\mathbf{b} = \begin{bmatrix} 4 \\ 1 \end{bmatrix}$$

The columns of **A** corresponding to the slack variables (i.e., $\mathbf{A}_{.4}$ and $\mathbf{A}_{.5}$) give rise to the identity matrix.

Although it is not reasonable to expect all of the original constraints to be $\leq$ constraints and $\mathbf{b} \geq \mathbf{0}$, still, for each such constraint, one less artificial variable is required in constructing the phase 1 LP because the corresponding slack variable serves the same purpose.

Taking the strategy of trying to reduce the number of artificial variables one step further, it is possible to develop a phase 1 procedure that requires no artificial variables at all. Instead, the method begins by arbitrarily selecting a potential basis matrix **B** from **A**. Assuming that **B** is nonsingular, one can compute the values of the corresponding basic variables by $\mathbf{x_B} = \mathbf{B}^{-1}\mathbf{b}$. In the event that $\mathbf{B}^{-1}\mathbf{b} \geq \mathbf{0}$, an initial bfs has been found, and phase 1 is not needed. On the other hand, if $\mathbf{B}^{-1}\mathbf{b} \geq \mathbf{0}$ does not hold, then one would like to "move" to another potential basis matrix $\mathbf{B}'$ in some guaranteed noncycling fashion. The desired movement is accomplished by a sequence of new phase 1 LPs. To construct the desired sequence, consider the current vector $\mathbf{x} = (\mathbf{x_B}, \mathbf{x_N}) = (\mathbf{B}^{-1}\mathbf{b}, \mathbf{0})$ that does not satisfy $\mathbf{B}^{-1}\mathbf{b} \geq \mathbf{0}$. Consequently, there is an i with $1 \leq i \leq m$, for which $(\mathbf{x_B})_i = (\mathbf{B}^{-1}\mathbf{b})_i < 0$. In other words, at least one of the basic variables has a negative value. The objective is to see if its value can be increased to 0. To find out, try to maximize $(\mathbf{x_B})_i$, subject, of course, to the constraints that $\mathbf{Ax} = \mathbf{b}$, $\mathbf{x_N} \geq 0$, and those components of $\mathbf{x_B}$ for which $\mathbf{B}^{-1}\mathbf{b} \geq 0$ should remain nonnegative. If $(\mathbf{x_B})_i$ cannot be increased to 0, then the original LP is infeasible; otherwise, when $(\mathbf{x_B})_i = 0$, a new negative-valued basic variable is selected for the objective function. Then the entire process is repeated.

There are various modifications to this basic strategy, and sometimes a good guess can be made for the initial $\mathbf{B}$ matrix. Nonetheless, any method that begins with a potential basis matrix and then attempts to move to a new one is referred to as a *crashing technique*. Most professional codes incorporate some form of crashing technique.

Coordinate Transformations

The next improvement comes from using the geometry of the phase 1 LP. The idea is to arrange for phase 1 to end with a feasible solution to the original LP that is geometrically very close to the optimal solution, thus reducing the number of iterations required in phase 2. Consider the following LP:

$$\begin{array}{ll}
\text{Minimize} & -x_1 - x_2 \\
\text{subject to} & 2x_1 + 3x_2 \leq 26 \\
& 2x_1 - x_2 \leq 10 \\
& x_1 - 2x_2 \leq 2 \\
& -x_1 + 2x_2 \leq 8 \\
& x_1 + 3x_2 \geq 7 \\
& 3x_1 - x_2 \geq 1 \\
& 0 \leq x_1 \leq 8 \\
& 0 \leq x_2 \leq 7
\end{array}$$

depicted in Figure 7.5. After putting the LP in standard form, one can apply the usual phase 1 procedure that would start with $x_1 = x_2 = 0$ and end with

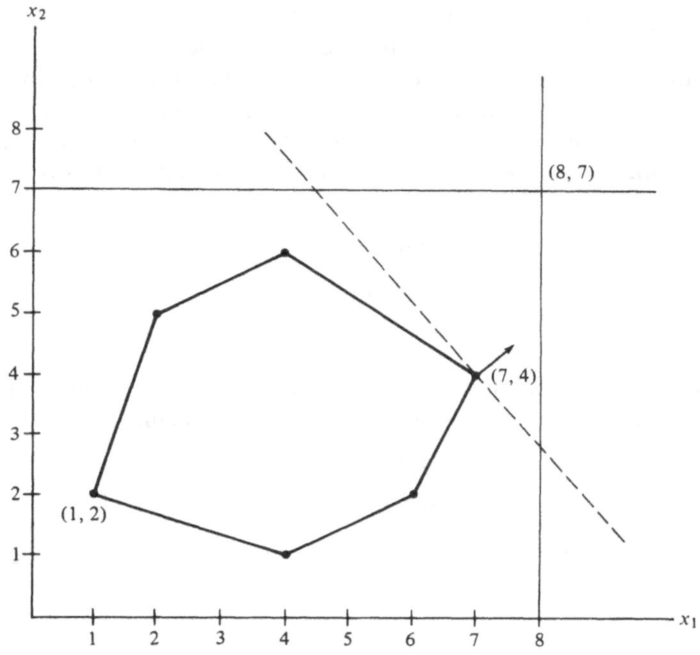

FIGURE 7.5. Geometry of phase 1.

$x_1 = 1$ and $x_2 = 2$, as seen in Figure 7.5. Unfortunately, this point is "far away" from the desired solution. One might therefore expect phase 2 to require many iterations before an optimal solution is obtained.

One way of remedying the problem would be to arrange for phase 1 to generate a feasible point on the "proper" side of the feasible region. Doing so can often be accomplished by changing the coordinate system of the LP before applying the simplex algorithm. To see how this would work on the example above, simply turn Figure 7.5 upside down and imagine that the point $x_1 = 8$ and $x_2 = 7$ is the origin of a new **w**-coordinate system. Now, when phase 1 is applied, it terminates with the optimal solution to the original problem, and hence phase 2 is not required.

The **w**-coordinate system is obtained based on the premise that phase 1, started at the origin, will terminate with a feasible solution that is geometrically close to the origin. Consequently, one should choose the **w**-coordinate system so that the optimal solution to the original LP is close to the new origin. From Figure 7.5, this can be accomplished by removing the feasible region and minimizing the original objective function over the box formed by the constraints $0 \le x_1 \le 8$, $0 \le x_2 \le 7$. The point thus obtained (in this case $x_1 = 8$, $x_2 = 7$) becomes the origin of the **w**-coordinate system.

This approach would run into difficulty if the upper bound constraints $x_1 \le 8$ and $x_2 \le 7$ did not exist, for then there would be no box over which to minimize the objective function. Even in the absence of upper bound constraints, some intelligent strategies can be tried, and the interested reader is referred to Chapter 10. For the moment, suppose that each variable x_i has a known finite upper bound u_i. Hence the original LP has the following form:

$$\begin{aligned} \text{Minimize} \quad & \mathbf{cx} \\ \text{subject to} \quad & \mathbf{Ax} = \mathbf{b} \\ & \mathbf{0} \le \mathbf{x} \le \mathbf{u} \end{aligned}$$

To obtain the origin for the **w**-coordinate system, one simply removes the constraints $\mathbf{Ax} = \mathbf{b}$ and solves a much simpler problem:

$$\begin{aligned} \text{Minimize} \quad & \mathbf{cx} \\ \text{subject to} \quad & \mathbf{0} \le \mathbf{x} \le \mathbf{u} \end{aligned}$$

It is indeed fortunate that a solution to this LP can be obtained with no work;

$$x_j^* = \begin{cases} u_j & \text{if} \quad c_j < 0 \\ 0 & \text{otherwise} \end{cases}$$

Thus $\mathbf{x}^*$ becomes the origin of the **w**-coordinate system, and it is now necessary to express the original data, consisting of **c**, **A**, and **b**, relative to the new coordinate system. To do so requires replacing each original variable x_j by a new variable w_j if $c_j \ge 0$, and by $u_j - w_j$ if $c_j < 0$; i.e.,

$$x_j = \begin{cases} u_j - w_j & \text{if} \quad c_j < 0 \\ w_j & \text{otherwise} \end{cases}$$

In the example of Figure 7.5, x_1 and x_2 are replaced by $8 - w_1$ and $7 - w_2$, respectively, thus producing the new LP:

$$\begin{aligned} \text{Minimize} \quad & w_1 + w_2 - 15 \\ \text{subject to} \quad & -2w_1 - 3w_2 \le -11 \\ & -2w_1 + w_2 \le 1 \\ & -w_1 + 2w_2 \le 8 \\ & w_1 - 2w_2 \le 2 \\ & -w_1 - 3w_2 \ge -22 \\ & -3w_1 + w_2 \ge -16 \\ & 0 \le w_1 \le 8 \\ & 0 \le w_2 \le 7 \end{aligned}$$

In general, a replacement of the original $\mathbf{x}$ variables by the new $\mathbf{w}$ variables results in a transformed LP of the following form:

$$\begin{aligned} \text{Minimize} \quad & \mathbf{c}'\mathbf{w} \\ \text{subject to} \quad & \mathbf{A}'\mathbf{w} = \mathbf{b}' \\ & \mathbf{0} \leq \mathbf{w} \leq \mathbf{u}' \end{aligned}$$

in which

$$c_j' = \begin{cases} -c_j & \text{if } c_j < 0 \\ c_j & \text{otherwise} \end{cases}$$

$$\mathbf{A}'_{\cdot j} = \begin{cases} -\mathbf{A}_{\cdot j} & \text{if } c_j < 0 \\ \mathbf{A}_{\cdot j} & \text{otherwise} \end{cases}$$

$$\mathbf{u}' = \mathbf{u}$$

$$\mathbf{b}' = \mathbf{b} - \sum_{j=1}^{n} \mathbf{A}_{\cdot j} \begin{cases} u_j & \text{if } c_j < 0 \\ 0 & \text{otherwise} \end{cases}$$

The transformed LP thus obtained is equivalent to the original one not only in form, but also in the sense that any solution $\mathbf{w}^*$ to the transformed LP can easily be used to create a solution $\mathbf{x}^*$ to the original LP. Similarly, if the transformed LP is infeasible, so is the original one.

It is also important to note that, in the transformed LP, one can make $\mathbf{b}' \geq \mathbf{0}$ because if $b_i' < 0$, then row i can be multiplied by -1. Since the transformed LP has the same basic form as the original one, all the other labor-saving ideas (including the crashing techniques) can also be applied to the transformed LP.

There is a second instructive way to implement the ideas of the coordinate transformation without actually having to transform the data. From Figure 7.5, it is reasonable to suggest developing a phase 1 procedure that has the capability of allowing the $\mathbf{x}$ variables to start off at values other than 0 as is currently done. To implement the ideas of the coordinate transformation, one starts phase 1 with the $\mathbf{x}$ variables at their desired values, i.e., at u_j if $c_j < 0$ and 0 otherwise. For Figure 7.5 this means starting phase 1 with $x_1 = 8$ and $x_2 = 7$. Such a method has been developed, and it offers several other advantages too. The interested reader is referred to Chapter 10.

Using the Original Objective Function

The final idea for improving computational efficiency arises from the observation that, in the phase 1 LP, the original objective function $\mathbf{cx}$ has been completely ignored. Perhaps there is some way to take it into account. For instance, in solving the phase 1 LP, consider an iteration in which more than one $\mathbf{x}$ variable can be chosen to enter the current basis. It might be best to choose the one that decreases the original objective function the most (or increases it the least). Yet another strategy for incorporating the original objective function into the phase 1 process is referred to as the *Big-M Method*. Recall that the phase 1 objective function **ey** was designed to drive the artificial variables toward the value 0. Another objective function which essentially accomplishes the same goal (while incorporating the original cost vector **c**) is to minimize

$$\mathbf{cx} + M(\mathbf{ey})$$

where M is chosen as a very large, positive real number. If M is very large, then the best way to minimize this new objective function is to make all the **y** variables have the value 0, if possible. The reason is that if a **y** variable is positive, the objective function value will be very large (because of M). In any event, most professional codes somehow incorporate the original objective function into the phase 1 process.

This section has described various ideas for improving the efficiency of the phase 1 procedure. The entire chapter has been devoted to suggesting methods for

1. reducing the amount of work that is involved in performing an iteration of the simplex algorithm,
2. reducing the number of iterations needed to solve both the phase 1 and phase 2 problems, and
3. maintaining numerical accuracy throughout the solution procedure.

It is now time to explore a whole new area of linear programming, known as duality theory.

EXERCISES

7.6.1. Consider the following LP:

$$\begin{array}{ll}
\text{Maximize} & x_1 + 3x_2 - 2x_3 - x_4 + 2x_5 - x_6 + 3x_7 \\
\text{subject to} & 2x_1 - x_3 + \frac{3}{2}x_5 + x_6 \le 5 \\
& x_1 + 3x_4 - 2x_5 + x_6 = 8 \\
& x_2 + x_3 - 4x_4 + 2x_7 \ge 7 \\
& -1 \le x_1 \le 1 \\
& 0 \le x_2 \le 3 \\
& -2 \le x_3 \le -1 \\
& 2 \le x_4 \le 4 \\
& -3 \le x_5 \le 0 \\
& 0 \le x_6 \le 2 \\
& 1 \le x_7 \le 3
\end{array}$$

Determine the transformed LP obtained by changing the coordinate system as indicated by the objective function.

DISCUSSION

Computational Issues

The computational issues related to obtaining by computer efficient and numerically accurate solutions fall in the area of numerical analysis, for which an appropriate reference would be Burden, Faires, and Reynolds [5]. The topic of solving large, sparse, linear systems of equations is treated by Reid [14], and the approach to scaling a matrix by Tomlin [16].

With all of the computational improvements that were presented in this chapter, one can still ask how efficient the simplex algorithm is. For instance, with a problem of n variables and m constraints, it would be nice to know, beforehand, the maximum number of bfs that the simplex algorithm would have to visit before finding an optimal one. This topic is referred to as the computational complexity of an algorithm, and it will be discussed briefly in Appendix B.

A more detailed description of the **LU** decomposition than was presented in this chapter is that of Bartels and Golub [2, 3]. Another advanced method for handling sparse basis matrices was developed by Forrest and Tomlin [9]. The revised simplex method was developed by Dantzig and Orchard-Hays [7, 8].

Computer Codes

Numerous professional computer codes for solving large LPs are available, the most notable being MPSX/370 [11] and APEX-III [1]. Some of the user manuals for these codes contain details on the implementation of partial/multiple pricing, crashing techniques, using the original objective function in phase 1, etc. For details on how to use the MAGEN system for data entry and matrix generation, see PDS/MAGEN [12]. It is worth mentioning that LP models having more than 20,000 variables are currently being solved routinely on the computer.

Special Structure

Special structure in the form of bound constraints and network flow problems is dealt with in Chapters 10 and 11, respectively. The principle of decomposition (and column generation), to take advantage of the special structure that can arise in the constraints of certain large LPs, is discussed by Dantzig and Wolfe [6], and staircase problems by Fourer [10]. Exploitation of degeneracy is described by Perold [13].

Phase 1

The Big-M method was originally developed by Charnes, and a detailed explanation is given by Bazaraa and Jarvis [4]. However, this method is not used in practice because it is difficult to know how large to make M and because if M is made too large, then numerical difficulties can arise on the computer.

A phase 1 procedure that uses only one artificial variable has been developed. Details can be found in the text by Bazaraa and Jarvis [4]. The concept of coordinate transformations was developed by Solow [15].

REFERENCES

1. APEX-III Reference Manual, Version 1.2, CDC Publication Number 76070000, Control Data Corporation, Minneapolis, MN, 1979.
2. R. H. Bartels and G. H. Golub, "The Simplex Method of Linear Programming Using **LU** Decomposition," *Communications of the ACM* 12, 266–268 (1969).
3. R. H. Bartels and G. H. Golub, "Algorithm 350: Simplex Method Procedure Employing **LU** Decomposition," *Communications of the ACM* 12, 275–278, 1969.
4. M. S. Bazaraa and J. J. Jarvis, *Linear Programming and Network Flows*, Wiley, New York, 1977.
5. R. L. Burden, J. D. Faires, and A. C. Reynolds, *Numerical Analysis*, Second Edition, Prindle, Weber and Schmidt, Boston, MA, 1981.
6. G. B. Dantzig and P. Wolfe, "The Decomposition Algorithm for Linear Programs," *Econometrica* 29(4), 767–778 (1961).
7. G. B. Dantzig and W. Orchard-Hays, "Notes on Linear Programming: Part V —Alternate Algorithm for the Revised Simplex Method Using Product Form for the Inverse," Research Memorandum RM-1268, The Rand Corporation, Santa Monica, CA, 1953.
8. G. B. Dantzig and W. Orchard-Hays, "The Product Form of the Inverse in the Simplex Method," *Mathematical Tables and Aids to Computation* 8(46), 64–67 (1954).
9. J. J. H. Forrest and J. A. Tomlin, "Updated Triangular Factors of the Basis to Maintain Sparsity in the Product Form Simplex Method," *Mathematical Programming* 2, 263–278 (1972).
10. R. Fourer, "Solving Staircase Linear Programs by the Simplex Method, 1: Inversion," *Mathematical Programming* 23(3), 274–313 (1982).
11. "Mathematical Programming System Extended/370 (MPSX/370)—Program Reference Manual," Third Edition, IBM Publication Number SH19-1096-1, International Business Machines, Inc., White Plains, NY, 1978.
12. "PDS/MAGEN—A General Purpose Problem Description System User Manual, Version 1.3," CDC Publication Number 84009900, developed by Haverly Systems, Inc., published by Control Data Corporation, Minneapolis, MN, 1978.
13. A. F. Perold, "A Degeneracy-Exploiting **LU** Factorization for the Simplex Method," *Mathematical Programming* 19(3), 239–254 (1980).

14. J. K. Reid (ed.), *Large Sparse Sets of Linear Equations*, Oxford Conference of 1970, Academic Press, London, 1971.
15. D. Solow, "The Special Theory of Coordinate Transformations in Linear Programming," Technical Memorandum No. 456, Department of Operations Research, Case Western Reserve University, Cleveland, OH, 1979.
16. J. A. Tomlin, "On Scaling Linear Programming Problems," *Mathematical Programming Study* 4, 146–166 (1975).

Chapter 8

Duality Theory

The simplex algorithm is capable of determining the *status* of an LP (i.e., if it is infeasible, optimal, or unbounded), but what can you do if a computer is not immediately available? The theory of duality, to be developed in this chapter, can often be used to determine the status of an LP without the need for a computer.

Duality theory can also be used to develop a new test for optimality that, in turn, leads to other finite improvement algorithms for solving LPs. Duality theory will play an important role throughout the remainder of this textbook.

8.1. THE DUAL LINEAR PROGRAMMING PROBLEM

The simplex algorithm of Chapter 5 is actually twice as efficient as it first appears. The reason is that it simultaneously solves two LPs: the original (or *primal*) LP and a closely related (or *dual*) LP. This section shows how the dual LP can be obtained from the primal LP and vice versa.

Recall that a primal LP in the standard form,

$$\begin{aligned} \text{Minimize} \quad & \mathbf{cx} \\ \text{subject to} \quad & \mathbf{Ax} = \mathbf{b} \\ & \mathbf{x} \geq \mathbf{0} \end{aligned}$$

has n variables and m equality constraints. The dual LP, on the other hand, has m variables and n constraints; i.e., there is one dual variable associated with each primal equality constraint and one dual constraint for each primal variable. There are other differences too. The **b** vector from the primal equality constraints becomes the cost vector for the dual LP, while the original cost row **c** enters the dual constraints. In matrix/vector notation, the dual LP is to

$$\begin{aligned} &\text{maximize} && \mathbf{ub} \\ &\text{subject to} && \mathbf{uA} \le \mathbf{c} \\ &&& \mathbf{u} \text{ unrestricted} \end{aligned}$$

Note that the vector **u** appears on the left of the **A** matrix and, as such, **u** is a $1 \times m$ row vector. Several other differences between the primal and dual problems are also evident. The primal objective is to minimize while the dual objective is to maximize. The primal has equality constraints, whereas the dual has inequality constraints. The primal variables are restricted to be greater than or equal to 0, but the dual variables are unrestricted. The following example demonstrates how to write the dual of an LP in standard form.

EXAMPLE 8.1: *Primal LP.*

$$\begin{aligned} &\text{Minimize} && -2x_1 - 3x_2 - 5x_3 \\ &\text{subject to} && 3x_1 + 4x_2 + x_3 + x_4 = 7 \\ &&& x_1 - x_2 + 2x_3 + x_5 = 5 \\ &&& x_1,\ x_2,\ x_3,\ x_4,\ x_5 \ge 0 \end{aligned}$$

so

$$\mathbf{c} = (-2, -3, -5, 0, 0)$$

$$\mathbf{A} = \begin{bmatrix} 3 & 4 & 1 & 1 & 0 \\ 1 & -1 & 2 & 0 & 1 \end{bmatrix}$$

$$\mathbf{b} = \begin{bmatrix} 7 \\ 5 \end{bmatrix}$$

Dual LP in matrix / vector notation.

$$\begin{aligned} &\text{Maximize} && (u_1, u_2)\begin{bmatrix} 7 \\ 5 \end{bmatrix} \\ &\text{subject to} && (u_1, u_2)\begin{bmatrix} 3 & 4 & 1 & 1 & 0 \\ 1 & -1 & 2 & 0 & 1 \end{bmatrix} \le (-2, -3, -5, 0, 0) \\ &&& u_1, u_2 \text{ unrestricted} \end{aligned}$$

Dual LP in equation form.

$$\begin{aligned} \text{Maximize}\quad & 7u_1 + 5u_2 \\ \text{subject to}\quad & 3u_1 + u_2 \le -2 \\ & 4u_1 - u_2 \le -3 \\ & u_1 + 2u_2 \le -5 \\ & u_1 \le 0 \\ & u_2 \le 0 \end{aligned}$$

It is also necessary to be able to find the dual of an LP that is not in standard form. One way to proceed is to convert the primal LP to standard form first and then to write the dual, as illustrated in the next example.

EXAMPLE 8.2: *Primal LP.*

$$\begin{aligned} \text{Minimize}\quad & -2x_1 + 3x_2 + 4x_3 \\ \text{subject to}\quad & x_1 + 2x_2 + 3x_3 \le 1 \\ & 5x_1 + 7x_2 - x_3 \ge 3 \\ & x_1 + 8x_2 - 7x_3 = 5 \\ & x_1,\ x_2,\ x_3 \ge 0 \end{aligned}$$

Primal LP in standard form.

$$\begin{aligned} \text{Minimize}\quad & -2x_1 + 3x_2 + 4x_3 \\ \text{subject to}\quad & x_1 + 2x_2 + 3x_3 + x_4 = 1 \\ & 5x_1 + 7x_2 - x_3 - x_5 = 3 \\ & x_1 + 8x_2 - 7x_3 = 5 \\ & x_1,\ x_2,\ x_3,\ x_4,\ x_5 \ge 0 \end{aligned}$$

so

$$\mathbf{c} = (-2, 3, 4, 0, 0)$$

$$\mathbf{A} = \begin{bmatrix} 1 & 2 & 3 & 1 & 0 \\ 5 & 7 & -1 & 0 & -1 \\ 1 & 8 & -7 & 0 & 0 \end{bmatrix}$$

$$\mathbf{b} = \begin{bmatrix} 1 \\ 3 \\ 5 \end{bmatrix}$$

Dual LP in matrix / vector notation.

$$\text{Maximize} \quad (u_1, u_2, u_3)\begin{bmatrix} 1 \\ 3 \\ 5 \end{bmatrix}$$

$$\text{subject to} \quad (u_1, u_2, u_3)\begin{bmatrix} 1 & 2 & 3 & 1 & 0 \\ 5 & 7 & -1 & 0 & -1 \\ 1 & 8 & -7 & 0 & 0 \end{bmatrix} \le (-2, 3, 4, 0, 0)$$

$$u_1, u_2, u_3 \quad \text{unrestricted}$$

Dual LP in equation form.

$$\begin{aligned} \text{Maximize} \quad & u_1 + 3u_2 + 5u_3 \\ \text{subject to} \quad & u_1 + 5u_2 + u_3 \le -2 \\ & 2u_1 + 7u_2 + 8u_3 \le 3 \\ & 3u_1 - u_2 - 7u_3 \le 4 \\ & u_1 \le 0 \\ & -u_2 \le 0 \\ & u_3 \text{ unrestricted} \end{aligned}$$

It is very useful to be able to go through these steps in matrix notation, as illustrated in the next example.

EXAMPLE 8.3:

$$\begin{aligned} \text{Minimize} \quad & \mathbf{cx} \\ \text{subject to} \quad & \mathbf{Ax} \ge \mathbf{b} \\ & \mathbf{x} \ge \mathbf{0} \end{aligned}$$

Subtract slacks:

$$\begin{aligned} \text{Minimize} \quad & \mathbf{cx} \\ \text{subject to} \quad & \mathbf{Ax} - \mathbf{Is} = \mathbf{b} \\ & \mathbf{x}, \quad \mathbf{s} \ge \mathbf{0} \end{aligned}$$

Rewrite

$$\begin{aligned} \text{Minimize} \quad & (\mathbf{c}, \mathbf{0})\begin{bmatrix} \mathbf{x} \\ \mathbf{s} \end{bmatrix} \\ \text{subject to} \quad & [\mathbf{A}, -\mathbf{I}]\begin{bmatrix} \mathbf{x} \\ \mathbf{s} \end{bmatrix} = \mathbf{b} \\ & (\mathbf{x}, \mathbf{s}) \ge \mathbf{0} \end{aligned}$$

Take the dual:

$$\begin{aligned} \text{Maximize} \quad & \mathbf{ub} \\ \text{subject to} \quad & \mathbf{u}[\mathbf{A}, -\mathbf{I}] \le (\mathbf{c}, \mathbf{0}) \\ & \mathbf{u} \text{ unrestricted} \end{aligned}$$

Finally, again rewrite:

$$\begin{aligned} \text{Maximize} \quad & \mathbf{ub} \\ \text{subject to} \quad & \mathbf{uA} \le \mathbf{c} \\ & \mathbf{u} \ge \mathbf{0} \end{aligned}$$

The particular pair of primal and dual problems in Example 8.3 are called *symmetric LPs*. These LPs illustrate a possible computational advantage of the dual problem. Once in standard form, the primal LP has m equality constraints and $m+n$ variables; the dual LP has n equality constraints and $m+n$ variables. The corresponding size of the primal basis is $m \times m$, as opposed to the $n \times n$ dual basis. If $m \ge n$, then, from a computational point of view, it would seem advantageous to consider solving the dual LP with its smaller basis rather than the primal LP with its larger basis, everything else being equal. As will be shown in Section 8.4, whichever LP is solved, the simplex algorithm automatically produces a solution to the other LP.

Another important property of the dual problem is that on writing its dual, one again obtains the primal LP; that is, the dual of the dual is the primal. For demonstration purposes, consider the dual LP obtained in Example 8.3. The next example shows that its dual is the primal.

EXAMPLE 8.4:

$$\begin{aligned} \text{Maximize} \quad & \mathbf{ub} \\ \text{subject to} \quad & \mathbf{uA} \le \mathbf{c} \\ & \mathbf{u} \ge \mathbf{0} \end{aligned}$$

Add slacks:

$$\begin{aligned} \text{Minimize} \quad & -\mathbf{ub} \\ \text{subject to} \quad & \mathbf{A}^{\mathrm{T}}\mathbf{u} + \mathbf{Is} = \mathbf{c} \\ & \mathbf{u}, \quad \mathbf{s} \ge \mathbf{0} \end{aligned}$$

Rewrite

$$\begin{aligned} \text{Minimize} \quad & (-\mathbf{b}, \mathbf{0})\begin{bmatrix} \mathbf{u} \\ \mathbf{s} \end{bmatrix} \\ \text{subject to} \quad & [\mathbf{A}^{\mathrm{T}}, \mathbf{I}]\begin{bmatrix} \mathbf{u} \\ \mathbf{s} \end{bmatrix} = \mathbf{c} \\ & (\mathbf{u}, \mathbf{s}) \ge \mathbf{0} \end{aligned}$$

Take the dual:

$$\begin{aligned} \text{Maximize} \quad & \mathbf{cw} \\ \text{subject to} \quad & \mathbf{wA}^{\mathrm{T}} \le -\mathbf{b} \\ & \mathbf{wI} \le \mathbf{0} \\ & \mathbf{w} \text{ unrestricted} \end{aligned}$$

Let $\mathbf{w} = -\mathbf{x}$,

$$\begin{aligned} \text{Maximize} \quad & -\mathbf{cx} \\ \text{subject to} \quad & -\mathbf{xA}^{\mathrm{T}} \le -\mathbf{b} \\ & -\mathbf{x} \le \mathbf{0} \end{aligned}$$

Finally, rewrite again:

$$\begin{aligned} \text{Minimize} \quad & \mathbf{cx} \\ \text{subject to} \quad & \mathbf{Ax} \ge \mathbf{b} \\ & \mathbf{x} \ge \mathbf{0} \end{aligned}$$

Although the method of first converting the given LP to standard form and then taking the dual always works, the algebraic manipulations are prone to careless mistakes. To avoid such problems and to save time, it is possible to develop specific rules for writing the dual of any LP quickly and accurately. The rules are based on the specific form of the primal LP, and they are summarized in Table 8.1. The next example demonstrates the use of Table 8.1.

TABLE 8.1. Converting from Primal to Dual

Primal minimization problem			
If the primal variables are	Then the dual constraints are	If the primal constraints are	Then the dual variables are
≥ 0	$\le$	$\ge$	≥ 0
≤ 0	$\ge$	$\le$	≤ 0
Unrestricted	$=$	$=$	Unrestricted
Primal maximization problem			
If the primal variables are	Then the dual constraints are	If the primal constraints are	Then the dual variables are
≥ 0	$\ge$	$\ge$	≤ 0
≤ 0	$\le$	$\le$	≥ 0
Unrestricted	$=$	$=$	Unrestricted

EXAMPLE 8.5: *Primal LP.*

$$\begin{aligned} \text{Maximize} \quad & 5x_1 + 3x_2 + 7x_3 \\ \text{subject to} \quad & x_1 + x_2 \leq 3 \\ & 5x_1 + 4x_3 \geq 4 \\ & 7x_1 - x_2 + x_3 = 5 \\ & x_1 \geq 0 \\ & x_2 \leq 0 \\ & x_3 \quad \text{unrestricted} \end{aligned}$$

Dual LP.

$$\begin{aligned} \text{Minimize} \quad & 3u_1 + 4u_2 + 5u_3 \\ \text{subject to} \quad & u_1 + 5u_2 + 7u_3 \geq 5 \\ & u_1 - u_3 \leq 3 \\ & 4u_2 + u_3 = 7 \\ & u_1 \geq 0 \\ & u_2 \leq 0 \\ & u_3 \quad \text{unrestricted} \end{aligned}$$

This section has shown how to find the dual of any LP. In any real-world situation the decision variables of the primal LP have a specific interpretation (e.g., how much of an item to produce). For such problems, it is natural to ask if the variables in the corresponding dual problem have a similar economic interpretation. The answer is yes, as will be discussed in the next section.

EXERCISES

8.1.1. Write the duals of the following LPs.

(a)

$$\begin{aligned} \text{Minimize} \quad & 3x_1 - x_2 + 4x_4 + 2x_5 + 5x_6 \\ \text{subject to} \quad & x_1 + x_2 - 5x_5 = 4 \\ & -6x_1 + 4x_3 + 6x_4 + 2x_5 = 7 \\ & 6x_2 - 5x_3 + 4x_5 - 5x_6 = 11 \\ & x_1, x_2, x_3, x_4, x_5, x_6 \geq 0 \end{aligned}$$

(b)

$$\begin{aligned} \text{Minimize} \quad & -x_1 + 3x_3 + 4x_5 + 5x_6 \\ \text{subject to} \quad & 4x_1 - 7x_2 + 3x_3 + x_5 - 2x_6 = 11 \\ & 2x_2 - x_3 + 4x_4 + 6x_6 = 7 \\ & x_1 - 2x_2 + 3x_4 + 3x_5 = 10 \\ & 7x_1 + x_3 - 4x_5 + 7x_6 = 12 \\ & x_1, x_2, x_3, x_4, x_5, x_6 \geq 0 \end{aligned}$$

8.1.2. Derive the matrix/vector dual for the following LPs. (Recall that **e** is the vector of all 1s.)

(a)

$$\begin{array}{ll} \text{Minimize} & \mathbf{ey} \\ \text{subject to} & \mathbf{Ax} + \mathbf{Iy} = \mathbf{b} \\ & \mathbf{x}, \quad \mathbf{y} \geq \mathbf{0} \end{array}$$

(b)

$$\begin{array}{ll} \text{Minimize} & \mathbf{0x} \\ \text{subject to} & \mathbf{Ax} = \mathbf{b} \\ & \mathbf{ex} = 1 \\ & \mathbf{x} \geq \mathbf{0} \end{array}$$

(c)

$$\begin{array}{ll} \text{Maximize} & \mathbf{cx} \\ \text{subject to} & \mathbf{Ax} = \mathbf{b} \\ & \mathbf{0} \leq \mathbf{x} \leq \mathbf{u} \quad (\mathbf{u} \text{ is an } n \text{ vector}) \end{array}$$

8.1.3. Solve the dual of the following LP by the graphical method of Chapter 2.

$$\begin{array}{ll} \text{Minimize} & x_1 + 2x_2 - x_3 \\ \text{subject to} & 4x_1 + 5x_2 + 3x_4 + x_5 = 10 \\ & 6x_1 + 3x_2 - 3x_3 + x_4 = 12 \\ & x_1, \quad x_2, \quad x_3, \quad x_4, \quad x_5 \geq 0 \end{array}$$

8.1.4. Use Table 8.1 to write the duals of the following LPs.

(a)

$$\begin{array}{ll} \text{Maximize} & x_1 - 2x_2 + 3x_4 \\ \text{subject to} & 2x_1 + 3x_2 - 4x_3 \geq 5 \\ & -x_1 + 2x_3 + 4x_4 \leq 7 \\ & x_1 + x_2 - 3x_3 + x_4 = 5 \\ & x_1, \quad x_2 \geq 0 \\ & x_3, \quad x_4 \leq 0 \end{array}$$

(b)

$$\begin{array}{lrrrrrrr}
\text{Minimize} & x_1 & +2x_2 & -3x_3 & & +4x_5 & & \\
\text{subject to} & x_1 & -2x_2 & & +3x_4 & & = & 5 \\
& & 2x_2 & -\ x_3 & & +4x_5 & \geq & -7 \\
& 2x_1 & -4x_2 & +5x_3 & -4x_4 & & \leq & 8 \\
& x_1 & +3x_2 & +4x_3 & & -5x_5 & \geq & 4 \\
& x_1, & x_2, & x_3 & & & \geq & 0 \\
& & & & x_4, & x_5 & \multicolumn{2}{l}{\text{unrestricted}}
\end{array}$$

8.2. ECONOMIC INTERPRETATION OF THE DUAL

When an LP arises from a real world problem, the decision variables represent certain physical quantities. As will be seen in this section, the dual variables also have an economic meaning that can provide valuable information to a decision maker.

The specific interpretation of the dual variables varies for each LP. One of the best ways to go about understanding what they represent is to determine what units they have. For instance, recall the first linear programming problem that was presented in this book. There, Mr. Miner is trying to decide how many pounds of gold and silver ore to take on board his spaceship so as to maximize the value of the cargo. The constraints of the LP represent the limited availability of pounds of cargo capacity and cubic feet of cargo space. Specifically, the primal problem is to

$$\begin{array}{lrrl}
\text{maximize} & 3x_1 & +2x_2 & \\
\text{subject to} & 1x_1 & +1x_2 & \leq 100 \\
& 2x_1 & +1x_2 & \leq 150 \\
& x_1, & x_2 & \geq\ \ 0
\end{array}$$

and its associated dual is to

$$\begin{array}{lrrl}
\text{minimize} & 100u_1 & +150u_2 & \\
\text{subject to} & 1u_1 & +\ \ 2u_2 & \geq 3 \\
& 1u_1 & +\ \ 1u_2 & \geq 2 \\
& u_1, & u_2 & \geq 0
\end{array}$$

To see what units the dual variables have, the primal and dual LPs will be rewritten with the units appearing where appropriate. For convenience, intergalactic credits will be replaced with dollars ($), and the word "ore" will be omitted. The primal LP is as follows:

$$\begin{aligned}
\text{Maximize}\quad & 3(\$/\text{lb of gold}) \times x_1(\text{lb of gold}) \\
& +2(\$/\text{lb of silver}) \times x_2(\text{lb of silver}) \\
\text{subject to}\quad & 1(\text{lb of cargo/lb of gold}) \times x_1(\text{lb of gold}) \\
& +1(\text{lb of cargo/lb of silver}) \times x_2(\text{lb of silver}) \\
& \qquad \leq 100(\text{lb of cargo}) \\
& 2(\text{ft}^3 \text{ of cargo/lb of gold}) \times x_1(\text{lb of gold}) \\
& +1(\text{ft}^3 \text{ of cargo/lb of silver}) \times x_2(\text{lb of silver}) \\
& \qquad \leq 150(\text{ft}^3 \text{ of cargo}) \\
& x_1, x_2 \geq 0
\end{aligned}$$

The dual LP is as follows:

$$\begin{aligned}
\text{Minimize}\quad & 100(\text{lb of cargo}) \times u_1 + 150(\text{ft}^3 \text{ of cargo}) \times u_2 \\
\text{subject to}\quad & 1(\text{lb of cargo/lb of gold}) \times u_1 \\
& +2(\text{ft}^3 \text{ of cargo/lb of gold}) \times u_2 \\
& \qquad \geq 3(\$/\text{lb of gold}) \\
& 1(\text{lb of cargo/lb of silver}) \times u_1 \\
& +1(\text{ft}^3 \text{ of cargo/lb of silver}) \times u_2 \\
& \qquad \geq 2(\$/\text{lb of silver}) \\
& u_1, u_2 \geq 0
\end{aligned}$$

The units of u_1 and u_2 should be chosen so that the resulting units on the left-hand side of the dual constraints are the same as those on the right-hand side. In the first constraint of the dual LP, it would seem that the units of u_1 should be \$/(lb of cargo), whereas those of u_2 should be \$/(ft^3 of space). Thus u_1 and u_2 represent some kind of price per pound of cargo capacity and cubic foot of cargo space, respectively. For this reason, the dual variables are often referred to as *shadow prices*. To gain further understanding of the economic meaning and uses of the shadow prices, it is helpful to consider the following interpretation of the dual LP.

Imagine that you are on the same asteroid with Mr. Miner and that you find yourself in need of buying his ship's 100 pounds of cargo capacity and 150 cubic feet of cargo space. If u_1 is the price per pound of cargo capacity and u_2 is the price per cubic foot of cargo space, then your total cost is going to be $100u_1 + 150u_2$, and hence your objective should be to choose values for u_1 and u_2 so as to minimize this cost.

On the other hand, you want to choose values of u_1 and u_2 that will entice Mr. Miner to sell his cargo resources to you rather than to use them himself. Midas knows that each pound of gold ore utilizes one pound of cargo capacity and two cubic feet of cargo space. At the prices u_1 and u_2, you would be paying him the equivalent of $u_1 + 2u_2$ dollars for each pound of gold ore. Yet he knows that the gold ore is worth \$3 per pound on the open market. Thus, to induce him to sell you his cargo resources, you must choose values for u_1 and u_2 so that $u_1 + 2u_2 \geq 3$. Similarly, for silver, u_1 and u_2 should satisfy $u_1 + u_2 \geq 2$. Of course, $u_1, u_2 \geq 0$, so the problem that you must solve is to

$$\begin{array}{lrrl} \text{minimize} & 100u_1 & +150u_2 & \text{(your cost)} \\ \text{subject to} & 1u_1 + & 2u_2 \geq 3 & \text{(motivation for Midas to sell)} \\ & 1u_1 + & 1u_2 \geq 2 & \\ & u_1, & u_2 \geq 0 & \end{array}$$

This is precisely the dual LP associated with the original primal problem.

The simplex algorithm can now be used to obtain the optimal primal and dual solutions $x_1^*, x_2^*, u_1^*, u_2^*$. The values of u_1^* and u_2^* provide further economic information that can be valuable to the decision maker. For instance, Midas might be thinking of leaving behind some of his equipment so that his spaceship can carry some additional pounds of gold or silver. The question to be answered is how much should Midas be willing to pay for s_1 additional pounds of cargo capacity and s_2 additional cubic feet of cargo space so that the extra cost is more than offset by the extra profit to be earned.

At the prices u_1^* and u_2^*, the additional cost for the cargo resources is $u_1^* s_1 + u_2^* s_2$. To compute the extra profit that Midas can earn with the additional resources, he must solve a new LP,

$$\begin{array}{lrrl} \text{Maximize} & 3x_1 & +2x_2 & \\ \text{subject to} & 1x_1 & +1x_2 \leq & 100 + s_1 \\ & 2x_1 & +1x_2 \leq & 150 + s_2 \\ & x_1, & x_2 \geq & 0 \end{array}$$

and then deduct the previous profit of $3x_1^* + 2x_2^*$. If the additional profit exceeds the cost, then Midas should be willing to pay for the extra cargo resources at the prices u_1^* and u_2^*. As a result of the theory to be developed later in this chapter, it can be shown that, as long as s_1 and s_2 are not chosen too large, the prices u_1^* and u_2^* have the property that the cost of the extra cargo resources is exactly offset by the additional profits that can be earned.

In other words, u_1^* and u_2^* can be interpreted as the price that Midas would be willing to pay for (small) additional amounts of cargo capacity and space. With this interpretation in mind, it is reasonable to conjecture that if Mr. Miner has not used up all 100 pounds of his cargo capacity when loading x_1^* and x_2^* pounds of the ore, then u_1^* should be 0; i.e., he should not be willing to pay anything for additional pounds of cargo capacity because there is some left over. Similarly, if there is excess cargo space left over, u_2^* should be 0.

Further information can be gleaned from the value of u_1^* and u_2^*. Midas can use these values to determine which ores are not profitable to take. At the prices u_1^* and u_2^*, the value of the cargo resources required by one pound of gold ore is $u_1^* + 2u_2^*$ and that of silver is $u_1^* + u_2^*$. The numbers $u_1^* + 2u_2^*$ and $u_1^* + u_2^*$ are referred to as the *opportunity cost* associated with x_1^* and x_2^* pounds of gold and silver ore, respectively. It should not be surprising to discover that if the opportunity cost of an item exceeds its profit per unit, then the item should not be chosen.

This section has described the economic interpretation and various uses of the dual variables. The next section begins the study of duality theory.

EXERCISES

8.2.1. For each of the formulation exercises in Section 1.2, write the dual LP. Then identify the units of the dual variables and briefly describe the economic significance and use of those variables.

8.3. WEAK DUALITY

The purpose of this section is to show how the dual problem can be used to test a current primal solution for optimality. To do so, consider the following primal and dual LPs:

Primal LP.

$$\begin{aligned} \text{Minimize} \quad & \mathbf{cx} \\ \text{subject to} \quad & \mathbf{Ax} = \mathbf{b} \\ & \mathbf{x} \geq \mathbf{0} \end{aligned}$$

Dual LP.

$$\begin{aligned} \text{Maximize} \quad & \mathbf{ub} \\ \text{subject to} \quad & \mathbf{uA} \leq \mathbf{c} \\ & \mathbf{u} \text{ unrestricted} \end{aligned}$$

Recall from Chapter 2 the analogy of drawing balls from a box to solve an LP, and imagine that one box corresponds to the primal LP and a second box corresponds to the dual LP. For the primal problem, the objective is to find a ball having the smallest number on it; the dual objective is to find a ball with the largest number on it.

There is a remarkable relationship between the dual and primal balls that enables one to use the numbers on the dual balls as a test for optimality in the primal problem (and vice versa). The property, known as *weak duality*, states that the number on every primal ball is greater than or equal to the number on every dual ball.

Proposition 8.1. *If* $\mathbf{x}$ *is feasible for the primal and* $\mathbf{u}$ *is feasible for the dual, then* $\mathbf{ub} \leq \mathbf{cx}$.

OUTLINE OF PROOF. The proof is done by working forward from the hypothesis by algebraic manipulations. Specifically, let $\mathbf{x}$ and $\mathbf{u}$ be primal and dual feasible solutions, respectively, so that $\mathbf{Ax} = \mathbf{b}$, $\mathbf{x} \geq \mathbf{0}$, and $\mathbf{uA} \leq \mathbf{c}$. These facts will be used to establish the desired conclusion that $\mathbf{ub} \leq \mathbf{cx}$ by first showing that $\mathbf{ub} = \mathbf{uAx}$ and then that $\mathbf{uAx} \leq \mathbf{cx}$, for then $\mathbf{ub} = \mathbf{uAx} \leq \mathbf{cx}$.

To obtain the equality $\mathbf{uAx} = \mathbf{ub}$, simply multiply both sides of the constraints $\mathbf{Ax} = \mathbf{b}$ on the left by the vector $\mathbf{u}$, obtaining $\mathbf{uAx} = \mathbf{ub}$. The other inequality, $\mathbf{uAx} \leq \mathbf{cx}$, is obtained in a similar fashion by multiplying both sides of the dual constraints $\mathbf{uA} \leq \mathbf{c}$ on the right by the vector $\mathbf{x}$. It is necessary to use $\mathbf{x} \geq \mathbf{0}$ to be able to conclude that $\mathbf{uAx} \leq \mathbf{cx}$.

PROOF OF PROPOSITION 8.1. Let $\mathbf{x}$ and $\mathbf{u}$ be primal and dual feasible vectors, respectively, and so $\mathbf{Ax} = \mathbf{b}$, $\mathbf{x} \geq \mathbf{0}$, and $\mathbf{uA} \leq \mathbf{c}$. On multiplying both sides of the equality $\mathbf{Ax} = \mathbf{b}$ by $\mathbf{u}$, one obtains $\mathbf{uAx} = \mathbf{ub}$. Similarly, on multiplying both sides of the inequality $\mathbf{uA} \leq \mathbf{c}$ by the nonnegative vector $\mathbf{x}$, one obtains $\mathbf{uAx} \leq \mathbf{cx}$. Combining these two results yields $\mathbf{ub} = \mathbf{uAx} \leq \mathbf{cx}$. □

Proposition 8.1 provides the mechanism for devising a new test for optimality. Imagine that you have a primal ball $\mathbf{x}$ whose number is $\mathbf{cx}$ and you wish to know if it is optimal. As a result of weak duality, you know that the number $\mathbf{cx}$ is greater than or equal to the number on each and every dual ball. What would happen if you could somehow find one dual ball $\mathbf{u}$ for which $\mathbf{cx}$ was equal to the number on the dual ball (namely, $\mathbf{ub}$)? Since you cannot hope to do better, the primal (and dual) ball must be optimal. This observation is formally stated in the next theorem.

Theorem 8.1. *If* $\mathbf{x}$ *is feasible for the primal,* $\mathbf{u}$ *is feasible for the dual, and* $\mathbf{cx} = \mathbf{ub}$, *then* $\mathbf{x}$ *and* $\mathbf{u}$ *are primal and dual optimal solutions, respectively.*

The proof is clear from the discussion preceding Theorem 8.1 and is therefore omitted.

Although the simplex algorithm does not use this test for optimality directly it is interesting to note that, when an optimal bfs $\mathbf{x} = (\mathbf{x}_\mathbf{B}, \mathbf{x}_\mathbf{N}) = (\mathbf{B}^{-1}\mathbf{b}, \mathbf{0}) \geq \mathbf{0}$ is found, the simplex algorithm has simultaneously produced a dual optimal solution $\mathbf{u}$ with $\mathbf{cx} = \mathbf{ub}$. Indeed, $\mathbf{u}$ is $\mathbf{c}_\mathbf{B}\mathbf{B}^{-1}$, as shown in the next proposition.

Proposition 8.2. *If* $\mathbf{x} = (\mathbf{x}_\mathbf{B}, \mathbf{x}_\mathbf{N}) = (\mathbf{B}^{-1}\mathbf{b}, \mathbf{0}) \geq \mathbf{0}$ *is a bfs that satisfies* $\mathbf{c}_\mathbf{N} - \mathbf{c}_\mathbf{B}\mathbf{B}^{-1}\mathbf{N} \geq \mathbf{0}$, *then the vector* $\mathbf{u} = \mathbf{c}_\mathbf{B}\mathbf{B}^{-1}$ *is optimal for the dual and satisfies* $\mathbf{cx} = \mathbf{ub}$.

OUTLINE OF PROOF. The forward–backward method gives rise to the abstraction question "How can I show that a vector (namely, $\mathbf{u} = \mathbf{c}_\mathbf{B}\mathbf{B}^{-1}$) is optimal for the dual LP?" One answer is provided by Theorem 8.1, whereby it must be shown that $\mathbf{x}$ is feasible for the primal, $\mathbf{u}$ is feasible for the dual, and $\mathbf{cx} = \mathbf{ub}$.

Clearly, $\mathbf{x}$ is feasible for the primal because, by hypothesis, $\mathbf{x}$ is a bfs. To show that $\mathbf{u}$ is feasible for the dual, it must be shown that $\mathbf{u} = \mathbf{c}_\mathbf{B}\mathbf{B}^{-1}$ satisfies the dual constraints, namely, $\mathbf{uA} \leq \mathbf{c}$. Working forward from the hypothesis that $\mathbf{x}$ satisfies $\mathbf{c}_\mathbf{N} - \mathbf{c}_\mathbf{B}\mathbf{B}^{-1}\mathbf{N} \geq \mathbf{0}$, one has

$$\begin{aligned}\mathbf{uA} &= (\mathbf{c}_\mathbf{B}\mathbf{B}^{-1})\mathbf{A} = \mathbf{c}_\mathbf{B}\mathbf{B}^{-1}[\mathbf{B},\mathbf{N}] = (\mathbf{c}_\mathbf{B}\mathbf{B}^{-1}\mathbf{B}, \mathbf{c}_\mathbf{B}\mathbf{B}^{-1}\mathbf{N})\\ &= (\mathbf{c}_\mathbf{B}\mathbf{I}, \mathbf{c}_\mathbf{B}\mathbf{B}^{-1}\mathbf{N}) = (\mathbf{c}_\mathbf{B}, \mathbf{c}_\mathbf{B}\mathbf{B}^{-1}\mathbf{N})\\ &\leq (\mathbf{c}_\mathbf{B}, \mathbf{c}_\mathbf{N}) = \mathbf{c}\end{aligned}$$

Thus $\mathbf{u}$ is feasible for the dual. All that remains is to show that $\mathbf{cx} = \mathbf{ub}$. Working forward from the fact that $\mathbf{x}_\mathbf{B} = \mathbf{B}^{-1}\mathbf{b}$ and $\mathbf{x}_\mathbf{N} = \mathbf{0}$, one has

$$\begin{aligned}\mathbf{ub} &= (\mathbf{c}_\mathbf{B}\mathbf{B}^{-1})\mathbf{b} = \mathbf{c}_\mathbf{B}(\mathbf{B}^{-1}\mathbf{b}) = \mathbf{c}_\mathbf{B}\mathbf{x}_\mathbf{B} = \mathbf{c}_\mathbf{B}\mathbf{x}_\mathbf{B} + \mathbf{c}_\mathbf{N}(\mathbf{0}) = \mathbf{c}_\mathbf{B}\mathbf{x}_\mathbf{B} + \mathbf{c}_\mathbf{N}\mathbf{x}_\mathbf{N}\\ &= \mathbf{cx}\end{aligned}$$

PROOF OF PROPOSITION 8.2. First it will be shown that $\mathbf{u} = \mathbf{c}_\mathbf{B}\mathbf{B}^{-1}$ is feasible for the dual. By hypothesis, $\mathbf{c}_\mathbf{B}\mathbf{B}^{-1}\mathbf{N} \leq \mathbf{c}_\mathbf{N}$, and so

$$\mathbf{uA} = \mathbf{c}_\mathbf{B}\mathbf{B}^{-1}[\mathbf{B},\mathbf{N}] = (\mathbf{c}_\mathbf{B}, \mathbf{c}_\mathbf{B}\mathbf{B}^{-1}\mathbf{N}) \leq (\mathbf{c}_\mathbf{B}, \mathbf{c}_\mathbf{N}) = \mathbf{c}$$

To establish that $\mathbf{u}$ is optimal for the dual, by Theorem 8.1, one need only show that $\mathbf{cx} = \mathbf{ub}$ since $\mathbf{x}$ is clearly feasible for the primal. But $\mathbf{x}_\mathbf{B} = \mathbf{B}^{-1}\mathbf{b}$ and $\mathbf{x}_\mathbf{N} = \mathbf{0}$, so

$$\mathbf{ub} = (\mathbf{c}_\mathbf{B}\mathbf{B}^{-1})\mathbf{b} = \mathbf{c}_\mathbf{B}(\mathbf{B}^{-1}\mathbf{b}) = \mathbf{c}_\mathbf{B}\mathbf{x}_\mathbf{B} + \mathbf{c}_\mathbf{N}\mathbf{x}_\mathbf{N} = \mathbf{cx}$$

and the proof is complete. □

It is important to realize that, even when the primal problem does have an optimal solution, the simplex algorithm might not produce an optimal bfs for that LP. The reason is that, if artificial variables are left in the final basis (at value 0) at the end of phase 1, then the simplex algorithm must be applied to the auxiliary LP instead of the primal LP (see Section 6.3). The next proposition establishes that, even in this case, if the primal LP has an optimal solution, then the simplex algorithm will produce an optimal solution to the dual LP.

Proposition 8.3. *If the primal LP has an optimal solution, then there are optimal solutions* $\mathbf{x}^*$ *and* $\mathbf{u}^*$ *to the primal and dual problems, respectively, for which* $\mathbf{cx}^* = \mathbf{u}^*\mathbf{b}$.

OUTLINE OF PROOF. The appearance of the quantifier "there are" suggests using the construction method to produce the desired solutions $\mathbf{x}^*$ and $\mathbf{u}^*$. Specifically, the simplex algorithm will be applied to the primal LP. To that end, the algorithm will first be applied to the phase 1 problem, and there are two cases that can arise.

Case 1: There are no artificial variables in the final phase 1 basis.

In this case, phase 2 can be implemented directly. By hypothesis, the simplex algorithm must then terminate with an optimal bfs $\mathbf{x}^* = (\mathbf{x}_\mathbf{B}^*, \mathbf{x}_\mathbf{N}^*) = (\mathbf{B}^{-1}\mathbf{b}, \mathbf{0}) \geq \mathbf{0}$ for the primal LP. Proposition 8.2 then shows that the vector $\mathbf{u}^* = \mathbf{c}_\mathbf{B}\mathbf{B}^{-1}$ is the desired optimal dual solution for which $\mathbf{cx}^* = \mathbf{u}^*\mathbf{b}$.

Case 2: There are artificial variables (all at value 0) in the final phase 1 basis.

In this case, the simplex algorithm is applied to the auxiliary LP:

$$\begin{array}{lllll} \text{Minimize} & \mathbf{cx} & & & \\ \text{subject to} & \mathbf{Ax} & +\mathbf{Iy} & & = \mathbf{b} \\ & & \mathbf{ey} & +z & = 0 \\ & \mathbf{x}, & \mathbf{y}, & z & \geq 0 \end{array}$$

Once again, the hypothesis that the primal problem has an optimal solution ensures that the simplex algorithm will produce an optimal bfs $(\mathbf{x}^*, \mathbf{y}^*, z^*)$ to the auxiliary LP. Of course, $\mathbf{x}^*$ is optimal for the primal problem (see Proposition 6.6); however, one has to be somewhat clever in producing the desired dual solution $\mathbf{u}^*$. It comes from considering the dual problem associated with the auxiliary LP:

$$\begin{array}{llll} \text{Maximize} & \mathbf{ub} & & \\ \text{subject to} & \mathbf{uA} & & \leq \mathbf{c} \\ & \mathbf{uI} & + v\mathbf{e} & \leq \mathbf{0} \\ & & v & \leq 0 \\ & \mathbf{u} & & \text{unrestricted} \end{array}$$

where v is a real number.

Since the simplex algorithm has produced an optimal bfs $(\mathbf{x}^*, \mathbf{y}^*, z^*)$ for the auxiliary LP, Proposition 8.2 ensures that one has also obtained a corresponding dual optimal solution, say, $(\mathbf{u}^*, v^*)$, for which $\mathbf{cx}^* = \mathbf{u}^*\mathbf{b}$. It remains only to show that $\mathbf{u}^*$ is optimal for the original dual LP, and this is left as an exercise.

PROOF OF PROPOSITION 8.3. To produce the desired solutions $\mathbf{x}^*$ and $\mathbf{u}^*$, consider what happens when the simplex algorithm is applied to the primal LP. At the end of phase 1 (which must produce a feasible solution to the primal), if no artificial variables appear in the final basis, then phase 2 can be implemented directly. The hypothesis then ensures that an optimal bfs $\mathbf{x}^* = (\mathbf{x}^*_{\mathbf{B}}, \mathbf{x}^*_{\mathbf{N}}) = (\mathbf{B}^{-1}\mathbf{b}, \mathbf{0}) \geq \mathbf{0}$ is obtained, but then Proposition 8.2 guarantees that the vector $\mathbf{u}^* = \mathbf{c}_{\mathbf{B}}\mathbf{B}^{-1}$ is optimal for the dual and satisfies $\mathbf{cx}^* = \mathbf{u}^*\mathbf{b}$.

In the event that phase 1 terminates feasibly with some artificial variables (all at value 0) in the final basis, phase 2 must be implemented on the auxiliary LP:

$$\begin{aligned} \text{Minimize} \quad & \mathbf{cx} \\ \text{subject to} \quad & \mathbf{Ax} + \mathbf{Iy} \qquad = \mathbf{b} \\ & \qquad\quad \mathbf{ey} + z = 0 \\ & \mathbf{x}, \quad \mathbf{y}, \quad z \geq 0 \end{aligned}$$

The hypothesis that the primal LP has an optimal solution ensures that phase 2 will produce an optimal bfs $(\mathbf{x}^*, \mathbf{y}^*, z^*)$ to the auxiliary LP. The desired dual solution $\mathbf{u}^*$ is found by considering the dual problem associated with the auxiliary LP:

$$\begin{aligned} \text{Maximize} \quad & \mathbf{ub} \\ \text{subject to} \quad & \mathbf{uA} \qquad \leq \mathbf{c} \\ & \mathbf{uI} + v\mathbf{e} \leq \mathbf{0} \\ & \qquad\quad v \leq 0 \\ & \mathbf{u} \qquad\quad \text{unrestricted} \end{aligned}$$

where v is a real number.

Since the simplex algorithm has produced an optimal bfs $(\mathbf{x}^*, \mathbf{y}^*, z^*)$ for the auxiliary LP, Proposition 8.2 ensures that one also has obtained a corresponding dual solution $(\mathbf{u}^*, v^*)$ for which $\mathbf{cx}^* = \mathbf{u}^*\mathbf{b}$. It remains to show that $\mathbf{u}^*$ is optimal for the dual LP, and this is left to the reader (see Exercise 8.3.1). □

To summarize weak duality, if **x** and **u** are feasible vectors for the primal and dual problems, respectively, then one way to test **x** (and **u**) for optimality is to see whether $\mathbf{cx} = \mathbf{ub}$. There is another completely equivalent way of stating this test for optimality, known as *complementary slackness*, which is particularly useful in designing other algorithms for solving linear programming problems. Instead of checking for $\mathbf{cx} = \mathbf{ub}$, one can replace **b** by **Ax** (since **x** is feasible), obtaining $\mathbf{cx} = \mathbf{u}(\mathbf{Ax})$ or, equivalently, $(\mathbf{c} - \mathbf{uA})\mathbf{x} = 0$. This, then, is the third test for optimality. It can be stated in a slightly different form, as shown in the next proposition.

Proposition 8.4. *If* **x** *and* **u** *are feasible for the primal and dual problems, respectively, and if, in addition, for each* j *with* $1 \le j \le n$ *and* $x_j > 0$, $(\mathbf{c} - \mathbf{uA})_j = 0$, *then* **x** *and* **u** *are optimal solutions.*

OUTLINE OF PROOF. The forward–backward method gives rise to the abstraction question "How can I show that two vectors (namely, **x** and **u**) are optimal solutions for the primal and dual problems?" The question is answered by means of Theorem 8.1, whereby it will be shown that **x** and **u** are feasible for the primal and dual, respectively, and that $\mathbf{cx} = \mathbf{ub}$. That **x** and **u** are feasible follows immediately from the hypothesis; the additional information in the hypothesis will be used to establish that $\mathbf{cx} = \mathbf{ub}$.

To see that $\mathbf{cx} = \mathbf{ub}$, note that $\mathbf{b} = \mathbf{Ax}$ (since **x** is feasible for the primal), and hence $\mathbf{ub} = \mathbf{uAx}$. Thus, to conclude that $\mathbf{cx} = \mathbf{ub}$, the forward process will be used to establish that $\mathbf{cx} = \mathbf{uAx}$, i.e., that $(\mathbf{c} - \mathbf{uA})\mathbf{x} = 0$. In other words, the hypothesis that, for each j with $1 \le j \le n$ and $x_j > 0$, $(\mathbf{c} - \mathbf{uA})_j = 0$ leads to the conclusion that $(\mathbf{c} - \mathbf{uA})\mathbf{x} = 0$ and hence that $\mathbf{ub} = \mathbf{cx}$.

Working backward, consider how one might be able to show that $(\mathbf{c} - \mathbf{uA})\mathbf{x} = 0$. Since $\mathbf{c} - \mathbf{uA}$ and **x** are both n vectors, the equation $(\mathbf{c} - \mathbf{uA})\mathbf{x} = 0$ can be written equivalently as $\sum_{i=1}^{n}(\mathbf{c} - \mathbf{uA})_i x_i = 0$. Thus, to show that $(\mathbf{c} - \mathbf{uA})\mathbf{x} = 0$, it will be shown that each of the terms in the summation is 0, i.e., that for each i with $1 \le i \le n$, $(\mathbf{c} - \mathbf{uA})_i x_i = 0$. The appearance of the quantifier "for all" in the backward process suggests using the choose method, so let i with $1 \le i \le n$. To see that $(\mathbf{c} - \mathbf{uA})_i x_i = 0$, note that, since **x** is feasible, either $x_i = 0$ or $x_i > 0$. If $x_i = 0$, then of course $(\mathbf{c} - \mathbf{uA})_i x_i = 0$. On the other hand, if $x_i > 0$, then, by specializing the hypothesis that, for each j with $1 \le j \le n$ and $x_j > 0$, $(\mathbf{c} - \mathbf{uA})_j = 0$ to the value of $j = i$, it follows that $(\mathbf{c} - \mathbf{uA})_i = 0$ and hence $(\mathbf{c} - \mathbf{uA})_i x_i = 0$.

PROOF OF PROPOSITION 8.4. Let **x** and **u** be feasible for the primal and dual problems, respectively. Theorem 8.1 can be used to show that they are both optimal by establishing that $\mathbf{cx} = \mathbf{ub}$. But $\mathbf{b} = \mathbf{Ax}$ (since **x** is feasible for the primal), and hence it will be shown that $\mathbf{cx} = \mathbf{uAx}$ or, equivalently, that $(\mathbf{c} - \mathbf{uA})\mathbf{x} = \sum_{i=1}^{n}(\mathbf{c} - \mathbf{uA})_i x_i = 0$. To this end, let i with $1 \le i \le n$. If $x_i = 0$, then of course $(\mathbf{c} - \mathbf{uA})_i x_i = 0$; if $x_i > 0$, then the hypothesis guarantees that $(\mathbf{c} - \mathbf{uA})_i = 0$, and once again $(\mathbf{c} - \mathbf{uA})_i x_i = 0$. □

Proposition 8.4 has created yet another test for optimality. The next example demonstrates its use.

EXAMPLE 8.6: *Primal LP.*

$$\begin{array}{lrrrrrl} \text{Minimize} & -2x_1 & -3x_2 & -\ x_3 & & & \\ \text{subject to} & x_1 & -\ x_2 & +2x_3 & +x_4 & & =1 \\ & 4x_1 & +2x_2 & -\ x_3 & & +x_5 & =2 \\ & x_1, & x_2, & x_3, & x_4, & x_5 & \geq 0 \end{array}$$

Dual LP.

$$\begin{array}{lrrl} \text{Maximize} & u_1 & +2u_2 & \\ \text{subject to} & u_1 & +4u_2 & \leq -2 \\ & -\ u_1 & +2u_2 & \leq -3 \\ & 2u_1 & -\ u_2 & \leq -1 \\ & u_1, & u_2 & \leq \ \ 0 \end{array}$$

To see that $\mathbf{x} = (0, \frac{5}{3}, \frac{4}{3}, 0, 0)$ is optimal for the primal, let $\mathbf{u} = (-\frac{5}{3}, -\frac{7}{3})$. It is easy to verify that $\mathbf{x}$ is feasible for the primal and $\mathbf{u}$ is feasible for the dual. Furthermore, $x_2 > 0$ and $x_3 > 0$; correspondingly,

$$(\mathbf{c} - \mathbf{uA})_2 = c_2 - \mathbf{uA}_{.2} = -3 - (-\tfrac{5}{3}, -\tfrac{7}{3})\begin{bmatrix} -1 \\ 2 \end{bmatrix} = 0$$

$$(\mathbf{c} - \mathbf{uA})_3 = c_3 - \mathbf{uA}_{.3} = -1 - (-\tfrac{5}{3}, -\tfrac{7}{3})\begin{bmatrix} 2 \\ -1 \end{bmatrix} = 0$$

Thus $\mathbf{x}$ passes the test for optimality given in Proposition 8.4.

After Example 8.6 it is reasonable to wonder why one would ever want to use the complementary slackness conditions as a test for optimality. The answer lies in computational considerations. When designing an algorithm to solve problems, whatever test for optimality is used should have the property that if the current vector $\mathbf{x}$ does not pass the test, then that fact should indicate how to move a new solution $\mathbf{x}'$. Indeed, that was one of the advantages of testing whether the reduced costs $\mathbf{c_N} - \mathbf{c_B B}^{-1}\mathbf{N} \geq \mathbf{0}$, for, if they are not, that information was used in the simplex algorithm to construct a direction in which to move. The complementary slackness test for optimality also exhibits the desirable property of indicating how to move to the new solution. More will be said about the complementary slackness conditions when the appropriate algorithm is developed.

This section has described the property of weak duality and has provided two new tests for optimality. The next section shows how the status of the dual (i.e., if it is infeasible, optimal, or unbounded) can be used to determine the status of the primal LP and vice versa.

EXERCISES

8.3.1. In the proof of Proposition 8.3, prove that if $(\mathbf{u}^*, v^*)$ is an optimal solution for the dual of the auxiliary LP, then $\mathbf{u}^*$ is optimal for the dual of the original LP.

8.3.2. Consider the following primal and dual LPs. The primal LP is to

$$\begin{array}{lrrrrrrl} \text{minimize} & -4x_1 & +6x_2 & +7x_3 & & +x_5 & +5x_6 & \\ \text{subject to} & -x_1 & +2x_2 & +x_3 & -x_4 & & +2x_6 & = 5 \\ & x_1 & & -2x_3 & & +x_5 & +x_6 & = 3 \\ & & x_2 & & +2x_4 & & +x_6 & = 3 \\ & x_1, & x_2, & x_3, & x_4, & x_5, & x_6 & \geq 0 \end{array}$$

The dual LP is to

$$\begin{array}{lrrrl} \text{maximize} & 5u_1 & +3u_2 & +3u_3 & \\ \text{subject to} & -u_1 & +u_2 & & \leq -4 \\ & 2u_1 & & +u_3 & \leq 6 \\ & u_1 & -2u_2 & & \leq 7 \\ & -u_1 & & +2u_3 & \leq 0 \\ & & u_2 & & \leq 1 \\ & 2u_1 & +u_2 & +u_3 & \leq 5 \\ & u_1, & u_2, & u_3 & \text{unrestricted} \end{array}$$

Use the complementary slackness conditions of Proposition 8.4 to show that the vectors $\mathbf{x}^* = (1,1,0,0,0,2)$ and $\mathbf{u}^* = (3,-1,0)$ are optimal for their respective problems.

8.3.3. **(a)** Derive the appropriate complementary slackness conditions for the symmetric LPs presented in Example 8.3.

(b) Use the complementary slackness conditions of part (a) to determine if the vectors $\mathbf{x}^* = (0,2,4)$ and $\mathbf{u}^* = (0,1,0)$ are optimal for the following primal and dual LPs, respectively. The primal LP is to

$$\begin{array}{lrrrl} \text{minimize} & 5x_1 & -x_2 & +x_3 & \\ \text{subject to} & x_1 & +2x_2 & +x_3 & \geq 5 \\ & 2x_1 & -x_2 & +x_3 & \geq 2 \\ & -x_1 & -x_2 & +3x_3 & \geq 9 \\ & x_1, & x_2, & x_3 & \geq 0 \end{array}$$

The dual LP is to

$$\begin{array}{llcccl} \text{maximize} & 5u_1 & +2u_2 & +9u_3 & & \\ \text{subject to} & u_1 & +2u_2 & -\ u_3 & \leq & 5 \\ & 2u_1 & -\ u_2 & -\ u_3 & \leq & -1 \\ & u_1 & +\ u_2 & +3u_3 & \leq & 1 \\ & u_1, & u_2, & u_3 & \geq & 0 \end{array}$$

8.4. STRONG DUALITY

The ultimate goal of this section is to be able to determine whether the primal LP is infeasible, optimal, or unbounded, by using the dual LP and vice versa. For instance, it will be shown that if both the primal and dual problems are known to be feasible, then, without using a computer, one can conclude that both problems have optimal solutions.

Imagine for the moment that you somehow know that the primal LP has a feasible solution. This means that, when the simplex algorithm is applied to the primal problem, either unboundedness or optimality is detected. The next proposition shows that if the dual is also known to have a feasible solution than the primal LP cannot be unbounded, and hence optimality must occur.

Proposition 8.5. *If the primal and dual problems both have feasible solutions, then the primal is not unbounded.*

OUTLINE OF PROOF. The appearance of the word "not" in the conclusion suggests using the contradiction method. Thus it can be assumed that the primal and dual problems have feasible solutions (say, $\mathbf{x}$ and $\mathbf{u}$, respectively), and, furthermore, that the primal LP is unbounded. Thus one has $\mathbf{Ax} = \mathbf{b}$, $\mathbf{x} \geq \mathbf{0}$, and $\mathbf{uA} \leq \mathbf{c}$; in addition, by the definition of unboundedness, there is a direction $\mathbf{d} \geq \mathbf{0}$ such that $\mathbf{Ad} = \mathbf{0}$ and $\mathbf{cd} < 0$. The strategy is to work forward from this information to establish a contradiction to $\mathbf{cd} < 0$, by showing that $\mathbf{cd} \geq 0$. Specifically, multiplying both sides of the inequality $\mathbf{uA} \leq \mathbf{c}$ by the nonnegative vector $\mathbf{d}$ and using $\mathbf{Ad} = \mathbf{0}$, one obtains

$$\mathbf{cd} \geq (\mathbf{uA})\mathbf{d} = \mathbf{u}(\mathbf{Ad}) = \mathbf{u}(\mathbf{0}) = 0$$

PROOF OF PROPOSITION 8.5. Assume, to the contrary, that the primal LP is unbounded. From the hypothesis, let $\mathbf{x}$ and $\mathbf{u}$ be feasible solutions for the primal and dual problems, respectively, so $\mathbf{Ax} = \mathbf{b}$, $\mathbf{x} \geq \mathbf{0}$, and $\mathbf{uA} \leq \mathbf{c}$.

Since the primal is assumed to be unbounded, there is a direction $\mathbf{d} \geq \mathbf{0}$ such that $\mathbf{Ad} = \mathbf{0}$ and $\mathbf{cd} < 0$. Multiplying both sides of the inequality $\mathbf{uA} \leq \mathbf{c}$ by the nonnegative vector $\mathbf{d}$ and using $\mathbf{Ad} = \mathbf{0}$, one obtains

$$\mathbf{cd} \geq (\mathbf{uA})\mathbf{d} = \mathbf{u}(\mathbf{Ad}) = \mathbf{u}(\mathbf{0}) = 0$$

which contradicts $\mathbf{cd} < 0$. □

With the help of Propositions 8.3 and 8.5, it is possible to establish a more profound relationship between the primal and dual problems, as is shown in the next theorem.

Theorem 8.2. *If the primal and dual problems both have feasible solutions, then there are optimal solutions* $\mathbf{x}^*$ *and* $\mathbf{u}^*$, *respectively, for which* $\mathbf{cx}^* = \mathbf{u}^*\mathbf{b}$.

OUTLINE OF PROOF. Since the conclusion of Theorem 8.2 is exactly the same as that of Proposition 8.3, one need only work forward from the hypothesis of this theorem to establish the hypothesis of Proposition 8.3.

In other words, the assumption that the primal and dual problems both have feasible solutions will be used to show that the primal LP has an optimal solution, for then, by Proposition 8.3, the desired solutions $\mathbf{x}^*$ and $\mathbf{u}^*$ exist. The forward–backward method now gives rise to the abstraction question "How can I show that an LP (namely, the primal) has an optimal solution?" One answer is to apply the simplex algorithm to the primal problem and somehow to rule out infeasible and unbounded termination. The hypothesis that the primal problem has a feasible solution clearly excludes infeasible termination. On the other hand, the assumption that the dual also has a feasible solution ensures that the hypothesis of Proposition 8.5 is true, and thus the conclusion of Proposition 8.5 rules out unbounded termination.

PROOF OF THEOREM 8.2. To derive the desired conclusion, it will be shown that the primal LP has an optimal solution, for then, by Proposition 8.3, there will be optimal solutions $\mathbf{x}^*$ and $\mathbf{u}^*$ for which $\mathbf{cx}^* = \mathbf{u}^*\mathbf{b}$. To see that the primal LP has an optimal solution, consider what happens when the simplex algorithm is applied to the primal problem. The hypothesis not only rules out infeasible termination, but also ensures that unbounded termination cannot occur because of Proposition 8.5. □

Theorem 8.2 is one version of what is known as the *fundamental theorem of duality*, and it has established a relationship between the status of the primal LP and that of the dual; if both are feasible, then both have optimal solutions.

With the use of Theorem 8.2, it is possible to establish other relationships between the status of the two LPs. There are four possible combinations, depending on whether the primal and dual are feasible or infeasible (see Table 8.2). Theorem 8.2 has supplied the information for the box in the upper left corner of Table 8.2. The remainder of this section is devoted to filling in the rest of the boxes. Indeed, the next proposition supplies the information for the box in the lower left corner of Table 8.2.

TABLE 8.2. Possible Primal and Dual Relationships

	Primal feasible	Primal infeasible
Dual feasible		
Dual infeasible		

Proposition 8.6. *If the primal LP has a feasible solution and the dual LP does not, then the primal LP is unbounded.*

OUTLINE OF PROOF. The forward–backward method gives rise to the abstraction question "How can I show that an LP (namely, the primal) is unbounded?" One answer is by the definition, but a second answer is to apply the simplex algorithm to the primal LP and somehow to rule out infeasible and optimal termination. The hypothesis that the primal has a feasible solution ensures that infeasible termination of the simplex algorithm cannot happen.

It remains to show that an optimal solution cannot be produced by the simplex algorithm. The appearance of the word "not" now suggests proceeding by the contradiction method, whereby it can be assumed that the simplex algorithm does produce an optimal solution to the primal. With this, a contradiction is reached by showing that the dual is both infeasible and feasible at the same time.

The hypothesis states that the dual is infeasible. On the other hand, since the primal has an optimal solution, the conclusion of Proposition 8.3 states that the dual has an optimal (and thus feasible) solution, hence establishing the contradiction.

PROOF OF PROPOSITION 8.6. To show that the primal is unbounded, consider what happens when the simplex algorithm is applied to the primal LP. The hypothesis ensures that infeasible termination is impossible. If an optimal solution is produced, then, by Proposition 8.3, the dual must also have an optimal solution. Since this contradicts the hypothesis that the dual is infeasible, it must be the case that the simplex algorithm terminates in an unbounded condition. □

Proposition 8.6 has filled in the information in the lower left corner of Table 8.2. The next proposition addresses the upper right corner.

Proposition 8.7. *If the dual LP is feasible and the primal LP is infeasible, then the dual LP is unbounded.*

OUTLINE OF PROOF. The forward–backward method gives rise to the abstraction question "How can I show that an LP (namely, the dual) is unbounded?" One answer is by the definition, whereby it is necessary to show that

1. the dual is feasible (which is true by hypothesis) and
2. there is a direction $\mathbf{d}$ such that $\mathbf{dA} \le \mathbf{0}$ and $\mathbf{db} > 0$.

The desired direction is constructed by working forward from the assumption that the primal is infeasible. To make use of this information, recall the phase 1 LP,

$$\begin{array}{ll} \text{Minimize} & \mathbf{ey} \\ \text{subject to} & \mathbf{Ax} + \mathbf{Iy} = \mathbf{b} \\ & \mathbf{x},\quad \mathbf{y} \ge \mathbf{0} \end{array}$$

and its associated dual:

$$\begin{array}{ll} \text{Maximize} & \mathbf{ub} \\ \text{subject to} & \mathbf{uA} \le \mathbf{0} \\ & \mathbf{uI} \le \mathbf{e} \\ & \mathbf{u} \text{ unrestricted} \end{array}$$

The direction $\mathbf{d}$ is going to be any optimal solution to the phase 1 dual, but how does one show that the phase 1 dual has an optimal solution? Here is where Theorem 8.2 is useful because, according to that theorem, one need only establish that both the phase 1 LP and its dual are feasible, for then both have optimal solutions. It is easy to see that $\mathbf{x} = \mathbf{0}$, $\mathbf{y} = \mathbf{b}$ is feasible for the phase 1 LP and that $\mathbf{u} = \mathbf{0}$ is feasible for the dual LP. Consequently, by Theorem 8.2, both have optimal solutions [say, $(\mathbf{x}^*, \mathbf{y}^*)$ and $\mathbf{d}$, respectively] and, furthermore, $\mathbf{ey}^* = \mathbf{db}$.

It remains to show that $\mathbf{d}$ has the desired properties that $\mathbf{dA} \le \mathbf{0}$ and $\mathbf{db} > 0$. Both of these statements are verified by using the fact that $\mathbf{d}$ is optimal for the phase 1 dual together with the hypothesis that the primal LP is infeasible, i.e., $\mathbf{y}^* \ne \mathbf{0}$. Since $\mathbf{d}$ is optimal (and hence feasible) for the phase 1 dual, it must be that $\mathbf{dA} \le \mathbf{0}$. Also, since $\mathbf{db} = \mathbf{ey}^*$ and $\mathbf{ey}^* > 0$ (because $\mathbf{y}^* \ne \mathbf{0}$), $\mathbf{db} > 0$.

PROOF OF PROPOSITION 8.7. By hypothesis, the dual is feasible, so to show that it is unbounded a direction $\mathbf{d}$ must be found with the property that $\mathbf{dA} \le \mathbf{0}$ and $\mathbf{db} > 0$. To that end, consider the phase 1 LP,

$$\begin{array}{ll} \text{Minimize} & \mathbf{ey} \\ \text{subject to} & \mathbf{Ax} + \mathbf{Iy} = \mathbf{b} \\ & \mathbf{x},\quad \mathbf{y} \ge \mathbf{0} \end{array}$$

and its associated dual.

$$\begin{aligned} \text{Maximize} \quad & \mathbf{ub} \\ \text{subject to} \quad & \mathbf{uA} \le \mathbf{0} \\ & \mathbf{uI} \le \mathbf{e} \\ & \mathbf{u} \text{ unrestricted} \end{aligned}$$

Clearly, $(\mathbf{x}, \mathbf{y}) = (\mathbf{0}, \mathbf{b})$ and $\mathbf{u} = \mathbf{0}$ are feasible for the phase 1 LP and its dual, respectively. Hence, by Theorem 8.2, both have optimal solutions, say, $(\mathbf{x}^*, \mathbf{y}^*)$ and $\mathbf{d}$, respectively, for which $\mathbf{db} = \mathbf{ey}^*$. Consequently, $\mathbf{dA} \le \mathbf{0}$, and since the primal is assumed to be infeasible, $\mathbf{ey}^* > 0$, and so $\mathbf{db} = \mathbf{ey}^* > 0$. □

Proposition 8.7 has supplied the information for the box in the upper right corner of Table 8.2. The next example will address the remaining box in the lower right corner by showing that it is possible for both the primal and dual problems to be infeasible at the same time.

EXAMPLE 8.7: *Primal LP.*

$$\begin{aligned} \text{Minimize} \quad & x_1 - 3x_2 \\ \text{subject to} \quad & x_1 - x_2 = 0 \\ & x_1 - x_2 = 1 \\ & x_1, \quad x_2 \ge 0 \end{aligned}$$

Dual LP.

$$\begin{aligned} \text{Maximize} \quad & u_2 \\ \text{subject to} \quad & u_1 + u_2 \le 1 \\ & -u_1 - u_2 \le -3 \\ & u_1, \quad u_2 \text{ unrestricted} \end{aligned}$$

Table 8.2 is now complete. All possible relationships between the status of the primal and dual problems have been examined. The complete information is summarized in Table 8.3. This table can be very useful in determin-

TABLE 8.3. Primal and Dual Relationships

	Primal feasible	Primal infeasible
Dual feasible	Primal optimal Dual optimal	Dual unbounded
Dual infeasible	Primal unbounded	Can happen (see Example 8.7)

ing the status of a given LP. For instance, to show that a particular LP has an optimal solution, you might check that both the LP and its dual are feasible. Similarly, to establish that a particular LP is unbounded, you need only show that it is feasible and that its dual is not. The next proposition demonstrates the use of Table 8.3.

Proposition 8.8. *If the dual LP has an optimal solution, then the primal LP has an optimal solution.*

OUTLINE OF PROOF. By looking at Table 8.3, the only box containing a dual optimal solution is the upper left corner, where it can be seen that the primal LP also has an optimal solution, as desired. To make the proof formal, one can argue as follows. By hypothesis, the dual has an optimal (and hence feasible solution), so one must be in either the upper left or upper right corner of Table 8.3. If one were in the upper right corner, the dual would be unbounded, which it is not by hypothesis. Thus one must be in the upper left corner of Table 8.3, and hence the primal problem has an optimal solution.

PROOF OF PROPOSITION 8.8. Since the dual is optimal, the primal LP cannot be infeasible; otherwise, by Proposition 8.7, the dual LP would be unbounded. Hence both the primal and dual LPs are feasible, and so, by Theorem 8.2, both have optimal solutions. □

Another typical use of Table 8.3 arises in the study of *linear inequalities*. Consider, for example, the linear system $\mathbf{Ax} = \mathbf{b}$, $\mathbf{x} \geq \mathbf{0}$, and suppose that you happen to know that there is no vector $\mathbf{x} \geq \mathbf{0}$ for which $\mathbf{Ax} = \mathbf{b}$. To derive some useful information from the fact that the system has no solution, consider the following associated primal LP.

$$\begin{aligned} \text{Minimize} \quad & \mathbf{0x} \\ \text{subject to} \quad & \mathbf{Ax} = \mathbf{b} \\ & \mathbf{x} \geq \mathbf{0} \end{aligned}$$

Its dual is to

$$\begin{aligned} \text{maximize} \quad & \mathbf{ub} \\ \text{subject to} \quad & \mathbf{uA} \leq \mathbf{0} \\ & \mathbf{u} \text{ unrestricted} \end{aligned}$$

Since the linear system $\mathbf{Ax} = \mathbf{b}$, $\mathbf{x} \geq \mathbf{0}$, has no solution, the primal LP is infeasible. The idea now is to use Table 8.3 to make some kind of statement regarding the dual LP. To do so requires knowing whether the dual is feasible or not. In this particular instance, the dual LP is feasible because

$\mathbf{u} = \mathbf{0}$ is a dual feasible solution. Therefore, since the primal problem is infeasible and the dual is feasible, one must be located in the upper right corner of Table 8.3, and hence the dual must be unbounded. By definition, this means that there must be a direction of unboundedness $\mathbf{d}$ for the dual such that $\mathbf{dA} \leq \mathbf{0}$ and $\mathbf{db} > 0$. In other words, if the system $\mathbf{Ax} = \mathbf{b}$, $\mathbf{x} \geq \mathbf{0}$, has no solution, then the system $\mathbf{uA} \leq \mathbf{0}$, $\mathbf{ub} > 0$, does have a solution.

In a similar fashion, if the system $\mathbf{uA} \leq \mathbf{0}$, $\mathbf{ub} > 0$, does not have a solution, then the LP,

$$\begin{aligned} \text{Maximize} \quad & \mathbf{ub} \\ \text{subject to} \quad & \mathbf{uA} \leq \mathbf{0} \\ & \mathbf{u} \text{ unrestricted} \end{aligned}$$

is not unbounded. Consequently, since $\mathbf{u} = \mathbf{0}$ is a feasible solution, the above LP and its dual,

$$\begin{aligned} \text{Minimize} \quad & \mathbf{0x} \\ \text{subject to} \quad & \mathbf{Ax} = \mathbf{b} \\ & \mathbf{x} \geq \mathbf{0} \end{aligned}$$

must both have optimal solutions (see Table 8.3). Therefore there is an $\mathbf{x} \geq \mathbf{0}$ such that $\mathbf{Ax} = \mathbf{b}$. Thus, if there is no solution to $\mathbf{uA} \leq \mathbf{0}$, $\mathbf{ub} > 0$, then there is a solution to $\mathbf{Ax} = \mathbf{b}$, $\mathbf{x} \geq \mathbf{0}$.

Finally, note that the two systems $\mathbf{Ax} = \mathbf{b}$, $\mathbf{x} \geq \mathbf{0}$ and $\mathbf{uA} \leq \mathbf{0}$, $\mathbf{ub} > 0$ cannot both have solutions, for otherwise the primal LP would be feasible while the dual LP would be unbounded, and, according to Table 8.3, this is not possible. The next proposition summarizes these results; it is known as *Farkas's lemma*. Its proof will already be clear.

Proposition 8.9. *Either the system* $\mathbf{Ax} = \mathbf{b}$, $\mathbf{x} \geq \mathbf{0}$, *has a solution, or else the system* $\mathbf{uA} \leq \mathbf{0}$, $\mathbf{ub} > 0$, *has a solution, but not both.*

Results such as Farkas's lemma are referred to as *theorems of the alternative* because they state that either one system or its associated alternative system has a solution, but not both. Statements regarding the feasibility of such systems can usually be phrased in terms of the feasibility or unboundedness of an associated LP (as was done for Farkas's lemma). Then, by using Table 8.3, one can make the appropriate alternative statement for the associated dual LP. Finally, one must translate the statement regarding the dual problem back into the context of the linear system.

This section has examined all possible relationships between the status of the primal and dual problems. The information is summarized in Table 8.3.

The next section uses many of the ideas presented here to develop a new algorithm for solving the dual of an LP in standard form.

EXERCISES

8.4.1. What can be said about the status of the following pair of primal and dual LPs? Explain. The primal LP is to

$$\begin{array}{ll} \text{minimize} & \mathbf{ex} \\ \text{subject to} & \mathbf{Ax} \geq -\mathbf{e} \\ & \mathbf{x} \geq \mathbf{0} \end{array}$$

The dual LP is to

$$\begin{array}{ll} \text{maximize} & -\mathbf{ue} \\ \text{subject to} & \mathbf{uA} \leq \mathbf{e} \\ & \mathbf{u} \geq \mathbf{0} \end{array}$$

8.4.2. For an LP in standard form, prove that if $\mathbf{x}^*$ is optimal for the primal and $\mathbf{u}^*$ is optimal for the dual, then $\mathbf{cx}^* = \mathbf{u}^*\mathbf{b}$.

8.4.3. Consider the following LP:

$$\begin{array}{lrrrrrr} \text{Minimize} & 2x_1 & +3x_2 & -\ x_3 & -7x_4 & -8x_5 & \\ \text{subject to} & 8x_1 & +3x_2 & -4x_3 & +\ x_4 & & = 6 \\ & 2x_1 & +6x_2 & -\ x_3 & & +\ x_5 & = 3 \\ & x_1, & x_2, & x_3, & x_4, & x_5 & \geq 0 \end{array}$$

Use the vectors $\mathbf{x} = (0,0,0,6,3)$ and $\mathbf{d} = (0,0,1,4,1)$ to show that the above LP is unbounded. What can you say about the status of the dual LP?

8.4.4. Suppose that the following system of equations has no feasible solution:

$$\begin{aligned} \mathbf{Ax} &= \mathbf{0} \\ \mathbf{ex} &= 1 \\ \mathbf{x} &\geq \mathbf{0} \end{aligned}$$

Derive the associated system of equations and use duality theory to explain why that system has a solution.

8.4.5. Suppose that the following system of equations has no feasible solution:

$$\begin{aligned} \mathbf{Ad} &= \mathbf{0} \\ \mathbf{cd} &< 0 \\ \mathbf{d} &\geq \mathbf{0} \end{aligned}$$

Derive the associated system of equations and use duality theory to explain why this system has a solution.

8.5. THE DUAL SIMPLEX ALGORITHM

For an LP in standard form,

$$\begin{aligned} \text{Minimize} \quad & \mathbf{cx} \\ \text{subject to} \quad & \mathbf{Ax} = \mathbf{b} \\ & \mathbf{x} \geq \mathbf{0} \end{aligned}$$

that has an optimal solution, the simplex algorithm can be used to compute it. In so doing, the algorithm simultaneously produces an optimal solution to the dual LP:

$$\begin{aligned} \text{Maximize} \quad & \mathbf{ub} \\ \text{subject to} \quad & \mathbf{uA} \leq \mathbf{c} \\ & \mathbf{u} \text{ unrestricted} \end{aligned}$$

The purpose of this section is to describe a different method for obtaining an optimal dual solution. The new procedure is referred to as the *dual simplex algorithm*, in contrast to the *primal simplex algorithm* of Chapter 5. Not surprisingly, when the dual simplex algorithm has computed an optimal solution to the dual problem, it simultaneously produces an optimal solution to the primal LP. The advantages and disadvantages of the dual simplex algorithm will also be discussed.

To understand how the dual simplex algorithm works, recall that the primal simplex algorithm starts with a (primal) basic feasible solution. After this solution is tested for optimality, a direction is computed, and a move is made to another primal basic feasible solution whose objective value has decreased. In a similar spirit, the dual simplex algorithm starts with a dual basic feasible solution (to be defined in a moment). After this solution is tested for optimality, a direction is computed, and a move is made to another dual basic feasible solution whose dual objective value has increased.

To describe the steps of such an algorithm, it is necessary to define formally a dual basic feasible solution (dbfs). Conveniently, a dbfs can be defined using an $m \times m$ submatrix $\mathbf{B}$ of $\mathbf{A}$ much as was done with a primal basic feasible solution.

Definition 8.1. An m vector $\mathbf{u}$ is a *dual basic feasible solution* if there is a nonsingular $m \times m$ submatrix $\mathbf{B}$ of $\mathbf{A}$ such that

1. $\mathbf{u} = \mathbf{c_B B}^{-1}$ and
2. $\mathbf{uA} \leq \mathbf{c}$.

As usual, the remaining $n - m$ columns of $\mathbf{A}$ will be denoted $\mathbf{N}$. Note that condition 2 of Definition 8.1 ensures that $\mathbf{u}$ is feasible for the dual problem. The next example illustrates a dbfs.

EXAMPLE 8.8: *Primal LP.*

$$\begin{array}{lrrrrrl} \text{Minimize} & -x_1 & + x_2 & - x_3 & & & \\ \text{subject to} & 7x_1 & +2x_2 & & +x_4 & & =5 \\ & 4x_1 & -2x_2 & +3x_3 & & +x_5 & =3 \\ & x_1, & x_2, & x_3, & x_4, & x_5 & \geq 0 \end{array}$$

so

$$\mathbf{c} = (-1, 1, -1, 0, 0)$$
$$\mathbf{A} = \begin{bmatrix} 7 & 2 & 0 & 1 & 0 \\ 4 & -2 & 3 & 0 & 1 \end{bmatrix}$$
$$\mathbf{b} = \begin{bmatrix} 5 \\ 3 \end{bmatrix}$$

Dual LP.

$$\begin{array}{lrrl} \text{Maximize} & 5u_1 & +3u_2 & \\ \text{subject to} & 7u_1 & +4u_2 & \leq -1 \\ & 2u_1 & -2u_2 & \leq 1 \\ & & 3u_2 & \leq -1 \\ & u_1, & u_2 & \leq 0 \end{array}$$

The vector $\mathbf{u} = (0, -\frac{1}{3})$ is a dbfs in which

$$\mathbf{B} = [\mathbf{A}_{.3}, \mathbf{A}_{.4}] = \begin{bmatrix} 0 & 1 \\ 3 & 0 \end{bmatrix}$$
$$\mathbf{N} = [\mathbf{A}_{.1}, \mathbf{A}_{.2}, \mathbf{A}_{.5}] = \begin{bmatrix} 7 & 2 & 0 \\ 4 & -2 & 1 \end{bmatrix}$$

because

$$\mathbf{c_B B}^{-1} = (-1, 0)\begin{bmatrix} 0 & \frac{1}{3} \\ 1 & 0 \end{bmatrix} = (0, -\tfrac{1}{3}) = \mathbf{u}$$

and

$$\begin{aligned} \mathbf{uA} &= (0, -\tfrac{1}{3})\begin{bmatrix} 7 & 2 & 0 & 1 & 0 \\ 4 & -2 & 3 & 0 & 1 \end{bmatrix} \\ &= (-\tfrac{4}{3}, \tfrac{2}{3}, -1, 0, -\tfrac{1}{3}) \\ &\leq (-1, 1, -1, 0, 0) \\ &= \mathbf{c} \end{aligned}$$

With the concept of a dbfs, it is possible to develop the dual simplex algorithm much as the primal simplex algorithm was developed in Chapter 5. The "phase 1" issue of how to find an initial dbfs for the dual LP will be addressed shortly; for now, assume that you have somehow found such a dbfs, say, **u**, together with the associated **B** matrix. The justification for the following steps of the dual simplex algorithm will be left to the exercises. The similarities and differences between these steps and the corresponding ones for the primal simplex algorithm are discussed after the steps are illustrated with an example.

Step 1. Test for optimality: If $\mathbf{B}^{-1}\mathbf{b} \geq \mathbf{0}$, then the current dbfs $\mathbf{u} = \mathbf{c_B}\mathbf{B}^{-1}$ is optimal for the dual LP. Otherwise, find a k^* with $1 \leq k^* \leq m$ such that $(\mathbf{B}^{-1}\mathbf{b})_{k^*} < 0$. Go to step 2.

Step 2. Computing the direction of movement: Compute the m (row) vector $\mathbf{d} = -(\text{row } k^* \text{ of } \mathbf{B}^{-1}) = -\mathbf{I}_{k^*.}\mathbf{B}^{-1}$. (This particular direction is chosen because it leads to an increase in the dual objective function since $\mathbf{db} > 0$.) Go to step 3.

Step 3. Computing the amount of movement: If $\mathbf{dN} \leq \mathbf{0}$, then the dual LP is unbounded. Otherwise, compute

$$\begin{aligned} t^* &= \min\{(\mathbf{c_N} - \mathbf{uN})_j/(\mathbf{dN})_j : 1 \leq j \leq n-m, (\mathbf{dN})_j > 0\} \\ &= (\mathbf{c_N} - \mathbf{uN})_{j^*}/(\mathbf{dN})_{j^*} \end{aligned}$$

(t^* is the largest value of t for which $\mathbf{u} + t\mathbf{d}$ remains feasible for the dual LP).

Step 4. Moving to the new dbfs $\mathbf{u}'$ *and pivoting:* Move from $\mathbf{u}$ to the new dbfs $\mathbf{u}'$ by computing $\mathbf{u}' = \mathbf{u} + t^*\mathbf{d}$. The associated $\mathbf{B}'$ matrix is obtained by replacing column k^* of $\mathbf{B}$ with column j^* of $\mathbf{N}$, just as in the primal simplex algorithm. The new inverse matrix $\mathbf{B}'^{-1}$ can then be computed.

These steps will be demonstrated in the next example. Observe that, when the current dbfs passes the test for optimality, a primal optimal solution has simultaneously been produced and that $\mathbf{B}^{-1}\mathbf{b}$ comprises the values of the optimal primal basic variables. The remaining nonbasic primal variables have the value 0.

EXAMPLE 8.9: *Primal LP.*

$$\begin{array}{lrrrrrl} \text{Minimize} & 2x_1 & +3x_2 & +\ x_3 & & & \\ \text{subject to} & x_1 & +\ x_2 & -2x_3 & +x_4 & & = -1 \\ & 4x_1 & -2x_2 & -\ x_3 & & +x_5 & = -2 \\ & x_1, & x_2, & x_3, & x_4, & x_5 & \geq\ 0 \end{array}$$

so

$$\mathbf{c} = (2, 3, 1, 0, 0)$$

$$\mathbf{A} = \begin{bmatrix} 1 & 1 & -2 & 1 & 0 \\ 4 & -2 & -1 & 0 & 1 \end{bmatrix}$$

$$\mathbf{b} = \begin{bmatrix} -1 \\ -2 \end{bmatrix}$$

Dual LP.

$$\begin{aligned}\text{Maximize} \quad & -u_1 - 2u_2 \\ \text{subject to} \quad & u_1 + 4u_2 \le 2 \\ & u_1 - 2u_2 \le 3 \\ & -2u_1 - u_2 \le 1 \\ & u_1 \le 0 \\ & u_2 \le 0\end{aligned}$$

Iteration 1

Step 0. The initial dbfs for this iteration:

$$\mathbf{B} = [\mathbf{A}_{.4}, \mathbf{A}_{.5}] = \begin{bmatrix} 1 & 0 \\ 0 & 1 \end{bmatrix}$$

$$\mathbf{N} = [\mathbf{A}_{.1}, \mathbf{A}_{.2}, \mathbf{A}_{.3}] = \begin{bmatrix} 1 & 1 & -2 \\ 4 & -2 & -1 \end{bmatrix}$$

$$\mathbf{u} = \mathbf{c_B B}^{-1} = (0,0)\begin{bmatrix} 1 & 0 \\ 0 & 1 \end{bmatrix} = (0,0)$$

Step 1. Testing for optimality:

$$\mathbf{B}^{-1}\mathbf{b} = \begin{bmatrix} 1 & 0 \\ 0 & 1 \end{bmatrix}\begin{bmatrix} -1 \\ -2 \end{bmatrix} = \begin{bmatrix} -1 \\ -2 \end{bmatrix}$$

So $k^* = 1$ or 2. Choose $k^* = 2$.

Step 2. Computing the direction of movement:

$$\mathbf{d} = -\text{row } k^* \text{ of } \mathbf{B}^{-1} = (0, -1)$$

Step 3. Computing the amount of movement:

$$\mathbf{dN} = (0, -1)\begin{bmatrix} 1 & 1 & -2 \\ 4 & -2 & -1 \end{bmatrix} = (-4, 2, 1)$$

$$\begin{aligned} t^* &= \min\{(\mathbf{c_N} - \mathbf{uN})_j / (\mathbf{dN})_j : (\mathbf{dN})_j > 0\} \\ &= \min\{(\mathbf{c_N} - \mathbf{uN})_2/(\mathbf{dN})_2, \quad (\mathbf{c_N} - \mathbf{uN})_3/(\mathbf{dN})_3\} \\ &= \min\{\tfrac{3}{2}, 1\} \\ &= 1 \qquad \text{so} \quad j^* = 3 \end{aligned}$$

Step 4. Pivoting: Column $k^* = 2$ of $\mathbf{B}$ is replaced with column $j^* = 3$ of $\mathbf{N}$, so

$$\mathbf{B}' = \begin{bmatrix} 1 & -2 \\ 0 & -1 \end{bmatrix}$$

$$\mathbf{N}' = \begin{bmatrix} 1 & 1 & 0 \\ 4 & -2 & 1 \end{bmatrix}$$

$$\mathbf{u}' = \mathbf{u} + t^*\mathbf{d} = (0,0) + 1(0,-1) = (0,-1)$$

Iteration 2

Step 0. The dbfs for this iteration:

$$\mathbf{B} = [\mathbf{A}_{.4}, \mathbf{A}_{.3}] = \begin{bmatrix} 1 & -2 \\ 0 & -1 \end{bmatrix}$$

$$\mathbf{N} = [\mathbf{A}_{.1}, \mathbf{A}_{.2}, \mathbf{A}_{.5}] = \begin{bmatrix} 1 & 1 & 0 \\ 4 & -2 & 1 \end{bmatrix}$$

$$\mathbf{u} = \mathbf{c_B}\mathbf{B}^{-1} = (0,1)\begin{bmatrix} 1 & -2 \\ 0 & -1 \end{bmatrix} = (0,-1)$$

Step 1. Testing for optimality:

$$\mathbf{B}^{-1}\mathbf{b} = \begin{bmatrix} 1 & 0 \\ 0 & 1 \end{bmatrix}\begin{bmatrix} -1 \\ -2 \end{bmatrix} = \begin{bmatrix} -1 \\ -2 \end{bmatrix}$$

Since $\mathbf{B}^{-1}\mathbf{b} \geq \mathbf{0}$, the current dbfs $\mathbf{u} = (0,-1)$ is optimal for the dual problem. The corresponding solution for the primal problem is $\mathbf{x_B} = (x_4, x_3) = \mathbf{B}^{-1}\mathbf{b} = (3, 2)$ and $\mathbf{x_N} = (x_1, x_2, x_5) = (0, 0, 0)$, so $\mathbf{x} = (0, 0, 2, 3, 0)$.

In order to complete the development of the dual simplex algorithm, a procedure must be developed to find an initial dbfs. One might hope for some kind of "phase 1" LP associated with the dual. Ideally, such a phase 1 problem should have two desirable properties:

1. There is an "obvious" dbfs from which to initiate the dual simplex algorithm.
2. An optimal solution to the phase 1 problem should indicate whether the dual LP has a feasible solution or not and, in the former case, indicate how to find an initial dbfs.

Unfortuantley, no such phase 1 LP has been found, and it is in this regard that the primal and dual simplex algorithms differ. The exercises will present several valid approaches for finding an initial dbfs, but none of them is as easy to implement computationally as the corresponding phase 1 procedure for the primal simplex algorithm. For this reason, the dual simplex algorithm is usually used only when the particular dual problem of

TABLE 8.4

Primal simplex algorithm	Dual simplex algorithm
1. An initial bfs is required to start. If none is available, a phase 1 LP with an obvious bfs is solved	1. An initial dbfs is required to start. If none is available, a phase 1 LP is solved
2. Movement is made from one bfs to another. At each step a primal feasible solution is available but a dual feasible solution is not	2. Movement is made from one dbfs to another. At each step a dual feasible solution is available but a primal feasible solution is not
3. Given a bfs $$\mathbf{x} = (\mathbf{x_B}, \mathbf{x_N}) = (\mathbf{B}^{-1}\mathbf{b}, \mathbf{0}) \geq \mathbf{0}$$ the test for optimality consists of using the $\mathbf{B}$ matrix to see if the "potential" dual solution $\mathbf{u} = \mathbf{c_B}\mathbf{B}^{-1}$ is feasible for the dual LP by checking whether $\mathbf{c_N} - \mathbf{c_B}\mathbf{B}^{-1}\mathbf{N} \geq \mathbf{0}$	3. Given a dbfs $$\mathbf{u} = \mathbf{c_B}\mathbf{B}^{-1}$$ the test for optimality consists of using the $\mathbf{B}$ matrix to see if the "potential" primal solution $\mathbf{x_B} = \mathbf{B}^{-1}\mathbf{b}$, $\mathbf{x_N} = \mathbf{0}$, is feasible for the primal LP by checking whether $\mathbf{B}^{-1}\mathbf{b} \geq \mathbf{0}$
4. If the bfs fails the test for optimality, this determines a column to enter the basis by providing a value of j^* with $1 \leq j^* \leq n - m$, for which $(\mathbf{c_N} - \mathbf{c_B}\mathbf{B}^{-1}\mathbf{N})_{j^*} < 0$	4. If the dbfs fails the test for optimality, this determines a column to leave the basis by providing a value of k^* with $1 \leq k^* \leq m$, for which $(\mathbf{B}^{-1}\mathbf{b})_{k^*} < 0$
5. The value of j^* in step 4 is used to create a direction of movement $$\mathbf{d} = (\mathbf{d_B}, \mathbf{d_N}) = (-\mathbf{B}^{-1}\mathbf{N}_{\cdot j^*}, \mathbf{I}_{\cdot j^*})$$ that decreases the primal objective value	5. The value of k^* in step 4 is used to create a direction of movement $$\mathbf{d} = -(\text{row } k^* \text{ of } \mathbf{B}^{-1}) = -\mathbf{I}_{k^*\cdot}\mathbf{B}^{-1}$$ that increases the dual objective value

interest happens to have an obvious dbfs from which to get started. Indeed, many useful problems have this desirable feature. Several such examples will be discussed here and in the next chapter.

Consider, for instance, a primal LP in the following form:

$$\begin{aligned} \text{Minimize} \quad & \mathbf{cx} \\ \text{subject to} \quad & \mathbf{Ax} \leq \mathbf{b} \\ & \mathbf{x} \geq \mathbf{0} \end{aligned}$$

or

$$\begin{aligned} \text{Minimize} \quad & (\mathbf{c}, \mathbf{0})\begin{bmatrix} \mathbf{x} \\ \mathbf{s} \end{bmatrix} \\ \text{subject to} \quad & [\mathbf{A}, \mathbf{I}]\begin{bmatrix} \mathbf{x} \\ \mathbf{s} \end{bmatrix} = \mathbf{b} \\ & \mathbf{x}, \mathbf{s} \geq \mathbf{0} \end{aligned}$$

TABLE 8.4 (*continued*)

Primal simplex algorithm	Dual simplex algorithm
6a. A min ratio test is used to determine how far one can move from $\mathbf{x}$ in the direction $\mathbf{d}$ and remain feasible for the primal. Only the negative components of $\mathbf{d_B}$ play a role in computing the value of t^*	6a. A min ratio test is used to determine how far one can move from $\mathbf{u}$ in the direction $\mathbf{d}$ and remain feasible for the dual. Only the positive components of $\mathbf{dN}$ play a role in computing the value of t^*
6b. If $\mathbf{d_B} \geq \mathbf{0}$, then $t^* = \infty$, and the primal LP is unbounded and the dual LP is infeasible	6b. If $\mathbf{dN} \leq \mathbf{0}$, then $t^* = \infty$ and the dual LP is unbounded and the primal LP is infeasible
6c. If $t^* = 0$, degeneracy has occurred because a primal basic variable has the value 0. Degeneracy can be resolved by Bland's rule	6c. If $t^* = 0$, degeneracy has occurred because $(\mathbf{c_N} - \mathbf{uN})_{j^*}$ has the value 0. Degeneracy can be resolved by Bland's rule
6d. If the primal is not unbounded, the min ratio test provides a value of k^* with $1 \leq k^* \leq m$, that determines the column to leave the basis	6d. If the dual is not unbounded, the min ratio test provides a value of j^* with $1 \leq j^* \leq n - m$, that determines the column to enter the basis
7. The pivot operation is performed by (a) moving from $\mathbf{x}$ to $\mathbf{x} + t^*\mathbf{d}$, (b) forming the new basis matrix $\mathbf{B}'$ by replacing column k^* of $\mathbf{B}$ with column j^* of $\mathbf{N}$, and (c) computing the matrix $\mathbf{B}'^{-1}$	7. The pivot operation is performed by (a) moving from $\mathbf{u}$ to $\mathbf{u} + t^*\mathbf{d}$, (b) forming the new basis matrix $\mathbf{B}'$ by replacing column k^* of $\mathbf{B}$ with column j^* of $\mathbf{N}$, and (c) computing the matrix $\mathbf{B}'^{-1}$

and its associated dual:

$$
\begin{aligned}
\text{Maximize} \quad & \mathbf{ub} \\
\text{subject to} \quad & \mathbf{uA} \leq \mathbf{c} \\
& \mathbf{uI} \leq \mathbf{0}
\end{aligned}
$$

In the event that $\mathbf{c} \geq \mathbf{0}$, then the vector $\mathbf{u} = \mathbf{0}$ is a dbfs for the dual LP. The associated $\mathbf{B}$ matrix is $\mathbf{I}$. Consequently, the dual simplex algorithm can be applied immediately to solve the dual LP, and, at optimality, a solution to the primal LP is also obtained. Notice that if one were to apply the primal simplex algorithm to the primal LP and $\mathbf{b} \geq \mathbf{0}$ were false, then both phase 1 and phase 2 would be needed, whereas only phase 2 is needed for the dual simplex algorithm.

Other uses of the dual simplex algorithm will be presented in the next chapter. For now, a step-by-step comparison of the primal and dual algorithms is in order. It will be found in Table 8.4.

EXERCISES

8.5.1. Solve the following problems using the dual simplex algorithm. Also find the optimal primal solution.

(a)

$$\begin{array}{lrcl} \text{Maximize} & 6u_1 + 12u_2 & & \\ \text{subject to} & 5u_1 + 2u_2 & \leq & 1 \\ & 3u_1 + 7u_2 & \leq & 7 \\ & -u_1 & \leq & 3 \\ & -u_2 & \leq & 2 \end{array}$$

The initial dbfs is $\mathbf{u} = (-3, -2)$.

(b)

$$\begin{array}{lrcl} \text{Maximize} & 3u_1 + 8u_2 & & \\ \text{subject to} & 4u_1 + 3u_2 & \leq & 2 \\ & u_1 + 2u_2 & \leq & 4 \\ & -u_1 & \leq & 3 \\ & -u_2 & \leq & 2 \\ & 2u_1 - u_2 & \leq & 1 \end{array}$$

The initial dbfs is $\mathbf{u} = (-3, -2)$.

8.5.2. Prove that the m vector $\mathbf{d} = -(\text{row } k^* \text{ of } \mathbf{B}^{-1})$ used in the dual simplex algorithm satisfies $\mathbf{db} > 0$.

8.5.3. Prove that if $\mathbf{u}$ is a dbfs that fails the test for optimality in the dual simplex algorithm and $\mathbf{d}$ and t^* are defined accordingly, then, for all t with $0 \leq t \leq t^*$, $\mathbf{u} + t\mathbf{d}$ is feasible for the dual LP.

8.5.4. Discuss how degeneracy arises in the dual simplex algorithm. Specifically, indicate how t^* might be 0. Also describe precisely how Bland's rule could be used in the dual simplex algorithm.

8.5.5. The following method is one way to find an initial dbfs for the dual of an LP in standard form. Let $\mathbf{B}$ be any nonsingular $m \times m$ submatrix of $\mathbf{A}$. Using the initial bfs $\mathbf{x} = (\mathbf{x_B}, \mathbf{x_N}) = (\mathbf{0}, \mathbf{e})$, apply the primal simplex algorithm to the LP:

$$\begin{array}{ll} \text{Minimize} & \mathbf{cx} \\ \text{subject to} & \mathbf{Ax} = \mathbf{Be} \\ & \mathbf{x} \geq \mathbf{0} \end{array}$$

(a) If this LP is unbounded, then what can be said about the status of the original dual (and primal) LP? Explain.

(b) If this LP is optimal, then show that the corresponding optimal dual solution $\mathbf{u^*}$ is an initial dbfs for the original dual LP.

DISCUSSION

Duality Theory

The first person to recognize duality theory in a formal way appears to have been the mathematician J. von Neumann, although he never published a paper on the subject. One of the initial papers on duality is by Gale, Kuhn, and Tucker [3].

Duality is now recognized as being one of the most important aspects of linear programming theory. A generalization of duality theory has led to a test for optimality in nonlinear programming known as the Kuhn–Tucker conditions (see Kuhn and Tucker [4], for example). Similar uses of duality theory have arisen in numerous combinatorial optimization problems, and the study of duality theory has led to the efficient development of numerous primal algorithms.

Dual Simplex and Related Algorithms

The dual simplex algorithm was developed by Lemke [5]. Another algorithm for solving LPs is called the primal–dual algorithm (developed by Dantzig, Ford, and Fulkerson [1]), so named because it works simultaneously on both the primal and dual problems. In any event, good professional simplex codes print out the values of the dual variables at optimality.

Linear Inequalities

Farkas's lemma [2] is one of many theorems of the alternative. A more extensive list and references can be found in the text by Mangasarian [6].

REFERENCES

1. G. B. Dantzig, L. R. Ford, and D. R. Fulkerson, "A Primal–Dual Algorithm for Linear Programs," in *Linear Inequalities and Related Systems* (H. W. Kuhn and A. W. Tucker, eds.), Annals of Mathematics Study No. 38, Princeton University Press, Princeton, NJ, 1956, pp. 171–181.
2. J. Farkas, "Uber die Theorie der Einfachen Ungleichugen," *Journal fur die Reine und Angewandte Mathematik*, 124, 1–27 (1901).
3. D. Gale, H. W. Kuhn, and A. W. Tucker, "Linear Programming and the Theory of Games," in *Activity Analysis of Production and Allocation* (T. C. Koopmans, ed.), Cowles Commission Monograph No. 13, Wiley, New York, 1951, pp. 317–329.
4. H. W. Kuhn and A. W. Tucker, "Nonlinear Programming," in *Proceedings of the Second Berkeley Symposium on Mathematical Statistics and Probability* (J. Neyman, ed.), University of California Press, Berkeley, CA, 1950, pp. 481–492.
5. C. E. Lemke, "The Dual Method of Solving the Linear Programming Problem," *Naval Research Logistics Quarterly* **1**(1), 48–54 (1954).
6. O. L. Mangasarian, *Nonlinear Programming*, McGraw-Hill, New York, 1969.

Chapter 9

Sensitivity and Parametric Analysis

After an LP has been formulated and solved, it is often the case that a change in the original problem has to be made; for example, a new constraint may have to be added, or a cost coefficient may need to be modified. It is, of course, possible to formulate the new LP and to solve it once again from scratch; in most cases, however, doing so is both wasteful and unnecessary. *Sensitivity analysis* (sometimes called *postoptimality analysis*) deals with the issue of using the optimal solution to the original problem as a starting point from which to solve the new problem. Sometimes the original primal solution is used (together with the primal simplex algorithm); at other times it is more convenient to use the original dual solution (together with the dual simplex algorithm).

This chapter will present sensitivity analysis for the following types of modifications to the original LP: (1) changes in the **c** vector, (2) changes in the **b** vector, (3) changes in the **A** matrix, (4) the addition of new variables, and (5) the addition of new constraints. Throughout this chapter, let $\mathbf{x}^* = (\mathbf{x}_B^*, \mathbf{x}_N^*) = (\mathbf{B}^{-1}\mathbf{b}, \mathbf{0}) \geq \mathbf{0}$ be an optimal primal bfs to the original LP and $\mathbf{u}^* = \mathbf{c}_B\mathbf{B}^{-1}$ be the corresponding optimal dual basic feasible solution (dbfs).

9.1. CHANGES IN THE c VECTOR

Changing the Cost Coefficients

After obtaining the optimal solution $\mathbf{x}^*$ to the original LP, it may happen that the decision maker needs to change one or more of the cost coefficients, so the **c** vector will change to **c**′. Hence the new LP to be solved is to

$$\begin{aligned} \text{minimize} \quad & \mathbf{c}'\mathbf{x} \\ \text{subject to} \quad & \mathbf{A}\mathbf{x} = \mathbf{b} \\ & \mathbf{x} \geq \mathbf{0} \end{aligned}$$

In order to use $\mathbf{x}^*$ to solve the new problem, observe that the constraints of both the original and the new problem are the same. Thus $\mathbf{x}^*$ is a bfs for the new problem from which the simplex algorithm can then be initiated to solve the new LP (see Figure 9.1). To solve the new problem starting from $\mathbf{x}^*$, the test for optimality at the first iteration would be to check if $\mathbf{c}'_N - \mathbf{c}'_B\mathbf{B}^{-1}\mathbf{N} \geq \mathbf{0}$. If so, then $\mathbf{x}^*$ is optimal for the new LP; otherwise, a direction of movement can be computed and the remaining steps of the algorithm carried out as usual. In a finite number of iterations, the simplex algorithm will either determine that the new LP is unbounded or else produce an optimal solution, as illustrated in the next example.

EXAMPLE 9.1: Consider Example 5.2:

$$\begin{aligned} \text{Minimize} \quad & -2x_1 - 3x_2 - x_3 \\ \text{subject to} \quad & x_1 - x_2 + 2x_3 + x_4 = 1 \\ & 4x_1 + 2x_2 - x_3 + x_5 = 2 \\ & x_1, \; x_2, \; x_3, \; x_4, \; x_5 \geq 0 \end{aligned}$$

The optimal solution was found to be $(0, \frac{5}{3}, \frac{4}{3}, 0, 0)$ with an objective function value of $-\frac{19}{3}$. Suppose that the **c** vector is changed to $\mathbf{c}' = (-12, -3, -1, 0, 0)$.

FIGURE 9.1. Changing the cost vector.

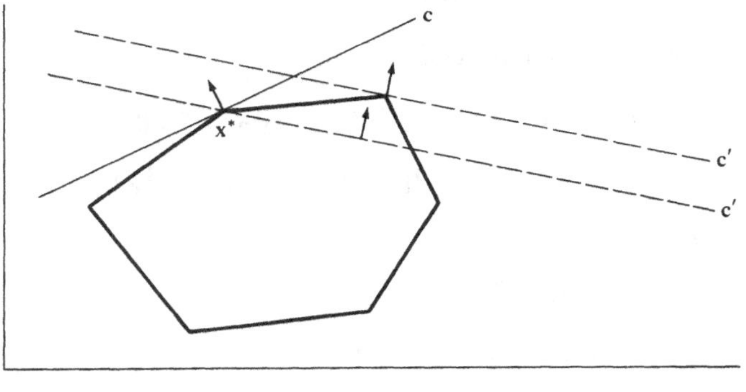

Iteration 1

Step 0. Initialization: Start from the optimal basis of the previous solution. The bfs for this iteration:

$$\mathbf{B} = [\mathbf{A}_{\cdot 3}, \mathbf{A}_{\cdot 2}] = \begin{bmatrix} 2 & -1 \\ -1 & 2 \end{bmatrix}$$

$$\mathbf{N} = [\mathbf{A}_{\cdot 1}, \mathbf{A}_{\cdot 4}, \mathbf{A}_{\cdot 5}] = \begin{bmatrix} 1 & 1 & 0 \\ 4 & 0 & 1 \end{bmatrix}$$

$$\mathbf{x_B} = \begin{bmatrix} x_3 \\ x_2 \end{bmatrix} = \mathbf{B}^{-1}\mathbf{b} = \begin{bmatrix} \frac{2}{3} & \frac{1}{3} \\ \frac{1}{3} & \frac{2}{3} \end{bmatrix}\begin{bmatrix} 1 \\ 2 \end{bmatrix} = \begin{bmatrix} \frac{4}{3} \\ \frac{5}{3} \end{bmatrix}$$

$$\mathbf{x_N} = (x_1, x_4, x_5) = (0,0,0)$$

Step 1. Testing for optimality:

$$\mathbf{c}'_{\mathbf{N}} - \mathbf{c}'_{\mathbf{B}}\mathbf{B}^{-1}\mathbf{N} = (-12,0,0) - (-1,-3)\begin{bmatrix} \frac{2}{3} & \frac{1}{3} \\ \frac{1}{3} & \frac{2}{3} \end{bmatrix}\begin{bmatrix} 1 & 1 & 0 \\ 4 & 0 & 1 \end{bmatrix}$$

$$= (-12,0,0) - (-\tfrac{5}{3}, -\tfrac{7}{3})\begin{bmatrix} 1 & 1 & 0 \\ 4 & 0 & 1 \end{bmatrix}$$

$$= (-12,0,0) - (-11, -\tfrac{5}{3}, -\tfrac{7}{3})$$

$$= (-1, \tfrac{5}{3}, \tfrac{7}{3}) \qquad \text{so} \quad j^* = 1$$

Step 2. Computing the direction of movement:

$$\mathbf{d_B} = \begin{bmatrix} d_3 \\ d_2 \end{bmatrix} = -\mathbf{B}^{-1}\mathbf{N}_{\cdot j^*} = -\begin{bmatrix} \frac{2}{3} & \frac{1}{3} \\ \frac{1}{3} & \frac{2}{3} \end{bmatrix}\begin{bmatrix} 1 \\ 4 \end{bmatrix} = \begin{bmatrix} -2 \\ -3 \end{bmatrix}$$

$$\mathbf{d_N} = (d_1, d_4, d_5) = \mathbf{I}_{\cdot j^*} = \mathbf{I}_{\cdot 1} = (1,0,0)$$

$$\mathbf{d} = (1, -3, -2, 0, 0)$$

Step 3. Computing the amount of movement:

$$t^* = \min\{-(\mathbf{x_B})_k/(\mathbf{d_B})_k : (\mathbf{d_B})_k < 0\}$$

$$= \min\{-(\mathbf{x_B})_1/(\mathbf{d_B})_1, -(\mathbf{x_B})_2/(\mathbf{d_B})_2\}$$

$$= \min\{-\tfrac{4}{3}/-2, -\tfrac{5}{3}/-3\}$$

$$= \min\{\tfrac{2}{3}, \tfrac{5}{9}\}$$

$$= \tfrac{5}{9} \qquad \text{so} \quad k^* = 2$$

Step 4. Pivoting: Column $k^* = 2$ of $\mathbf{B}$ is replaced with column $j^* = 1$ of $\mathbf{N}$, so

$$\mathbf{B}' = \begin{bmatrix} 2 & 1 \\ -1 & 4 \end{bmatrix}$$

$$\mathbf{N}' = \begin{bmatrix} -1 & 1 & 0 \\ 2 & 0 & 1 \end{bmatrix}$$

$$\mathbf{x}' = \mathbf{x} + t^*\mathbf{d} = (0, \tfrac{5}{3}, \tfrac{4}{3}, 0, 0) + \tfrac{5}{9}(1, -3, -2, 0, 0)$$
$$= (\tfrac{5}{9}, 0, \tfrac{2}{9}, 0, 0)$$

Iteration 2

Step 0. The bfs for this iteration:

$$\mathbf{B} = [\mathbf{A}_{.3}, \mathbf{A}_{.1}] = \begin{bmatrix} 2 & 1 \\ -1 & 4 \end{bmatrix}$$

$$\mathbf{N} = [\mathbf{A}_{.2}, \mathbf{A}_{.4}, \mathbf{A}_{.5}] = \begin{bmatrix} -1 & 1 & 0 \\ 2 & 0 & 1 \end{bmatrix}$$

$$\mathbf{x_B} = \begin{bmatrix} x_3 \\ x_1 \end{bmatrix} = \mathbf{B}^{-1}\mathbf{b} = \begin{bmatrix} \frac{4}{9} & -\frac{1}{9} \\ \frac{1}{9} & \frac{2}{9} \end{bmatrix} \begin{bmatrix} 1 \\ 2 \end{bmatrix} = \begin{bmatrix} \frac{2}{9} \\ \frac{5}{9} \end{bmatrix}$$

$$\mathbf{x_N} = (x_2, x_4, x_5) = (0, 0, 0)$$
$$\mathbf{x} = (\tfrac{5}{9}, 0, \tfrac{2}{9}, 0, 0)$$

Step 1. Testing for optimality:

$$\mathbf{c}'_\mathbf{N} - \mathbf{c}'_\mathbf{B}\mathbf{B}^{-1}\mathbf{N} = (-3, 0, 0) - (-1, -12)\begin{bmatrix} \frac{4}{9} & -\frac{1}{9} \\ \frac{1}{9} & \frac{2}{9} \end{bmatrix}\begin{bmatrix} -1 & 1 & 0 \\ 2 & 0 & 1 \end{bmatrix}$$
$$= (-3, 0, 0) - (-\tfrac{16}{9}, -\tfrac{23}{9})\begin{bmatrix} -1 & 1 & 0 \\ 2 & 0 & 1 \end{bmatrix}$$
$$= (-3, 0, 0) - (-\tfrac{30}{9}, -\tfrac{16}{9}, -\tfrac{23}{9})$$
$$= (\tfrac{3}{9}, \tfrac{16}{9}, \tfrac{23}{9})$$

Since $\mathbf{c}'_\mathbf{N} - \mathbf{c}'_\mathbf{B}\mathbf{B}^{-1}\mathbf{N} \geq \mathbf{0}$, the current solution $\mathbf{x} = (\frac{5}{9}, 0, \frac{2}{9}, 0, 0)$ is optimal. The objective value is $\mathbf{cx} = -\frac{62}{9}$.

Sensitivity of the Cost Coefficients

A related question might be asked by a decision maker after the original LP has been solved: "By how much can I change one particular cost coefficient without changing the current optimal solution $\mathbf{x}^*$?" In other words, how sensitive is the current bfs to changes in the cost vector? To answer this question, think of the particular cost coefficient of interest as a variable instead of a fixed number. As its value changes, so do the reduced costs associated with $\mathbf{x}^*$. As long as the reduced costs remain nonnegative, $\mathbf{x}^*$ remains optimal. In determining the range of values over which a cost coefficient can be varied without affecting the optimal solution $\mathbf{x}^*$, two cases arise.

Case 1: The cost coefficient under consideration corresponds to a nonbasic component, say, $(\mathbf{c_N})_j$.

In this case, the basic components $\mathbf{c_B}$ do not change, and hence neither does $\mathbf{c_B B^{-1} N}$. Think of $(\mathbf{c_N})_j$ as a variable. As long as $(\mathbf{c_N})_j \geq (\mathbf{c_B B^{-1} N})_j = (\mathbf{c_B B^{-1}})\mathbf{N}_{.j}$, the reduced costs remain nonnegative. In other words, for any value $(\mathbf{c_N})_j \geq (\mathbf{c_B B^{-1}})\mathbf{N}_{.j}$, the solution $\mathbf{x}^*$ remains optimal.

Case 2: The cost coefficient under consideration corresponds to a basic component, say, $(\mathbf{c_B})_k$.

In this case, $\mathbf{c_N}$ does not change, so, in order for the reduced costs to remain nonnegative, it is necessary to determine the values of $(\mathbf{c_B})_k$ for which $\mathbf{c_N} \geq \mathbf{c_B}(\mathbf{B}^{-1}\mathbf{N})$. As you will show (Exercise 9.1.2), as long as $(\mathbf{c_B})_k$ is between some lower bound t_1 and some upper bound t_2, the reduced costs remain nonnegative. In other words, if $t_1 \leq (\mathbf{c_B})_k \leq t_2$, the current bfs $\mathbf{x}^*$ remains optimal.

The Parametric Cost Problem

The final problem to be discussed in this section is known as the *parametric cost problem*. For a given original cost vector $\mathbf{c}$ together with another n vector $\mathbf{c}'$, the goal is to solve the following parametric LP for each value of $s \geq 0$:

$$\begin{aligned} \text{Minimize} \quad & (\mathbf{c} + s\mathbf{c}')\mathbf{x} \\ \text{subject to} \quad & \mathbf{Ax} = \mathbf{b} \\ & \mathbf{x} \geq \mathbf{0} \end{aligned}$$

Since there are an infinite number of values for s, there are an infinite number of problems to solve; yet, because there are only a finite number of bfs associated with the $\mathbf{A}$ matrix, it is possible to develop a finite solution procedure. Specifically, consider the set consisting of all nonnegative real numbers as representing the possible values for s. The solution procedure "partitions" this interval into a finite number of subintervals of the form

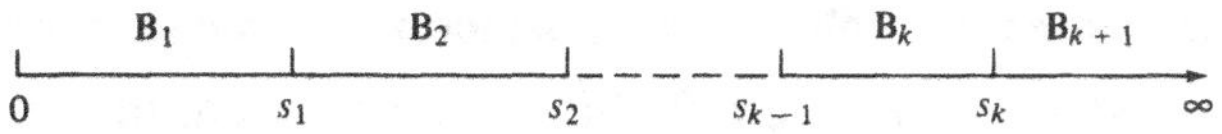

FIGURE 9.2. Intervals for the parametric LP.

$0 < s_1 < \cdots < s_k < \infty$. Associated with each interval $[s_{i-1}, s_i]$ is a basis matrix $\mathbf{B}_i$ such that, for all values of s between s_{i-1} and s_i, the optimal values of the basic variables for the parametric LP can be computed as $\mathbf{B}_i^{-1}\mathbf{b}$ (see Figure 9.2).

Consider the following finite improvement algorithm for finding the values for $s_1, \ldots, s_k$ and the associated basis matrices $\mathbf{B}_1, \ldots, \mathbf{B}_{k+1}$.

Step 0. Initialization

Set $s = 0$ and solve the original LP:

$$\begin{aligned} \text{Minimize} \quad & \mathbf{cx} \\ \text{subject to} \quad & \mathbf{Ax} = \mathbf{b} \\ & \mathbf{x} \geq \mathbf{0} \end{aligned}$$

As with any LP, there are three possibilities:

Case 1: The LP is infeasible.

In this case, the parametric LP is infeasible for all values of s because, as s changes, the constraints of the parametric LP do not change.

Case 2: The LP is unbounded.

In this case, the simplex algorithm has produced a current bfs, $\mathbf{x}$, together with a direction of unboundedness $\mathbf{d}$ that satisfies $\mathbf{d} \geq \mathbf{0}$, $\mathbf{Ad} = \mathbf{0}$, and $\mathbf{cd} < 0$. As s increases from its current value of 0, the corresponding cost vector changes from $\mathbf{c}$ to $\mathbf{c} + s\mathbf{c}'$. It might happen that if s is made large enough, then $\mathbf{d}$ is no longer a direction of unboundedness. In other words, it may happen that if s is increased enough, then the condition of unboundedness for the parametric LP no longer applies. The details of computing the appropriate values for s are left to Exercise 9.1.4.

Case 3: An optimal bfs $\mathbf{x}^* = (\mathbf{x}_\mathbf{B}^*, \mathbf{x}_\mathbf{N}^*) = (\mathbf{B}^{-1}\mathbf{b}, \mathbf{0}) \geq \mathbf{0}$ is obtained.

In this case, a basis matrix has been obtained for the first interval. The next step of the procedure determines the right endpoint for the interval.

Step 1. Determining the Value of s for Which $\mathbf{x}^*$ Remains Optimal

At the end of step 0, if the optimal bfs $\mathbf{x}^*$ is produced, then the associated reduced costs satisfy $\mathbf{c_N} - \mathbf{c_B B^{-1} N} \geq \mathbf{0}$. As s increases from its current value of 0, the cost vector of the parametric LP changes from $\mathbf{c}$ to $\mathbf{c} + s\mathbf{c}'$, and hence the reduced costs also change. In particular, $\mathbf{c_N}$ changes to $(\mathbf{c} + s\mathbf{c}')_\mathbf{N}$ and $\mathbf{c_B}$ changes to $(\mathbf{c} + s\mathbf{c}')_\mathbf{B}$, so the reduced costs change from $\mathbf{c_N} - \mathbf{c_B B^{-1} N}$ to $(\mathbf{c_N} + s\mathbf{c'_N}) - (\mathbf{c_B} + s\mathbf{c'_B})\mathbf{B^{-1} N}$. Therefore, on setting $\mathbf{y} = \mathbf{c_N} - \mathbf{c_B B^{-1} N}$ and $\mathbf{y}' = \mathbf{c'_N} - \mathbf{c'_B B^{-1} N}$, as long as s is chosen so that $\mathbf{y} + s\mathbf{y}' \geq \mathbf{0}$, the current bfs, $\mathbf{x}^*$, must remain optimal. In other words, the right endpoint, say, s^*, of the interval is the largest value of s for which $\mathbf{y} + s\mathbf{y}' \geq \mathbf{0}$.

To compute the value for s^*, observe that if $\mathbf{y}' \geq \mathbf{0}$, then, for all values of $s \geq 0$, $\mathbf{y} + s\mathbf{y}' \geq \mathbf{0}$, and hence $s^* = \infty$. On the other hand, if it is not the case that $\mathbf{y}' \geq \mathbf{0}$, then s^* can be computed by the formula

$$s^* = \min\left\{ -y_j/y'_j : 1 \leq j \leq n - m \text{ and } y'_j < 0 \right\} = -y_{j*}/y'_{j*}$$

Notice the similarity to the min ratio test.

Thus, for all values of $0 \leq s \leq s^*$, the bfs $\mathbf{x}^*$ is optimal for the parametric LP. The final step of the procedure describes how to "move into" the next interval and how to find the new basis matrix.

Step 2. Moving to a New bfs

For values of s "slightly" larger than s^*, component j^* of the reduced costs becomes negative (because $s^* = -y_{j*}/y'_{j*}$ was chosen in this way). Hence it would seem logical to attempt to bring $\mathbf{N}_{.j*}$ into the basis. To do so, one would compute $\mathbf{d} = (\mathbf{d_B}, \mathbf{d_N}) = (-\mathbf{B}^{-1}\mathbf{N}_{.j*}, \mathbf{I}_{.j*})$ as usual. If $\mathbf{d_B} \geq \mathbf{0}$, then, for all values of $s \geq s^*$, the parametric LP is unbounded (see Exercise 9.1.5). Otherwise, a column k^* can be determined to leave the basis by the usual min ratio test. After the pivot operation is performed, the $\mathbf{c}$ vector must be modified to $\mathbf{c} + s^*\mathbf{c}'$, and the entire process can then be repeated by returning to step 1, thus determining each successive interval and the corresponding bfs. As long as $s^* > 0$ at each step, the finiteness of the procedure is ensured because each basis matrix can be optimal over at most one interval (see Exercise 9.1.6). The process of solving the parametric LP will now be demonstrated with an example.

EXAMPLE 9.2: Consider Example 5.2 with the $\mathbf{c}$ vector modified as follows:

$$\begin{array}{lrl}
\text{Minimize} & & (-2x_1 - 3x_2 - x_3) + s(-10x_1 - 3x_2 + x_3 - x_4) \\
\text{subject to} & & x_1 - x_2 + 2x_3 + x_4 = 1 \\
& & 4x_1 + 2x_2 - x_3 + x_5 = 2 \\
& & x_1,\ x_2,\ x_3,\ x_4,\ x_5 \geq 0
\end{array}$$

That is,

$$\mathbf{c} = (-2, -3, -1, 0, 0)$$
$$\mathbf{c}' = (-10, -3, 1, -1, 0)$$
$$\mathbf{A} = \begin{bmatrix} 1 & -1 & 2 & 1 & 0 \\ 4 & 2 & -1 & 0 & 1 \end{bmatrix}$$
$$\mathbf{b} = \begin{bmatrix} 1 \\ 2 \end{bmatrix}$$

Iteration 1

Step 0. Initialization: At $s = 0$ the LP is precisely that of Example 5.2, and the optimal bfs for that problem is

$$\mathbf{B} = [\mathbf{A}_{.3}, \mathbf{A}_{.2}] = \begin{bmatrix} 2 & -1 \\ -1 & 2 \end{bmatrix}$$
$$\mathbf{N} = [\mathbf{A}_{.1}, \mathbf{A}_{.5}, \mathbf{A}_{.4}] = \begin{bmatrix} 1 & 0 & 1 \\ 4 & 1 & 0 \end{bmatrix}$$
$$\mathbf{x_B} = \begin{bmatrix} x_3 \\ x_2 \end{bmatrix} = \mathbf{B}^{-1}\mathbf{b} = \begin{bmatrix} \frac{2}{3} & \frac{1}{3} \\ \frac{1}{3} & \frac{2}{3} \end{bmatrix} \begin{bmatrix} 1 \\ 2 \end{bmatrix} = \begin{bmatrix} \frac{4}{3} \\ \frac{5}{3} \end{bmatrix}$$
$$\mathbf{x_N} = (x_1, x_5, x_4) = (0, 0, 0)$$
$$\mathbf{x} = (0, \tfrac{5}{3}, \tfrac{4}{3}, 0, 0)$$

Step 1. Determining the values of s for which the current solution is optimal:

$$\mathbf{y} = \mathbf{c_N} - \mathbf{c_B}\mathbf{B}^{-1}\mathbf{N} = (-2, 0, 0) - (-1, -3)\begin{bmatrix} \frac{2}{3} & \frac{1}{3} \\ \frac{1}{3} & \frac{2}{3} \end{bmatrix}\begin{bmatrix} 1 & 0 & 1 \\ 4 & 1 & 0 \end{bmatrix}$$
$$= (-2, 0, 0) - (-\tfrac{5}{3}, -\tfrac{7}{3})\begin{bmatrix} 1 & 0 & 1 \\ 4 & 1 & 0 \end{bmatrix}$$
$$= (-2, 0, 0) - (-11, -\tfrac{7}{3}, -\tfrac{5}{3})$$
$$= (9, \tfrac{7}{3}, \tfrac{5}{3})$$

$$\mathbf{y}' = \mathbf{c}'_{\mathbf{N}} - \mathbf{c}'_{\mathbf{B}}\mathbf{B}^{-1}\mathbf{N} = (-10, 0, -1) - (1, -3)\begin{bmatrix} \frac{2}{3} & \frac{1}{3} \\ \frac{1}{3} & \frac{2}{3} \end{bmatrix}\begin{bmatrix} 1 & 0 & 1 \\ 4 & 1 & 0 \end{bmatrix}$$
$$= (-10, 0, -1) - (-\tfrac{1}{3}, -\tfrac{5}{3})\begin{bmatrix} 1 & 0 & 1 \\ 4 & 1 & 0 \end{bmatrix}$$
$$= (-10, 0, -1) - (-7, -\tfrac{5}{3}, -\tfrac{1}{3})$$
$$= (-3, \tfrac{5}{3}, -\tfrac{2}{3})$$

$$s^* = \min\{-y_j/y'_j : y'_j < 0\} = \min\{-y_1/y'_1, -y_3/y'_3\}$$
$$= \min\{-9/-3, -\tfrac{5}{3}/-\tfrac{2}{3}\} = \min\{3, \tfrac{5}{2}\}$$
$$= \tfrac{5}{2} \qquad \text{so} \quad j^* = 3$$

Thus, for all values of $0 \le s \le \frac{5}{2}$, the current solution $\mathbf{x}^* = (0, \frac{5}{3}, \frac{4}{3}, 0, 0)$ is optimal.

Step 2. Moving to a new bfs:

a. Computing the direction of movement:

$$\mathbf{d_B} = \begin{bmatrix} d_3 \\ d_2 \end{bmatrix} = -\mathbf{B}^{-1}\mathbf{N}_{.j^*} = -\begin{bmatrix} \frac{2}{3} & \frac{1}{3} \\ \frac{1}{3} & \frac{2}{3} \end{bmatrix}\begin{bmatrix} 1 \\ 0 \end{bmatrix} = \begin{bmatrix} -\frac{2}{3} \\ -\frac{1}{3} \end{bmatrix}$$

$$\mathbf{d_N} = (d_1, d_5, d_4) = \mathbf{I}_{.j^*} = \mathbf{I}_{.3} = (0,0,1)$$

$$\mathbf{d} = (0, -\tfrac{1}{3}, -\tfrac{2}{3}, 1, 0)$$

b. Computing the amount of movement t^*:

$$\begin{aligned} t^* &= \min\{-(\mathbf{x_B})_k/(\mathbf{d_B})_k : (\mathbf{d_B})_k < 0\} \\ &= \min\{-\tfrac{4}{3}/-\tfrac{2}{3}, -\tfrac{5}{3}/-\tfrac{1}{3}\} \\ &= \min\{2,5\} \\ &= 2 \qquad \text{so} \quad k^* = 1 \end{aligned}$$

c. Pivoting: Column $k^* = 1$ of $\mathbf{B}$ is replaced with column $j^* = 3$ of $\mathbf{N}$, so

$$\mathbf{B}' = [\mathbf{A}_{.4}, \mathbf{A}_{.2}] = \begin{bmatrix} 1 & -1 \\ 0 & 2 \end{bmatrix}$$

$$\mathbf{N}' = [\mathbf{A}_{.1}, \mathbf{A}_{.5}, \mathbf{A}_{.3}] = \begin{bmatrix} 1 & 0 & 2 \\ 4 & 1 & -1 \end{bmatrix}$$

$$\begin{aligned} \mathbf{x}' &= \mathbf{x} + t^*\mathbf{d} \\ &= (0, \tfrac{5}{3}, \tfrac{4}{3}, 0, 0) + 2(0, -\tfrac{1}{3}, -\tfrac{2}{3}, 1, 0) \\ &= (0,1,0,2,0) \end{aligned}$$

The new $\mathbf{c}$ vector is

$$\begin{aligned} \mathbf{c} &= \mathbf{c} + s^*\mathbf{c}' \\ &= (-2, -3, -1, 0, 0) + \tfrac{5}{2}(-10, -3, 1, -1, 0) \\ &= (-27, -\tfrac{21}{2}, \tfrac{3}{2}, -\tfrac{5}{2}, 0) \end{aligned}$$

Iteration 2

Step 1. Determining the values of s for which the current solution is optimal:

$$\mathbf{y} = \mathbf{c_N} - \mathbf{c_B}\mathbf{B}^{-1}\mathbf{N} = (-27, 0, \tfrac{3}{2}) - (-\tfrac{5}{2}, -\tfrac{21}{2})\begin{bmatrix} 1 & \frac{1}{2} \\ 0 & \frac{1}{2} \end{bmatrix}\begin{bmatrix} 1 & 0 & 2 \\ 4 & 1 & -1 \end{bmatrix}$$

$$\begin{aligned} &= (-27, 0, \tfrac{3}{2}) - (-\tfrac{5}{2}, -\tfrac{13}{2})\begin{bmatrix} 1 & 0 & 2 \\ 4 & 1 & -1 \end{bmatrix} \\ &= (-27, 0, \tfrac{3}{2}) - (-\tfrac{57}{2}, -\tfrac{13}{2}, \tfrac{3}{2}) \\ &= (\tfrac{3}{2}, \tfrac{13}{2}, 0) \end{aligned}$$

$$\mathbf{y}' = \mathbf{c}'_{\mathbf{N}} - \mathbf{c}'_{\mathbf{B}}\mathbf{B}^{-1}\mathbf{N} = (-10,0,1) - (-1,-3)\begin{bmatrix} 1 & \frac{1}{2} \\ 0 & \frac{1}{2} \end{bmatrix}\begin{bmatrix} 1 & 0 & 2 \\ 4 & 1 & -1 \end{bmatrix}$$

$$= (-10,0,1) - (-1,-2)\begin{bmatrix} 1 & 0 & 2 \\ 4 & 1 & -1 \end{bmatrix}$$

$$= (-10,0,1) - (-9,-2,0)$$

$$= (-1,2,1)$$

$$s^* = \min\{-y_j/y'_j : y'_j < 0\} = \min\{-y_1/y'_1\} = \min\{-\tfrac{3}{2}/-1\}$$

$$= \tfrac{3}{2} \qquad \text{so} \quad j^* = 1$$

Thus, for all values of $\frac{5}{2} \le s \le \frac{5}{2} + \frac{3}{2}$, or $\frac{5}{2} \le s \le 4$, the current solution $\mathbf{x}^* = (0,1,0,2,0)$ is optimal.

Step 2. Moving to a new bfs:

a. Computing the direction of movement:

$$\mathbf{d}_{\mathbf{B}} = \begin{bmatrix} d_4 \\ d_2 \end{bmatrix} = -\mathbf{B}^{-1}\mathbf{N}_{\cdot j^*} = -\begin{bmatrix} 1 & \frac{1}{2} \\ 0 & \frac{1}{2} \end{bmatrix}\begin{bmatrix} 1 \\ 4 \end{bmatrix} = \begin{bmatrix} -3 \\ -2 \end{bmatrix}$$

$$\mathbf{d}_{\mathbf{N}} = (d_1, d_5, d_3) = \mathbf{I}_{\cdot j^*} = \mathbf{I}_{\cdot 1} = (1,0,0)$$

$$\mathbf{d} = (1,-2,0,-3,0)$$

b. Computing the amount of movement t^*:

$$t^* = \min\{-(\mathbf{x}_{\mathbf{B}})_k/(\mathbf{d}_{\mathbf{B}})_k : (\mathbf{d}_{\mathbf{B}})_k < 0\}$$

$$= \min\{-2/-3, -1/-2\}$$

$$= \min\{\tfrac{2}{3}, \tfrac{1}{2}\}$$

$$= \tfrac{1}{2} \qquad \text{so} \quad k^* = 2$$

c. Pivoting: Column $k^* = 2$ of **B** is replaced with column $j^* = 1$ of **N**, so

$$\mathbf{B}' = [\mathbf{A}_{\cdot 4}, \mathbf{A}_{\cdot 1}] = \begin{bmatrix} 1 & 1 \\ 0 & 4 \end{bmatrix}$$

$$\mathbf{N}' = [\mathbf{A}_{\cdot 2}, \mathbf{A}_{\cdot 5}, \mathbf{A}_{\cdot 3}] = \begin{bmatrix} -1 & 0 & 2 \\ 2 & 1 & -1 \end{bmatrix}$$

$$\mathbf{x}' = \mathbf{x} + t^*\mathbf{d}$$

$$= (0,1,0,2,0) + \tfrac{1}{2}(1,-2,0,-3,0)$$

$$= (\tfrac{1}{2},0,0,\tfrac{1}{2},0)$$

The new **c** vector is equal to the previous **c** plus $s^*\mathbf{c}'$:

$$\mathbf{c} = (-27, -\tfrac{21}{2}, \tfrac{3}{2}, -\tfrac{5}{2}, 0) + \tfrac{3}{2}(-10,-3,1,-1,0)$$

$$= (-42,-15,3,-4,0)$$

Iteration 3

Step 1. Determining the values of s for which the current solution is optimal:

$$\mathbf{y} = \mathbf{c}_\mathbf{N} - \mathbf{c}_\mathbf{B}\mathbf{B}^{-1}\mathbf{N} = (-15,0,3) - (-4,-42)\begin{bmatrix} 1 & -\frac{1}{4} \\ 0 & \frac{1}{4} \end{bmatrix}\begin{bmatrix} -1 & 0 & 2 \\ 2 & 1 & -1 \end{bmatrix}$$

$$= (-15,0,3) - (-4, -\tfrac{19}{2})\begin{bmatrix} -1 & 0 & 2 \\ 2 & 1 & -1 \end{bmatrix}$$

$$= (-15,0,3) - (-15, -\tfrac{19}{2}, \tfrac{3}{2})$$

$$= (0, \tfrac{19}{2}, \tfrac{3}{2})$$

$$\mathbf{y}' = \mathbf{c}'_\mathbf{N} - \mathbf{c}'_\mathbf{B}\mathbf{B}^{-1}\mathbf{N} = (-3,0,1) - (-1,-10)\begin{bmatrix} 1 & -\frac{1}{4} \\ 0 & \frac{1}{4} \end{bmatrix}\begin{bmatrix} -1 & 0 & 2 \\ 2 & 1 & -1 \end{bmatrix}$$

$$= (-3,0,1) - (-1, -\tfrac{9}{4})\begin{bmatrix} -1 & 0 & 2 \\ 2 & 1 & -1 \end{bmatrix}$$

$$= (-3,0,1) - (-\tfrac{7}{2}, -\tfrac{9}{4}, \tfrac{1}{4})$$

$$= (\tfrac{1}{2}, \tfrac{9}{4}, \tfrac{3}{4})$$

Since $\mathbf{y}' \geq \mathbf{0}$, for all values of $s \geq 4$, the current bfs $\mathbf{x} = (\frac{1}{2},0,0,\frac{1}{2},0)$ is optimal.

This section has shown how to use the optimal bfs for an LP in standard form, to solve a new LP in which some (or all) of the cost coefficients are modified. In each case, the primal simplex algorithm is initiated from $\mathbf{x}^*$. In the next section, similar problems associated with changes in the $\mathbf{b}$ vector will be discussed. There, however, the dual simplex algorithm will be initiated from the dual solution $\mathbf{u}^*$.

EXERCISES

9.1.1. Consider the problem given in Exercise 5.1.2,

$$\begin{array}{ll} \text{Minimize} & -2x_1 - x_2 + x_3 + 5x_4 - 3x_5 \\ \text{subject to} & 4x_1 + 2x_2 + x_3 + x_4 = 3 \\ & 2x_1 + 2x_2 + 3x_4 + x_5 = 2 \\ & x_1, \quad x_2, \quad x_3, \quad x_4, \quad x_5 \geq 0 \end{array}$$

for which the optimal solution $\mathbf{x}^* = (0,0,3,0,2)$. If the objective function is changed to $\mathbf{c}' = (-4,-3,2,5,-1)$, then find the new optimal solution using $\mathbf{x}^*$ as an initial bfs.

9.1.2. **(a)** In the original LP of Exercise 9.1.1, by how much can $c_2 = -1$ be varied before the optimal solution is changed?

(b) Use trial and error to determine how much $c_5 = -3$ can be increased and decreased in the original LP of Exercise 9.1.1 before the optimal bfs $\mathbf{x}^* = (0, 0, 3, 0, 2)$ needs to be changed.

(c) Given an optimal bfs $\mathbf{x}^*$ to an LP, and given k with $1 \le k \le m$, determine the values of t_1 and t_2 so that, as long as the value of $(\mathbf{c_B})_k$ is between t_1 and t_2, $\mathbf{x}^*$ remains optimal.

9.1.3. Solve the following parametric LP:

$$\begin{aligned} \text{Minimize} \quad & 2x_1 - x_2 - 4x_3 + x_4 + x_5 + s(-3x_1 + 2x_2 - 4x_3 - x_4 - x_5) \\ \text{subject to} \quad & 2x_1 + 3x_2 + 3x_3 + x_4 = 23 \\ & 3x_1 + 4x_2 - x_3 + x_5 = 31 \\ & x_1, \quad x_2, \quad x_3, \quad x_4, \quad x_5 \ge 0 \end{aligned}$$

It has the optimal solution $\mathbf{x}^* = (0, \frac{23}{3}, 0, 0, \frac{1}{3})$ when $s = 0$.

9.1.4. Consider the following parametric LP:

$$\begin{aligned} \text{Minimize} \quad & (\mathbf{c} + s\mathbf{c}')\mathbf{x} \\ \text{subject to} \quad & \mathbf{A}\mathbf{x} = \mathbf{b} \\ & \mathbf{x} \ge \mathbf{0} \end{aligned}$$

Suppose that at $s = 0$ the LP is unbounded. Derive a formula for determining how large s can be made before the parametric LP is no longer unbounded.

9.1.5. In the solution procedure for the parametric LP, show that if, in step 2, $\mathbf{d_B} \ge \mathbf{0}$, then, for all $s \ge s^*$, the parametric LP is unbounded.

9.1.6. Suppose that the parametric LP in Exercise 9.1.4 is bounded when $s = 0$. Let $\mathbf{B}$ be the optimal basis matrix for the interval $[0, s^*]$, and let $\mathbf{B}_1$ be the optimal basis matrix for the next interval. Show that, for all $s > s^*$, $\mathbf{B}$ is not optimal.

9.2. CHANGES IN THE b VECTOR

Changing the Right-Hand-Side Coefficients

As in the previous section, suppose that $\mathbf{x}^* = (\mathbf{x}_\mathbf{B}^*, \mathbf{x}_\mathbf{N}^*) = (\mathbf{B}^{-1}\mathbf{b}, \mathbf{0}) \ge \mathbf{0}$ is an optimal bfs for an LP in standard form. Also, let $\mathbf{u}^* = \mathbf{c_B}\mathbf{B}^{-1}$ be the associated optimal dbfs (dual basic feasible solution—see Section 8.5). After one has obtained these optimal solutions, it may happen that one or more of the components of $\mathbf{b}$ need to be changed. If $\mathbf{b}'$ is the modified vector, then the new LP to be solved is to

$$\begin{aligned} \text{minimize} \quad & \mathbf{cx} \\ \text{subject to} \quad & \mathbf{Ax} = \mathbf{b}' \\ & \mathbf{x} \geq \mathbf{0} \end{aligned}$$

One of the difficulties that can arise in using $\mathbf{x}^*$ as a starting point from which to solve the new problem is that $\mathbf{x}^*$ might not be feasible for the new LP since the constraints have changed. To avoid the necessity of having to use the phase 1 procedure to reestablish feasibility, an alternative method has been found using the dual simplex algorithm. Consider the original dual LP,

$$\begin{aligned} \text{Maximize} \quad & \mathbf{ub} \\ \text{subject to} \quad & \mathbf{uA} \leq \mathbf{c} \\ & \mathbf{u} \text{ unrestricted} \end{aligned}$$

and the associated new dual LP,

$$\begin{aligned} \text{Maximize} \quad & \mathbf{ub}' \\ \text{subject to} \quad & \mathbf{uA} \leq \mathbf{c} \\ & \mathbf{u} \text{ unrestricted} \end{aligned}$$

It is important to note that the constraints of both of the dual problems are identical. Thus, if $\mathbf{u}^*$ is an optimal dbfs for the original dual LP, then $\mathbf{u}^*$ is certainly a dbfs for the new dual LP. Thus the dual simplex algorithm can be applied to the new dual LP by starting with the initial dbfs $\mathbf{u}^*$. At the first iteration, the test for optimality would be to check whether $\mathbf{B}^{-1}\mathbf{b}' \geq \mathbf{0}$. If so, then $\mathbf{u}^*$ is optimal for the new dual problem (and hence $\mathbf{B}^{-1}\mathbf{b}'$ are the values of the optimal basic variables for the new primal LP). Otherwise, a direction of movement can be computed and the remaining steps of the procedure carried out. In a finite number of iterations, the dual simplex algorithm will either determine that the new dual LP is unbounded (in which case, the new primal LP is infeasible—see Table 8.3), or else produce an optimal dual solution (in which case the new primal LP also has an optimal solution). These steps will now be illustrated with the following example.

EXAMPLE 9.3: *Primal LP.*

$$\begin{aligned} \text{Minimize} \quad & 2x_1 + 3x_2 + x_3 \\ \text{subject to} \quad & x_1 + x_2 - 2x_3 + x_4 = -1 \\ & 4x_1 - 2x_2 - x_3 + x_5 = -2 \\ & x_1, \ x_2, \ x_3, \ x_4, \ x_5 \geq 0 \end{aligned}$$

Dual LP.

$$\begin{array}{ll} \text{Maximize} & -u_1 - 2u_2 \\ \text{subject to} & u_1 + 4u_2 \leq 2 \\ & u_1 - 2u_2 \leq 3 \\ & -2u_1 - u_2 \leq 1 \\ & u_1 \leq 0 \\ & u_2 \leq 0 \end{array}$$

The dbfs $\mathbf{u} = (0, -1)$ is optimal for this dual problem. If $\mathbf{b}'$ is now given as $(-5, -2)$ the new problem to be solved is the new dual LP

$$\begin{array}{ll} \text{Maximize} & -5u_1 - 2u_2 \\ \text{subject to} & u_1 + 4u_2 \leq 2 \\ & u_1 - 2u_2 \leq 3 \\ & -2u_1 - u_2 \leq 1 \\ & u_1 \leq 0 \\ & u_2 \leq 0 \end{array}$$

Iteration 1

Step 0. The initial dbfs for this iteration:

$$\mathbf{B} = [\mathbf{A}_{\cdot 4}, \mathbf{A}_{\cdot 3}] = \begin{bmatrix} 1 & -2 \\ 0 & -1 \end{bmatrix}$$

$$\mathbf{N} = [\mathbf{A}_{\cdot 1}, \mathbf{A}_{\cdot 2}, \mathbf{A}_{\cdot 5}] = \begin{bmatrix} 1 & 1 & 0 \\ 4 & -2 & 1 \end{bmatrix}$$

$$\mathbf{u} = \mathbf{c_B}\mathbf{B}^{-1} = (0, 1)\begin{bmatrix} 1 & -2 \\ 0 & -1 \end{bmatrix} = (0, -1)$$

Step 1. Testing for optimality:

$$\mathbf{B}^{-1}\mathbf{b}' = \begin{bmatrix} 1 & -2 \\ 0 & -1 \end{bmatrix}\begin{bmatrix} -5 \\ -2 \end{bmatrix} = \begin{bmatrix} -1 \\ 2 \end{bmatrix} \qquad \text{so} \quad k^* = 1$$

Step 2. Computing the direction of movement:

$$\mathbf{d} = -(\text{row } k^* \text{ of } \mathbf{B}^{-1}) = (-1, 2)$$

Step 3. Computing the amount of movement:

$$\mathbf{dN} = (-1,2)\begin{bmatrix} 1 & 1 & 0 \\ 4 & -2 & 1 \end{bmatrix} = (7,-5,2)$$

$$\begin{aligned} t^* &= \min\{(\mathbf{c_N} - \mathbf{uN})_j/(\mathbf{dN})_j : (\mathbf{dN})_j > 0\} \\ &= \min\{(\mathbf{c_N} - \mathbf{uN})_1/(\mathbf{dN})_1, (\mathbf{c_N} - \mathbf{uN})_3/(\mathbf{dN})_3\} \\ &= \min\{[2-(-4)]/7, [0-(-1)]/2\} \\ &= \min\{\tfrac{6}{7}, \tfrac{1}{2}\} \\ &= \tfrac{1}{2} \quad \text{so} \quad j^* = 3 \end{aligned}$$

Step 4. Pivoting: Column $k^* = 1$ of $\mathbf{B}$ is replaced with column $j^* = 3$ of $\mathbf{N}$, so

$$\mathbf{B}' = \begin{bmatrix} 0 & -2 \\ 1 & -1 \end{bmatrix}$$

$$\mathbf{N}' = \begin{bmatrix} 1 & 1 & 1 \\ 4 & -2 & 0 \end{bmatrix}$$

$$\begin{aligned} \mathbf{u}' &= \mathbf{u} + t^*\mathbf{d} = (0,-1) + (\tfrac{1}{2})(-1,2) \\ &= (-\tfrac{1}{2}, 0) \end{aligned}$$

Iteration 2

Step 0. The dbfs for this iteration is:

$$\mathbf{B} = [\mathbf{A}_{.5}, \mathbf{A}_{.3}] = \begin{bmatrix} 0 & -2 \\ 1 & -1 \end{bmatrix}$$

$$\mathbf{N} = [\mathbf{A}_{.1}, \mathbf{A}_{.2}, \mathbf{A}_{.4}] = \begin{bmatrix} 1 & 1 & 1 \\ 4 & -2 & 0 \end{bmatrix}$$

$$\mathbf{u} = \mathbf{c_B}\mathbf{B}^{-1} = (0,1)\begin{bmatrix} -\tfrac{1}{2} & 1 \\ -\tfrac{1}{2} & 0 \end{bmatrix} = (-\tfrac{1}{2}, 0)$$

Step 1. Testing for optimality:

$$\mathbf{B}^{-1}\mathbf{b}' = \begin{bmatrix} -\tfrac{1}{2} & 1 \\ -\tfrac{1}{2} & 0 \end{bmatrix}\begin{bmatrix} -5 \\ -2 \end{bmatrix} = \begin{bmatrix} \tfrac{1}{2} \\ \tfrac{5}{2} \end{bmatrix}$$

Since $\mathbf{B}^{-1}\mathbf{b}' \geq 0$, the current dbfs $= (-\tfrac{1}{2}, 0)$ is optimal for the new dual problem. The corresponding solution for the new primal problem is $\mathbf{x_B} = (x_5, x_3) = \mathbf{B}^{-1}\mathbf{b}' = (\tfrac{1}{2}, \tfrac{5}{2})$ and $\mathbf{x_N} = (x_1, x_2, x_4) = (0,0,0)$ so $\mathbf{x} = (0, 0, \tfrac{5}{2}, 0, \tfrac{1}{2})$.

Sensitivity of the Right-Hand Side

A related question of interest might be asked by the decision maker after solving the original primal LP: "By how much can one particular component of $\mathbf{b}$ be changed without changing the current set of basic variables?" In other words, how sensitive is the set of basic variables to changes in $\mathbf{b}$? Recall that when $\mathbf{b}$ changes to $\mathbf{b}'$ the basic variables change in value from $\mathbf{B}^{-1}\mathbf{b}$ to $\mathbf{B}^{-1}\mathbf{b}'$, while the reduced costs $\mathbf{c_N} - \mathbf{c_B}\mathbf{B}^{-1}\mathbf{N}$ (which are nonnegative) do not change at all. Hence, as long as $\mathbf{B}^{-1}\mathbf{b}'$ remains nonnegative, the current basis matrix remains optimal.

To determine the range of values over which one particular component of $\mathbf{b}$ (say, b_k) can be varied without affecting the fact that $\mathbf{B}$ is an optimal basis matrix, think of b_k as a variable instead of a fixed number. As will be shown in Exercise 9.2.2, as long as b_k is between some lower bound t_1 and some upper bound t_2, the values of the basic variables remain nonnegative. In other words, for $t_1 \leq b_k \leq t_2$ the current $\mathbf{B}$ matrix is always optimal (although the values of the basic variables change as b_k varies).

The Parametric Right-Hand-Side Problem

The final problem to be discussed in this section is known as the *parametric right-hand-side problem*. Given the original right-hand-side vector $\mathbf{b}$ together with another m vector $\mathbf{b}'$, the objective is to solve the following parametric LP for each value of $s \geq 0$:

$$\begin{array}{ll} \text{Minimize} & \mathbf{cx} \\ \text{subject to} & \mathbf{Ax} = \mathbf{b} + s\mathbf{b}' \\ & \mathbf{x} \geq \mathbf{0} \end{array}$$

Since there are an infinite number of values for s, there are an infinite number of problems to solve. However, as in the parametric problem of the previous section, a finite procedure can be developed. To that end, consider the following parametric dual LP:

$$\begin{array}{ll} \text{Maximize} & \mathbf{u}(\mathbf{b} + s\mathbf{b}') \\ \text{subject to} & \mathbf{uA} \leq \mathbf{c} \\ & \mathbf{u} \text{ unrestricted} \end{array}$$

As s varies, the objective function changes, but the constraints do not. Thus the approach is similar to that of the previous section, namely, to divide the interval from 0 to ∞ into a finite number of subintervals of the form $0 < s_1 < \cdots < s_k < \infty$. Associated with each subinterval, an optimal basis matrix $\mathbf{B}_i$ will be sought with the property that, for all values of s such that

$s_{i-1} \leq s \leq s_i$, $\mathbf{B}_i$ is optimal for the parametric dual LP. This basis matrix can then be used to compute the corresponding solution for the parametric primal LP. The detailed steps of the procedure are given next. For simplicity, it will be assumed that $\mathbf{u}^* = \mathbf{c_B B}^{-1}$ is an optimal dbfs for the parametric dual LP in which $s = 0$.

Step 1. Determining the Values of s for Which u* Remains Optimal

Because $\mathbf{B}^{-1}\mathbf{b} \geq \mathbf{0}$, $\mathbf{u}^*$ is optimal for the dual LP. As s increases from its current value of 0, the values of the primal basic variables change from $\mathbf{B}^{-1}\mathbf{b}$ to $\mathbf{B}^{-1}(\mathbf{b} + s\mathbf{b}')$, so, as long as $\mathbf{B}^{-1}(\mathbf{b} + s\mathbf{b}')$ remains nonnegative, the current basis matrix is optimal for the dual and primal problems. Thus, if s^* is the largest such value for s, it can be computed explicitly by the formula

$$s^* = \min\{-(\mathbf{B}^{-1}\mathbf{b})_k/(\mathbf{B}^{-1}\mathbf{b}')_k : (\mathbf{B}^{-1}\mathbf{b}')_k < 0\}$$

If $s^* = \infty$ (which happens when $\mathbf{B}^{-1}\mathbf{b}' \geq \mathbf{0}$), then the current basis matrix remains optimal for all nonnegative values of s; otherwise,

$$s^* = -(\mathbf{B}^{-1}\mathbf{b})_{k^*}/(\mathbf{B}^{-1}\mathbf{b}')_{k^*}$$

Thus, for all values of $0 \leq s \leq s^*$ the current basis matrix remains optimal. The next step of the procedure describes how to "move into" the next subinterval.

Step 2. Moving to a New dbfs

For all values of s "slightly" larger than s^*, component k^* of the primal basic variables becomes negative because k^* was chosen in this way. Thus one may compute the direction of movement $\mathbf{d} = -(\text{row } k^* \text{ of } \mathbf{B}^{-1})$. If $\mathbf{dN} \leq \mathbf{0}$ then, for all values of $s \geq s^*$, the parametric dual LP is unbounded and the parametric primal LP is infeasible. Otherwise, a column, say, j^*, can be determined to enter the basis matrix by the usual min ratio test of the dual simplex algorithm. After the pivot operation is performed, the $\mathbf{b}$ vector must be modified to $\mathbf{b} + s^*\mathbf{b}'$, and the entire process can be repeated by returning to step 1, thus determining each successive interval and the corresponding optimal basis matrix. As long as $s^* > 0$, the finiteness of the procedure is ensured because each basis matrix can be optimal over at most one interval, as was the case in the previous section.

The process of solving the parametric dual LP will now be demonstrated with an example.

EXAMPLE 9.4: For the vectors $\mathbf{b} = (-1, -2)$ and $\mathbf{b}' = (-4, 0)$, the dual LP in Example 8.9 becomes a parametric dual LP:

$$\begin{array}{ll} \text{Maximize} & \mathbf{u}[(-1,-2) + s(-4,0)] \\ \text{subject to} & u_1 + 4u_2 \leq 2 \\ & u_1 - 2u_2 \leq 3 \\ & -2u_1 - u_2 \leq 1 \\ & u_1 \leq 0 \\ & u_2 \leq 0 \end{array}$$

Iteration 1

Step 0. The initial dbfs for this iteration:

$$\mathbf{B} = [\mathbf{A}_{.4}, \mathbf{A}_{.3}] = \begin{bmatrix} 1 & -2 \\ 0 & -1 \end{bmatrix}$$

$$\mathbf{N} = [\mathbf{A}_{.1}, \mathbf{A}_{.2}, \mathbf{A}_{.5}] = \begin{bmatrix} 1 & 1 & 0 \\ 4 & -2 & 1 \end{bmatrix}$$

$$\mathbf{u} = \mathbf{c_B}\mathbf{B}^{-1} = (0,1)\begin{bmatrix} 1 & -2 \\ 0 & -1 \end{bmatrix} = (0, -1)$$

Step 1. Determining the value of s for which the current dual solution is optimal:

$$\mathbf{B}^{-1}\mathbf{b} = \begin{bmatrix} 1 & -2 \\ 0 & -1 \end{bmatrix}\begin{bmatrix} -1 \\ -2 \end{bmatrix} = \begin{bmatrix} 3 \\ 2 \end{bmatrix}$$

$$\mathbf{B}^{-1}\mathbf{b}' = \begin{bmatrix} 1 & -2 \\ 0 & -1 \end{bmatrix}\begin{bmatrix} -4 \\ 0 \end{bmatrix} = \begin{bmatrix} -4 \\ 0 \end{bmatrix}$$

$$\begin{aligned} s^* &= \min\{-(\mathbf{B}^{-1}\mathbf{b})_k/(\mathbf{B}^{-1}\mathbf{b}')_k : (\mathbf{B}^{-1}\mathbf{b}')_k < 0\} \\ &= \min\{-(\mathbf{B}^{-1}\mathbf{b})_1/(\mathbf{B}^{-1}\mathbf{b}')_1\} \\ &= \tfrac{3}{4} \qquad \text{so} \quad k^* = 1 \end{aligned}$$

This means that the current basis is optimal for all values of s for which $0 \leq s \leq \frac{3}{4}$. For values of $s > \frac{3}{4}$, a new bfs will provide the optimal solution.

Step 2. Moving to a new dbfs:

a. Computing the direction of movement:

$$\mathbf{d} = -(\text{row } k^* \text{ of } \mathbf{B}^{-1}) = (-1, 2)$$

$$\mathbf{dN} = (-1,2)\begin{bmatrix} 1 & 1 & 0 \\ 4 & -2 & 1 \end{bmatrix} = (7, -5, 2)$$

b. Computing the amount of movement t^*:

$$\begin{aligned} t^* &= \min\{(\mathbf{c_N}-\mathbf{uN})_j/(\mathbf{dN})_j : (\mathbf{dN})_j > 0\} \\ &= \min\{(\mathbf{c_N}-\mathbf{uN})_1/(\mathbf{dN})_1, (\mathbf{c_N}-\mathbf{uN})_3/(\mathbf{dN})_3\} \\ &= \min\{[2-(-4)]/7, [0-(-1)]/2\} \\ &= \min\{\tfrac{6}{7}, \tfrac{1}{2}\} \\ &= \tfrac{1}{2} \qquad \text{so} \quad j^* = 3 \end{aligned}$$

c. Pivoting: Column $k^* = 1$ of $\mathbf{B}$ is replaced with column $j^* = 3$ of $\mathbf{N}$, so

$$\mathbf{B}' = \begin{bmatrix} 0 & -2 \\ 1 & -1 \end{bmatrix}$$

$$\mathbf{N}' = \begin{bmatrix} 1 & 1 & 1 \\ 4 & -2 & 0 \end{bmatrix}$$

$$\begin{aligned} \mathbf{u}' &= \mathbf{u} + t^*\mathbf{d} \\ &= (0,1) + (\tfrac{1}{2})(-1,2) \\ &= (-\tfrac{1}{2},0) \end{aligned}$$

d. Updating the $\mathbf{b}$ vector: The new $\mathbf{b}$ vector is

$$\mathbf{b} = (-1,-2) + (\tfrac{3}{4})(-4,0) = (-4,-2)$$

Iteration 2

Step 1. Determining the value of s for which the current solution is optimal:

$$\mathbf{B}^{-1}\mathbf{b}' = \begin{bmatrix} -\frac{1}{2} & 1 \\ -\frac{1}{2} & 0 \end{bmatrix} \begin{bmatrix} -4 \\ 0 \end{bmatrix} = \begin{bmatrix} 2 \\ 2 \end{bmatrix}$$

Since $(\mathbf{B}^{-1}\mathbf{b}') \geq \mathbf{0}$, the current dbfs $(-\tfrac{1}{2},0)$ remains optimal for all values of $s \geq \tfrac{3}{4}$.

This section has shown how to use the optimal dbfs $\mathbf{u}^*$ to solve a new LP in which some (or all) of the $\mathbf{b}$ vector is changed. In each case, the dual simplex algorithm is initiated from $\mathbf{u}^*$. The next section addresses the issue of handling changes in the $\mathbf{A}$ matrix.

EXERCISES

9.2.1. Consider the following LP:

$$\begin{array}{llllll} \text{Minimize} & 2x_1 & +4x_2 & +3x_3 & +2x_4 & + \ x_5 \\ \text{subject to} & 4x_1 & + \ x_2 & - \ x_3 & & +2x_5 = 3 \\ & 3x_1 & +2x_2 & & - \ x_4 & - \ x_5 = 8 \\ & x_1, & x_2, & x_3, & x_4, & x_5 \geq 0 \end{array}$$

This LP has an optimal solution of $\mathbf{x}^* = (0,4,1,0,0)$. Find the new optimal solution if the $\mathbf{b}$ vector is changed to $(5,5)$.

9.2.2. For the optimal bfs $\mathbf{x}^* = (0,4,1,0,0)$ to the original LP given in Exercise 9.2.1 (before the $\mathbf{b}$ vector was changed), do the following:

(a) Find by trial and error how much b_1 can vary such that $\mathbf{x}^*$ remains optimal.

(b) Find by trial and error how much b_2 can vary such that $\mathbf{x}^*$ remains optimal.

(c) Given an optimal bfs $\mathbf{x}^*$ to an LP and given k with $1 \le k \le m$, determine the values of t_1 and t_2 so that, as long as the value of b_k is between t_1 and t_2, the current basis matrix remains optimal.

9.2.3. Solve the parametric LP

$$\begin{aligned} \text{Minimize} \quad & 4x_1 + 3x_2 + 2x_3 + x_4 \\ \text{subject to} \quad & x_1 - x_2 \qquad\quad + 2x_4 = 3 + s(1) \\ & 2x_1 \qquad\quad - x_3 - x_4 = 8 + s(-3) \\ & x_1, \quad x_2, \quad x_3, \quad x_4 \ge 0 \end{aligned}$$

starting with the initial bfs $(4,1,0,0)$.

9.2.4. Suppose that $\mathbf{x}^*$ is an optimal bfs to the following parametric LP at $s = 0$:

$$\begin{aligned} \text{Minimize} \quad & \mathbf{cx} \\ \text{subject to} \quad & \mathbf{Ax} = \mathbf{b} + s\mathbf{b}' \\ & \mathbf{x} \ge \mathbf{0} \end{aligned}$$

Find how large s can be made before the problem becomes unbounded. [*Hint*: Find how large s can be made before the dual becomes infeasible.]

9.3. CHANGES IN THE A MATRIX

The previous two sections have addressed the issue of solving an LP in which either the $\mathbf{c}$ vector or $\mathbf{b}$ vector has been changed. This section will discuss changes in the $\mathbf{A}$ matrix. Once again, suppose that $\mathbf{x}^* = (\mathbf{x}^*_{\mathbf{B}}, \mathbf{x}^*_{\mathbf{N}}) = (\mathbf{B}^{-1}\mathbf{b}, \mathbf{0}) \ge \mathbf{0}$ is an optimal bfs for an LP in standard form. Modifications in the $\mathbf{A}$ matrix fall into two categories.

Case 1: The $\mathbf{N}$ matrix changes to $\mathbf{N}'$.

In this case, the $\mathbf{B}$ matrix and $\mathbf{b}$ vector are not affected, and hence $\mathbf{x}^*$ is a bfs from which the simplex algorithm can be initiated to solve the new LP:

$$\begin{aligned} \text{Minimize} \quad & \mathbf{c_B x_B} + \mathbf{c_N x_N} \\ \text{subject to} \quad & \mathbf{Bx_B} + \mathbf{N}'\mathbf{x_N} = \mathbf{b} \\ & \mathbf{x_B}, \quad \mathbf{x_N} \ge \mathbf{0} \end{aligned}$$

In a finite number of iterations, the algorithm will either produce an optimal bfs for the new problem, or else determine that it is unbounded.

Case 2: The $\mathbf{B}$ matrix changes to $\mathbf{B}'$.

In this case, $\mathbf{x}^*$ might not be a bfs for the new LP:

$$\begin{aligned} \text{Minimize} \quad & \mathbf{c_B x_B} + \mathbf{c_N x_N} \\ \text{subject to} \quad & \mathbf{B' x_B} + \mathbf{N x_N} = \mathbf{b} \\ & \mathbf{x_B}, \quad \mathbf{x_N} \geq \mathbf{0} \end{aligned}$$

In fact, there is no guarantee that $\mathbf{B}'$ is even nonsingular. Furthermore, $\mathbf{u}^*$ might not even be a dbfs for the new dual LP.

One approach for handling this difficulty involves a two-phase procedure. The first phase attempts to replace the columns of the current basis matrix $\mathbf{B}$ with those of $\mathbf{B}'$ and $\mathbf{N}$. If successful, then the result is an initial bfs, say, $\mathbf{x}'$, for the new LP. The second phase is then to use $\mathbf{x}'$ as an initial bfs from which the primal simplex algorithm can be applied to solve the new LP.

In order to implement the first phase, a method is needed for replacing the columns of the current matrix $\mathbf{B}$ with those of $\mathbf{B}'$ and $\mathbf{N}$. This objective can be accomplished by attempting to make the current basic variables $\mathbf{x_B}$ all decrease to 0 and simultaneously become nonbasic. To that end, consider the "phase 1" LP

$$\begin{aligned} \text{Minimize} \quad & \mathbf{e x_B} \\ \text{subject to} \quad & \mathbf{B x_B} + \mathbf{N x_N} + \mathbf{B' y} = \mathbf{b} \\ & \mathbf{x_B}, \quad \mathbf{x_N}, \quad \mathbf{y} \geq \mathbf{0} \end{aligned}$$

The objective function serves to force the current basic variables to zero, and the $\mathbf{y}$ variables (and $\mathbf{B}'$ matrix) have been added so as to allow the columns of $\mathbf{B}'$ to enter the basis.

Indeed, the vector $(\mathbf{x_B}, \mathbf{x_N}, \mathbf{y}) = (\mathbf{x_B^*}, \mathbf{0}, \mathbf{0})$ is an initial bfs for this "phase 1" LP. Thus the simplex algorithm can be initiated to solve the problem, and the possible outcomes are similar to those of the phase 1 procedure of Chapter 6. For instance, the above "phase 1" LP cannot be unbounded because the value of the objective function is always nonnegative. Consequently, in a finite number of iterations, an optimal bfs, say, $(\mathbf{x'_B}, \mathbf{x'_N}, \mathbf{y}')$, must be produced. In the event that $\mathbf{x'_B} \neq \mathbf{0}$, the new LP is infeasible, as will be shown in Exercise 9.3.3. On the other hand, if $\mathbf{x'_B} = \mathbf{0}$, then there are two cases.

Subcase a: All the $\mathbf{x'_B}$ variables are nonbasic. In this case, $(\mathbf{x'_N}, \mathbf{y}')$ is a bfs for the new LP. Hence the simplex algorithm can be initiated there to determine if the new LP is optimal or unbounded.

Subcase b: Some of the $\mathbf{x'_B}$ variables are basic at value 0. In this case, an artificial constraint (and variable) can be added to the new LP in much the same way as was done for phase 1 in Section 6.3. The new LP then has the following form:

$$\begin{array}{llllll} \text{Minimize} & \mathbf{0x_B} & +\mathbf{c_N x_N} & +\mathbf{c'_B y} & & \\ \text{subject to} & \mathbf{Bx_B} & +\mathbf{Nx_N} & +\mathbf{B'y} & & =\mathbf{b} \\ & \mathbf{ex_B} & & & +z & =0 \\ & \mathbf{x_B}, & \mathbf{x_N}, & \mathbf{y}, & z & \geq \mathbf{0} \end{array}$$

Note that z is the new variable, and the second constraint is the artificial constraint. The vector $(\mathbf{x_B}, \mathbf{x_N}, \mathbf{y}, z) = (\mathbf{x'_B}, \mathbf{x'_N}, \mathbf{y}', 0)$ is an initial bfs for the new LP.

This section has addressed the problem of using $\mathbf{x}^*$ to solve a new LP in which the A matrix changes. The remaining section of this chapter discusses similar solution procedures when a new variable or constraint is to be added to the original problem.

EXERCISES

9.3.1. Consider the following LP:

$$\begin{array}{lrrrrrl} \text{Minimize} & -x_1 & -2x_2 & +x_3 & & & \\ \text{subject to} & 4x_1 & +5x_2 & & +3x_4 & +x_5 & =10 \\ & 6x_1 & +3x_2 & -3x_3 & +x_4 & & =12 \\ & x_1, & x_2, & x_3, & x_4, & x_5 & \geq 0 \end{array}$$

This LP has an optimal bfs at $\mathbf{x}^* = (\frac{5}{3}, \frac{2}{3}, 0, 0, 0)$. Using $\mathbf{x}^*$ as an initial bfs, find the optimal solution to the LP:

$$\begin{array}{lrrrrrl} \text{Minimize} & -x_1 & -2x_2 & +x_3 & & & \\ \text{subject to} & 4x_1 & +5x_2 & & +3x_4 & -2x_5 & =10 \\ & 6x_1 & +3x_2 & +3x_3 & +2x_4 & & =12 \\ & x_1, & x_2, & x_3, & x_4, & x_5 & \geq 0 \end{array}$$

9.3.2. Suppose the original LP given in Exercise 9.3.1 is changed to the following LP:

$$\begin{array}{lrrrrrl} \text{Minimize} & -x_1 & -2x_2 & +x_3 & & & \\ \text{subject to} & 2x_1 & +x_2 & & +3x_4 & +x_5 & =10 \\ & 3x_1 & -x_2 & -3x_3 & +x_4 & & =12 \\ & x_1, & x_2, & x_3, & x_4, & x_5 & \geq 0 \end{array}$$

Note that the optimal solution $(\frac{5}{3},\frac{2}{3},0,0,0)$ to the original LP is no longer feasible. Formulate an appropriate "phase 1" LP and provide an initial bfs for that problem.

9.3.3. Suppose $(\mathbf{x}'_\mathbf{B},\mathbf{x}'_\mathbf{N},\mathbf{y}')$ solves the following phase 1 LP:

$$\begin{array}{ll} \text{Minimize} & \mathbf{ex_B} \\ \text{subject to} & \mathbf{Bx_B} + \mathbf{Nx_N} + \mathbf{B'y} = \mathbf{b} \\ & \mathbf{x_B}, \quad \mathbf{x_N}, \quad \mathbf{y} \geq \mathbf{0} \end{array}$$

Show that, when $\mathbf{x}'_\mathbf{B} \neq \mathbf{0}$, the new LP is infeasible.

9.4. THE ADDITION OF A NEW VARIABLE OR CONSTRAINT

Adding a New Variable

As before, suppose that $\mathbf{x}^* = (\mathbf{x}^*_\mathbf{B},\mathbf{x}^*_\mathbf{N}) = (\mathbf{B}^{-1}\mathbf{b},\mathbf{0}) \geq \mathbf{0}$ is an optimal bfs for an LP in standard form. After obtaining the solution, it may become necessary to add a new variable, x_{n+1}, together with its cost coefficient c_{n+1}, and its associated column of the $\mathbf{A}$ matrix $\mathbf{A}_{\cdot n+1}$. Thus the new LP is to

$$\begin{array}{ll} \text{minimize} & \mathbf{cx} + c_{n+1}x_{n+1} \\ \text{subject to} & \mathbf{Ax} + \mathbf{A}_{\cdot n+1}x_{n+1} = \mathbf{b} \\ & \mathbf{x}, \quad x_{n+1} \geq \mathbf{0} \end{array}$$

Think of the variable x_{n+1} as a nonbasic variable at value 0; then the vector $(\mathbf{x}, x_{n+1}) = (\mathbf{x}^*,0)$ is an initial bfs for the new LP. As such, the simplex algorithm can be initiated there to solve the new LP. Specifically, the reduced costs of all of the nonbasic variables, except (possibly) x_{n+1}, are nonnegative because $\mathbf{x}^*$ is optimal for the original LP. Hence only the reduced cost corresponding to x_{n+1} needs to be checked for nonnegativity. Specifically, if $c_{n+1} - (\mathbf{c_B}\mathbf{B}^{-1})\mathbf{A}_{\cdot n+1} \geq 0$, then $(\mathbf{x}^*,0)$ is optimal for the new LP. Otherwise, the remaining steps of the simplex algorithm can be carried out (i.e., x_{n+1} can be made basic) and further iterations performed, if necessary. In a finite number of iterations, either an optimal solution to the new LP will be produced, or else unboundedness will be detected.

Adding a New Constraint

The addition of a new constraint requires more care than the addition of a new variable. To be specific, suppose that a new constraint of the form $\mathbf{ax} \leq b_{m+1}$ (where $\mathbf{a}$ is an n vector) is to be added to the original problem.

After adding a slack variable, the new LP to be solved is to

$$\begin{aligned} \text{minimize} \quad & \mathbf{cx} \\ \text{subject to} \quad & \mathbf{Ax} \qquad = \mathbf{b} \\ & \mathbf{ax} + s = b_{m+1} \\ & \mathbf{x}, \quad s \geq \mathbf{0} \end{aligned}$$

or, equivalently, if $\mathbf{x}^*$ is an optimal bfs for the original LP, to

$$\begin{aligned} \text{minimize} \quad & \mathbf{c_B x_B} + \mathbf{c_N x_N} \\ \text{subject to} \quad & \mathbf{Bx_B} + \mathbf{Nx_N} \qquad = \mathbf{b} \\ & \mathbf{a_B x_B} + \mathbf{a_N x_N} + s = b_{m+1} \\ & \mathbf{x_B}, \quad \mathbf{x_N}, \quad s \geq \mathbf{0} \end{aligned}$$

One of the difficulties that can arise in using $\mathbf{x}^*$ as a starting point from which to solve the new problem is that $\mathbf{x}^*$ might not be feasible for the new LP since the new constraints may "exclude" $\mathbf{x}^*$, as shown in Figure 9.3. Furthermore, a bfs for the new LP would, of necessity, consist of $m+1$ variables. Perhaps $\mathbf{x}^*_\mathbf{B}$ and the slack variable s are the desired $m+1$ basic variables. To find out, consider the corresponding nonsingular $(m+1) \times (m+1)$ matrix

$$\hat{\mathbf{B}} = \begin{bmatrix} \mathbf{B} & \mathbf{0} \\ \mathbf{a_B} & 1 \end{bmatrix}$$

and its inverse

$$\hat{\mathbf{B}}^{-1} = \begin{bmatrix} \mathbf{B}^{-1} & \mathbf{0} \\ -\mathbf{a_B B}^{-1} & 1 \end{bmatrix}$$

In the event that

$$\hat{\mathbf{B}}^{-1} \begin{bmatrix} \mathbf{b} \\ b_{m+1} \end{bmatrix} \geq \mathbf{0}$$

FIGURE 9.3. The geometry of adding a new constraint.

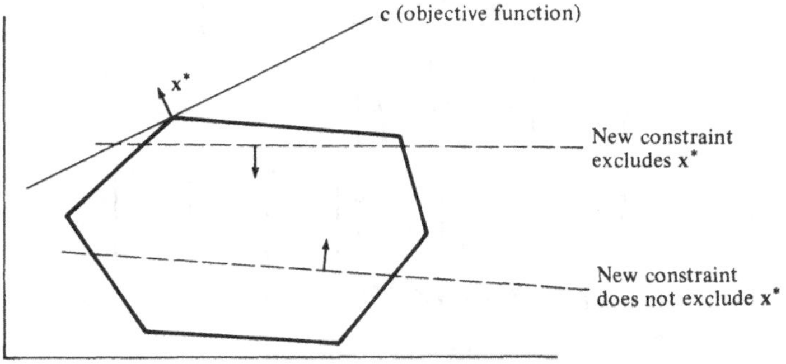

then indeed an initial bfs for the new LP has been found; moreover, the bfs is optimal, as will be shown in Exercise 9.4.2. Otherwise, the dual simplex algorithm can be initiated from the dbfs $(\mathbf{u}, v) = (\mathbf{c_B B}^{-1}, 0)$, as illustrated in the next example.

EXAMPLE 9.5: Consider Example 8.9:

$$\begin{array}{llllllll} \text{Minimize} & 2x_1 & +3x_2 & + x_3 & & & & \\ \text{subject to} & x_1 & + x_2 & -2x_3 & +x_4 & & = & -1 \\ & 4x_1 & -2x_2 & - x_3 & & +x_5 & = & -2 \\ & x_1, & x_2, & x_3, & x_4, & x_5 & \geq & 0 \end{array}$$

This problem was shown to have the optimal solution $\mathbf{x}^* = (0, 0, 2, 3, 0)$. The dual problem has the optimal solution $\mathbf{u}^* = (0, -1)$. Suppose now that one were to add the constraint

$$x_2 + 2x_3 \leq 3$$

The new problem can be formulated as follows:

$$\begin{array}{lllllllll} \text{Minimize} & 2x_1 & +3x_2 & + x_3 & & & & & \\ \text{subject to} & x_1 & + x_2 & -2x_3 & +x_4 & & & = & -1 \\ & 4x_1 & -2x_2 & - x_3 & & +x_5 & & = & -2 \\ & & x_2 & +2x_3 & & & +x_6 & = & 3 \\ & x_1, & x_2, & x_3, & x_4, & x_5, & x_6 & \geq & 0 \end{array}$$

Iteration 1

Step 0. Check if the previous bfs is optimal for the new LP. If not, find a new dbfs for the dual:

a. Determine the current bfs for the new LP:

$$\hat{\mathbf{B}} = \begin{bmatrix} \mathbf{B} & \mathbf{0} \\ \mathbf{a_B} & 1 \end{bmatrix} = \begin{bmatrix} 1 & -2 & 0 \\ 0 & -1 & 0 \\ 0 & 2 & 1 \end{bmatrix}$$

$$\hat{\mathbf{N}} = \begin{bmatrix} 1 & 1 & 0 \\ 4 & -2 & 1 \\ 0 & 1 & 0 \end{bmatrix}$$

$$\hat{\mathbf{B}}^{-1} = \begin{bmatrix} \mathbf{B}^{-1} & \mathbf{0} \\ -\mathbf{a_B B}^{-1} & 1 \end{bmatrix} = \begin{bmatrix} 1 & -2 & 0 \\ 0 & -1 & 0 \\ 0 & 2 & 1 \end{bmatrix}$$

$$\mathbf{x_B} = (x_4, x_3, x_6)$$

$$\mathbf{x_N} = (x_1, x_2, x_5)$$

b. Testing for optimality:

$$\hat{\mathbf{B}}^{-1}\begin{bmatrix}\mathbf{b}\\ b_{m+1}\end{bmatrix}=\begin{bmatrix}1 & -2 & 0\\ 0 & -1 & 0\\ 0 & 2 & 1\end{bmatrix}\begin{bmatrix}-1\\ -2\\ 3\end{bmatrix}=\begin{bmatrix}3\\ 2\\ -1\end{bmatrix}$$

Since this vector is not $\geq \mathbf{0}$, the current bfs is not optimal.

c. Find the dbfs for the dual problem:

$$(\mathbf{u}, v)=(\mathbf{c}_{\mathbf{B}}\mathbf{B}^{-1},0)=\left[(0,1)\begin{bmatrix}1 & -2\\ 0 & -1\end{bmatrix}, 0\right]$$

$$=(0,-1,0)$$

The matrices $\hat{\mathbf{B}}$ and $\hat{\mathbf{B}}^{-1}$ are now denoted $\mathbf{B}$ and $\mathbf{B}^{-1}$.

Step 1. Test for optimality:

$$\mathbf{B}^{-1}\begin{bmatrix}\mathbf{b}\\ b_{m+1}\end{bmatrix}=\begin{bmatrix}1 & -2 & 0\\ 0 & -1 & 0\\ 0 & 2 & 1\end{bmatrix}\begin{bmatrix}-1\\ -2\\ 3\end{bmatrix}$$

$$=\begin{bmatrix}3\\ 2\\ -1\end{bmatrix} \qquad \text{so} \quad k^*=3$$

Step 2. Computing the direction of movement:

$$\mathbf{d}=-(\text{row } k^* \text{ of } \mathbf{B}^{-1})=(0,-2,-1)$$

Step 3. Computing the amount of movement:

$$\mathbf{dN}=(0,-2,-1)\begin{bmatrix}1 & 1 & 0\\ 4 & -2 & 1\\ 0 & 1 & 0\end{bmatrix}=\begin{bmatrix}-8\\ 3\\ -2\end{bmatrix}$$

$$t^*=\min\{(\mathbf{c_N}-\mathbf{uN})_j/(\mathbf{dN})_j : (\mathbf{dN})_j>0\}$$

$$=\min\{(\mathbf{c_N}-\mathbf{uN})_2/(\mathbf{dN})_2\}$$

$$=\tfrac{1}{3} \qquad \text{so} \quad j^*=2$$

Step 4. Pivoting: Column $k^*=3$ of $\mathbf{B}$ is replaced by column $j^*=2$ of $\mathbf{N}$, so

$$\mathbf{B}'=\begin{bmatrix}1 & -2 & 1\\ 0 & -1 & -2\\ 0 & 2 & 1\end{bmatrix}$$

$$\mathbf{N}'=\begin{bmatrix}1 & 0 & 0\\ 4 & 0 & 1\\ 0 & 1 & 0\end{bmatrix}$$

$$\mathbf{u}'=\mathbf{u}+t^*\mathbf{d}=(0,-1,0)+(\tfrac{1}{3})(0,-2,-1)$$

$$=(0,-\tfrac{5}{3},-\tfrac{1}{3})$$

Iteration 2

Step 0. The dbfs for this iteration:

$$\mathbf{B} = [\mathbf{A}_{.4}, \mathbf{A}_{.3}, \mathbf{A}_{.2}] = \begin{bmatrix} 1 & -2 & 1 \\ 0 & -1 & -2 \\ 0 & 2 & 1 \end{bmatrix}$$

$$\mathbf{N} = [\mathbf{A}_{.1}, \mathbf{A}_{.6}, \mathbf{A}_{.5}] = \begin{bmatrix} 1 & 0 & 0 \\ 4 & 0 & 1 \\ 0 & 1 & 0 \end{bmatrix}$$

$$\mathbf{u} = \mathbf{c_B B}^{-1} = (0,1,3)\begin{bmatrix} 1 & \frac{4}{3} & \frac{5}{3} \\ 0 & \frac{1}{3} & \frac{2}{3} \\ 0 & -\frac{2}{3} & -\frac{1}{3} \end{bmatrix} = (0, -\tfrac{5}{3}, -\tfrac{1}{3})$$

Step 1. Testing for optimality:

$$\mathbf{B}^{-1}\mathbf{b} = \begin{bmatrix} 1 & \frac{4}{3} & \frac{5}{3} \\ 0 & \frac{1}{3} & \frac{2}{3} \\ 0 & -\frac{2}{3} & -\frac{1}{3} \end{bmatrix}\begin{bmatrix} -1 \\ -2 \\ 3 \end{bmatrix} = \begin{bmatrix} \frac{4}{3} \\ \frac{4}{3} \\ \frac{1}{3} \end{bmatrix}$$

Since $\mathbf{B}^{-1}\mathbf{b} \geq \mathbf{0}$, the current dbfs $\mathbf{u} = (0, -\frac{5}{3}, -\frac{1}{3})$ is optimal for the dual problem. The corresponding solution for the primal problem is $\mathbf{x_B} = (x_4, x_3, x_2) = (\frac{4}{3}, \frac{4}{3}, \frac{1}{3})$ and $\mathbf{x_N} = (x_1, x_6, x_5) = (0,0,0)$, so $x = (0, \frac{1}{3}, \frac{4}{3}, \frac{4}{3}, 0, 0)$. The value of the objective function is $\frac{7}{3}$.

This section has addressed the issue of adding a new variable or constraint to the original LP. The remaining chapters of this text will examine some special topics in linear programming.

EXERCISES

9.4.1. Consider the LP given in Exercise 9.3.1:

$$\begin{array}{llllllll} \text{Minimize} & -\ x_1 & -2x_2 & +\ x_3 & & & & \\ \text{subject to} & 4x_1 & +5x_2 & & +3x_4 & +x_5 & = 10 \\ & 6x_1 & +3x_2 & -3x_3 & +\ x_4 & & = 12 \\ & x_1, & x_2, & x_3, & x_4, & x_5 \geq & 0 \end{array}$$

It has the optimal bfs $\mathbf{x}^* = (\frac{5}{3}, \frac{2}{3}, 0, 0, 0)$. Suppose another decision variable has also been added to the LP, and the new LP looks as follows:

$$\begin{array}{lllllllll} \text{Minimize} & -\ x_1 & -2x_2 & +\ x_3 & & & -\ 3x_6 & \\ \text{subject to} & 4x_1 & +5x_2 & & +3x_4 & +x_5 & +10x_6 & = 10 \\ & 6x_1 & +3x_2 & -3x_2 & +2x_4 & & +15x_6 & = 12 \\ & x_1, & x_2, & x_3, & x_4, & x_5, & x_6 \geq & 0 \end{array}$$

Find an optimal solution to the new LP.

9.4.2. Show that if a new constraint of the form $\mathbf{ax} \le b_{m+1}$ is added to an LP and if

$$\hat{\mathbf{B}}^{-1}\begin{bmatrix}\mathbf{b}\\ b_{m+1}\end{bmatrix} \ge \mathbf{0}$$

then the bfs is optimal for the new LP.

9.4.3. Consider the LP given in Exercise 9.1.1:

$$\begin{array}{lllllll} \text{Minimize} & 2x_1 & - \;x_2 & +x_3 & +5x_4 & -3x_5 & \\ \text{subject to} & 4x_1 & +2x_2 & +x_3 & +\;x_4 & & =3 \\ & 2x_1 & +2x_2 & & +3x_4 & +\;x_5 & =2 \\ & x_1, & x_2, & x_3, & x_4, & x_5 & \ge 0 \end{array}$$

It has the optimal bfs $\mathbf{x}^* = (0,0,3,0,2)$. Suppose that the new constraint

$$x_1 + 2x_2 + 4x_3 + x_4 + 2x_5 \le 12$$

is added to the LP. Find an initial bfs for the new LP and determine if that bfs is optimal.

9.4.4. Develop a solution procedure when a constraint of the form $\mathbf{ax} = b_{m+1}$ is added instead of $\mathbf{ax} \le b_{m+1}$.

DISCUSSION

Parametric Linear Programming and Sensitivity Analysis

Some of the original developments in the area of parametric programming are due to Manne [3], and Gass and Saaty [2]. For more details on the topics of this chapter, see Gal [1]. It is worth noting that, because of the usefulness of postoptimality analysis, most professional codes have the capability of printing information pertaining to sensitivity analysis.

Continuous Time Linear Programming

A more difficult problem than parametric programming is to allow all of data (**A** matrix, **b** and **c** vectors) to change values at each point in some time interval, with the objective of optimizing the integral of the cost functions over time. The resulting problem is referred to as continuous time linear programming. Some appropriate references on this topic would be Grinold [4] and Perold [5].

REFERENCES

1. T. Gal, *Postoptimal Analyses, Parametric Programming and Related Topics*, McGraw-Hill, New York, 1979.
2. S. I. Gass and T. L. Saaty, "The Parametric Objective Function, Part 1," *Operations Research* 2, 316–319 (1954).
3. A. S. Manne, "Notes on Parametric Programming," RAND Report P-468, The Rand Corporation, Santa Monica CA, 1953.
4. G. C. Grinold, "Continuous Programming Part One: Linear Objectives," *Journal of Mathematical Analysis and Applications* 28, 32–51 (1969).
5. A. F. Perold, "Extreme Points and Basic Feasible Solutions in Continuous Time Linear Programming," *SIAM Journal on Control and Optimization* 19(1), 52–63 (1981).

Chapter 10

Techniques for Handling Bound Constraints

Recall from Chapter 1 that a bound constraint is a constraint involving only one variable, such as $l_i \le x_i$ or $x_i \le u_i$ (where l_i and u_i are given real numbers). It is of course possible to convert each such constraint into an equality constraint by adding or subtracting a slack variable; however, doing so increases the number of variables in the problem (one for each bound constraint). Moreover, by increasing the number of equality constraints, the size of a basis matrix will, of necessity, be increased accordingly (i.e., by 1 for each new equality constraint). Because of the inherent simplicity of bound constraints, it is possible (and computationally expedient) to modify the simplex algorithm so as to avoid the necessity of increasing the number of variables and constraints, as will be described in this chapter.

10.1. FINITE BOUNDS

For the purposes of this section, suppose that each variable x_i of an LP has a known finite lower bound l_i and upper bound u_i. Thus, the original LP has the following form:

$$\begin{aligned} \text{Minimize} \quad & \mathbf{cx} \\ \text{subject to} \quad & \mathbf{Ax} = \mathbf{b} \\ & \mathbf{l} \le \mathbf{x} \le \mathbf{u} \end{aligned}$$

It is left as an exercise to show that such an LP cannot be unbounded. The components of the vectors **l** and **u** can be negative, positive, or zero;

furthermore, it will be assumed that $\mathbf{l} \le \mathbf{u}$. In this section, a modification will be made to the simplex algorithm of Chapter 5 so as to enable it to solve the above LP without increasing the size of the basis matrix or the number of variables. Doing so requires the development of a simplex procedure that allows the nonbasic variables to assume values other than 0.

In handling bound constraints it is desirable to allow each nonbasic variable to have a value equal to either its lower bound or its upper bound, i.e., for each j with $1 \le j \le n - m$, either $(\mathbf{x_N})_j = (\mathbf{l_N})_j$ or $(\mathbf{x_N})_j = (\mathbf{u_N})_j$. As a result, it then becomes necessary to change the formula for computing the values of the basic variables from $\mathbf{x_B} = \mathbf{B}^{-1}\mathbf{b}$ to $\mathbf{x_B} = \mathbf{B}^{-1}(\mathbf{b} - \mathbf{N}\mathbf{x_N})$. To see why this new formula is correct, recall that the original equality constraints $\mathbf{Ax} = \mathbf{b}$ can be rewritten $\mathbf{B}\mathbf{x_B} + \mathbf{N}\mathbf{x_N} = \mathbf{b}$. In Chapter 5, when $\mathbf{x_N} = \mathbf{0}$, the equality constraints become $\mathbf{B}\mathbf{x_B} = \mathbf{b}$, and since $\mathbf{B}$ is nonsingular, $\mathbf{x_B} = \mathbf{B}^{-1}\mathbf{b}$. In the present situation, however, $\mathbf{x_N}$ might not be equal to 0. Nonetheless, the equality constraints can be written $\mathbf{B}\mathbf{x_B} = \mathbf{b} - \mathbf{N}\mathbf{x_N}$. Once again, it is possible to multiply both sides of this equality by $\mathbf{B}^{-1}$, thus giving rise to the formula

$$\mathbf{x_B} = \mathbf{B}^{-1}(\mathbf{b} - \mathbf{N}\mathbf{x_N})$$

In other words, if the nonbasic components of a vector are allowed to assume nonzero values, then, in order to satisfy the equality constraints, it had better be the case that the values of the basic variables can be computed by the formula $\mathbf{x_B} = \mathbf{B}^{-1}(\mathbf{b} - \mathbf{N}\mathbf{x_N})$. However, do not forget that the bound constraints $\mathbf{l} \le \mathbf{x} \le \mathbf{u}$ must also be satisfied. The nonbasic components $\mathbf{x_N}$ do satisfy $\mathbf{l_N} \le \mathbf{x_N} \le \mathbf{u_N}$ because for each j, $(\mathbf{x_N})_j = (\mathbf{l_N})_j$ or $(\mathbf{u_N})_j$. In order for the basic components of $\mathbf{x}$ to satisfy the bound constraints, it had better be the case that $\mathbf{l_B} \le \mathbf{x_B} \le \mathbf{u_B}$; otherwise, $\mathbf{x}$ will not be feasible.

These observations are put together in the next definition, which formalizes the concept of a basic feasible solution for an LP having bound constraints.

Definition 10.1. A vector $\mathbf{x}$ is a *bounded basic feasible solution* (bbfs) for an LP in the form,

$$\begin{aligned} \text{Minimize} \quad & \mathbf{cx} \\ \text{subject to} \quad & \mathbf{Ax} = \mathbf{b} \\ & \mathbf{l} \le \mathbf{x} \le \mathbf{u} \end{aligned}$$

if there is a nonsingular $m \times m$ submatrix $\mathbf{B}$ of $\mathbf{A}$ such that, together with the remaining columns of $\mathbf{A}$, denoted $\mathbf{N}$,

1. $\mathbf{x_B} = \mathbf{B}^{-1}(\mathbf{b} - \mathbf{N}\mathbf{x_N})$;
2. for every j with $1 \le j \le n - m$, either $(\mathbf{x_N})_j = (\mathbf{l_N})_j$ or $(\mathbf{x_N})_j = (\mathbf{u_N})_j$; and
3. $\mathbf{l_B} \le \mathbf{x_B} \le \mathbf{u_B}$.

The next example illustrates a bbfs.

EXAMPLE 10.1:

$$\begin{aligned} \text{Minimize} \quad & x_1 - x_2 + 3x_3 - 2x_4 + 6x_5 \\ \text{subject to} \quad & x_1 + x_3 - x_4 + 2x_5 = 2 \\ & x_2 + 2x_3 + x_4 + x_5 = 4 \\ & 2 \le x_1 \le 7 \\ & -3 \le x_2 \le 9 \\ & -4 \le x_3 \le 1 \\ & 1 \le x_4 \le 5 \\ & -2 \le x_5 \le 3 \end{aligned}$$

For this LP the vector $\mathbf{x} = (x_1, x_2, x_3, x_4, x_5) = (6, 3, 1, 1, -2)$ is a bbfs in which $\mathbf{x_B} = (x_1, x_2) = (6, 3)$ and $\mathbf{x_N} = (x_3, x_4, x_5) = (1, 1, -2)$. To see why, let

$$\mathbf{B} = [\mathbf{A}_{.1}, \mathbf{A}_{.2}] = \begin{bmatrix} 1 & 0 \\ 0 & 1 \end{bmatrix}, \qquad \mathbf{N} = [\mathbf{A}_{.3}, \mathbf{A}_{.4}, \mathbf{A}_{.5}] = \begin{bmatrix} 1 & -1 & 2 \\ 2 & 1 & 1 \end{bmatrix}$$

so

$$\mathbf{N}\mathbf{x_N} = \begin{bmatrix} 1 & -1 & 2 \\ 2 & 1 & 1 \end{bmatrix} \begin{bmatrix} 1 \\ 1 \\ -2 \end{bmatrix} = \begin{bmatrix} -4 \\ 1 \end{bmatrix}$$

$$\mathbf{B}^{-1}(\mathbf{b} - \mathbf{N}\mathbf{x_N}) = \begin{bmatrix} 1 & 0 \\ 0 & 1 \end{bmatrix} \begin{bmatrix} 2 \\ 4 \end{bmatrix} - \begin{bmatrix} -4 \\ 1 \end{bmatrix} = \begin{bmatrix} 6 \\ 3 \end{bmatrix} = \mathbf{x_B}$$

Also, $x_3 = 1$ is a nonbasic variable at its upper bound, $x_4 = 1$ is a nonbasic variable at its lower bound, $x_5 = -2$ is a nonbasic variable at its lower bound. Finally, $x_1 = 6$, so $2 \le x_1 \le 7$, and $x_2 = 3$, so $-3 \le x_2 \le 9$. Thus the above solution is a bbfs.

It is important to note from Definition 10.1 that there are only a finite number of bbfs for a given LP. Indeed, there are at most $2^{n-m}\binom{n}{m}$ bbfs (why?). To solve the bounded LP problem, a finite improvement algorithm will be developed for moving from one bbfs to another while strictly decreasing the objective value at each step. The phase 1 issue of how to find an initial bbfs will be left to the reader (see Exercise 10.1.3); for now, assume that $\mathbf{x}$ is an initial bbfs from which to get started. To perform the test for optimality, one first computes the reduced costs $\mathbf{c_N} - \mathbf{c_B}\mathbf{B}^{-1}\mathbf{N}$, as usual, but then it is necessary to check both the following:

1. for all j with $1 \le j \le n - m$ and $(\mathbf{x_N})_j = (\mathbf{l_N})_j$, $(\mathbf{c_N} - \mathbf{c_B}\mathbf{B}^{-1}\mathbf{N})_j \ge 0$ and
2. for all j with $1 \le j \le n - m$ and $(\mathbf{x_N})_j = (\mathbf{u_N})_j$, $(\mathbf{c_N} - \mathbf{c_B}\mathbf{B}^{-1}\mathbf{N})_j \le 0$.

It is left as Exercise 10.1.5 to show that if statements 1 and 2 both hold for a bbfs $\mathbf{x}$, then $\mathbf{x}$ is optimal.

Suppose that $\mathbf{x}$ is a bbfs that fails the test for optimality, i.e., either

1. there is a j with $1 \le j \le n-m$ and $(\mathbf{x_N})_j = (\mathbf{l_N})_j$ such that $(\mathbf{c_N} - \mathbf{c_B}\mathbf{B}^{-1}\mathbf{N})_j < 0$; or
2. there is a j with $1 \le j \le n-m$ and $(\mathbf{x_N})_j = (\mathbf{u_N})_j$ such that $(\mathbf{c_N} - \mathbf{c_B}\mathbf{B}^{-1}\mathbf{N})_j > 0$.

In either case, it is possible to construct a direction leading to a new bbfs with a strictly better objective value. Corresponding to the two cases are the respective directions

1. $\mathbf{d} = (\mathbf{d_B}, \mathbf{d_N}) = (-\mathbf{B}^{-1}\mathbf{N}_{\cdot j}, \mathbf{I}_{\cdot j})$ and
2. $\mathbf{d} = (\mathbf{d_B}, \mathbf{d_N}) = (\mathbf{B}^{-1}\mathbf{N}_{\cdot j}, -\mathbf{I}_{\cdot j})$

Again, it is left to the reader (Exercise 10.1.6) to show that, in either case, the direction leads to improvement, i.e., $\mathbf{cd} < 0$. Nonetheless, it is worth noting that the direction in the first case corresponds to increasing a nonbasic variable from a lower bound (when $\mathbf{d_N} = \mathbf{I}_{\cdot j}$) and, in the second case, the direction corresponds to decreasing a nonbasic variable from an upper bound (when $\mathbf{d_N} = -\mathbf{I}_{\cdot j}$).

It remains to determine t^*—how far one can move from $\mathbf{x}$ in the direction $\mathbf{d}$ and remain feasible. As before, the equality constraints are always satisfied in the direction $\mathbf{d}$, as will be shown in Exercise 10.1.6. Thus only a bound constraint can become infeasible. Three cases can arise:

1. A basic variable decreases to a lower bound.
2. A basic variable increases to an upper bound.
3. The value of the nonbasic variable $(\mathbf{x_N})_j$ moves from its lower bound to its upper bound, or vice versa.

Through a min ratio formula, it is possible to compute the value for t^*. Specifically, corresponding to each of the three cases above, one would compute

1. $t_1 = \min\{[(\mathbf{x_B})_k - (\mathbf{l_B})_k]/-(\mathbf{d_B})_k : (\mathbf{d_B})_k < 0\}$,
2. $t_2 = \min\{[(\mathbf{u_B})_k - (\mathbf{x_B})_k]/(\mathbf{d_B})_k : (\mathbf{d_B})_k > 0\}$, and
3. $t_3 = (\mathbf{u_N})_j - (\mathbf{l_N})_j$

Finally, t^* is given by $t^* = \min\{t_1, t_2, t_3\}$.

When $t^* = t_1$ or $t^* = t_2$, the pivot operation is performed in the usual manner; i.e., the nonbasic variable $(\mathbf{x_N})_j$ becomes basic and $(\mathbf{x_B})_k$ becomes a nonbasic variable whose value is either $(\mathbf{l_B})_k$ if $t^* = t_1$ or $(\mathbf{u_B})_k$ if $t^* = t_2$. Also, we can use the new basis matrix to compute the values of the new basic variables by the formula $\mathbf{x_B} = \mathbf{B}^{-1}(\mathbf{b} - \mathbf{N}\mathbf{x_N})$. Finally, when $t^* = t_3$, no pivot operation is needed. Instead, the nonbasic variable $(\mathbf{x_N})_j$ changes its value from one bound to another (either lower to upper, or vice versa). The new values of the (old) basic variables are $\mathbf{x_B} + t^*\mathbf{d_B}$.

The steps of this algorithm will now be applied to solve the LP in Example 10.1.

EXAMPLE 10.2: Recall the LP in Example 10.1.

$$\begin{array}{ll}\text{Minimize} & x_1 - x_2 + 3x_3 - 2x_4 + 6x_5 \\ \text{subject to} & x_1 \qquad + x_3 - x_4 + 2x_5 = 2 \\ & \qquad x_2 + 2x_3 + x_4 + x_5 = 4 \\ & 2 \le x_1 \le 7 \\ & -3 \le x_2 \le 9 \\ & -4 \le x_3 \le 1 \\ & 1 \le x_4 \le 5 \\ & -2 \le x_5 \le 3 \end{array}$$

Iteration 1

Step 0. The bbfs for this iteration:

$$\mathbf{x_B} = (x_1, x_2) = (6, 3), \qquad \mathbf{B} = \begin{bmatrix} 1 & 0 \\ 0 & 1 \end{bmatrix}$$

$$\mathbf{x_N} = (x_3, x_4, x_5) = (1, 1, -2), \qquad \mathbf{N} = \begin{bmatrix} 1 & -1 & 2 \\ 2 & 1 & 1 \end{bmatrix}$$

Step 1. Testing for optimality:

$$\mathbf{c_B} = (1, -1), \qquad \mathbf{c_N} = (3, -2, 6)$$

$$\mathbf{c_B B^{-1} N} = (1, -1)\begin{bmatrix} 1 & 0 \\ 0 & 1 \end{bmatrix}\begin{bmatrix} 1 & -1 & 2 \\ 2 & 1 & 1 \end{bmatrix} = (-1, -2, 1)$$

$$\mathbf{c_N} - \mathbf{c_B B^{-1} N} = (3, -2, 6) - (-1, -2, 1) = (4, 0, 5)$$

Choose $j^* = 1$ since $(\mathbf{x_N})_1 = 1$ is a nonbasic variable at its upper bound and $(\mathbf{c_N} - \mathbf{c_B B^{-1} N})_1 = 4 > 0$.

Step 2. Computing the direction of movement:

$$\mathbf{d} = (\mathbf{d_B}, \mathbf{d_N}) = \left(\mathbf{B}^{-1}\mathbf{N}_{.1}, -\mathbf{I}_{.1}\right)$$

$$\mathbf{d_B} = \begin{bmatrix} 1 & 0 \\ 0 & 1 \end{bmatrix}\begin{bmatrix} 1 \\ 2 \end{bmatrix} = \begin{bmatrix} 1 \\ 2 \end{bmatrix} = \begin{bmatrix} d_1 \\ d_2 \end{bmatrix}$$

$$\mathbf{d_N} = \begin{bmatrix} -1 \\ 0 \\ 0 \end{bmatrix} = \begin{bmatrix} d_3 \\ d_4 \\ d_5 \end{bmatrix}$$

Step 3. Computing the amount of movement:

$$t_1 = \infty \qquad \text{(because } \mathbf{d_B} \text{ has no negative components)}$$

$$t_2 = \min\{[(\mathbf{u_B})_1 - (\mathbf{x_B})_1]/(\mathbf{d_B})_1, [(\mathbf{u_B})_2 - (\mathbf{x_B})_2]/(\mathbf{d_B})_2\}$$

$$= \min\{(7-6)/1, (9-3)/2\} = 1$$

$$t_3 = (\mathbf{u_N})_1 - (\mathbf{l_N})_1 = 1 - (-4) = 5$$

$$t^* = \min\{t_1, t_2, t_3\} = t_2 = 1 \qquad \text{so} \quad k^* = 1$$

Step 4. The pivot operation: Column $k^* = 1$ of $\mathbf{B}$ is replaced with column $j^* = 1$ of $\mathbf{N}$. Also, $(\mathbf{x_B})_1$ leaves the basis and becomes nonbasic at its upper bound of 7, and $(\mathbf{x_N})_1$ becomes basic at value 0. So the new bbfs is $\mathbf{x}' = \mathbf{x} + t^*\mathbf{d} = (6,3,1,1,-2) + 1(1,2,-1,0,0) = (7,5,0,1,-2)$.

Iteration 2

Step 0. The bbfs for this iteration:

$$\mathbf{x_B} = (x_3, x_2) = (0,5), \qquad \mathbf{B} = \begin{bmatrix} 1 & 0 \\ 2 & 1 \end{bmatrix}$$

$$\mathbf{x_N} = (x_1, x_4, x_5) = (7,1,-2), \qquad \mathbf{N} = \begin{bmatrix} 1 & -1 & 2 \\ 0 & 1 & 1 \end{bmatrix}$$

$$\mathbf{x_B} = \mathbf{B}^{-1}(\mathbf{b} - \mathbf{N}\mathbf{x_N}) = \begin{bmatrix} 1 & 0 \\ -2 & 1 \end{bmatrix}\left[\begin{bmatrix} 2 \\ 4 \end{bmatrix} - \begin{bmatrix} 1 & -1 & 2 \\ 0 & 1 & 1 \end{bmatrix}\begin{bmatrix} 7 \\ 1 \\ -2 \end{bmatrix}\right] = \begin{bmatrix} 0 \\ 5 \end{bmatrix}$$

Step 1. Testing for optimality:

$$\mathbf{c_B} = (3,-1), \qquad \mathbf{c_N} = (1,-2,6)$$

$$\mathbf{c_B}\mathbf{B}^{-1}\mathbf{N} = (3,-1)\begin{bmatrix} 1 & 0 \\ -2 & 1 \end{bmatrix}\begin{bmatrix} 1 & -1 & 2 \\ 0 & 1 & 1 \end{bmatrix} = (5,-6,9)$$

$$\mathbf{c_N} - \mathbf{c_B}\mathbf{B}^{-1}\mathbf{N} = (1,-2,6) - (5,-6,9) = (-4,4,-3)$$

Choose $j^* = 3$ because $(\mathbf{x_N})_3 = -2$ is a nonbasic variable at its lower bound and $(\mathbf{c_N} - \mathbf{c_B}\mathbf{B}^{-1}\mathbf{N})_3 = -3 < 0$.

Step 2. Computing the direction of movement:

$$\mathbf{d} = (\mathbf{d_B}, \mathbf{d_N}) = (-\mathbf{B}^{-1}\mathbf{N}_{.3}, \mathbf{I}_{.3})$$

$$\mathbf{d_B} = -\mathbf{B}^{-1}\mathbf{N}_{.3} = -\begin{bmatrix} 1 & 0 \\ -2 & 1 \end{bmatrix}\begin{bmatrix} 2 \\ 1 \end{bmatrix} = \begin{bmatrix} -2 \\ 3 \end{bmatrix} = \begin{bmatrix} d_3 \\ d_2 \end{bmatrix}$$

$$\mathbf{d_N} = \begin{bmatrix} 0 \\ 0 \\ 1 \end{bmatrix} = \begin{bmatrix} d_1 \\ d_4 \\ d_5 \end{bmatrix}$$

Step 3. Computing the amount of movement:

$$t_1 = \min\{[(\mathbf{x_B})_1 - (\mathbf{l_B})_1]/-(\mathbf{d_B})_1]\} = [0-(-4)]/-(-2) = 2$$

$$t_2 = \min\{[(\mathbf{u_B})_2 - (\mathbf{x_B})_2]/(\mathbf{d_B})_2\} = (9-5)/3 = \tfrac{4}{3}$$

$$t_3 = (\mathbf{u_N})_3 - (\mathbf{l_N})_3 = 3-(-2) = 5$$

$$t^* = \min\{t_1, t_2, t_3\} = t_2 = \tfrac{4}{3} \qquad \text{so} \quad k^* = 2$$

Step 4. The pivot operation: Column $k^* = 2$ of $\mathbf{B}$ is replaced with column $j^* = 3$ of $\mathbf{N}$. Also, $(\mathbf{x_B})_2$ leaves the basis and becomes nonbasic at its upper bound of 9, and $(\mathbf{x_N})_3$ becomes basic at value $-\tfrac{2}{3}$. The new bbfs is $\mathbf{x}' = \mathbf{x} + t^*\mathbf{d} = (7,5,0,1,-2) + \tfrac{4}{3}(0,3,-2,0,1) = (7,9,-\tfrac{8}{3},1,-\tfrac{2}{3})$.

Iteration 3

Step 0. The bbfs for this iteration:

$$\mathbf{x_B} = (x_3, x_5) = (-\tfrac{8}{3}, -\tfrac{2}{3}), \qquad \mathbf{B} = \begin{bmatrix} 1 & 2 \\ 2 & 1 \end{bmatrix}$$

$$\mathbf{x_N} = (x_1, x_4, x_2) = (7,1,9), \qquad \mathbf{N} = \begin{bmatrix} 1 & -1 & 0 \\ 0 & 1 & 1 \end{bmatrix}$$

Step 1. Testing for optimality:

$$\mathbf{c_B} = (3,6), \qquad \mathbf{c_N} = (1,-2,-1)$$

$$\mathbf{c_B}\mathbf{B}^{-1}\mathbf{N} = (3,6)\begin{bmatrix} -\frac{1}{3} & \frac{2}{3} \\ \frac{2}{3} & -\frac{1}{3} \end{bmatrix}\begin{bmatrix} 1 & -1 & 0 \\ 0 & 1 & 1 \end{bmatrix} = (3,-3,0)$$

$$\mathbf{c_N} - \mathbf{c_B}\mathbf{B}^{-1}\mathbf{N} = (1,-2,-1) - (3,-3,0) = (-2,1,-1)$$

Since $(\mathbf{c_N} - \mathbf{c_B}\mathbf{B}^{-1}\mathbf{N})_1 = -2 < 0$ and $x_1 = 7$ is at its upper bound, and $(\mathbf{c_N} - \mathbf{c_B}\mathbf{B}^{-1}\mathbf{N})_2 = 1 > 0$ and $x_4 = 1$ is at its lower bound, and $(\mathbf{c_N} - \mathbf{c_B}\mathbf{B}^{-1}\mathbf{N})_3 = -1 < 0$ and $x_2 = 9$ is at its upper bound, the test for optimality is satisfied. Therefore, the current solution $\mathbf{x}^* = (x_1, x_2, x_3, x_4, x_5) = (7, 9, -\tfrac{8}{3}, 1, -\tfrac{2}{3})$ is optimal.

This section has described a modified simplex algorithm that, without increasing the number of variables or the size of the basis matrix, is capable of solving an LP in which each variable has a known finite lower and upper bound. The next section extends this algorithm to allow the nonbasic variables to start with any values between their lower and upper bounds.

EXERCISES

10.1.1. Consider an LP in the following form.

$$\begin{aligned} \text{Minimize} \quad & \mathbf{cx} \\ \text{subject to} \quad & \mathbf{Ax} = \mathbf{b} \\ & \mathbf{l} \le \mathbf{x} \le \mathbf{u} \end{aligned}$$

(where $\mathbf{l}$ and $\mathbf{u}$ are finite lower and upper bounds). Show that the LP cannot be unbounded.

10.1.2. Show that the vector $\mathbf{x} = (2,1,1,1,0)$ is a bbfs for Example 10.1.

10.1.3. Consider the following phase 1 LP associated with the LP in Exercise 10.1.1:

$$\begin{aligned} \text{Minimize} \quad & \mathbf{ey} \\ \text{subject to} \quad & \mathbf{Ax} + \mathbf{Iy} = \mathbf{b} \\ & \mathbf{l} \le \mathbf{x} \le \mathbf{u} \\ & \mathbf{0} \le \mathbf{y} \end{aligned}$$

(a) Show that if $(\mathbf{x}, \mathbf{y})$ is feasible for the phase 1 LP and $\mathbf{y} = \mathbf{0}$, then $\mathbf{x}$ is feasible for the LP in Exercise 10.1.1.
(b) Show that if $\mathbf{b} - \mathbf{Al} \ge \mathbf{0}$, then the vector $(\mathbf{x}, \mathbf{y}) = (\mathbf{l}, \mathbf{b} - \mathbf{Al})$ is an initial bbfs for the phase 1 LP.
(c) Explain how to create an initial bbfs (in which $\mathbf{x} = \mathbf{l}$) for the phase 1 problem if it is not the case that $\mathbf{b} - \mathbf{Al} \ge \mathbf{0}$.

10.1.4. For the following LP:

$$\begin{aligned} \text{Minimize} \quad & x_1 + 3x_2 - 6x_3 - x_4 \\ \text{subject to} \quad & -x_1 + 2x_2 + x_3 + x_4 = 6 \\ & -3x_1 + x_2 + x_3 - 3x_4 = 5 \\ & 1 \le x_1 \le 2 \\ & 1 \le x_2 \le 6 \\ & -2 \le x_3 \le 6 \\ & -1 \le x_4 \le 0 \end{aligned}$$

(a) Create the phase 1 LP.
(b) Solve the phase 1 problem using the initial bbfs $(\mathbf{x}, \mathbf{y}) = (1, 1, -2, -1, 8, 6)$.
(c) Using the results obtained from phase 1, solve the original LP.

10.1.5. Prove that if $\mathbf{x}$ is a bbfs that satisfies

1. for all j with $1 \le j \le n - m$ and $(\mathbf{x_N})_j = (\mathbf{l_N})_j$, $(\mathbf{c_N} - \mathbf{c_B B^{-1} N})_j \ge 0$, and
2. for all j with $1 \le j \le n - m$ and $(\mathbf{x_N})_j = (\mathbf{u_N})_j$, $(\mathbf{c_N} - \mathbf{c_B B^{-1} N})_j \le 0$,

then $\mathbf{x}$ is optimal.

10.1.6. (a) Show that if $(\mathbf{c_N} - \mathbf{c_B B^{-1} N})_j > 0$, then the direction $\mathbf{d} = (\mathbf{d_B}, \mathbf{d_N}) = (\mathbf{B^{-1} N}_{.j}, -\mathbf{I}_{.j})$ satisfies $\mathbf{cd} < 0$.
(b) Show that if $\mathbf{x}$ is a bbfs and $\mathbf{d}$ is the direction given in part (a), then, for all $t \ge 0$, $\mathbf{x} + t\mathbf{d}$ satisfies the equality constraints.

10.1.7. For the algorithm of this section, explain under what conditions $t^* = 0$.

10.2. HANDLING NONBASIC VALUES BETWEEN THEIR BOUNDS

In the previous section a finite improvement algorithm was developed for dealing with an LP in which each variable had a known finite lower and upper bound. The approach was to allow each nonbasic variable to take on a value of either its lower or upper bound. In this section a more general procedure will be developed that allows the nonbasic variables to assume any arbitrary values between their bounds. One of the primary advantages of this added flexibility is that it enables the development of a phase 1 procedure that is capable of being initiated from any starting point. Also, the ability to handle unrestricted variables will be incorporated into the new procedure. Consequently, for this section, consider an LP in this form:

$$\begin{aligned} \text{Minimize} \quad & \mathbf{cx} \\ \text{subject to} \quad & \mathbf{Ax} = \mathbf{b} \\ & \mathbf{l} \le \mathbf{x} \le \mathbf{u} \end{aligned}$$

in which $\mathbf{l} \le \mathbf{u}$, but it could be that $l_i = -\infty$, or $u_i = +\infty$ or both.

To allow for arbitrary values of the nonbasic variables, yet another concept of a basic feasible solution is required. It is provided in the next definition.

Definition 10.2. A vector $\mathbf{x}$ is said to be a *basic feasible solution* (bfs) *of order p* (where $0 \le p \le n - m$) if there is a nonsingular $m \times m$ submatrix $\mathbf{B}$ of $\mathbf{A}$ such that, together with the remaining $n - m$ columns of $\mathbf{A}$ denoted by $\mathbf{N}$,

1. $\mathbf{x_B} = \mathbf{B}^{-1}(\mathbf{b} - \mathbf{N}\mathbf{x_N})$,
2. $\mathbf{l} \le \mathbf{x} \le \mathbf{u}$, and
3. precisely p of the nonbasic variables have values equal to one of their bounds.

The nonbasic variables whose values are strictly between their bounds will be called *superbasic variables*. The next example illustrates a bfs of order $p = 2$.

EXAMPLE 10.3:

$$\begin{aligned} \text{Minimize} \quad & x_1 + 2x_2 - 4x_3 - 5x_4 + x_5 \\ \text{subject to} \quad & x_1 - x_3 + 2x_4 - x_5 = 2 \\ & x_2 + 2x_3 - 3x_4 + 2x_5 = 6 \\ & -3 \le x_1 \le 5 \\ & 1 \le x_2 \le 6 \\ & -1 \le x_3 \le 3 \\ & -4 \le x_4 \le 7 \\ & 2 \le x_5 \le 6 \end{aligned}$$

For this LP the vector $\mathbf{x} = (x_1, x_2, x_3, x_4, x_5) = (3, 2, -1, 2, 6)$ is a bfs of order 2 in which $\mathbf{x_B} = (x_1, x_2) = (3, 2)$ and $\mathbf{x_N} = (x_3, x_4, x_5) = (-1, 2, 6)$. To see why, let

$$\mathbf{B} = \begin{bmatrix} 1 & 0 \\ 0 & 1 \end{bmatrix}, \qquad \mathbf{N} = \begin{bmatrix} -1 & 2 & -1 \\ 2 & -3 & 2 \end{bmatrix}$$

$$\mathbf{Nx_N} = \begin{bmatrix} -1 & 2 & -1 \\ 2 & -3 & 2 \end{bmatrix} \begin{bmatrix} -1 \\ 2 \\ 6 \end{bmatrix} = \begin{bmatrix} -1 \\ 4 \end{bmatrix}$$

$$\mathbf{b} - \mathbf{Nx_N} = \begin{bmatrix} 2 \\ 6 \end{bmatrix} - \begin{bmatrix} -1 \\ 4 \end{bmatrix} = \begin{bmatrix} 3 \\ 2 \end{bmatrix}$$

$$\mathbf{B}^{-1}(\mathbf{b} - \mathbf{Nx_N}) = \begin{bmatrix} 1 & 0 \\ 0 & 1 \end{bmatrix} \begin{bmatrix} 3 \\ 2 \end{bmatrix} = \begin{bmatrix} 3 \\ 2 \end{bmatrix} = \mathbf{x_B}$$

Also, $-3 \le x_1 = 3 \le 5$ is a basic variable, $1 \le x_2 = 2 \le 6$ is a basic variable, $x_3 = -1$ is a nonbasic variable at its lower bound; $-4 \le x_4 = 2 \le 7$ is a superbasic variable, and $x_5 = 6$ is a nonbasic variable at its upper bound, so precisely two of the nonbasic variables are at their bounds.

It is important to note that when $p = n - m$ Definition 10.2 becomes precisely that of a bbfs given in Definition 10.1. Hence there are a finite number of bfs of order $n - m$. However, for $p < n - m$, there are not a finite number of bfs of order p (why?).

In designing a finite improvement algorithm to solve this problem, suppose that $\mathbf{x}$ is a given bfs of some order, say, p. The phase 1 issue of how to find such a vector will be left to the reader (Exercise 10.2.3). At each iteration, the proposed procedure will produce a new bfs of order p or $p + 1$, but never of order $p - 1$. As long as the order of each successive bfs strictly increases by one, progress is being made toward a bfs of order $n - m$, of which there are then a finite number. On the other hand, when the new bfs is also of order p (instead of $p + 1$), there is a potential danger of generating an infinite number of such bfs and never reaching a bfs of order $p + 1$. Fortunately, for the proposed procedure, it can be shown that, under nondegeneracy, a finite number of bfs or order p can be visited before moving to a bfs of order $p + 1$. The steps of the finite improvement algorithm follow.

Step 0. Initialization: Find an initial bfs of some order, say, p (see the exercises). Go to step 1.

Step 1. Testing for optimality: Compute the reduced costs, $\mathbf{c_N} - \mathbf{c_B B}^{-1}\mathbf{N}$, and define the three sets

$$J = \{1 \le j \le n - m : (\mathbf{x_N})_j = (\mathbf{l_N})_j\}$$

$$K = \{1 \le j \le n - m : (\mathbf{x_N})_j = (\mathbf{u_N})_j\}$$

$$L = \{1 \le j \le n - m : (\mathbf{l_N})_j < (\mathbf{x_N})_j < (\mathbf{u_N})_j\}$$

If (1) for all j in J, $(\mathbf{c_N} - \mathbf{c_B}\mathbf{B}^{-1}\mathbf{N})_j \geq 0$; (2) for all j in K, $(\mathbf{c_N} - \mathbf{c_B}\mathbf{B}^{-1}\mathbf{N})_j \leq 0$; and (3) for all j in L, $(\mathbf{c_N} - \mathbf{c_B}\mathbf{B}^{-1}\mathbf{N})_j = 0$; then $\mathbf{x}$ is optimal. Otherwise, choose an "appropriate" j^* and go to step 2a, 2b, or 2c depending on whether j^* is in J, K, or L, respectively.

Step 2. Computing the direction of movement:

a. Set $\mathbf{d} = (\mathbf{d_B}, \mathbf{d_N}) = (-\mathbf{B}^{-1}\mathbf{N}_{.j^*}, \mathbf{I}_{.j^*})$.
b. Set $\mathbf{d} = (\mathbf{d_B}, \mathbf{d_N}) = (\mathbf{B}^{-1}\mathbf{N}_{.j^*}, -\mathbf{I}_{.j^*})$.
c. Set $\mathbf{d} = (\mathbf{d_B}, \mathbf{d_N}) = (-\mathbf{B}^{-1}\mathbf{N}_{.j^*}, \mathbf{I}_{.j^*})$ if $(\mathbf{c_N} - \mathbf{c_B}\mathbf{B}^{-1}\mathbf{N})_{j^*} < 0$; otherwise, set $\mathbf{d} = (\mathbf{d_B}, \mathbf{d_N}) = (\mathbf{B}^{-1}\mathbf{N}_{.j^*}, -\mathbf{I}_{.j^*})$.

Go to step 3. (*Note*: $\mathbf{d_N} = \mathbf{I}_{.j^*}$ corresponds to increasing a nonbasic variable, and $\mathbf{d_N} = -\mathbf{I}_{.j^*}$ corresponds to decreasing a nonbasic variable.)

Step 3. Computing the amount of movement: Compute

$$t_1 = \min\{[(\mathbf{x_B})_k - (\mathbf{l_B})_k]/-(\mathbf{d_B})_k : (\mathbf{d_B})_k < 0\}$$

$$t_2 = \min\{[(\mathbf{u_B})_k - (\mathbf{x_B})_k]/(\mathbf{d_B})_k : (\mathbf{d_B})_k > 0\}$$

$$t_3 = \begin{cases} (\mathbf{u_N})_{j^*} - (\mathbf{x_N})_{j^*} & \text{if } \mathbf{d_N} = \mathbf{I}_{.j^*} \\ (\mathbf{x_N})_{j^*} - (\mathbf{l_N})_{j^*} & \text{if } \mathbf{d_N} = -\mathbf{I}_{.j^*} \end{cases}$$

$$t^* = \min\{t_1, t_2, t_3\}$$

Note that t_1, t_2, t_3, and t^* can be computed even if a lower and/or upper bound is not finite. Thus the values of t_1, t_2, t_3, and t^* might not be finite. If t^* is not finite, then the LP is unbounded. Otherwise, go to step 4a if $t^* = t_1$ or $t^* = t_2$. If $t^* = t_3$, then go to step 4b.

Step 4. Pivoting:

a. In this case, a basic variable, say, $(\mathbf{x_B})_{k^*}$, has either increased to a finite upper bound or else decreased to a finite lower bound. Make $(\mathbf{x_N})_{j^*}$ basic and $(\mathbf{x_B})_{k^*}$ nonbasic. Compute the new values of the variables by $\mathbf{x}' = \mathbf{x} + t^*\mathbf{d}$. Find the inverse of the new basis matrix and go to step 1.
b. In this case, the nonbasic variable $(\mathbf{x_N})_{j^*}$ has moved to a finite bound. Compute the values of the new variables by $\mathbf{x}' = \mathbf{x} + t^*\mathbf{d}$. Do not change the basis matrix. Go to step 1.

It is worth noting that if j^* is in L (see step 1), then the order of the new bfs $\mathbf{x}'$ will be $p+1$ (why?), but if j^* is in J or K, then the order of $\mathbf{x}'$ will be p. However, in the latter case, all of the superbasic variables (which are necessarily nonbasic) remain fixed in value (only the nonbasic variable $(\mathbf{x_N})_{j^*}$ changes its value). As you will show (Exercise 10.2.2) as long as the superbasic variables are fixed in value, there are only a finite number of bfs of order p. It is this property, together with the nondegeneracy assumption, that ensures the finiteness of the algorithm. The steps will now be demonstrated with the LP in Example 10.2.

EXAMPLE 10.4: Consider the LP in Example 10.2,

$$\begin{aligned} \text{Minimize} \quad & x_1 + 2x_2 - 4x_3 - 5x_4 + x_5 \\ \text{subject to} \quad & x_1 \qquad - x_3 + 2x_4 - x_5 = 2 \\ & \qquad x_2 + 2x_3 - 3x_4 + 2x_5 = 6 \\ & -3 \le x_1 \le 5 \\ & 1 \le x_2 \le 6 \\ & -1 \le x_3 \le 3 \\ & -4 \le x_4 \le 7 \\ & 2 \le x_5 \le 6 \end{aligned}$$

Iteration 1

Step 0. The bfs for this iteration:

$$\mathbf{x_B} = (x_1, x_2) = (3, 2), \qquad \mathbf{B} = \begin{bmatrix} 1 & 0 \\ 0 & 1 \end{bmatrix}$$

$$\mathbf{x_N} = (x_3, x_4, x_5) = (-1, 2, 6), \qquad \mathbf{N} = \begin{bmatrix} -1 & 2 & -1 \\ 2 & -3 & 2 \end{bmatrix}$$

Step 1. Testing for optimality:

$$\mathbf{c_B} = (1, 2), \qquad \mathbf{c_N} = (-4, -5, 1)$$

$$\mathbf{c_B B^{-1} N} = (1, 2)\begin{bmatrix} -1 & 2 & -1 \\ 2 & -3 & 2 \end{bmatrix} = (3, -4, 3)$$

$$\mathbf{c_N} - \mathbf{c_B B^{-1} N} = (-4, -5, 1) - (3, -4, 3) = (-7, -1, -2)$$

$$J = \{1\}, \qquad K = \{3\}, \qquad L = \{2\}$$

Choose $j^* = 2$ since $(\mathbf{x_N})_2$ is a superbasic variable and $(\mathbf{c_N} - \mathbf{c_B B^{-1} N})_2 = -1 \neq 0$. One could also have chosen $j^* = 1$ since $(\mathbf{x_N})_1$ is a nonbasic variable at a lower bound whose reduced cost is strictly negative.

Step 2. Computing the direction of movement: Note that $(\mathbf{x_N})_2$ could be either increased or decreased since it is a superbasic variable. However, because its reduced cost is less than 0, $(\mathbf{x_N})_2$ is increased in order to improve the objective function:

$$\mathbf{d} = (\mathbf{d_B}, \mathbf{d_N}) = \left(-\mathbf{B}^{-1}\mathbf{N}_{.2}, \mathbf{I}_{.2}\right)$$

$$\mathbf{d_B} = -\begin{bmatrix} 1 & 0 \\ 0 & 1 \end{bmatrix}\begin{bmatrix} 2 \\ -3 \end{bmatrix} = \begin{bmatrix} -2 \\ 3 \end{bmatrix} = \begin{bmatrix} d_1 \\ d_2 \end{bmatrix}$$

$$\mathbf{d_N} = \begin{bmatrix} 0 \\ 1 \\ 0 \end{bmatrix} = \begin{bmatrix} d_3 \\ d_4 \\ d_5 \end{bmatrix}$$

Step 3. Computing the amount of movement:

$$\begin{aligned} t_1 &= \min\{[(\mathbf{x_B})_1 - (\mathbf{l_B})_1]/-(\mathbf{d_B})_1\} \\ &= [3-(-3)]/-(-2) = 3 \\ t_2 &= \min\{[(\mathbf{u_B})_2 - (\mathbf{x_B})_2]/(\mathbf{d_B})_2\} \\ &= (6-2)/3 = \tfrac{4}{3} \\ t_3 &= (\mathbf{u_N})_2 - (\mathbf{x_N})_2 = 7 - 2 = 5 \\ t^* &= \min\{t_1, t_2, t_3\} = t_2 = \tfrac{4}{3} \qquad \text{so} \quad k^* = 2 \end{aligned}$$

Step 4. The pivot operation: Column $k^* = 2$ of **B** is replaced with column $j^* = 2$ of **N**. Also, $(\mathbf{x_B})_2$ leaves the basis and becomes nonbasic at its upper bound, and $(\mathbf{x_N})_2$ becomes basic. The new bfs (or order 3) is $\mathbf{x}' = \mathbf{x} + t^*\mathbf{d} = (3, 2, -1, 2, 6) + \frac{4}{3}(-2, 3, 0, 1, 0) = (\frac{1}{3}, 6, -1, \frac{10}{3}, 6)$.

Iteration 2

Step 0. The bfs for this iteration:

$$\mathbf{x_B} = (x_1, x_4) = (\tfrac{1}{3}, \tfrac{10}{3}), \qquad \mathbf{B} = \begin{bmatrix} 1 & 2 \\ 0 & -3 \end{bmatrix}$$

$$\mathbf{x_N} = (x_3, x_2, x_5) = (-1, 6, 6), \qquad \mathbf{N} = \begin{bmatrix} -1 & 0 & -1 \\ 2 & 1 & 2 \end{bmatrix}$$

$$\mathbf{N}\mathbf{x_N} = \begin{bmatrix} -1 & 0 & -1 \\ 2 & 1 & 2 \end{bmatrix}\begin{bmatrix} -1 \\ 6 \\ 6 \end{bmatrix} = \begin{bmatrix} -5 \\ 16 \end{bmatrix}$$

$$\mathbf{b} - \mathbf{N}\mathbf{x_N} = \begin{bmatrix} 2 \\ 6 \end{bmatrix} - \begin{bmatrix} -5 \\ 16 \end{bmatrix} = \begin{bmatrix} 7 \\ -10 \end{bmatrix}$$

$$\mathbf{x_B} = \mathbf{B}^{-1}(\mathbf{b} - \mathbf{N}\mathbf{x_N}) = \begin{bmatrix} 1 & \frac{2}{3} \\ 0 & -\frac{1}{3} \end{bmatrix}\begin{bmatrix} 7 \\ -10 \end{bmatrix} = \begin{bmatrix} \frac{1}{3} \\ \frac{10}{3} \end{bmatrix} = \begin{bmatrix} x_1 \\ x_4 \end{bmatrix}$$

Step 1. Testing for optimality:

$$\mathbf{c_B} = (1, -5), \qquad \mathbf{c_N} = (-4, 2, 1)$$

$$\mathbf{c_B B}^{-1} = (1, -5)\begin{bmatrix} 1 & \frac{2}{3} \\ 0 & -\frac{1}{3} \end{bmatrix} = (1, \tfrac{7}{3})$$

$$\mathbf{c_N} - \mathbf{c_B B}^{-1}\mathbf{N} = (-4, 2, 1) - (1, \tfrac{7}{3})\begin{bmatrix} -1 & 0 & -1 \\ 2 & 1 & 2 \end{bmatrix}$$

$$= (-4, 2, 1) - (\tfrac{11}{3}, \tfrac{7}{3}, \tfrac{11}{3})$$

$$= (-\tfrac{23}{3}, -\tfrac{1}{3}, -\tfrac{8}{3})$$

$$J = \{1\}, \qquad K = \{2, 3\}, \qquad L = \varnothing$$

Choose $j^* = 1$ since $(\mathbf{x_N})_1$ is nonbasic at its lower bound and $(\mathbf{c_N} - \mathbf{c_B B}^{-1}\mathbf{N})_1 < 0$.

Step 2. Computing the direction of movement:

$$\mathbf{d} = (\mathbf{d_B}, \mathbf{d_N}) = (-\mathbf{B}^{-1}\mathbf{N}_{.1}, \mathbf{I}_{.1})$$

$$\mathbf{d_B} = \begin{bmatrix} -1 & -\frac{2}{3} \\ 0 & \frac{1}{3} \end{bmatrix}\begin{bmatrix} -1 \\ 2 \end{bmatrix} = \begin{bmatrix} -\frac{1}{3} \\ \frac{2}{3} \end{bmatrix}$$

$$\mathbf{d_N} = \begin{bmatrix} 1 \\ 0 \\ 0 \end{bmatrix} = \begin{bmatrix} d_3 \\ d_2 \\ d_5 \end{bmatrix}$$

Step 3. Computing the amount of movement:

$$t_1 = \min\{[(\mathbf{x_B})_1 - (\mathbf{l_B})_1]/-(\mathbf{d_B})_1\}$$

$$= [\tfrac{1}{3} - (-3)]/-(-\tfrac{1}{3}) = 10$$

$$t_2 = \min\{[(\mathbf{u_B})_2 - (\mathbf{x_B})_2]/(\mathbf{d_B})_2\}$$

$$= (7 - \tfrac{10}{3})/\tfrac{2}{3} = \tfrac{11}{2}$$

$$t_3 = (\mathbf{u_N})_1 - (\mathbf{x_N})_1 = 3 - (-1) = 4$$

$$t^* = \min\{t_1, t_2, t_3\} = t_3 = 4$$

Step 4. The pivot operation: Since $t^* = t_3$, no pivot operation is needed. The new bfs (of order 3) is $\mathbf{x}' = \mathbf{x} + t^*\mathbf{d} = (\tfrac{1}{3}, 6, -1, \tfrac{10}{3}, 6) + 4(-\tfrac{1}{3}, 0, 1, \tfrac{2}{3}, 0) = (-1, 6, 3, 6, 6)$.

Iteration 3

Step 0. The bfs for this iteration:

$$\mathbf{x_B} = (x_1, x_4) = (-1, 6)$$

$$\mathbf{x_N} = (x_3, x_2, x_5) = (3, 6, 6)$$

The **B** and **N** matrices are as in the previous iteration.

Step 1. Test for optimality: Since **B** and **N** have not changed, the reduced costs for this iteration are the same as in the previous iteration, so

$$\mathbf{c_N} - \mathbf{c_B B^{-1} N} = \left(-\tfrac{23}{3}, -\tfrac{1}{3}, -\tfrac{8}{3}\right)$$

However, $J = \varnothing$, $K = \{1,2,3\}$, $L = \varnothing$, and since for all j in K, $(\mathbf{c_N} - \mathbf{c_B B^{-1} N})_j \le 0$, the current solution $\mathbf{x}^* = (-1,6,3,6,6)$ is optimal.

This section has described a finite improvement algorithm that is capable of working with values of the nonbasic variables that are strictly between their bounds. This procedure is capable of handling unrestricted variables. It also provides a phase 1 procedure that can be initiated from any starting point, as will be shown in Exercise 10.2.3. The final section of this chapter will present a state-of-the-art method for handling bound constraints.

EXERCISES

10.2.1. What is the order of the bfs $\mathbf{x} = (4,2,0,0,2)$ for the LP in Example 10.3? Prove that your answer is correct.

10.2.2. **(a)** Explain why there are not a finite number of bfs of order p, where $p < n - m$.
(b) Explain why, at each iteration, the order of the bfs is either p or $p+1$, but never $p-1$. Specifically, explain when the new bfs will have order p, and when it will have order $p+1$.
(c) Explain why there are a finite number of bfs of order p $(< n - m)$ when the superbasic variables are fixed in value.

10.2.3. For a given starting point $\mathbf{x}'$, explain how to create an initial bfs (of some order) for the following phase 1 LP:

$$\begin{aligned} \text{Minimize} \quad & \mathbf{ey} \\ \text{subject to} \quad & \mathbf{Ax} + \mathbf{Iy} = \mathbf{b} \\ & \mathbf{l} \le \mathbf{x} \le \mathbf{u} \\ & \mathbf{0} \le \mathbf{y} \end{aligned}$$

in which the $\mathbf{x}$ variables are nonbasic at value $\mathbf{x}'$.

10.2.4. Solve the following LP:

$$\begin{aligned} \text{Minimize} \quad & -x_1 + 2x_2 + x_3 - x_4 \\ \text{subject to} \quad & x_1 + 3x_2 + 2x_3 = 5 \\ & x_2 + 2x_3 + x_4 = 6 \\ & 0 \le x_2 \le 9 \\ & 0 \le x_3 \le 8 \\ & x_1, x_4 \text{ unrestricted} \end{aligned}$$

Start with the initial bfs $\mathbf{x} = (0,1,1,3)$ in which x_2 and x_3 are the basic variables.

10.3. GENERALIZED UPPER BOUNDING

In this section, a state-of-the-art procedure will be presented for handling LPs in which some of the constraints have a special structure. To be specific, consider an LP:

$$\begin{aligned} \text{Minimize} \quad & \mathbf{cx} \\ \text{subject to} \quad & \mathbf{Ax} = \mathbf{b} \quad (m \text{ rows}) \\ & \mathbf{A'x} = \mathbf{b'} \quad (p \text{ rows}) \\ & \mathbf{x} \geq \mathbf{0} \end{aligned}$$

where $\mathbf{A}'$ is a $p \times n$ matrix and $\mathbf{b}'$ is a p vector. If $\mathbf{A}'$ satisfies the property that each element is -1, 0, or $+1$ and each column of $\mathbf{A}'$ has at most one nonzero entry, then the technique of *generalized upper bounding* (GUB) can be applied. Observe that when bound constraints of the form $\mathbf{l} \leq \mathbf{x} \leq \mathbf{u}$ are put in standard form by adding and subtracting slack variables, the resulting constraints do possess the desired requirements for GUB. The following is an example of an LP that can be solved by GUB.

EXAMPLE 10.5:

$$\begin{array}{lllllllll} \text{Minimize} & & & & & -3x_5 & -x_6 & -2x_7 & \\ \text{subject to} & x_1 & & & & +2x_5 & +x_6 & +\ x_7 & = 9 \\ & & x_2 & & & +\ x_5 & -x_6 & +\ x_7 & = 3 \\ & & & x_3 & & +\ x_5 & & & = 2 \\ & & & & x_4 & & +x_6 & +\ x_7 & = 4 \\ & x_1, & x_2, & x_3, & x_4, & x_5, & x_6, & x_7 & \geq 0 \end{array}$$

$$\mathbf{c} = (0, 0, 0, 0, -3, -1, -2)$$

$$\mathbf{A} = \begin{bmatrix} 1 & 0 & 0 & 0 & 2 & 1 & 1 \\ 0 & 1 & 0 & 0 & 1 & -1 & 1 \end{bmatrix}, \quad \mathbf{b} = \begin{bmatrix} 9 \\ 3 \end{bmatrix}$$

$$\mathbf{A}' = \begin{bmatrix} 0 & 0 & 1 & 0 & 1 & 0 & 0 \\ 0 & 0 & 0 & 1 & 0 & 1 & 1 \end{bmatrix}, \quad \mathbf{b}' = \begin{bmatrix} 2 \\ 4 \end{bmatrix}$$

In essence GUB is a technique that is designed to solve an LP in the appropriate form by working with an $m \times m$ basis matrix instead of an $(m+p) \times (m+p)$ basis matrix, as would be done with the revised simplex algorithm. The approach is based on the observation that all basis matrices for these types of LPs have a very special form that can be exploited.

For ease of exposition, it will be assumed that each element of $\mathbf{A}'$ is either 0 or 1. The case in which an element of $\mathbf{A}'$ could be -1 will be left to Exercise 10.3.1. Under this hypothesis, it can be further assumed that each component of $\mathbf{b}'$ is positive, for if $b_i' < 0$, then the LP must be infeasible since row i of $\mathbf{A}'$ contains only 0s and 1s, and also $\mathbf{x} \geq \mathbf{0}$. Also, if $b_i' = 0$, then, in order for $\mathbf{x}$ to satisfy $\mathbf{A}'_{i.}\mathbf{x} = b_i' = 0$, it must necessarily be the case that if $\mathbf{A}_{ij} = 1$, then $x_j = 0$, and those variables and constraints can be eliminated from the problem. The result of assuming that $\mathbf{b}' > \mathbf{0}$ is that each of the last p rows of an $(m+p)\times(m+p)$ basis matrix $\mathbf{B}$ must contain at least one element equal to 1. Otherwise, a row of the $\mathbf{B}$ matrix would contain all 0s, and hence $\mathbf{B}$ would be singular. Finally, since each of the last p rows of $\mathbf{B}$ contains at least one element equal to 1, it is possible to rearrange the columns of $\mathbf{B}$ so that a $p \times p$ identity matrix appears in the lower right corner. Specifically, $\mathbf{B}$ can be written

$$\begin{array}{cc} & \begin{array}{cc} m & p \end{array} \\ \mathbf{B} = & \begin{bmatrix} \mathbf{R} & \mathbf{S} \\ \mathbf{T} & \mathbf{I} \end{bmatrix} \end{array} \begin{array}{l} \\ m \text{ rows} \\ p \text{ rows} \end{array}$$

Recall that the $\mathbf{B}$ matrix will be used to solve the two linear systems $\mathbf{uB} = \mathbf{c_B}$ and $\mathbf{Bd_B} = -\mathbf{N}_{.j}$. Because of the special structure of $\mathbf{B}$, these two systems can be solved efficiently with the use of the lower triangular matrix

$$\begin{array}{cc} & \begin{array}{cc} m & p \end{array} \\ \mathbf{L} = & \begin{bmatrix} \mathbf{I} & \mathbf{0} \\ -\mathbf{T} & \mathbf{I} \end{bmatrix} \end{array} \begin{array}{l} \\ m \text{ rows} \\ p \text{ rows} \end{array}$$

in a manner similar to the $\mathbf{LU}$ decomposition of Section 7.4.

To solve the system $\mathbf{uB} = \mathbf{c_B}$, one will actually solve the equivalent system $\mathbf{u(BL)} = \mathbf{c_B L}$ because the matrix $\mathbf{BL}$ has the following useful structure:

$$\begin{array}{cc} & \begin{array}{cc} m & p \end{array} \\ \mathbf{BL} = & \begin{bmatrix} \mathbf{R}-\mathbf{ST} & \mathbf{S} \\ \mathbf{0} & \mathbf{I} \end{bmatrix} \end{array} \begin{array}{l} \\ m \text{ rows} \\ p \text{ rows} \end{array}$$

The matrix $\mathbf{Z} = \mathbf{R} - \mathbf{ST}$ is the desired $m \times m$ "basis" matrix that will be stored, updated, and used in the computations. In other words, formulas will now be developed for using the $\mathbf{Z}$ matrix to solve the systems $\mathbf{u(BL)} = \mathbf{c_B L}$ and to perform all the steps of the simplex algorithm.

To specify the details, it will be convenient to let $\mathbf{y}^m$ and $\mathbf{y}^p$ denote the first m and last p components of an $m+p$ vector $\mathbf{y}$, respectively. Then, to find $\mathbf{u}$ satisfying $\mathbf{u(BL)} = \mathbf{c_B L}$, it is necessary to solve

$$(\mathbf{u}^m, \mathbf{u}^p)\begin{bmatrix} \mathbf{Z} & \mathbf{S} \\ \mathbf{0} & \mathbf{I} \end{bmatrix} = ((\mathbf{c_B})^m, (\mathbf{c_B})^p)\begin{bmatrix} \mathbf{I} & \mathbf{0} \\ -\mathbf{T} & \mathbf{I} \end{bmatrix}$$

or

$$\mathbf{u}^m\mathbf{Z} = (\mathbf{c_B})^m - (\mathbf{c_B})^p\mathbf{T}$$

$$\mathbf{u}^m\mathbf{S} + \mathbf{u}^p\mathbf{I} = (\mathbf{c_B})^p$$

or

$$\mathbf{u}^m = \left[(\mathbf{c_B})^m - (\mathbf{c_B})^p\mathbf{T}\right]\mathbf{Z}^{-1}$$

$$\mathbf{u}^p = (\mathbf{c_B})^p - \mathbf{u}^m\mathbf{S}$$

In other words, knowing $\mathbf{Z}^{-1}$, one can compute $\mathbf{u}^m$, $\mathbf{u}^p$, and hence $\mathbf{u}$.

After $\mathbf{u}$ is known, the reduced costs can be determined and a nonbasic column can be selected to enter the basis matrix. For convenience, let that column be denoted by the $m+p$ vector $\mathbf{a} = (\mathbf{a}^m, \mathbf{a}^p)$. It is now necessary to solve the linear system $\mathbf{Bd_B} = -\mathbf{a}$. Once again this task will be accomplished by using the matrix $\mathbf{BL}$. Specifically, one first solves the system $(\mathbf{BL})\mathbf{w} = -\mathbf{a}$ for $\mathbf{w}$, for then $\mathbf{d_B}$ can be computed by $\mathbf{d_B} = \mathbf{Lw}$. To find $\mathbf{w}$, one must solve $(\mathbf{BL})\mathbf{w} = -\mathbf{a}$, or

$$\begin{bmatrix} \mathbf{Z} & \mathbf{S} \\ \mathbf{0} & \mathbf{I} \end{bmatrix}\begin{bmatrix} \mathbf{w}^m \\ \mathbf{w}^p \end{bmatrix} = \begin{bmatrix} -\mathbf{a}^m \\ -\mathbf{a}^p \end{bmatrix}$$

or

$$\mathbf{Zw}^m + \mathbf{Sw}^p = -\mathbf{a}^m$$

$$\mathbf{Iw}^p = -\mathbf{a}^p$$

or

$$\mathbf{w}^p = -\mathbf{a}^p$$

$$\mathbf{w}^m = \mathbf{Z}^{-1}(-\mathbf{a}^m + \mathbf{Sa}^p)$$

Finally, $\mathbf{d_B} = \mathbf{Lw}$, so

$$(\mathbf{d_B})^m = \mathbf{w}^m = \mathbf{Z}^{-1}(-\mathbf{a}^m + \mathbf{Sa}^p)$$

$$(\mathbf{d_B})^p = \mathbf{w}^p - \mathbf{Tw}^m = -\mathbf{a}^p - \mathbf{T}(\mathbf{d_B})^m$$

The final issue that must be addressed is that of updating $\mathbf{R}, \mathbf{S}, \mathbf{T}$, and $\mathbf{Z}^{-1}$ when the new column enters the basis. As usual, let k^* with $1 \le k^* \le p+m$ be the column of $\mathbf{B}$ that is leaving. There are two cases.

Case 1. $1 \le k^* \le m$

In this case, the $p \times p$ identity matrix in the lower right corner of the basis matrix will remain. The net result of this fact is that $\mathbf{Z}^{-1}$ can be updated with an $m \times m$ eta matrix $\mathbf{E}$ obtained by replacing column k^* of the $m \times m$ identity matrix with $-(\mathbf{d_B})^m$. In Exercise 10.3.2 it will be shown that when $\mathbf{E}$ is constructed in this way, the new matrix $\mathbf{Z}'$ satisfies $\mathbf{Z}' = \mathbf{ZE}$, and so $\mathbf{Z}'^{-1} = \mathbf{E}^{-1}\mathbf{Z}^{-1}$. Of course, $\mathbf{R}$ and $\mathbf{T}$ must also be updated.

Case 2. $m+1 \leq k^* \leq m+p$

In this case, the $p \times p$ identity matrix in the lower right corner of the basis matrix might be destroyed, depending on the form of the entering column. There is one fortunate circumstance in which the identity matrix remains unchanged.

Subcase a: Row k^* of $\mathbf{B}$ = row k^* of $\mathbf{I}$: In this subcase, component k^* of the entering column must equal 1, for if it contained a 0, then row k^* of the new basis matrix would be the zero vector, which is impossible. Moreover, as will be shown in Exercise 10.3.3, the $\mathbf{Z}$ matrix does not change at all, i.e., $\mathbf{Z}' = \mathbf{Z}$, and so no updating needs to be performed. However, the $\mathbf{S}$ matrix must be updated.

Subcase b: Row k^* of $\mathbf{B} \neq$ row k^* of $\mathbf{I}$: In this subcase, the $p \times p$ identity matrix in $\mathbf{B}$ might be lost. To circumvent this problem, prior to performing the pivot operation, column k^* of $\mathbf{B}$ can be exchanged with some other column of $\mathbf{B}$, say i, in such a way that a $p \times p$ identity matrix still appears in the lower right corner. Then, when the pivot operation is performed, the identity matrix is not destroyed. As will be shown in Exercise 10.3.4, it is always possible to find an appropriate column of $\mathbf{B}$ to exchange with column k^*.

When column i of $\mathbf{B}$ is exchanged with column k^*, the $\mathbf{Z}$ matrix changes to $\tilde{\mathbf{Z}}$. Thus, prior to pivoting, it is first necessary to update $\mathbf{Z}^{-1}$ into $\tilde{\mathbf{Z}}^{-1}$. To do so, create the $m \times m$ matrix $\mathbf{F}$ by replacing row i of the $m \times m$ identity matrix with the negative of the first m components of row k^* of $\mathbf{B}$. It can then be shown that $\mathbf{F}^{-1} = \mathbf{F}$, and $\tilde{\mathbf{Z}}^{-1} = \mathbf{F}^{-1}\mathbf{Z}^{-1}$. Also, $\mathbf{S}$ must be changed to $\tilde{\mathbf{S}}$, and component i and k^* of $\mathbf{d_B}$ must be interchanged. Finally, $\mathbf{Z}$ can be updated as in case 1.

These computations will now be illustrated for the LP in Example 10.5.

EXAMPLE 10.6: Consider the LP in Example 10.5:

$$
\begin{array}{llllllll}
\text{Minimize} & & & & & -3x_5 & -x_6 & -2x_7 \\
\text{subject to} & x_1 & & & & +2x_5 & +x_6 & +\ x_7 = 9 \\
 & & x_2 & & & x_5 & -x_6 & +\ x_7 = 3 \\
 & & & x_3 & & +\ x_5\ + & & \quad\ \ = 2 \\
 & & & & x_4 & & +x_6 & +\ x_7 = 4 \\
 & x_1, & x_2, & x_3, & x_4, & x_5, & x_6, & x_7 \geq 0
\end{array}
$$

Iteration 1

Step 0. The bfs for this iteration:

$$\mathbf{x_B} = (x_1, x_2, x_3, x_4) = (9,3,2,4), \qquad \mathbf{x_N} = (x_5, x_6, x_7) = (0,0,0)$$

$$\mathbf{B} = \begin{bmatrix} 1 & 0 & 0 & 0 \\ 0 & 1 & 0 & 0 \\ 0 & 0 & 1 & 0 \\ 0 & 0 & 0 & 1 \end{bmatrix}, \qquad \mathbf{N} = \begin{bmatrix} 2 & 1 & 1 \\ 1 & -1 & 1 \\ 1 & 0 & 0 \\ 0 & 1 & 1 \end{bmatrix}$$

$$\mathbf{R} = \begin{bmatrix} 1 & 0 \\ 0 & 1 \end{bmatrix}, \qquad \mathbf{S} = \begin{bmatrix} 0 & 0 \\ 0 & 0 \end{bmatrix}, \qquad \mathbf{T} = \begin{bmatrix} 0 & 0 \\ 0 & 0 \end{bmatrix}$$

$$\mathbf{Z} = \mathbf{R} - \mathbf{ST} = \begin{bmatrix} 1 & 0 \\ 0 & 1 \end{bmatrix}$$

$$\mathbf{L} = \begin{bmatrix} 1 & 0 & 0 & 0 \\ 0 & 1 & 0 & 0 \\ 0 & 0 & 1 & 0 \\ 0 & 0 & 0 & 1 \end{bmatrix}, \qquad m = 2, \qquad p = 2$$

Step 1. Testing for optimality [by solving the system $\mathbf{u}(\mathbf{BL}) = \mathbf{c_B L}$ for $\mathbf{u}$]:

$$\mathbf{u}^m = \left[(\mathbf{c_B})^m - (\mathbf{c_B})^p \mathbf{T}\right]\mathbf{Z}^{-1} = [(0,0) - (0,0)\mathbf{T}]\mathbf{I} = (0,0)$$

$$\mathbf{u}^p = (\mathbf{c_B})^p - \mathbf{u}^m \mathbf{S} = (0,0) - (0,0)\mathbf{S} = (0,0)$$

$$\mathbf{u} = (\mathbf{u}^m, \mathbf{u}^p) = (0,0,0,0)$$

Computing the reduced costs:

$$\mathbf{c_N} - \mathbf{uN} = \mathbf{c_N} = (-3, -1, -2), \qquad \text{so} \quad j^* = 1$$

Step 2. Computing the direction of movement [by solving the system $\mathbf{Bd_B} = -\mathbf{N}_{.j^*} = -\mathbf{a}$ for $\mathbf{d_B}$):

$$(\mathbf{d_B})^m = \mathbf{Z}^{-1}(-\mathbf{a}^m + \mathbf{S}\mathbf{a}^p) = \mathbf{I}[(-2,-1) + (\mathbf{0})\mathbf{a}^p] = (-2,-1)$$

$$(\mathbf{d_B})^p = -\mathbf{a}^p - \mathbf{T}(\mathbf{d_B})^m = (-1,0) - (\mathbf{0})(\mathbf{d_B})^m = (-1,0)$$

$$\mathbf{d_B} = \left((\mathbf{d_B})^m, (\mathbf{d_B})^p\right) = (-2,-1,-1,0)$$

Step 3. Computing the amount of movement:

$$\begin{aligned} t^* &= \min\{(\mathbf{x_B})_k / -(\mathbf{d_B})_k : (\mathbf{d_B})_k < 0\} \\ &= \min\{\tfrac{9}{2}, \tfrac{3}{1}, \tfrac{2}{1}\} \\ &= 2 \qquad \text{so} \quad k^* = 3 \end{aligned}$$

Step 4. The pivot operation (case 2a): Column $k^* = 3$ of $\mathbf{B}$ is replaced with column $j^* = 1$ of $\mathbf{N}$ and, in this case, $\mathbf{Z}$ and $\mathbf{Z}^{-1}$ do not change, so

$$\mathbf{B}' = \begin{bmatrix} 1 & 0 & 2 & 0 \\ 0 & 1 & 1 & 0 \\ 0 & 0 & 1 & 0 \\ 0 & 0 & 0 & 1 \end{bmatrix}, \qquad \mathbf{N}' = \begin{bmatrix} 0 & 1 & 1 \\ 0 & -1 & 1 \\ 1 & 0 & 0 \\ 0 & 1 & 1 \end{bmatrix}$$

$$\begin{aligned} \mathbf{x}' &= \mathbf{x} + t^*\mathbf{d} \\ &= (9,3,2,4,0,0,0) + 2(-2,-1,-1,0,1,0,0) \\ &= (5,1,0,4.2,0,0) \end{aligned}$$

Iteration 2

Step 0. The bfs for this iteration:

$$\mathbf{x_B} = (x_1, x_2, x_5, x_4) = (5,1,2,4), \qquad \mathbf{x_N} = (x_3, x_6, x_7) = (0,0,0)$$

$$\mathbf{B} = \begin{bmatrix} 1 & 0 & 2 & 0 \\ 0 & 1 & 1 & 0 \\ 0 & 0 & 1 & 0 \\ 0 & 0 & 0 & 1 \end{bmatrix}, \qquad \mathbf{N} = \begin{bmatrix} 0 & 1 & 1 \\ 0 & -1 & 1 \\ 1 & 0 & 0 \\ 0 & 1 & 1 \end{bmatrix}$$

$$\mathbf{R} = \begin{bmatrix} 1 & 0 \\ 0 & 1 \end{bmatrix}, \qquad \mathbf{S} = \begin{bmatrix} 2 & 0 \\ 1 & 0 \end{bmatrix}, \qquad \mathbf{T} = \begin{bmatrix} 0 & 0 \\ 0 & 0 \end{bmatrix}$$

$$\mathbf{Z} = \begin{bmatrix} 1 & 0 \\ 0 & 1 \end{bmatrix}, \qquad \mathbf{L} = \mathbf{I}$$

Step 1. Testing for optimality [by solving the system $\mathbf{u(BL)} = \mathbf{c_B L}$ for $\mathbf{u}$]:

$$\mathbf{u}^m = \left[(\mathbf{c_B})^m - (\mathbf{c_B})^p\mathbf{T}\right]\mathbf{Z}^{-1} = [(0,0) - (-3,0)(\mathbf{0})]\mathbf{I} = (0,0)$$

$$\mathbf{u}^p = (\mathbf{c_B})^p - \mathbf{u}^m\mathbf{S} = (-3,0) - (0,0)\mathbf{S} = (-3,0)$$

$$\mathbf{u} = (\mathbf{u}^m, \mathbf{u}^p) = (0,0,-3,0)$$

Computing the reduced costs:

$$\mathbf{c_N} - \mathbf{uN} = (0,-1,-2) - (0,0,-3,0)\begin{bmatrix} 0 & 1 & 1 \\ 0 & -1 & 1 \\ 1 & 0 & 0 \\ 0 & 1 & 1 \end{bmatrix}$$

$$= (0,-1,-2) - (-3,0,0)$$

$$= (3,-1,-2) \qquad \text{so} \quad j^* = 3$$

Step 2. Computing the direction of movement (by solving the system $\mathbf{Bd_B} = -\mathbf{N}_{\cdot j^*} = -\mathbf{a}$ for $\mathbf{d_B}$):

$$(\mathbf{d_B})^m = \mathbf{Z}^{-1}(-\mathbf{a}^m + \mathbf{Sa}^p)$$

$$= \mathbf{I}\left[\begin{bmatrix} -1 \\ -1 \end{bmatrix} + \begin{bmatrix} 2 & 0 \\ 1 & 0 \end{bmatrix}\begin{bmatrix} 0 \\ 1 \end{bmatrix}\right] = \begin{bmatrix} -1 \\ -1 \end{bmatrix}$$

$$(\mathbf{d_B})^p = -\mathbf{a}^p - \mathbf{T}(\mathbf{d_B})^m$$

$$= \begin{bmatrix} 0 \\ -1 \end{bmatrix} - (\mathbf{0})(\mathbf{d_B})^m = \begin{bmatrix} 0 \\ -1 \end{bmatrix}$$

$$\mathbf{d_B} = \left((\mathbf{d_B})^m, (\mathbf{d_B})^p\right) = (-1,-1,0,-1)$$

Step 3. Computing the amount of movement:

$$t^* = \min\left\{(\mathbf{x_B})_k / -(\mathbf{d_B})_k : (\mathbf{d_B})_k < 0\right\}$$

$$= \min\left\{\tfrac{5}{1}, \tfrac{1}{1}, \tfrac{4}{1}\right\}$$

$$= 1 \qquad \text{so} \quad k^* = 2$$

Step 4. The pivot operation (case 1): Column $k^* = 2$ of $\mathbf{B}$ is replaced with column $j^* = 3$ of $\mathbf{N}$. Also, $\mathbf{Z}$ and $\mathbf{Z}^{-1}$ change, so

$$\mathbf{B}' = \begin{bmatrix} 1 & 1 & 2 & 0 \\ 0 & 1 & 1 & 0 \\ 0 & 0 & 1 & 0 \\ 0 & 1 & 0 & 1 \end{bmatrix}, \qquad \mathbf{N}' = \begin{bmatrix} 0 & 1 & 0 \\ 0 & -1 & 1 \\ 1 & 0 & 0 \\ 0 & 1 & 0 \end{bmatrix}$$

$$\begin{aligned} \mathbf{x}' &= \mathbf{x} + t^*\mathbf{d} \\ &= (5,1,0,4,2,0,0) + 1(-1,-1,0,-1,0,0,1) \\ &= (4,0,0,3,2,0,1) \end{aligned}$$

Iteration 3

Step 0. The bfs for this iteration:

$$\mathbf{x_B} = (x_1, x_7, x_5, x_4) = (4,1,2,3), \qquad \mathbf{x_N} = (x_3, x_6, x_2) = (0,0,0)$$

$$\mathbf{B} = \begin{bmatrix} 1 & 1 & 2 & 0 \\ 0 & 1 & 1 & 0 \\ 0 & 0 & 1 & 0 \\ 0 & 1 & 0 & 1 \end{bmatrix}, \qquad \mathbf{N} = \begin{bmatrix} 0 & 1 & 0 \\ 0 & -1 & 1 \\ 1 & 0 & 0 \\ 0 & 1 & 0 \end{bmatrix}$$

$$\mathbf{R} = \begin{bmatrix} 1 & 1 \\ 0 & 1 \end{bmatrix}, \qquad \mathbf{S} = \begin{bmatrix} 2 & 0 \\ 1 & 0 \end{bmatrix}, \qquad \mathbf{T} = \begin{bmatrix} 0 & 0 \\ 0 & 1 \end{bmatrix}$$

$$\mathbf{Z} = \mathbf{R} - \mathbf{ST} = \begin{bmatrix} 1 & 1 \\ 0 & 1 \end{bmatrix}$$

$$\mathbf{L} = \begin{bmatrix} 1 & 0 & 0 & 0 \\ 0 & 1 & 0 & 0 \\ 0 & 0 & 1 & 0 \\ 0 & -1 & 0 & 1 \end{bmatrix}$$

Step 1. Testing for optimality [by solving the system $\mathbf{u(BL)} = \mathbf{c_B L}$ for $\mathbf{u}$]:

$$\begin{aligned} \mathbf{u}^m &= \left[(\mathbf{c_B})^m - (\mathbf{c_B})^p\mathbf{T}\right]\mathbf{Z}^{-1} \\ &= \left[(0,-2) - (-3,0)\begin{bmatrix} 0 & 0 \\ 0 & 1 \end{bmatrix}\right]\begin{bmatrix} 1 & -1 \\ 0 & 1 \end{bmatrix} = (0,-2)\begin{bmatrix} 1 & -1 \\ 0 & 1 \end{bmatrix} \\ &= (0,-2) \end{aligned}$$

$$\begin{aligned} \mathbf{u}^p &= (\mathbf{c_B})^p - \mathbf{u}^m\mathbf{S} \\ &= (-3,0) - (0,-2)\begin{bmatrix} 2 & 0 \\ 1 & 0 \end{bmatrix} = (-3,0) - (-2,0) \\ &= (-1,0) \end{aligned}$$

$$\mathbf{u} = (\mathbf{u}^m, \mathbf{u}^p) = (0,-2,-1,0)$$

Computing the reduced costs:

$$\mathbf{c_N} - \mathbf{uN} = (0,-1,0) - (0,-2,-1,0)\begin{bmatrix} 0 & 1 & 0 \\ 0 & -1 & 1 \\ 1 & 0 & 0 \\ 0 & 1 & 0 \end{bmatrix}$$

$$= (0,-1,0) - (-1,2,-2)$$

$$= (1,-3,2) \qquad \text{so} \quad j^* = 2$$

Step 2. Computing the direction of movement (by solving the system $\mathbf{Bd_B} = -\mathbf{N}_{.j^*} = -\mathbf{a}$ for $\mathbf{d_B}$):

$$(\mathbf{d_B})^m = \mathbf{Z}^{-1}(-\mathbf{a}^m + \mathbf{Sa}^p)$$

$$= \begin{bmatrix} 1 & -1 \\ 0 & 1 \end{bmatrix}\left[\begin{bmatrix} -1 \\ 1 \end{bmatrix} + \begin{bmatrix} 2 & 0 \\ 1 & 0 \end{bmatrix}\begin{bmatrix} 0 \\ 1 \end{bmatrix}\right]$$

$$= \begin{bmatrix} 1 & -1 \\ 0 & 1 \end{bmatrix}\begin{bmatrix} -1 \\ 1 \end{bmatrix} = \begin{bmatrix} -2 \\ 1 \end{bmatrix}$$

$$(\mathbf{d_B})^p = -\mathbf{a}^p - \mathbf{T}(\mathbf{d_B})^m$$

$$= \begin{bmatrix} 0 \\ -1 \end{bmatrix} - \begin{bmatrix} 0 & 0 \\ 0 & 1 \end{bmatrix}\begin{bmatrix} -2 \\ 1 \end{bmatrix} = \begin{bmatrix} 0 \\ -2 \end{bmatrix}$$

$$\mathbf{d_B} = \left((\mathbf{d_B})^m, (\mathbf{d_B})^p\right) = (-2,1,0,-2)$$

Step 3. Computing the amount of movement:

$$t^* = \min\{(\mathbf{x_B})_k / -(\mathbf{d_B})_k : (\mathbf{d_B})_k < 0\} = \min\{\tfrac{4}{2}, \tfrac{3}{2}\}$$

$$= \tfrac{3}{2} \qquad \text{so} \quad k^* = 4$$

Step 4. The pivot operation (case 2b): In this case, prior to performing the pivot operation, column $k^* = 4$ of $\mathbf{B}$ is exchanged with column $i = 2$ of $\mathbf{B}$. The new matrices are

$$\mathbf{B} = \begin{bmatrix} 1 & 0 & 2 & 1 \\ 0 & 0 & 1 & 1 \\ 0 & 0 & 1 & 0 \\ 0 & 1 & 0 & 1 \end{bmatrix}$$

$$\mathbf{R} = \begin{bmatrix} 1 & 0 \\ 0 & 0 \end{bmatrix}, \qquad \mathbf{S} = \begin{bmatrix} 2 & 1 \\ 1 & 1 \end{bmatrix}, \qquad \mathbf{T} = \begin{bmatrix} 0 & 0 \\ 0 & 1 \end{bmatrix}$$

$$\mathbf{Z} = \mathbf{R} - \mathbf{ST} = \begin{bmatrix} 1 & -1 \\ 0 & -1 \end{bmatrix}$$

The result of interchanging these columns in the **B** matrix is that those corresponding components of $\mathbf{d_B}$ must also interchange (i.e., components 2 and 4), so

$$(\mathbf{d_B})^m = \begin{bmatrix} -2 \\ -2 \end{bmatrix}, \qquad (\mathbf{d_B})^p = \begin{bmatrix} 0 \\ 1 \end{bmatrix}$$

Using the new data, we can now perform the pivot operation by replacing column $k^* = 2$ of **B** with column $j^* = 2$ of **N** (note that k^* has changed from 4 to 2):

$$\mathbf{B}' = \begin{bmatrix} 1 & 1 & 2 & 1 \\ 0 & -1 & 1 & 1 \\ 0 & 0 & 1 & 0 \\ 0 & 1 & 0 & 1 \end{bmatrix}, \qquad \mathbf{N}' = \begin{bmatrix} 0 & 0 & 0 \\ 0 & 0 & 1 \\ 1 & 0 & 0 \\ 0 & 1 & 1 \end{bmatrix}$$

$$\begin{aligned} \mathbf{x}' &= \mathbf{x} + t^*\mathbf{d} \\ &= (4,0,0,3,2,0,1) + \tfrac{3}{2}(-2,0,0,-2,0,1,1) \\ &= (1,0,0,0,2,\tfrac{3}{2},\tfrac{5}{2}) \end{aligned}$$

Iteration 4

Step 0. The bfs for this iteration:

$$\mathbf{x_B} = (x_1, x_6, x_5, x_7) = (1, \tfrac{3}{2}, 2, \tfrac{5}{2}), \qquad \mathbf{x_N} = (x_3, x_4, x_2) = (0,0,0)$$

$$\mathbf{B} = \begin{bmatrix} 1 & 1 & 2 & 1 \\ 0 & -1 & 1 & 1 \\ 0 & 0 & 1 & 0 \\ 0 & 1 & 0 & 1 \end{bmatrix}, \qquad \mathbf{N} = \begin{bmatrix} 0 & 0 & 0 \\ 0 & 0 & 1 \\ 1 & 0 & 0 \\ 0 & 1 & 1 \end{bmatrix}$$

$$\mathbf{R} = \begin{bmatrix} 1 & 1 \\ 0 & -1 \end{bmatrix}, \qquad \mathbf{S} = \begin{bmatrix} 2 & 1 \\ 1 & 1 \end{bmatrix}, \qquad \mathbf{T} = \begin{bmatrix} 0 & 0 \\ 0 & 1 \end{bmatrix}$$

$$\mathbf{Z} = \mathbf{R} - \mathbf{ST} = \begin{bmatrix} 1 & 0 \\ 0 & -2 \end{bmatrix}$$

$$\mathbf{L} = \begin{bmatrix} 1 & 0 & 0 & 0 \\ 0 & 1 & 0 & 0 \\ 0 & 0 & 1 & 0 \\ 0 & -1 & 0 & 1 \end{bmatrix}$$

Step 1. Testing for optimality [by solving the system $\mathbf{u}(\mathbf{BL}) = \mathbf{c_B L}$ for $\mathbf{u}$]:

$$\begin{aligned}
\mathbf{u}^m &= \left[(\mathbf{c_B})^m - (\mathbf{c_B})^p \mathbf{T}\right]\mathbf{Z}^{-1} \\
&= \left[(0,-1) - (-3,-2)\begin{bmatrix} 0 & 0 \\ 0 & 1 \end{bmatrix}\right]\begin{bmatrix} 1 & 0 \\ 0 & -\frac{1}{2} \end{bmatrix} \\
&= [(0,-1) - (0,-2)]\begin{bmatrix} 1 & 0 \\ 0 & -\frac{1}{2} \end{bmatrix} \\
&= (0,1)\begin{bmatrix} 1 & 0 \\ 0 & -\frac{1}{2} \end{bmatrix} \\
&= (0,-\tfrac{1}{2}) \\
\mathbf{u}^p &= (\mathbf{c_B})^p - \mathbf{u}^m \mathbf{S} \\
&= (-3,-2) - (0,-\tfrac{1}{2})\begin{bmatrix} 2 & 1 \\ 1 & 1 \end{bmatrix} \\
&= (-3,-2) - (-\tfrac{1}{2},-\tfrac{1}{2}) \\
&= (-\tfrac{5}{2},-\tfrac{3}{2}) \\
\mathbf{u} &= (\mathbf{u}^m,\mathbf{u}^p) = (0,-\tfrac{1}{2},-\tfrac{5}{2},-\tfrac{3}{2})
\end{aligned}$$

Computing the reduced costs:

$$\begin{aligned}
\mathbf{c_N} - \mathbf{uN} &= (0,0,0) - (0,-\tfrac{1}{2},-\tfrac{5}{2},-\tfrac{3}{2})\begin{bmatrix} 0 & 0 & 0 \\ 0 & 0 & 1 \\ 1 & 0 & 0 \\ 0 & 1 & 0 \end{bmatrix} \\
&= (0,0,0) - (\tfrac{5}{2},-\tfrac{3}{2},-\tfrac{1}{2}) \\
&= (\tfrac{5}{2},\tfrac{3}{2},\tfrac{1}{2})
\end{aligned}$$

Since $\mathbf{c_N} - \mathbf{uN} \geq \mathbf{0}$, the current bfs $\mathbf{x}^* = (1,0,0,0,2,\frac{3}{2},\frac{5}{2})$ is optimal.

This section has presented GUB, a state-of-the-art procedure for exploiting a special structure that often arises in the constraints of an LP. The larger the value of p in comparison to m, the more effective is GUB. Computationally, it is advantageous to use GUB when p constitutes at least one-third of the constraints. The next chapter deals with yet another special structure that can arise in the constraints of certain LPs.

EXERCISES

10.3.1. Develop a GUB algorithm that is capable of handling problems in which elements of $\mathbf{A}'$ could be -1.

10.3.2. Show that if $1 \le k^* \le m$, and $\mathbf{E}$ is the eta matrix obtained by replacing column k^* of $\mathbf{I}$ with $-(\mathbf{d_B})^m$, then $\mathbf{Z}' = \mathbf{ZE}$.

10.3.3. Show that if $m + 1 \le k^* \le m + p$ and row k^* of $\mathbf{B}$ = row k^* of $\mathbf{I}$, then $\mathbf{Z}' = \mathbf{Z}$.

10.3.4. Show that if $m + 1 \le k^* \le m + p$ and row k^* of $\mathbf{B} \ne$ row k^* of $\mathbf{I}$, then it is always possible to find a column i of $\mathbf{B}$ to exchange with k^* so as to retain an identity matrix in the lower right corner.

10.3.5. Solve the following LP with GUB:

$$\begin{array}{lllllll} \text{Minimize} & & & & -5x_5 & -7x_6 & -3x_7 \\ \text{subject to} & x_1 & & & +\ x_5 & +\ x_6 & -\ x_7 = 6 \\ & & x_2 & & +\ x_5 & +2x_6 & +\ x_7 = 14 \\ & & & x_3 & +\ x_5 & & +\ x_7 = 8 \\ & & & x_4 & & +\ x_6 & = 4 \\ & x_1, x_2, x_3, x_4, & & & x_5, & x_6, & x_7 \ge 0 \end{array}$$

starting with the initial bfs $\mathbf{x} = (6, 14, 8, 4, 0, 0, 0)$.

DISCUSSION

Bounding Techniques

The first attempt to handle bound constraints was by Dantzig [1]. The generalized upper bounding approach was presented by Dantzig and Van Slyke [2].

The concept of superbasic variables first appeared in a nonlinear programming algorithm called the convex simplex method developed by Zangwill [4]. The use of superbasic variables, as described in this chapter, was developed by Solow [3].

REFERENCES

1. G. B. Dantzig, "Upper Bounds, Secondary Constraints, and Block Triangularity," *Econometrica* 23(2), 174–183, (1955); also in Research Memorandum RM-1367, The Rand Corporation, Santa Monica, CA, 1954.
2. G. B. Dantzig and R. M. Van Slyke, "Generalized Upper Bounded Techniques for Linear Programming I, II," ORC Report 64-17 (1964) and 64-18 (1965), Operations Research Center, University of California, Berkeley, CA.
3. D. Solow, "Development and Applications of a Simplex Algorithm for Handling Values of the Nonbasic Variables Between Their Lower and Upper Bounds," Working Paper, Department of Operations Research, Case Western Reserve University, Cleveland, OH, 1980.
4. W. I. Zangwill, "The Convex Simplex Method," *Management Science* 14(3), 221–238 (1967).

Chapter 11

Network Flow Problems

In the previous chapter, special structure in the constraints of certain LPs was exploited so as to reduce the computational effort and storage space needed to solve the problem. *Network flow problems* are a class of linear programming problems that also have a special structure, enabling one to perform the steps of the simplex algorithm with no multiplications or divisions. Because of this, it is possible to solve extremely large network problems in a very reasonable amount of computer time, as will be described in this chapter.

11.1. THE NETWORK LP AND ITS PROPERTIES

Consider a factory-to-market distribution problem in which a chocolate company has two factories, two intermediate transshipment warehouses, and three retail stores. The two factories have produced a known quantity of chocolate, say, 50 and 70 pounds, respectively. They wish to ship their goods to the three retail stores where there is a known demand of 30, 40, and 50 pounds of chocolate, respectively. The goods may be shipped along certain routes, and through the two transshipment warehouses if necessary, as depicted in Figure 11.1.

Each of the circles in Figure 11.1 is called a *node*. Nodes 1 and 2 are referred to as *sources*, because they are the sources of the supplies. Nodes 5–7 are the *demand nodes* or *sinks*. Nodes 3 and 4 are referred to as *intermediate*, or *transshipment*, nodes. The lines connecting the nodes in Figure 11.1 are called *arcs*, and they represent the possible routings between the nodes. The arrow on each arc indicates the allowable direction of

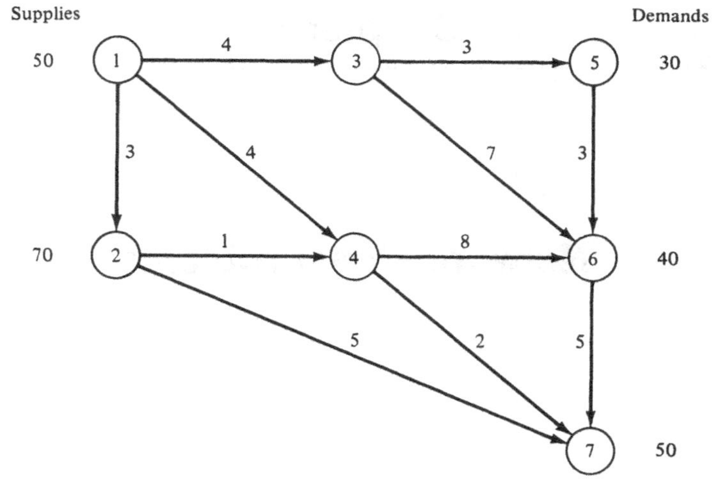

FIGURE 11.1. The chocolate company problem.

movement. The number associated with each arc in Figure 11.1 is a cost (in dollars per pound of chocolate) for shipping along that arc. The objective of the company is to determine how much chocolate should be shipped through each arc so as to transport all of the supplies from the sources to the sinks (leaving nothing at the transshipment nodes) while simultaneously minimizing the total shipping cost.

Following the three steps of problem formulation (see Chapter 1), the variables must first be specified. To do so, let *ij* represent the arc from node number *i* pointing toward node number *j*. Then, let x_{ij} be the variable representing the amount (in pounds) that is to be shipped along arc *ij*. For the network in Figure 11.1, you can see that the vector **x** consists of the variables $x_{12}, x_{13}, x_{14}, x_{24}, x_{27}, x_{35}, x_{36}, x_{46}, x_{47}, x_{56}, x_{67}$.

The second step of problem formulation is to identify the objective function to minimize, which in this case is

$$3x_{12} + 4x_{13} + 4x_{14} + x_{24} + 5x_{27} + 3x_{35}$$

$$+ 7x_{36} + 8x_{46} + 2x_{47} + 3x_{56} + 5x_{67}$$

Finally, it is necessary to identify the constraints. Naturally, each $x_{ij} \geq 0$. The more interesting constraints are the *balance* constraints for each node. Specifically, for each of the sources, the total amount leaving minus the total amount entering should equal the supply at the source, so

$$x_{12} + x_{13} + x_{14} = 50 \quad \text{(source 1)}$$

$$x_{24} + x_{27} - x_{12} = 70 \quad \text{(source 2)}$$

For each of the transshipment nodes, the total amount leaving minus the total amount entering should be 0 (i.e., no goods are to be left at these nodes), so

$$x_{35} + x_{36} - x_{13} = 0 \quad \text{(warehouse 3)}$$

$$x_{46} + x_{47} - x_{14} - x_{24} = 0 \quad \text{(warehouse 4)}$$

In an analogous manner, for each of the sinks, the total amount entering minus the total amount leaving should equal the demand at the sink, so

$$x_{35} - x_{56} = 30 \quad \text{(sink 5)}$$

$$x_{36} + x_{46} + x_{56} - x_{67} = 40 \quad \text{(sink 6)}$$

$$x_{27} + x_{47} + x_{67} = 50 \quad \text{(sink 7)}$$

As can be seen from this problem, a network LP with m nodes requires m balance equations. For algebraic reasons that will soon become evident, it is convenient to multiply each of the sink constraints by -1; thus one can think of demand as a negative supply. In so doing, the network LP becomes:

$$\begin{array}{ll}
\text{Minimize} & 3x_{12} + 4x_{13} + 4x_{14} + x_{24} + 5x_{27} + 3x_{35} \\
& + 7x_{36} + 8x_{46} + 2x_{47} + 3x_{56} + 5x_{67} \\
\text{subject to} & x_{12} + x_{13} + x_{14} = 50 \\
& -x_{12} + x_{24} + x_{27} = 70 \\
& -x_{13} + x_{35} + x_{36} = 0 \\
& -x_{14} - x_{24} + x_{46} + x_{47} = 0 \\
& -x_{35} + x_{56} = -30 \\
& -x_{36} - x_{46} - x_{56} + x_{67} = -40 \\
& -x_{27} - x_{47} - x_{67} = -50 \\
& x_{ij} \geq 0
\end{array}$$

If $\mathbf{x}$ is a feasible solution to this LP, then each node is said to be balanced. Also, in order for there to be a feasible solution, it had better be the case that the total supply equals the total demand. The reason is that once the total supply has left the supply nodes, it must all end up at the demand nodes. The $\mathbf{A}$ matrix obtained from a network LP is called the *node–arc incidence matrix*. It consists of m rows (one for each node), and n columns (one for each arc). The column corresponding to arc ij contains a $+1$ in row i and a -1 in row j, and 0 elsewhere. In other words, each column of the node-arc incidence matrix contains exactly one 1, one -1, and zeros elsewhere. It is this structure that will ultimately be exploited. In this example the node-arc incidence matrix is

$$
\begin{array}{c} \\ 1 \\ 2 \\ 3 \\ 4 \\ 5 \\ 6 \\ 7 \end{array}
\begin{array}{c}
\begin{array}{ccccccccccc} 12 & 13 & 14 & 24 & 27 & 35 & 36 & 46 & 47 & 56 & 67 \end{array} \\
\left[\begin{array}{ccccccccccc}
1 & 1 & 1 & & & & & & & & \\
-1 & & & 1 & 1 & & & & & & \\
 & -1 & & & & 1 & 1 & & & & \\
 & & -1 & -1 & & & & 1 & 1 & & \\
 & & & & & -1 & & & & 1 & \\
 & & & & & & -1 & -1 & & -1 & 1 \\
 & & & & -1 & & & & -1 & & -1
\end{array}\right]
\end{array}
$$

One interesting property of the node–arc incidence matrix is that it does not contain a nonsingular $m \times m$ submatrix, and consequently it will not be possible to find an initial bfs. The proof of this statement is left as Exercise 11.1.2. Nonetheless, this difficulty can be overcome easily by making the observation that only $m-1$ of the m balance constraints are needed to describe the LP. The reason is that if there is balance at $m-1$ nodes, then the remaining node must also be balanced. In other words, any one of the constraints can be deleted from the problem. The remaining $(m-1) \times n$ matrix will be referred to as the *reduced node–arc incidence matrix*. When working with this matrix, an $(m-1) \times (m-1)$ basis matrix will be sought to initialize the simplex algorithm. One approach for doing so will be presented in Section 11.3. The next section will describe the details of the simplex algorithm, assuming that an initial bfs has been found.

EXERCISES

11.1.1. The following problem is a special type of network LP referred to as the assignment problem. A company wishes to assign three jobs to three machines in such a way that each job is assigned to some machine and no machine works on more than one job. The number in row i and column j of the following matrix represents the cost of assigning job i to machine j, respectively:

$$\begin{bmatrix} 8 & 7 & 6 \\ 5 & 7 & 8 \\ 6 & 8 & 7 \end{bmatrix}$$

(a) Draw the associated network. Identify the sources and sinks and their respective supplies and demands.

(b) Formulate the network LP problem so as to minimize the total cost of making the assignment.

11.1.2. Prove that any $m \times m$ submatrix $\mathbf{B}$ of a node–arc incidence matrix is singular.

11.2. THE SIMPLEX ALGORITHM FOR NETWORK FLOW PROBLEMS

In this section, it will be shown how the steps of the simplex algorithm can be performed without multiplications or divisions when applied to the reduced network LP in which one constraint has been deleted. To describe the steps, consider the following LP:

$$\begin{aligned} \text{Minimize} \quad & \mathbf{cx} \\ \text{subject to} \quad & \mathbf{Ax} = \mathbf{b} \\ & \mathbf{x} \geq \mathbf{0} \end{aligned}$$

in which $\mathbf{A}$ is the $(m-1) \times n$ reduced node–arc incidence matrix. A key observation enables the efficient implementation of the simplex algorithm (see Exercise 11.2.1):

Fact. For every nonsingular $(m-1) \times (m-1)$ basis matrix $\mathbf{B}$ of the above LP, there are permutation matrices $\mathbf{P}$ and $\mathbf{Q}$ such that $\mathbf{PBQ}$ is upper triangular.

The importance of this observation is that the two linear systems $\mathbf{uB} = \mathbf{c_B}$ and $\mathbf{Bd_B} = -\mathbf{N}_{.j}$ can be solved efficiently by the process of back substitution (see Section 4.4).

To solve the system $\mathbf{uB} = \mathbf{c_B}$, one can solve the equivalent system $\mathbf{uBQ} = \mathbf{c_B Q}$. To do so, one can first use forward substitution to solve the upper triangular system $\mathbf{w(PBQ)} = \mathbf{c_B Q}$ for $\mathbf{w}$, then compute $\mathbf{u} = \mathbf{wP}$.

In an analogous manner, one can solve the system $\mathbf{Bd_B} = -\mathbf{N}_{.j}$ by solving the equivalent system $\mathbf{(PB)d_B} = -\mathbf{PN}_{.j}$. To do so, one can first use back substitution to solve the upper triangular system $\mathbf{(PBQ)y} = -\mathbf{PN}_{.j}$ for $\mathbf{y}$ and then compute $\mathbf{d_B} = \mathbf{Qy}$.

It is worth noting that since each element of $\mathbf{B}$ is -1, 0, or $+1$, the process of back substitution can be performed without multiplications or divisions. Moreover, it is possible to show (although the proof will not be given here) that the vector $\mathbf{d_B}$ consists of components whose values are -1, 0, or $+1$. Consequently, when performing the ratios $(\mathbf{x_B})_k / -(\mathbf{d_B})_k$ to compute a value for t^*, one need perform no divisions. The following is a summary of the steps of the simplex algorithm when applied to the reduced network LP.

Step 0. Initialization: Find an initial bfs $\mathbf{x} = (\mathbf{x_B}, \mathbf{x_N})$ and a basis matrix $\mathbf{B}$, together with permutation matrices $\mathbf{P}$ and $\mathbf{Q}$, such that $\mathbf{PBQ}$ is upper triangular.

Step 1. Test for optimality: Using forward substitution, solve the upper triangular system $\mathbf{w(PBQ)} = \mathbf{c_B Q}$ for $\mathbf{w}$, and then compute $\mathbf{u} = \mathbf{wP}$. Finally, compute the reduced costs $\mathbf{c_N} - \mathbf{uN}$ (without performing multiplications). If these reduced costs are ≥ 0, then the current bfs is optimal.

Otherwise, select j^* with $1 \leq j^* \leq (n-m)$ and $(\mathbf{c_N} - \mathbf{uN})_{j^*} < 0$ by a rule that is computationally efficient (see Section 7.2).

Step 2. Computing the direction of movement: Using the value of j^* obtained in step 1, solve the upper triangular system $(\mathbf{PBQ})\mathbf{y} = -\mathbf{PN}_{.j^*}$ for $\mathbf{y}$, and then compute $\mathbf{d_B} = \mathbf{Qy}$.

Step 3. Computing the amount of movement: If $\mathbf{d_B} \geq \mathbf{0}$, then the LP is unbounded. Otherwise, using the fact that each component of $\mathbf{d_B}$ is either -1, 0, or $+1$, compute $t^* = \min\{(\mathbf{x_B})_k : (\mathbf{d_B})_k = -1\} = (\mathbf{x_B})_{k^*}$.

Potential cycling problems can arise if $t^* = 0$, i.e., if $(\mathbf{x_B})_{k^*} = 0$. These difficulties can be overcome by using Bland's rule. However, a special noncycling rule, called *Cunnigham's rule*, has been developed for this particular problem. The details will not be given here.

Step 4. Pivoting: Compute the values of the new variables by $\mathbf{x}' = \mathbf{x} + t^*\mathbf{d}$. Replace column k^* of $\mathbf{B}$ with column j^* of $\mathbf{N}$ to obtain $\mathbf{B}'$. Finally, find permutation matrices $\mathbf{P}'$ and $\mathbf{Q}'$ such that $\mathbf{P}'\mathbf{B}'\mathbf{Q}'$ is upper triangular, and return to step 1.

These steps will now be illustrated with the example of the previous section. It should be noted that, when solving "small" network problems by hand, the $\mathbf{P}$ and $\mathbf{Q}$ matrices are not needed explicitly. Rather, the two linear systems can be solved (using back substitution) by inspection. However, for large network problems that must be solved by computer, the $\mathbf{P}$ and $\mathbf{Q}$ matrices are needed, and they must be updated at the end of each iteration. In the following example, only the first iteration will use the $\mathbf{P}$ and $\mathbf{Q}$ matrices explicitly. In subsequent iterations, the linear systems will be solved by inspection.

EXAMPLE 11.1

$$\begin{aligned}
\text{Minimize}\quad & 3x_{12} + 4x_{13} + 4x_{14} + x_{24} + 5x_{27} + 3x_{35} \\
& + 7x_{36} + 8x_{46} + 2x_{47} + 3x_{56} + 5x_{67} \\
\text{subject to}\quad & x_{13} + x_{12} + x_{14} = 50 \\
& -x_{12} + x_{24} + x_{27} = 70 \\
& -x_{13} + x_{35} + x_{36} = 0 \\
& -x_{14} - x_{24} + x_{46} + x_{47} = 0 \\
& -x_{35} + x_{56} = -30 \\
& -x_{36} - x_{46} - x_{56} + x_{67} = -40 \\
& -x_{27} - x_{47} - x_{67} = -50 \\
& \mathbf{x} \geq \mathbf{0}
\end{aligned}$$

Iteration 1

Step 0. The bfs for this iteration:

$$\begin{aligned}\mathbf{x_B} &= (x_{13}, x_{24}, x_{27}, x_{35}, x_{36}, x_{46}) \\ &= (50, 20, 50, 30, 20, 20) \\ \mathbf{x_N} &= (x_{12}, x_{14}, x_{47}, x_{56}, x_{67}) \\ &= (0, 0, 0, 0, 0) \\ \mathbf{c_B} &= (c_{13}, c_{24}, c_{27}, c_{35}, c_{36}, c_{46}) \\ &= (4, 1, 5, 3, 7, 8) \\ \mathbf{c_N} &= (c_{12}, c_{14}, c_{47}, c_{56}, c_{67}) \\ &= (3, 4, 2, 3, 5)\end{aligned}$$

The initial basis matrix **B** is

$$\mathbf{B} = \begin{array}{c} \begin{matrix} 13 & 24 & 27 & 35 & 36 & 46 \end{matrix} \\ \begin{bmatrix} 1 & 0 & 0 & 0 & 0 & 0 \\ 0 & 1 & 1 & 0 & 0 & 0 \\ -1 & 0 & 0 & 1 & 1 & 0 \\ 0 & -1 & 0 & 0 & 0 & 1 \\ 0 & 0 & 0 & -1 & 0 & 0 \\ 0 & 0 & 0 & 0 & -1 & -1 \end{bmatrix} \end{array}$$

In what follows, $\mathbf{P}_{ij}$ is the permutation matrix obtained by exchanging rows i and j of the identity matrix. The permutation matrix $\mathbf{P} = \mathbf{P}_{35}\mathbf{P}_{36}\mathbf{P}_{56}\mathbf{P}_{14}\mathbf{P}_{24}\mathbf{P}_{24}\mathbf{P}_{13}$ so

$$\mathbf{P} = \begin{bmatrix} 0 & 1 & 0 & 0 & 0 & 0 \\ 0 & 0 & 0 & 1 & 0 & 0 \\ 0 & 0 & 0 & 0 & 0 & 1 \\ 0 & 0 & 1 & 0 & 0 & 0 \\ 0 & 0 & 0 & 0 & 1 & 0 \\ 1 & 0 & 0 & 0 & 0 & 0 \end{bmatrix}$$

The permutation matrix $\mathbf{Q} = \mathbf{P}_{13}\mathbf{P}_{36}\mathbf{P}_{45}$ so

$$\mathbf{Q} = \begin{bmatrix} 0 & 0 & 0 & 0 & 0 & 1 \\ 0 & 1 & 0 & 0 & 0 & 0 \\ 1 & 0 & 0 & 0 & 0 & 0 \\ 0 & 0 & 0 & 0 & 1 & 0 \\ 0 & 0 & 0 & 1 & 0 & 0 \\ 0 & 0 & 1 & 0 & 0 & 0 \end{bmatrix}$$

Step 1. Testing for optimality: To solve $\mathbf{uB} = \mathbf{c_B}$ for $\mathbf{u}$, first solve $\mathbf{w}(\mathbf{PBQ}) = \mathbf{c_B Q}$ for $\mathbf{w}$, so

$$\mathbf{w}\begin{bmatrix} 1 & 1 & 0 & 0 & 0 & 0 \\ 0 & -1 & 1 & 0 & 0 & 0 \\ 0 & 0 & -1 & -1 & 0 & 0 \\ 0 & 0 & 0 & 1 & 1 & -1 \\ 0 & 0 & 0 & 0 & -1 & 0 \\ 0 & 0 & 0 & 0 & 0 & 1 \end{bmatrix}$$

$$= (4,1,5,3,7,8)\begin{bmatrix} 0 & 0 & 0 & 0 & 0 & 1 \\ 0 & 1 & 0 & 0 & 0 & 0 \\ 1 & 0 & 0 & 0 & 0 & 0 \\ 0 & 0 & 0 & 0 & 1 & 0 \\ 0 & 0 & 0 & 1 & 0 & 0 \\ 0 & 0 & 1 & 0 & 0 & 0 \end{bmatrix}$$

$$\begin{aligned} w_1 &= 5 \\ w_1 - w_2 &= 1 \\ w_2 - w_3 &= 8 \\ -w_3 + w_4 &= 7 \\ w_4 - w_5 &= 3 \\ -w_4 + w_6 &= 4 \end{aligned}$$

yielding $\mathbf{w} = (5,4,-4,3,0,7)$. To find $\mathbf{u}$, compute

$$\mathbf{u} = \mathbf{wP}$$

$$= (5,4,-4,3,0,7)\begin{bmatrix} 0 & 1 & 0 & 0 & 0 & 0 \\ 0 & 0 & 0 & 1 & 0 & 0 \\ 0 & 0 & 0 & 0 & 0 & 1 \\ 0 & 0 & 1 & 0 & 0 & 0 \\ 0 & 0 & 0 & 0 & 1 & 0 \\ 1 & 0 & 0 & 0 & 0 & 0 \end{bmatrix}$$

$$= (7,5,3,4,0,-4)$$

Finally, the reduced costs are

$$\mathbf{c_N} - \mathbf{uN} = (3,4,2,3,5) - (7,5,3,4,0,-4)\begin{bmatrix} 1 & 1 & 0 & 0 & 0 \\ -1 & 0 & 0 & 0 & 0 \\ 0 & 0 & 0 & 0 & 0 \\ 0 & -1 & 1 & 0 & 0 \\ 0 & 0 & 0 & 1 & 0 \\ 0 & 0 & 0 & -1 & 1 \end{bmatrix}$$

$$= (3,4,2,3,5) - (2,3,4,4,-4)$$

$$= (1,1,-2,-1,9) \qquad \text{so } j^* = 3 \ (x_{47} \text{ enters})$$

Step 2. Computing the direction of movement: To solve $\mathbf{B}\mathbf{d}_{\mathbf{B}} = -\mathbf{N}_{.j*}$ for $\mathbf{d}_{\mathbf{B}}$, first solve $(\mathbf{PBQ})\mathbf{y} = -\mathbf{PN}_{.j*}$ for $\mathbf{y}$, so

$$(\mathbf{PBQ})\mathbf{y} = -\mathbf{PN}_{.j*}$$

$$\begin{bmatrix} 1 & 1 & 0 & 0 & 0 & 0 \\ 0 & -1 & 1 & 0 & 0 & 0 \\ 0 & 0 & -1 & -1 & 0 & 0 \\ 0 & 0 & 0 & 1 & 1 & -1 \\ 0 & 0 & 0 & 0 & -1 & 0 \\ 0 & 0 & 0 & 0 & 0 & 1 \end{bmatrix} \begin{bmatrix} y_1 \\ y_2 \\ y_3 \\ y_4 \\ y_5 \\ y_6 \end{bmatrix}$$

$$= \begin{bmatrix} 0 & 1 & 0 & 0 & 0 & 0 \\ 0 & 0 & 0 & 1 & 0 & 0 \\ 0 & 0 & 0 & 0 & 0 & 1 \\ 0 & 0 & 1 & 0 & 0 & 0 \\ 0 & 0 & 0 & 0 & 1 & 0 \\ 1 & 0 & 0 & 0 & 0 & 0 \end{bmatrix} \begin{bmatrix} 0 \\ 0 \\ 0 \\ -1 \\ 0 \\ 0 \end{bmatrix}$$

$$\begin{aligned} y_1 + y_2 \qquad &= 0 \\ -y_2 + y_3 \qquad &= -1 \\ -y_3 - y_4 \qquad &= 0 \\ y_4 + y_5 - y_6 &= 0 \\ -y_5 \qquad &= 0 \\ y_6 \qquad &= 0 \end{aligned}$$

yielding

$$\mathbf{y} = \begin{bmatrix} -1 \\ 1 \\ 0 \\ 0 \\ 0 \\ 0 \end{bmatrix}$$

To find $\mathbf{d}_{\mathbf{B}}$, compute

$$\mathbf{d}_{\mathbf{B}} = \mathbf{Q}\mathbf{y}$$

$$= \begin{bmatrix} 0 & 0 & 0 & 0 & 0 & 1 \\ 0 & 1 & 0 & 0 & 0 & 0 \\ 1 & 0 & 0 & 0 & 0 & 0 \\ 0 & 0 & 0 & 0 & 1 & 0 \\ 0 & 0 & 0 & 1 & 0 & 0 \\ 0 & 0 & 1 & 0 & 0 & 0 \end{bmatrix} \begin{bmatrix} -1 \\ 1 \\ 0 \\ 0 \\ 0 \\ 0 \end{bmatrix} = \begin{bmatrix} 0 \\ 1 \\ -1 \\ 0 \\ 0 \\ 0 \end{bmatrix}$$

Step 3. Computing the amount of movement:

$$t^* = \min\{(\mathbf{x_B})_k : (\mathbf{d_B})_k = -1\} = (\mathbf{x_B})_3$$
$$= 50 \qquad \text{so } k^* = 3 \ (x_{27} \text{ leaves})$$

Step 4. Pivoting: The new bfs is

$$\mathbf{x}' = \mathbf{x} + t^*\mathbf{d}$$
$$= \begin{bmatrix} 50 \\ 20 \\ 50 \\ 30 \\ 20 \\ 20 \\ 0 \\ 0 \\ 0 \\ 0 \\ 0 \end{bmatrix} + 50 \begin{bmatrix} 0 \\ 1 \\ -1 \\ 0 \\ 0 \\ 0 \\ 0 \\ 0 \\ 1 \\ 0 \\ 0 \end{bmatrix} = \begin{bmatrix} 50 \\ 70 \\ 0 \\ 30 \\ 20 \\ 20 \\ 0 \\ 0 \\ 50 \\ 0 \\ 0 \end{bmatrix} = \begin{bmatrix} x_{13} \\ x_{24} \\ x_{27} \\ x_{35} \\ x_{36} \\ x_{46} \\ x_{12} \\ x_{14} \\ x_{47} \\ x_{56} \\ x_{67} \end{bmatrix}$$

The basic variables corresponding to $\mathbf{x}'$ are:

$$\mathbf{x}'_{\mathbf{B}} = \begin{bmatrix} x_{13} \\ x_{24} \\ x_{47} \\ x_{35} \\ x_{36} \\ x_{46} \end{bmatrix} = \begin{bmatrix} 50 \\ 70 \\ 50 \\ 30 \\ 20 \\ 20 \end{bmatrix}$$

$$\mathbf{c}'_{\mathbf{B}} = (c_{13}, c_{24}, c_{47}, c_{35}, c_{36}, c_{46})$$
$$= (4, 1, 2, 3, 7, 8)$$

$$\mathbf{c}'_{\mathbf{N}} = (c_{12}, c_{14}, c_{27}, c_{56}, c_{67})$$
$$= (3, 4, 5, 3, 5)$$

Column $k^* = 3$ of $\mathbf{B}$ is exchanged with column $j^* = 3$ of $\mathbf{N}$, so

$$\mathbf{B}' = \begin{array}{c} \begin{matrix} 13 & 24 & 47 & 35 & 36 & 46 \end{matrix} \\ \begin{bmatrix} 1 & 0 & 0 & 0 & 0 & 0 \\ 0 & 1 & 0 & 0 & 0 & 0 \\ -1 & 0 & 0 & 1 & 1 & 0 \\ 0 & -1 & 1 & 0 & 0 & 1 \\ 0 & 0 & 0 & -1 & 0 & 0 \\ 0 & 0 & 0 & 0 & -1 & -1 \end{bmatrix} \end{array}$$

$$\mathbf{N}' = \begin{array}{c} \begin{array}{ccccc} 12 & 14 & 27 & 56 & 67 \end{array} \\ \begin{bmatrix} 1 & 1 & 0 & 0 & 0 \\ -1 & 0 & 1 & 0 & 0 \\ 0 & 0 & 0 & 0 & 0 \\ 0 & -1 & 0 & 0 & 0 \\ 0 & 0 & 0 & 1 & 0 \\ 0 & 0 & 0 & -1 & 1 \end{bmatrix} \end{array}$$

Starting with the next iteration, all linear systems will be solved (using back substitution) by inspection. The **P** and **Q** matrices will not be shown explicitly.

Iteration 2

Step 1. Testing for optimality: Solve the system $\mathbf{uB} = \mathbf{c_B}$ for $\mathbf{u}$,

$$\begin{aligned} u_1 - u_3 &= 4 \\ u_2 - u_4 &= 1 \\ u_4 &= 2 \\ u_3 - u_5 &= 3 \\ u_3 - u_6 &= 7 \\ u_4 - u_6 &= 8 \end{aligned}$$

yielding $\mathbf{u} = (5, 3, 1, 2, -2, -6)$.

Computing the reduced costs:

$$\mathbf{c_N} - \mathbf{uN} = (3,4,5,3,5) - (5,3,1,2,-2,-6)\begin{bmatrix} 1 & 1 & 0 & 0 & 0 \\ -1 & 0 & 1 & 0 & 0 \\ 0 & 0 & 0 & 0 & 0 \\ 0 & -1 & 0 & 0 & 0 \\ 0 & 0 & 0 & 1 & 0 \\ 0 & 0 & 0 & -1 & 1 \end{bmatrix}$$

$$= (3,4,5,3,5) - (2,3,3,4,-6)$$

$$= (1,1,2,-1,11) \qquad \text{so} \quad j^* = 4 \ (x_{56} \text{ enters})$$

Step 2. Computing the direction of movement: Solve the system $\mathbf{Bd_B} = -\mathbf{N}_{\cdot j^*}$ for $\mathbf{d_B}$,

$$\begin{aligned} (\mathbf{d_B})_1 &= 0 \\ (\mathbf{d_B})_2 &= 0 \\ -(\mathbf{d_B})_1 + (\mathbf{d_B})_4 + (\mathbf{d_B})_5 &= 0 \\ -(\mathbf{d_B})_2 + (\mathbf{d_B})_3 + (\mathbf{d_B})_6 &= 0 \\ -(\mathbf{d_B})_4 &= -1 \\ -(\mathbf{d_B})_5 - (\mathbf{d_B})_6 &= 1 \end{aligned}$$

Thus

$$\mathbf{d_B} = \begin{bmatrix} 0 \\ 0 \\ 0 \\ 1 \\ -1 \\ 0 \end{bmatrix}$$

Step 3. Computing the amount of movement:

$$\begin{aligned} t^* &= \min\{(\mathbf{x_B})_k : (\mathbf{d_B})_k = -1\} \\ &= (\mathbf{x_B})_5 \\ &= 20 \qquad \text{so} \quad k^* = 5 \;\; (x_{36} \text{ leaves}) \end{aligned}$$

Step 4. Pivoting: The new bfs is

$$\mathbf{x}' = \mathbf{x} + t^*\mathbf{d}$$

$$= \begin{bmatrix} 50 \\ 70 \\ 50 \\ 30 \\ 20 \\ 20 \\ 0 \\ 0 \\ 0 \\ 0 \\ 0 \end{bmatrix} + 20 \begin{bmatrix} 0 \\ 0 \\ 0 \\ 1 \\ -1 \\ 0 \\ 0 \\ 0 \\ 0 \\ 1 \\ 0 \end{bmatrix} = \begin{bmatrix} 50 \\ 70 \\ 50 \\ 50 \\ 0 \\ 20 \\ 0 \\ 0 \\ 0 \\ 20 \\ 0 \end{bmatrix} = \begin{bmatrix} x_{13} \\ x_{24} \\ x_{47} \\ x_{35} \\ x_{36} \\ x_{46} \\ x_{12} \\ x_{14} \\ x_{27} \\ x_{56} \\ x_{67} \end{bmatrix}$$

The basic variables corresponding to $\mathbf{x}'$ are:

$$\mathbf{x}'_\mathbf{B} = \begin{bmatrix} x_{13} \\ x_{24} \\ x_{47} \\ x_{35} \\ x_{56} \\ x_{46} \end{bmatrix} = \begin{bmatrix} 50 \\ 70 \\ 50 \\ 50 \\ 20 \\ 20 \end{bmatrix}$$

$$\begin{aligned} \mathbf{c}'_\mathbf{B} &= (c_{13}, c_{24}, c_{47}, c_{35}, c_{56}, c_{46}) \\ &= (4, 1, 2, 3, 3, 8) \end{aligned}$$

$$\begin{aligned} \mathbf{c}'_\mathbf{N} &= (c_{12}, c_{14}, c_{27}, c_{36}, c_{67}) \\ &= (3, 4, 5, 7, 5) \end{aligned}$$

Column $k^* = 5$ of $\mathbf{B}$ is exchanged with column $j^* = 4$ of $\mathbf{N}$, so

$$\mathbf{B}' = \begin{array}{c} \begin{matrix} 13 & 24 & 47 & 35 & 56 & 46 \end{matrix} \\ \begin{bmatrix} 1 & 0 & 0 & 0 & 0 & 0 \\ 0 & 1 & 0 & 0 & 0 & 0 \\ -1 & 0 & 0 & 1 & 0 & 0 \\ 0 & -1 & 1 & 0 & 0 & 1 \\ 0 & 0 & 0 & -1 & 1 & 0 \\ 0 & 0 & 0 & 0 & -1 & -1 \end{bmatrix} \end{array}$$

$$\mathbf{N}' = \begin{array}{c} \begin{matrix} 12 & 14 & 27 & 36 & 67 \end{matrix} \\ \begin{bmatrix} 1 & 1 & 0 & 0 & 0 \\ -1 & 0 & 1 & 0 & 0 \\ 0 & 0 & 0 & 1 & 0 \\ 0 & -1 & 0 & 0 & 0 \\ 0 & 0 & 0 & 0 & 0 \\ 0 & 0 & 0 & -1 & 1 \end{bmatrix} \end{array}$$

Iteration 3

Step 1. Testing for optimality (by solving the system $\mathbf{uB} = \mathbf{c_B}$ for $\mathbf{u}$):

$$\begin{aligned} u_1 - u_3 &= 4 \\ u_2 - u_4 &= 1 \\ u_4 &= 2 \\ u_3 - u_5 &= 3 \\ u_5 - u_6 &= 3 \\ u_4 - u_6 &= 8 \end{aligned}$$

yielding $\mathbf{u} = (4, 3, 0, 2, -3, -6)$. Computing the reduced costs:

$$\mathbf{c_N} - \mathbf{uN} = (3,4,5,7,5) - (4,3,0,2,-3,-6) \begin{bmatrix} 1 & 1 & 0 & 0 & 0 \\ -1 & 0 & 1 & 0 & 0 \\ 0 & 0 & 0 & 1 & 0 \\ 0 & -1 & 0 & 0 & 0 \\ 0 & 0 & 0 & 0 & 0 \\ 0 & 0 & 0 & -1 & 1 \end{bmatrix}$$

$$= (3,4,5,7,5) - (1,2,3,6,-6)$$

$$= (2,2,2,1,11)$$

Since $\mathbf{c_N} - \mathbf{uN} \geq \mathbf{0}$, the current solution is optimal with

$$\begin{aligned}\mathbf{x_B} &= (x_{13}, x_{24}, x_{47}, x_{35}, x_{56}, x_{46})\\ &= (50, 70, 50, 50, 20, 20)\\ \mathbf{x_N} &= (x_{12}, x_{14}, x_{27}, x_{36}, x_{67})\\ &= (0,0,0,0,0)\end{aligned}$$

The objective function value is

$$\begin{aligned}\mathbf{c_B x_B} &= (4,1,2,3,3,8)\begin{bmatrix}50\\70\\50\\50\\20\\20\end{bmatrix}\\ &= 740\end{aligned}$$

In order to be able to solve network problems with thousands of nodes and many thousands of arcs, as is done routinely in practice, a sophisticated computational implementation of these steps is required, including an updating procedure for the permutation matrices **P** and **Q**. That implementation pays particular attention to data structures, but it will not be described here. The next section will address the phase 1 issue of how to find an initial bfs for the reduced network LP or determine that none exists.

EXERCISES

11.2.1. **(a)** Show that for every basis matrix **B** of a reduced node–arc incidence matrix, there are permutation matrices **P** and **Q** such that **PBQ** is upper triangular.

(b) Using part (a), describe how to find the permutation matrices in the pivot step of the network simplex algorithm.

11.2.2. Apply the network simplex algorithm to

$$\begin{array}{lrl}\text{Minimize} & \multicolumn{2}{l}{2x_{12} + 2x_{13} + 3x_{24} + 5x_{25} + 4x_{34} + 3x_{35} + 5x_{32}}\\ \text{subject to} & x_{12} + x_{13} & = 100\\ & -x_{12} + x_{24} + x_{25} - x_{32} & = 0\\ & -x_{13} + x_{32} + x_{34} + x_{35} & = 0\\ & -x_{24} - x_{34} & = -60\\ & -x_{25} - x_{35} & = -40\\ & \mathbf{x} & \geq \mathbf{0}\end{array}$$

starting with the initial bfs in which $x_{13} = 100$, $x_{25} = 40$, $x_{34} = 60$, and $x_{32} = 40$. Solve all linear systems (using back substitution) by inspection.

11.3. THE PHASE 1 PROCEDURE FOR NETWORK FLOW PROBLEMS

In this section a phase 1 procedure will be developed that either produces an initial bfs for the reduced network LP or else determines that none exists. The approach is similar in spirit to the phase 1 method that was presented in Chapter 6. Before developing the details, however, it is worth noting that, in order for the reduced network LP to have any chance of being feasible, the total supply must be equal to the total demand. The reason is that, once the supplies are shipped from the sources, they must end up at the sinks (since nothing can be left at the transshipment nodes). Hence the total supply should equal the total demand, as is the case in Example 11.1.

Under this assumption, there is one instance when an initial bfs for the reduced network LP is readily available. Suppose that, in the original network, there is a node, say, 1, that has (a) an arc connected to each of the transshipment nodes, (b) an arc directed from node 1 to each of the sinks, and (c) an arc directed from each of the sources into node 1. In this case, an initial bfs can be constructed using these $m - 1$ arcs as follows:

1. Along each such arc of the form $1j$, ship an amount equal to the demand at node j (i.e., $x_{1j} = |b_j|$).
2. Along each such arc of the form $i1$, ship an amount equal to the supply at node i (i.e., $x_{i1} = |b_i|$).
3. Through each of the remaining arcs, ship nothing (i.e., for all other arcs ij, $x_{ij} = 0$).

In order to show that the resulting vector $\mathbf{x}$ is a bfs for the reduced network LP, it is necessary to identify the $(m - 1) \times (m - 1)$ basis matrix $\mathbf{B}$, and then to show that $\mathbf{x_B} = \mathbf{B}^{-1}\mathbf{b}$ and $\mathbf{x_N} = \mathbf{0}$.

The $\mathbf{B}$ matrix is constructed by selecting those columns of the reduced node–arc incidence matrix that correspond to the $m - 1$ arcs connected to node 1. For the network depicted in Figure 11.2, assuming node 1 to be the only source, and node 7 to be the only sink, the resulting $\mathbf{B}$ matrix is

$$
\begin{array}{c} \\ 1 \\ 2 \\ 3 \\ 4 \\ 5 \\ 6 \end{array}
\begin{array}{c}
\begin{array}{cccccc} 12 & 31 & 41 & 15 & 16 & 17 \end{array} \\
\left[\begin{array}{cccccc}
1 & -1 & -1 & 1 & 1 & 1 \\
-1 & & & & & \\
 & 1 & & & & \\
 & & 1 & & & \\
 & & & -1 & & \\
 & & & & -1 &
\end{array}\right]
\end{array}
$$

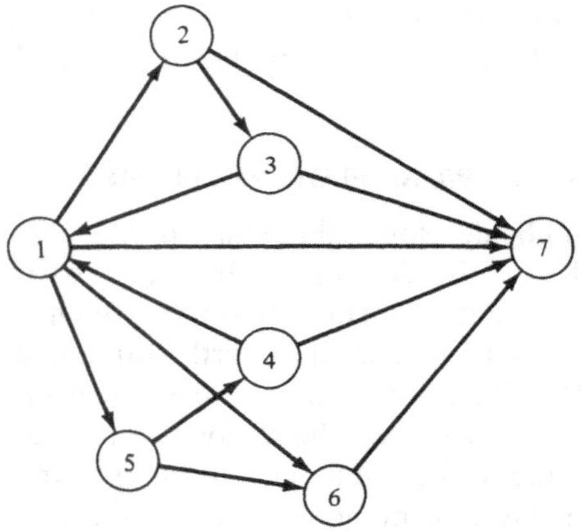

FIGURE 11.2. A network having an obvious initial bfs.

To see that this **B** matrix is nonsingular, observe what happens when the first row is moved to the last row of the matrix:

$$\begin{array}{c} \\ 2 \\ 3 \\ 4 \\ 5 \\ 6 \\ 1 \end{array} \begin{array}{c} \begin{array}{cccccc} 12 & 31 & 41 & 15 & 16 & 17 \end{array} \\ \left[\begin{array}{cccccc} -1 & & & & & \\ & 1 & & & & \\ & & 1 & & & \\ & & & -1 & & \\ & & & & -1 & \\ 1 & -1 & -1 & 1 & 1 & 1 \end{array}\right] \end{array}$$

As you can see, the result is a lower triangular matrix in which each diagonal element is either $+1$ or -1. Consequently, this matrix and the original **B** matrix are nonsingular (see Section 4.3). The formal proof that the vector **x** is a bfs will be left to Exercise 11.3.1.

Now it is time to address the issue of how to proceed when an initial bfs to the reduced network LP is not readily available, as is the case in Example 11.1. In this event, one chooses a fixed node, say, 1, and then one adds as many artificial arcs as necessary to connect node 1 to all of the other nodes in the manner described above. In practice, one might consider choosing, as the fixed node, a node requiring the addition of few artificial arcs. Observe the similarity of the current situation to that of adding artificial variables to form the phase 1 LP as described in Chapter 6.

The process of adding artificial arcs has been carried out for the network in Example 11.1, and the result is illustrated in Figure 11.3, with the artificial arcs indicated by dotted lines. The phase 1 network problem thus

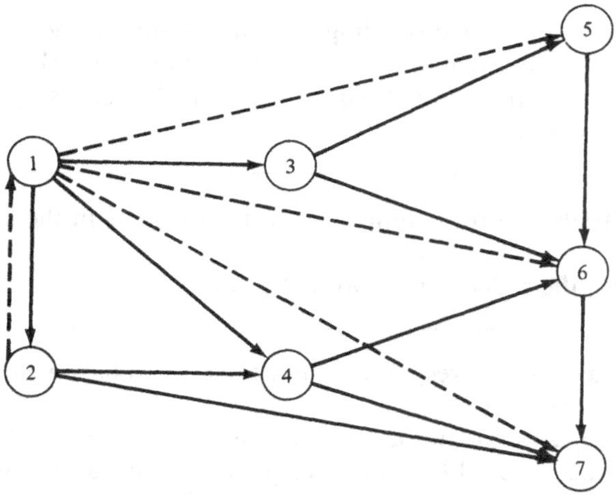

FIGURE 11.3. Adding artificial arcs.

obtained has the property that its reduced node–arc incidence matrix has an obvious initial basis matrix from which to initiate the simplex algorithm of the previous section. The initial bfs is constructed as was done above; namely, goods will be shipped through the arcs connected to node 1 and through no other arcs. Consequently, in the initial bfs one will be shipping goods along the artificial nonexisting arcs that were added to the original network. To reduce the amount of goods shipped along these artificial arcs to 0, a phase 1 objective (identical to the one used in Chapter 6) will be constructed. Specifically, one assigns a cost of 1 to each artificial arc. For Example 11.1 the phase 1 reduced network LP becomes:

$$\begin{array}{lll}
\text{Minimize} & x_{15} + x_{16} + x_{17} + x_{21} & \\
\text{subject to} & x_{12} + x_{13} + x_{14} + x_{15} + x_{16} + x_{17} - x_{21} = & 50 \\
& x_{21} + x_{24} + x_{27} - x_{12} = & 70 \\
& x_{35} + x_{36} - x_{13} = & 0 \\
& x_{46} + x_{47} - x_{14} - x_{24} = & 0 \\
& -x_{15} - x_{35} + x_{56} = & -30 \\
& -x_{16} - x_{36} - x_{46} - x_{56} + x_{67} = & -40 \\
& \mathbf{x} \geq & \mathbf{0}
\end{array}$$

As in the analysis of the phase 1 LP of Chapter 6, there are three possible outcomes when an optimal bfs $\mathbf{x}^*$ of the reduced network LP is obtained.

Case 1. All variables corresponding to artificial arcs are nonbasic.

In this case, the optimal bfs obtained at the end of the phase 1 procedure is an initial bfs for the original reduced network LP, and so phase 2 can be initiated immediately.

Case 2. A variable corresponding to an artificial arc is in the final basis at a positive value.

In this case, the reduced network LP is infeasible.

Case 3. All variables corresponding to artificial arcs have value 0, but some are in the final phase 1 basis.

In this case, several alternatives are available. One could attempt to replace those basic variables that correspond to artificial arcs with nonbasic variables corresponding to arcs in the original network. Such a procedure has been developed, but it will not be described here.

A second alternative for handling this case is similar in spirit to the one that was presented in Section 6.3, namely, adding a new constraint that ensures that the amount of goods shipped through the artificial arcs is always 0. The specific details, however, are a bit different. The desired goal is accomplished by considering the set S of nodes (including node 1) that are connected to node 1 with artificial arcs that appear in the optimal phase 1 basis. Suppose that there are k such arcs. To create the auxiliary LP, add an artificial transshipment node z; then

1. Eliminate all of the k artificial arcs (and variables) and
2. Create $k+1$ basic arcs (and variables) that point from each node in S to z.

Suppose, for example, that in Figure 11.3 all of the arcs indicated with dotted lines happen to end up in the final phase 1 basis at value 0. The result of adding the artificial node (and arcs) is shown in Figure 11.4.

Observe that since each of the new arcs points into z, and since z is a transshipment node, the amount shipped through these artificial arcs must be 0 because, if something were shipped into z, then there would be no way to ship the goods out of z. It takes a small amount of effort to use the optimal $(m-1)\times(m-1)$ basis matrix obtained at the end of phase 1 to create the initial $m \times m$ basis matrix for the "auxiliary" network LP. The details will be left as Exercises 11.3.3.

What is perhaps the most efficient computational procedure for handling Case 3 arose from the observation that, when artificial variables end up in the optimal phase 1 basis at value 0, it is possible to break the original network into two completely separate network problems, each of which can

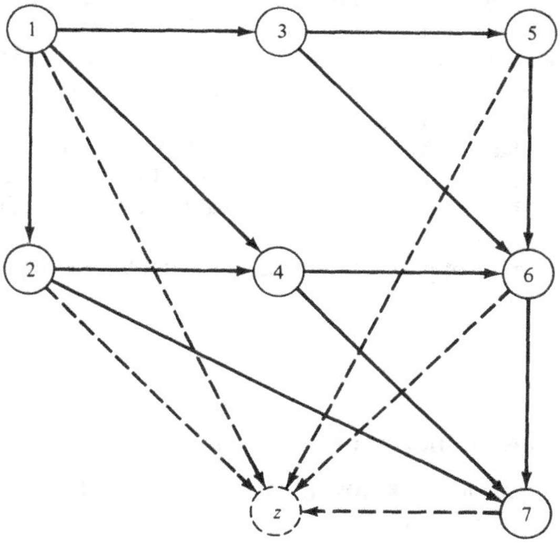

FIGURE 11.4. Adding an artificial node.

be solved independently of the other. Moreover, the optimal bfs obtained at the end of phase 1 indicates how to construct the two networks. The details will not be discussed here.

This chapter has shown how to exploit the special structure that arises in network problems so as to perform all of the steps of the simplex algorithm without using any multiplications or divisions. The message of the last two chapters should be clear: Whenever there is special structure in an LP, look for ways to enhance the performance of the simplex algorithm based on that special structure.

EXERCISES

11.3.1. Suppose that node 1 is connected to every other node with arcs directed from node 1 to the sinks and with arcs directed from the sources toward node 1. Show that the vector **x** obtained by shipping the appropriate amounts along those (and only those) arcs is a bfs.

11.3.2. Consider a network having arcs 12, 13, 14, 25, 35, 36, 45, 46, and 56. Suppose that node 1 is the source with a supply of 100. Also suppose that nodes 5 and 6 are sinks with demands of 40 and 60, respectively. Using node 3 as the fixed node, formulate and solve the phase 1 network problem.

11.3.3. Consider a network problem with six nodes in which node 1 has a supply of 30, node 6 has a demand of 30, and all other nodes are transshipment nodes. Suppose that the network has arcs 12, 13, 23, 25, 36, 41, 46, and 56.

(a) Write the phase 1 network LP using node 3 as the fixed node and adding the artificial arcs 34 and 35.

(b) Show that the bfs in which the basic variables are $x_{12} = 30$, $x_{23} = 30$, $x_{34} = 0$, $x_{35} = 0$, and $x_{36} = 30$ is optimal for the phase 1 LP.

(c) Draw the auxiliary network obtained by adding an artificial node z and the appropriate arcs.

(d) Use the bfs of part (b) to obtain an initial bfs for the auxiliary network in part (c).

DISCUSSION

On the History of Network and Related Problems

The development of network flow theory began with a very special problem known as the *transportation problem*, first studied by Hitchcock [9] and Kantorovitch [11]. It is interesting to note that the transportation problem was solved even before the simplex algorithm was invented, and it was Dantzig [4] who actually showed that the simplex method could be specialized to solve the transportation problem. Subsequently, Orden [16] extended that approach to handle the more general transshipment problem. Also, Charnes and Cooper [2] gave an intuitively appealing version of the simplex algorithm for solving the transportation problem that became known as the *stepping-stone method.* Koopmans [13] was the first to note the correspondence between basic feasible solutions in network problems and trees in graphs. Papers by Glicksman, Johnson, and Eselson [7], Scoins [17], and Johnson [10] discuss the use of data structures and indicate how the simplex algorithm can be implemented efficiently to solve large network problems.

Another special class of network problems is known as the *assignment problem*, and Kuhn [14] developed a primal–dual method for solving that problem. Subsequently, Ford and Fulkerson [5] extended the method to handle the transshipment problem. Fulkerson [6] and Minty [15] independently developed another approach, known as the *out-of-kilter algorithm*, for the same problem. The out-of-kilter algorithm was considered in the 1960s to be superior to the primal simplex algorithm. However, continued research by Srinivasan and Thompson [18], Glover, Karney, and Klingman [8], and Bradley, Brown, and Graves [1] has indicated that efficient implementations of the primal simplex algorithm are from about two to five times faster than the fastest out-of-kilter code. Currently, problems with tens of thousands of arcs can be solved quite routinely, and the specialized primal algorithms are 50–200 times faster than corresponding large-scale LP codes.

For Cunningham's pivoting rule [3] and a more detailed discussion of the topics of this chapter by Kennington and Helgason [12], the reader should consult the references.

REFERENCES

1. G. H. Bradley, G. G. Brown and G. W. Graves, "Design and Implementation of Large-Scale Transshipment Algorithms," *Management Science* 24(1), 1–34 (1977).
2. A. Charnes and W. W. Cooper, "The Stepping Stone Method of Explaining Linear Programming Calculations in Transportation Problems," *Management Science* 1(1), 1954.
3. W. H. Cunningham, "A Network Simplex Method," *Mathematical Programming* 11(2), 105–116 (1976).
4. G. B. Dantzig, "Application of the Simplex Method to a Transportation Problem," in *Activity Analysis and Allocation* (T. C. Koopmans, ed.), Cowles Commission Monograph No. 13, Wiley, New York, 1951.
5. L. R. Ford and D. R. Fulkerson, "A Primal–Dual Algorithm for the Capacitated Hitchcock Problem," *Naval Research Logistics Quarterly* 4(1), 47–54 (1957).
6. D. R. Fulkerson, "An Out-of-Kilter Method for Minimal Cost Flow Problems," *Journal of the Society for Industrial and Applied Mathematics* 9(1), 18–27 (1961).
7. S. Glicksman, L. Johnson, and L. Eselson, "Coding the Transportation Problem," *Naval Research Logistics Quarterly* 7(2), 169–183 (1960).
8. F. Glover, D. Karney, and D. Klingman, "Implementation and Computational Comparison of Primal, Dual, and Primal–Dual Computer Codes for Minimum Cost Network Flow Problems," *Networks* 4(3), 191–212 (1974).
9. F. L. Hitchcock, "Distribution of a Product from Several Sources to Numerous Localities," *Journal of Mathematical Physics* 20, 224–230 (1941).
10. E. L. Johnson, "Networks and Basic Solutions," *Operations Research* 14(4), 619–623 (1966).
11. L. Kantorovitch, "On the Translocation of Masses," *Comptes Rendus de l'Academie des Science de l'USSR*, 37, 199–201 (1942); English translation in *Management Science* 5(1), 1–8 (1958).
12. J. Kennington and R. Helgason, *Algorithms for Network Programming*, Wiley, New York, 1981.
13. T. C. Koopmans, "Optimum Utilization of the Transportation System," *Econometrica* 17(3–4), 136–146 (1949).
14. H. W. Kuhn, "The Hungarian Method for the Assignment Problem," *Naval Research Logistics Quarterly* 2(1–2), 83–97 (1955).
15. G. J. Minty, "Monotone Networks," *Proceedings of the Royal Society of London Series A* 257, 194–212 (1960).
16. A. Orden, "The Transshipment Problem," *Management Science*, 2(3), 276–285 (1956).
17. H. I. Scoins, "The Compact Representation of a Rooted Tree and the Transportation Problem," *International Symposium on Mathematical Programming*, London, 1964.
18. V. Srinivasan and G. L. Thompson, "Accelerated Algorithms for Labelling and Relabelling of Trees with Application for Distribution Problems," *Journal of the Association for Computing Machinery* 19, 712–726 (1972).

Appendix A

The Tableau Method

The tableau method has, historically, been the most common method for presenting and teaching the algebraic steps of the simplex algorithm. The primary advantage of the tableau method is that the desired computations can be carried out by keeping all of the columns of the **A** matrix in their original order, so it is not actually necessary to form the **B** and **N** matrices by rearranging the columns of **A**. It may at times be convenient to use the tableau method for solving problems having three or four equality constraints, but rarely for those with more. The primary disadvantage of the tableau method is that there is very little direct correlation between the algebraic steps and the associated geometric steps that were described in Chapter 2.

When using the tableau method to

$$\begin{aligned} \text{minimize} \quad & \mathbf{cx} \\ \text{subject to} \quad & \mathbf{Ax} = \mathbf{b} \\ & \mathbf{x} \geq \mathbf{0} \end{aligned}$$

it is algebraically convenient to modify the LP so as to minimize a single new variable z whose value will be that of the objective function at **x**. Thus the tableau method is designed to solve the following LP:

$$\begin{aligned} \text{Minimize} \quad & z \\ \text{subject to} \quad & z - \mathbf{cx} = 0 \\ & \mathbf{Ax} = \mathbf{b} \\ & \mathbf{x} \geq \mathbf{0} \end{aligned}$$

Given a set of basic and nonbasic variables for the above LP, the tableau method solves the necessary linear systems of equations by a process referred to as row operations, rather than by using the inverse of the basis matrix explicitly. Specifically, corresponding to each bfs $(z, \mathbf{x})$ of this LP is an $(m+1)\times(n+2)$ matrix called a *tableau*. The top row of the tableau (referred to as row 0) contains information pertaining to the reduced costs and to the value of the objective function at the bfs $\mathbf{x}$. The leftmost column of the tableau (referred to as column 0) corresponds to the additional variable z, which is always basic and whose value is that of the objective function at the current bfs $\mathbf{x}$. The next n columns of the tableau correspond to the original variables $x_1, \ldots, x_n$. The final column of the tableau contains the values of the basic variables.

Very specifically, if one were to rearrange the columns of $\mathbf{A}$ so that the basic ones appear first, followed by the nonbasic ones, then the information contained in the tableau would be as follows:

	z	$\mathbf{x_B}$	$\mathbf{x_N}$	RHS	Row
z	1	$\mathbf{0}$	$\mathbf{c_B B^{-1} N - c_N}$	$\mathbf{c_B B^{-1} b}$	0
$\mathbf{x_B}$	$\mathbf{0}$	$\mathbf{I}$	$\mathbf{B^{-1} N}$	$\mathbf{B^{-1} b}$	$1, \ldots, m$
Column	0	$1,\ldots,m$	$m+1,\ldots,n$	$n+1$	

The following observations are worth noting:

1. In row 0, the numbers appearing in the nonbasic columns are $\mathbf{c_B B^{-1} N - c_N}$, not $\mathbf{c_N - c_B B^{-1} N}$. Thus, when testing the tableau for optimality, one must check if these numbers (specifically, $\mathbf{c_B B^{-1} N - c_N}$) are all *nonpositive*, not nonnegative. These numbers in the tableau are often denoted $z_j - c_j$, where j is the column number in the tableau $(1 \le j \le n)$ and $z_j = \mathbf{c_B B^{-1} A}_{\cdot j}$.
2. The number in the last column of row 0 (namely, $\mathbf{c_B B^{-1} b}$) is the value of the objective function at $\mathbf{x}$ because $\mathbf{cx} = \mathbf{c_B x_B} + \mathbf{c_N x_N} = \mathbf{c_B x_B} = \mathbf{c_B B^{-1} b}$.
3. The matrix $\mathbf{B}^{-1}$ is not explicitly available from the tableau. Instead, under each nonbasic column is found the column vector $\mathbf{B}^{-1}\mathbf{N}_{\cdot j}$. Thus the associated direction vector $\mathbf{d_B} = -\mathbf{B}^{-1}\mathbf{N}_{\cdot j}$.
4. When the tableau is used for computational purposes, the columns remain in their original order $(1, 2, \ldots, n)$, and are not rearranged so that the basic columns appear first and are followed by the nonbasic columns.

Now it is possible to present the steps of the tableau method using the rule of steepest descent:

Step 1. Testing for optimality. See if, for each nonbasic variable x_j, the number in row 0 of the tableau under that variable is nonpositive (i.e., check if the reduced cost $z_j - c_j \le 0$). If so, then the current tableau (and bfs) is optimal. Otherwise, select, to enter the basis, a nonbasic variable x_k such that $z_k - c_k$ is as large a positive number as possible, and go to step 2.

Step 2. Computing the direction of movement. For the value of k obtained in step 1, the numbers appearing in that column of the tableau (below row 0) are $-\mathbf{d_B}$. Thus no computation needs to be done in this step. It is often the case that the variable $\mathbf{y}_k$ is used to denote the vector $\mathbf{B}^{-1}\mathbf{A}_{.k}$. Thus $\mathbf{d_B} = -\mathbf{y}_k$. Also, component i of $\mathbf{y}_k$ is denoted $\mathbf{y}_{ik}$ instead of $(\mathbf{y}_k)_i$.

Step 3. Computing the amount of movement. In this step it is necessary to perform the min ratio test. If $\mathbf{y}_k \leq \mathbf{0}$, i.e., if $\mathbf{d_B} \geq \mathbf{0}$, then the LP is unbounded. Otherwise, one computes, for each i with $1 \leq i \leq m$ and $\mathbf{y}_{ik} > 0$, the ratio

$$\bar{b}_i/\mathbf{y}_{ik}$$

where $\bar{b}_i = (\mathbf{B}^{-1}\mathbf{b})_i$, which is the number appearing in row i of the last column of the tableau. The subscript that provides the smallest ratio is generally denoted r instead of k^*; thus

$$t^* = \bar{b}_r/\mathbf{y}_{rk} = -(\mathbf{x_B})_{k^*}/(\mathbf{d_B})_{k^*}$$

It is common practice to circle, in the tableau, the element $\mathbf{y}_{rk}$, which is commonly referred to as the *pivot element*.

Step 4. The pivot operation. In this step, it is necessary to perform certain algebraic manipulations for creating all of the numbers needed to form the new tableau that corresponds to the new bfs $\mathbf{x}' = \mathbf{x} + t^*\mathbf{d}$. Although the necessary justification is omitted, the rows for the new tableau can be obtained from those of the old tableau using the following row operations:

1. Divide the old row r by $\mathbf{y}_{rk}$ to obtain the new row r.
2. To obtain the new row 0, subtract $z_k - c_k$ times the new row r from the old row 0.
3. For each other row $i = 1, \ldots, m$ (and $i \neq r$), add $-\mathbf{y}_{ik}$ times the new row r to the old row i.

To verify your arithmetic, at the end of the pivot operation, you should see the number 1 in the circled pivot position, and zeros in the rest of that column.

These steps will now be illustrated with the LP in Example 5.2:

EXAMPLE: Recall the LP in Example 5.2,

$$\begin{array}{llllllll}
\text{Minimize} & z & & & & & & \\
\text{subject to} & z + 2x_1 & + 3x_2 & + \;\; x_3 & & & = 0 \\
& \quad\;\; x_1 & - \;\; x_2 & + 2x_3 & + x_4 & & = 1 \\
& \quad 4x_1 & + 2x_2 & - \;\; x_3 & & + x_5 & = 2 \\
& \quad\;\; x_1, & x_2, & x_3, & x_4, & x_5 & \geq 0
\end{array}$$

Iteration 1

The tableau method will be initiated from the bfs in which $\mathbf{x} = (0,0,0,1,2)$ whose objective value is $z = 0$. In order to obtain the numbers in row 0 of the initial tableau, it is necessary to compute $\mathbf{c_B B^{-1} N} - \mathbf{c_N}$. In this case,

$$\mathbf{B} = [\mathbf{A}_{.4}, \mathbf{A}_{.5}] = \begin{bmatrix} 1 & 0 \\ 0 & 1 \end{bmatrix}$$

$$\mathbf{N} = [\mathbf{A}_{.1}, \mathbf{A}_{.2}, \mathbf{A}_{.3}] = \begin{bmatrix} 1 & -1 & 2 \\ 4 & 2 & -1 \end{bmatrix}$$

$$\begin{aligned} \mathbf{c_B B^{-1} N} - \mathbf{c_N} &= (0,0)\begin{bmatrix} 1 & 0 \\ 0 & 1 \end{bmatrix}\begin{bmatrix} 1 & -1 & 2 \\ 4 & 2 & -1 \end{bmatrix} - (-2, -3, -1) \\ &= (2,3,1) \end{aligned}$$

Then, for each column i of the $\mathbf{A}$ matrix, it is necessary to compute $\mathbf{B}^{-1}\mathbf{A}_{.i}$, but, since $\mathbf{B}^{-1} = \mathbf{I}$, $\mathbf{B}^{-1}\mathbf{A}_{.i} = \mathbf{A}_{.i}$. Finally, for the last column of the tableau one must compute $\bar{\mathbf{b}} = \mathbf{B}^{-1}\mathbf{b} = \mathbf{I}^{-1}\mathbf{b} = \mathbf{b}$. Thus the initial tableau is

z	x_1	x_2	x_3	x_4	x_5	RHS	
1	2	3	1	0	0	0	z
0	1	-1	2	1	0	1	x_4
0	4	(2)	-1	0	1	2	x_5

Step 1. Testing for optimality: Look in row 0 of the tableau under the nonbasic variables x_1, x_2, and x_3; the numbers 2, 3, and 1 are not all less than or equal to 0. Thus one chooses the value $k = 2$ corresponding to the most positive component (under x_2).

Step 2. Computing the direction of movement: Since $k = 2$, the corresponding column (under x_2) in the tableau is $\mathbf{y}_2$, so

$$\mathbf{y}_2 = \begin{bmatrix} -1 \\ 2 \end{bmatrix}$$

and the direction of movement is

$$\mathbf{d_B} = -\mathbf{y}_2 = \begin{bmatrix} 1 \\ -2 \end{bmatrix}$$

Step 3. Computing the amount of movement: Since $\mathbf{y}_2 \le \mathbf{0}$ is not true, the LP is not unbounded at this step. Thus one must compute the ratios $\bar{b}_i/\mathbf{y}_{i2}$, where $1 \le i \le m$ and $\mathbf{y}_{i2} > 0$. Since only $\mathbf{y}_{22} > 0$, it follows that

$$t^* = \bar{b}_2/\mathbf{y}_{22} = 2/2 = 1$$

and the value for r is 2 (i.e., $k^* = 2$). The pivot element $\mathbf{y}_{22}$ has been circled in the preceding tableau.

Step 4. The pivot operation ($k = 2$, $r = 2$): The new rows of the tableau are computed as

$$\begin{aligned}
\text{row } r &= (\text{old row } r)/\mathbf{y}_{rk} \\
&= (0,4,2,-1,0,1,2)/2 \\
&= (0,2,1,-\tfrac{1}{2},0,\tfrac{1}{2},1) \\
&= \text{new row 2} \\
\text{row } 0 &= \text{old row } 0 - (z_2 - c_2)(\text{new row } r) \\
&= (1,2,3,1,0,0,0) - 3(0,2,1,-\tfrac{1}{2},0,\tfrac{1}{2},1) \\
&= (1,-4,0,\tfrac{5}{2},0,-\tfrac{3}{2},-3) \\
\text{row } 1 &= \text{old row } 1 - \mathbf{y}_{1k}(\text{new row } r) \\
&= (0,1,-1,2,1,0,1) - (-1)(0,2,1,-\tfrac{1}{2},0,\tfrac{1}{2},1) \\
&= (0,3,0,\tfrac{3}{2},1,\tfrac{1}{2},2)
\end{aligned}$$

The new tableau is

z	x_1	x_2	x_3	x_4	x_5	RHS	
1	-4	0	$\frac{5}{2}$	0	$-\frac{3}{2}$	-3	z
0	3	0	($\frac{3}{2}$)	1	$\frac{1}{2}$	2	x_4
0	2	1	$-\frac{1}{2}$	0	$\frac{1}{2}$	1	x_2

Iteration 2

The bfs for this iteration is $\mathbf{x} = (0,1,0,2,0)$.

Step 1. Testing for optimality: Looking in row 0 of the tableau under the nonbasic variables x_1, x_3, and x_5, the numbers -4, $\frac{5}{2}$, and $-\frac{3}{2}$ are not all less than or equal to zero. Thus one chooses the value $k = 3$ corresponding to the most positive component (under x_3).

Step 2. Computing the direction of movement: Since $k = 3$, the corresponding column (under x_3) in the tableau is $\mathbf{y}_3$, so

$$\mathbf{y}_3 = \begin{bmatrix} \frac{3}{2} \\ -\frac{1}{2} \end{bmatrix}$$

and the direction of movement is

$$\mathbf{d}_{\mathbf{B}} = -\mathbf{y}_3 = \begin{bmatrix} -\frac{3}{2} \\ \frac{1}{2} \end{bmatrix}$$

Step 3. Computing the amount of movement: Since it is not the case that $\mathbf{y}_3 \leq \mathbf{0}$, the LP is not unbounded at this step. Thus one must compute the ratios $\bar{b}_i/\mathbf{y}_{i3}$, where $1 \le i \le m$ and $\mathbf{y}_{i3} > 0$. Since only $\mathbf{y}_{13} > 0$, it follows that

$$t^* = \bar{b}_1/\mathbf{y}_{13} = 2/\tfrac{3}{2} = \tfrac{4}{3}$$

and the value for r is 1 (i.e., $k^* = 1$). The pivot element $\mathbf{y}_{13}$ has been circled in the preceding tableau.

Step 4. The pivot operation ($k = 3$, $r = 1$): The new rows of the tableau are computed as

$$\begin{aligned}
\text{row } r &= (\text{old row } r)/\mathbf{y}_{rk}\\
&= (0,3,0,\tfrac{3}{2},1,\tfrac{1}{2},2)/\tfrac{3}{2}\\
&= (0,2,0,1,\tfrac{2}{3},\tfrac{1}{3},\tfrac{4}{3})\\
&= \text{new row } 1\\
\text{row } 0 &= \text{old row } 0 - (z_3 - c_3)(\text{new row } r)\\
&= (1,-4,0,\tfrac{5}{2},0,-\tfrac{3}{2},-3) - \tfrac{5}{2}(0,2,0,1,\tfrac{2}{3},\tfrac{1}{3},\tfrac{4}{3})\\
&= (1,-9,0,0,-\tfrac{5}{3},-\tfrac{7}{3},-\tfrac{19}{3})\\
\text{row } 2 &= \text{old row } 2 - \mathbf{y}_{2k}(\text{new row } r)\\
&= (0,2,1,-\tfrac{1}{2},0,\tfrac{1}{2},1) - (-\tfrac{1}{2})(0,2,0,1,\tfrac{2}{3},\tfrac{1}{3},\tfrac{4}{3})\\
&= (0,3,1,0,\tfrac{1}{3},\tfrac{2}{3},\tfrac{5}{3})
\end{aligned}$$

The new tableau is

z	x_1	x_2	x_3	x_4	x_5	RHS	
1	-9	0	0	$-\frac{5}{3}$	$-\frac{7}{3}$	$-\frac{19}{3}$	z
0	2	0	1	$\frac{2}{3}$	$\frac{1}{3}$	$\frac{4}{3}$	x_3
0	3	1	0	$\frac{1}{3}$	$\frac{2}{3}$	$\frac{5}{3}$	x_2

Iteration 3

The bfs for this iteration is $\mathbf{x} = (0, \tfrac{5}{3}, \tfrac{4}{3}, 0, 0)$.

Step 1. Testing for optimality: Look in row 0 of the tableau under the nonbasic variables x_1, x_4, and x_5. The numbers -9, $-\tfrac{5}{3}$, and $-\tfrac{7}{3}$ are all nonpositive. Thus the current bfs $\mathbf{x} = (0, \tfrac{5}{3}, \tfrac{4}{3}, 0, 0)$ is optimal, and the objective function value at $\mathbf{x}$ is $-\tfrac{19}{3}$.

When using the tableau method for both phase 1 and phase 2, one first formulates the phase 1 LP:

$$\begin{aligned}
&\text{Minimize} && \mathbf{ey}\\
&\text{subject to} && \mathbf{Ax} + \mathbf{Iy} = \mathbf{b}\\
& && \mathbf{x},\quad \mathbf{y} \ge \mathbf{0}
\end{aligned}$$

The corresponding initial tableau is

z	$\mathbf{x}$	$\mathbf{y}$	RHS	
1	$\mathbf{eA}$	$\mathbf{0}$	$\mathbf{eb}$	z
$\mathbf{0}$	$\mathbf{A}$	$\mathbf{I}$	$\mathbf{b}$	$\mathbf{y}$

It is convenient to carry the original cost vector $\mathbf{c}$ as an additional row (just under row 0). As the computations proceed, the original cost row is continually updated, just as would be done to any of the other rows in the tableau; however, one never uses the original cost row when performing the min ratio test in phase 1.

At the end of phase 1, if all of the $\mathbf{y}$ variables are nonbasic, then, to initiate phase 2, one simply crosses off the top row of the final phase 1 tableau (which corresponds to the phase 1 objective function) and all of the columns corresponding to the $\mathbf{y}$ variables. Finally, the original cost row (having been appropriately updated throughout phase 1) serves as row 0. The process of handling artificial variables in the final phase 1 basis at value 0 will not be addressed here.

DISCUSSION

The Tableau Method

The tableau method is the most common presentation of the algebraic steps of the simplex algorithm. In addition, many other procedures that rely on the simplex method, such as in integer and nonlinear programming, are presented in the tableau format. A complete presentation of the method is that of Bazaraa and Jarvis [1].

REFERENCES

1. M. S. Bazaraa and J. J. Jarvis, *Linear Programming and Network Flows*, Wiley, New York, 1977.

Appendix B

How Efficiently Can We Solve LP Problems?

In Chapter 7 many ideas were presented for improving the computational efficiency (and numerical accuracy) of the simplex algorithm. Even with all of these improvements, there is still the unanswered question of how efficient the simplex algorithm really is.

From a practical point of view, the answer is determined by examining the computational results of applying the algorithm on a large number of real world problems. In that regard, the simplex method has been an overwhelming success. It appears that, for real world problems with m constraints and n variables (in standard form), the simplex algorithm takes, on the average, about $3m$ iterations to obtain an optimal solution.

One might ask if there is some theoretical justification for these empirical results. None has been found so far. One question of interest would be to determine as a function of m and n the expected number of iterations needed to solve a random LP starting from an initial basic feasible solution. Such questions are currently being examined, but specific results are difficult to interpret because it is not always clear that random LPs are reflective of real world problems.

Another related question of theoretical interest is to determine, as a function of m and n, the maximum number of iterations that will be required to solve an LP. One can surely state that the absolute maximum number of iterations is $\binom{n}{m}$ because there are at most that many basic feasible solutions (see Section 5.1). Unfortunately, the function $f(m, n) = \binom{n}{m}$ is considered to be very large in terms of m and n. Perhaps there is a better function of m and n that could be used to predict the maximum number of bfs that will be visited by the simplex algorithm. The question therefore arises as to what types of functions are considered to be "good." Mathematicians and computer scientists have come to believe that polynomial functions such as $3m^2 + 5n^2$ or $300(mn)^2$ are good and that exponential functions such as 2^{mn} or $\binom{n}{m}$ are not. Part of the reason for this rather arbitrary decision is that, if you consider a fixed polynomial function such as $10n^2$ and a fixed exponential function such as 2^n, then there is an integer n' such that, for all values of $n > n'$, the exponential function is larger than the polynomial function. In this instance, for all values of $n > 9$, one has $2^n > 10n^2$.

Loosely speaking, an algorithm is said to be *polynomial* if there is a polynomial function p, in terms of the parameters describing the size of the problem, such that a solution to any specific problem of that size can be obtained with "computational effort" that is less than or equal to that predicted by the function p. With this concept of a polynomial algorithm in mind, one can ask the following:

1. Is the simplex algorithm polynomial?
2. If it is not, then does there exist some other algorithm for solving LP problems that is polynomial?

To answer the first question, it is important to note that, when discussing the simplex algorithm, a specific rule must be stated for actually choosing which nonbasic variable is to become basic in the event that more than one such variable has a negative reduced cost and, similarly, for choosing which variable is to become nonbasic (in case of ties in the minimum ratio test). Consider, for instance, the rule of steepest descent, which was presented in Section 5.3. The following LP was specifically designed to force the simplex algorithm (using the rule of steepest descent) to visit an exponential number of bfs. In this example, t can be chosen as any real number satisfying $0 < t < \frac{1}{2}$.

$$
\begin{array}{ll}
\text{Minimize} & -x_s \\
\text{subject to} & x_1 \geq 0 \\
& x_1 \leq 1 \\
& x_2 \geq t x_1 \\
& x_2 \leq 1 - t x_1 \\
& x_3 \geq t x_2 \\
& x_3 \leq 1 - t x_2 \\
& \vdots \\
& x_s \geq t x_{s-1} \\
& x_s \leq 1 - t x_{s-1}
\end{array}
$$

When slack variables are added and the unrestricted **x** variables are eliminated by using the equality constraints, the resulting LP has $m=s$ constraints and $n=2s$ variables. The simplex algorithm, when started at $\mathbf{x}=\mathbf{0}$, will then visit $f(s)=2^s-1$ bfs, and this function is larger than any polynomial in s. Consequently, with the rule of steepest descent the simplex algorithm is not polynomial. Indeed, for most known pivoting rules, an LP has been constructed that forces the simplex algorithm to visit an exponential number of bfs before obtaining an optimal solution. It would appear that no rule leads to a polynomial version of the simplex algorithm, although this statement has not been proved formally.

Until 1979 the answer to the second question was unknown. Then, a completely new approach to solving LPs was proposed with which, it was shown theoretically, a solution could be obtained to an LP with a polynomial amount of computational effort. That method and all of its variants are referred to as the *ellipsoid algorithms* because they work with ellipses that "shrink" toward a solution.

Given an $m \times n$ matrix **A** and an m vector **b**, the ellipsoid algorithms are designed to determine if there is a n vector **x** satisfying the linear system of inequalities

$$\mathbf{Ax} \leq \mathbf{b}$$

Suppose that such an algorithm were available. To see how that algorithm could be used to solve an LP, consider the following pair of symmetric LPs (see Section 8.2): The primal LP,

$$\begin{aligned} \text{Minimize} \quad & \mathbf{c}'\mathbf{x} \\ \text{subject to} \quad & \mathbf{A}'\mathbf{x} \geq \mathbf{b}' \\ & \mathbf{x} \geq \mathbf{0} \end{aligned}$$

and the dual LP,

$$\begin{aligned} \text{Maximize} \quad & \mathbf{ub}' \\ \text{subject to} \quad & \mathbf{uA}' \leq \mathbf{c}' \\ & \mathbf{u} \geq \mathbf{0} \end{aligned}$$

According to the duality theory presented in Chapter 8, optimal solutions to both LPs can be obtained by finding feasible vectors **x** and **u** that satisfy $\mathbf{c}'\mathbf{x} \leq \mathbf{ub}'$. Equivalently, one wants to find a solution to the system of inequalities $\mathbf{Ax} \leq \mathbf{b}$ in which

$$\mathbf{A} = \begin{bmatrix} \mathbf{c}' & -\mathbf{b}' \\ -\mathbf{A}' & \mathbf{0} \\ \mathbf{0} & \mathbf{A}'^{\mathrm{T}} \\ -\mathbf{I} & \mathbf{0} \\ \mathbf{0} & -\mathbf{I} \end{bmatrix}, \qquad \mathbf{b} = \begin{bmatrix} \mathbf{0} \\ -\mathbf{b}' \\ \mathbf{c}' \\ \mathbf{0} \\ \mathbf{0} \end{bmatrix}$$

In order to attempt finding a solution to the system of inequalities $\mathbf{Ax} \leq \mathbf{b}$, the ellipsoid algorithms generate a sequence of shrinking ellipses each of which contains at least one point $\mathbf{x}$ satisfying $\mathbf{Ax} \leq \mathbf{b}$, provided that such a point exists. For a given ellipse, the center is tested to see if it satisfies the system of inequalities. If not, then at least one inequality is violated. One can use any such inequality to construct a new ellipse that is substantially smaller in volume than the previous one, and the process is then repeated. It has been shown that, under suitable conditions, in a polynomial number of iterations, either the algorithm stops with a vector satisfying the system of inequalities or else the current ellipse will have shrunk to a size that is sufficiently small that it may be concluded the system of inequalities has no solution.

Before specifying the details of the ellipsoid algorithm, it is important to point out that, although the algorithm is theoretically polynomial, so far no implementation of the algorithm has proven to be computationally efficient on real world problems. Consequently, this algorithm is not used in practice.

The equation for an ellipse can be determined if its center $\mathbf{x}'$ is known together with a special type of nonsingular $n \times n$ matrix $\mathbf{B}$ that satisfies the property that, for all n vectors $\mathbf{x} \neq \mathbf{0}$, $\mathbf{x}^{\mathrm{T}}\mathbf{Bx} > 0$. Such matrices are called *positive definite* matrices. Using $\mathbf{x}'$ and $\mathbf{B}$, the ellipse is

$$E = \left\{\mathbf{x} : (\mathbf{x} - \mathbf{x}')^{\mathrm{T}}\mathbf{B}^{-1}(\mathbf{x} - \mathbf{x}') \leq 1\right\}$$

The steps of the ellipsoid algorithm follow:

Step 0. Initialization: Choose $\mathbf{x}' = \mathbf{0}$ and a positive definite matrix $\mathbf{B}$ such that the ellipse

$$E = \left\{\mathbf{x} : (\mathbf{x} - \mathbf{x}')^{\mathrm{T}}\mathbf{B}^{-1}(\mathbf{x} - \mathbf{x}') \leq 1\right\}$$

contains a vector $\mathbf{x}$ satisfying $\mathbf{Ax} \leq \mathbf{b}$ if such a vector exists. Go to step 1.

Step 1. Testing for optimality: If the center $\mathbf{x}'$ of the current ellipse satisfies $\mathbf{Ax}' \leq \mathbf{b}$, then stop—a feasible solution has been found. Otherwise, select any value of i with $1 \leq i \leq m$ such that $\mathbf{A}_{i.}\mathbf{x}' > b_i$. Go to step 2.

Step 2. Moving: From use of the column vector $\mathbf{a} = (\mathbf{A}_{i.})^{\mathrm{T}}$ and the real numbers $t = 1/(n+1)$, and $v = n^2/(n^2 - 1)$, the center of the new ellipse is found to be

$$\mathbf{x}' - \left(t/\sqrt{\mathbf{a}^{\mathrm{T}}\mathbf{Ba}}\,\right)\mathbf{Ba}$$

and the new positive definite matrix is

$$v\left[\mathbf{B} - 2t\left(\mathbf{Ba}(\mathbf{Ba})^{\mathrm{T}}\right)/\mathbf{a}^{\mathrm{T}}\mathbf{Ba}\right]$$

[Note that the $n \times 1$ column vector $\mathbf{Ba}$ times the $1 \times n$ row vector $(\mathbf{Ba})^{\mathrm{T}}$ is an $n \times n$ matrix.]

Use these data for the new ellipse, and return to step 1.

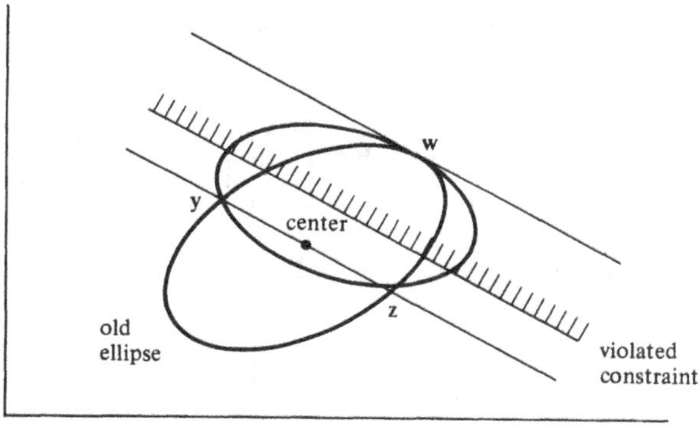

FIGURE B.1. Geometry of the ellipsoid algorithm.

Figure B.1 illustrates geometrically how the new ellipse is constructed from the original one. Start with any constraint for which the current center is infeasible. Move that constraint line parallel to itself (away from the center) until it is just tangent to the current ellipse at the point **w**. Then move that line parallel to itself in the opposite direction until the line passes through the current center, intersecting the ellipse at **y** and **z**. The new ellipse is the (unique) one that passes through **y** and **z** and is tangent to the line at **w**.

These steps will be demonstrated with the following example.

EXAMPLE: The ellipsoid algorithm will be used to find a feasible solution for the following system of inequalities:

$$-x_1 - x_2 \le -1$$
$$x_1 - 2x_2 \le -1$$
$$x_1 + 3x_2 \le 5$$
$$-5x_1 + 3x_2 \le 1$$

That is,

$$\mathbf{A} = \begin{bmatrix} -1 & -1 \\ 1 & -2 \\ 1 & 3 \\ -5 & 3 \end{bmatrix}, \qquad \mathbf{b} = \begin{bmatrix} -1 \\ -1 \\ 5 \\ 1 \end{bmatrix}$$

Here, $n = 2$, so $t = 1/(n+1) = \frac{1}{3}$ and $v = n^2/(n^2 - 1) = \frac{4}{3}$.

Iteration 1

Step 0. Initialization: Start with $\mathbf{x}' = (0, 0)$ and the positive definite matrix

$$\mathbf{B} = \begin{bmatrix} 5 & 0 \\ 0 & 5 \end{bmatrix}$$

Step 1. Testing for optimality (i.e., is $\mathbf{x}'$ feasible?):

$$\mathbf{Ax} = \mathbf{0} \nleq \begin{bmatrix} -1 \\ -1 \\ 5 \\ 1 \end{bmatrix}$$

Thus i can be chosen as either 1 or 2. Let $i = 1$.

Step 2. Moving:

$$\mathbf{a} = (\mathbf{A}_{i,\cdot})^{\mathrm{T}} = \begin{bmatrix} -1 \\ -1 \end{bmatrix}$$

$$\mathbf{Ba} = \begin{bmatrix} 5 & 0 \\ 0 & 5 \end{bmatrix}\begin{bmatrix} -1 \\ -1 \end{bmatrix} = \begin{bmatrix} -5 \\ -5 \end{bmatrix}$$

$$\mathbf{a}^{\mathrm{T}}\mathbf{Ba} = (-1, -1)\begin{bmatrix} -5 \\ -5 \end{bmatrix} = 5 + 5 = 10$$

$$\mathbf{Ba}(\mathbf{Ba})^{\mathrm{T}} = \begin{bmatrix} -5 \\ -5 \end{bmatrix}[-5, -5] = \begin{bmatrix} 25 & 25 \\ 25 & 25 \end{bmatrix}$$

Thus the center of the new ellipse is

$$\begin{aligned} \mathbf{x}'_{\text{new}} &= \mathbf{x}' - \left(t/\sqrt{\mathbf{a}^{\mathrm{T}}\mathbf{Ba}}\right)\mathbf{Ba} \\ &= \begin{bmatrix} 0 \\ 0 \end{bmatrix} - (1/3)(1/\sqrt{10})\begin{bmatrix} -5 \\ -5 \end{bmatrix} \\ &= 5\sqrt{10}/30\begin{bmatrix} 1 \\ 1 \end{bmatrix} = \begin{bmatrix} \sqrt{10}/6 \\ \sqrt{10}/6 \end{bmatrix} \end{aligned}$$

The new positive definite matrix is

$$\begin{aligned} \mathbf{B}_{\text{new}} &= v\left[\mathbf{B} - 2t\left(\mathbf{Ba}(\mathbf{Ba})^{\mathrm{T}}\right)/\mathbf{a}^{\mathrm{T}}\mathbf{Ba}\right] \\ &= \frac{4}{3}\left[\begin{bmatrix} 5 & 0 \\ 0 & 5 \end{bmatrix} - \frac{2}{3}\left(\frac{1}{10}\right)\begin{bmatrix} 25 & 25 \\ 25 & 25 \end{bmatrix}\right] \\ &= \frac{4}{3}\begin{bmatrix} \frac{10}{3} & -\frac{5}{3} \\ -\frac{5}{3} & \frac{10}{3} \end{bmatrix} \\ &= \begin{bmatrix} \frac{40}{9} & -\frac{20}{9} \\ -\frac{20}{9} & \frac{40}{9} \end{bmatrix} \end{aligned}$$

Iteration 2

Step 1. Testing for optimality (i.e., is $\mathbf{x}'$ feasible?):

$$\mathbf{Ax}' = \begin{bmatrix} -1 & -1 \\ 1 & -2 \\ 1 & 3 \\ -5 & 3 \end{bmatrix} \begin{bmatrix} \sqrt{10}/6 \\ \sqrt{10}/6 \end{bmatrix} = \begin{bmatrix} -\sqrt{10}/3 \\ -\sqrt{10}/6 \\ 2\sqrt{10}/3 \\ -\sqrt{10}/3 \end{bmatrix}$$

$$= \begin{bmatrix} -1.054 \\ -0.527 \\ 2.108 \\ -1.054 \end{bmatrix} \nleq \begin{bmatrix} -1 \\ -1 \\ 5 \\ 1 \end{bmatrix}$$

so $i = 2$.

Step 2. Moving:

$$\mathbf{a} = (\mathbf{A}_{2\cdot})^{\mathrm{T}} = \begin{bmatrix} 1 \\ -2 \end{bmatrix}$$

$$\mathbf{Ba} = \begin{bmatrix} \frac{40}{9} & -\frac{20}{9} \\ -\frac{20}{9} & \frac{40}{9} \end{bmatrix} \begin{bmatrix} 1 \\ -2 \end{bmatrix} = \begin{bmatrix} \frac{80}{9} \\ -\frac{100}{9} \end{bmatrix}$$

$$\mathbf{a}^{\mathrm{T}}\mathbf{Ba} = (1, -2) \begin{bmatrix} \frac{80}{9} \\ -\frac{100}{9} \end{bmatrix} = \frac{280}{9}$$

$$\mathbf{Ba}(\mathbf{Ba})^{\mathrm{T}} = \begin{bmatrix} \frac{80}{9} \\ -\frac{100}{9} \end{bmatrix} \left[\frac{80}{9}, -\frac{100}{9}\right] = 100 \begin{bmatrix} \frac{64}{81} & -\frac{80}{81} \\ -\frac{80}{81} & \frac{100}{81} \end{bmatrix}$$

Thus the center of the new ellipse is

$$\mathbf{x}'_{\text{new}} = \mathbf{x}' - \left(t/\sqrt{\mathbf{a}^{\mathrm{T}}\mathbf{Ba}}\right)\mathbf{Ba}$$

$$= \begin{bmatrix} \sqrt{10}/6 \\ \sqrt{10}/6 \end{bmatrix} - \frac{1}{3}\left(\sqrt{9/280}\right) \begin{bmatrix} \frac{80}{9} \\ -\frac{100}{9} \end{bmatrix}$$

$$= \begin{bmatrix} \sqrt{10}/6 \\ \sqrt{10}/6 \end{bmatrix} - \frac{1}{2\sqrt{70}} \begin{bmatrix} \frac{80}{9} \\ -\frac{100}{9} \end{bmatrix}$$

$$= \begin{bmatrix} -0.0042 \\ 1.1911 \end{bmatrix}$$

The new positive definite matrix is

$$\mathbf{B}_{\text{new}} = v\left[\mathbf{B} - 2t\left(\mathbf{Ba}(\mathbf{Ba})^{\mathrm{T}}\right)/\mathbf{a}^{\mathrm{T}}\mathbf{Ba}\right]$$

$$= \frac{4}{3}\left[\begin{bmatrix} \frac{40}{9} & -\frac{20}{9} \\ -\frac{20}{9} & \frac{40}{9} \end{bmatrix} - \frac{2}{3}\left(\frac{9}{280}\right)100\begin{bmatrix} \frac{64}{81} & -\frac{80}{81} \\ -\frac{80}{81} & \frac{100}{81} \end{bmatrix}\right]$$

$$= \begin{bmatrix} 3.6684 & -0.1411 \\ -0.1411 & 2.3986 \end{bmatrix}$$

Iteration 3

Step 1. Testing for optimality (i.e., is $\mathbf{x}'$ feasible?):

$$\mathbf{Ax}' = \begin{bmatrix} -1 & -1 \\ 1 & -2 \\ 1 & 3 \\ -5 & 3 \end{bmatrix}\begin{bmatrix} -0.0042 \\ 1.1911 \end{bmatrix} = \begin{bmatrix} -1.1869 \\ -2.3863 \\ 3.57 \\ 3.59 \end{bmatrix} \not\leq \begin{bmatrix} -1 \\ -1 \\ 5 \\ 1 \end{bmatrix}$$

so $i = 4$.

Step 2. Moving:

$$\mathbf{a} = (\mathbf{A}_{4\cdot})^{\mathrm{T}} = \begin{bmatrix} -5 \\ 3 \end{bmatrix}$$

$$\mathbf{Ba} = \begin{bmatrix} 3.6684 & -0.1411 \\ -0.1411 & 2.3986 \end{bmatrix}\begin{bmatrix} -5 \\ 3 \end{bmatrix} = \begin{bmatrix} -18.7654 \\ 7.9012 \end{bmatrix}$$

$$\mathbf{a}^{\mathrm{T}}\mathbf{Ba} = (-5, 3)\begin{bmatrix} -18.7654 \\ 7.9012 \end{bmatrix} = 117.531$$

$$\mathbf{Ba}(\mathbf{Ba})^{\mathrm{T}} = \begin{bmatrix} -18.7654 \\ 7.9012 \end{bmatrix}[-18.7654, 7.9012]$$

$$= \begin{bmatrix} 352.14 & -148.269 \\ -148.269 & 62.429 \end{bmatrix}$$

Thus the center of the new ellipse is

$$\mathbf{x}'_{\text{new}} = \mathbf{x}' - \left(t/\sqrt{\mathbf{a}^{\mathrm{T}}\mathbf{Ba}}\right)\mathbf{Ba}$$

$$= \begin{bmatrix} -0.0042 \\ 1.1911 \end{bmatrix} - \frac{1}{3}\left(\frac{1}{\sqrt{117.531}}\right)\begin{bmatrix} -18.7654 \\ 7.9012 \end{bmatrix}$$

$$= \begin{bmatrix} 0.5728 \\ 0.9481 \end{bmatrix}$$

The new positive definite matrix is

$$\mathbf{B}_{\text{new}} = v\left[\mathbf{B} - 2t\left(\mathbf{Ba}(\mathbf{Ba})^{\mathrm{T}}\right)/\mathbf{a}^{\mathrm{T}}\mathbf{Ba}\right]$$

$$= \frac{4}{3}\left[\begin{bmatrix} 3.6684 & -0.1411 \\ -0.1411 & 2.3986 \end{bmatrix} - \frac{2}{3}\left(\frac{1}{117.531}\right)\begin{bmatrix} 352.14 & -148.269 \\ -148.269 & 62.429 \end{bmatrix}\right]$$

$$= \begin{bmatrix} 2.2280 & 0.9332 \\ 0.9332 & 2.7260 \end{bmatrix}$$

Iteration 4

Step 1. Testing for optimality (i.e., is $\mathbf{x}'$ feasible?):

$$\mathbf{Ax}' = \begin{bmatrix} -1 & -1 \\ 1 & -2 \\ 1 & 3 \\ -5 & 3 \end{bmatrix}\begin{bmatrix} 0.5728 \\ 0.9481 \end{bmatrix} = \begin{bmatrix} -1.5209 \\ -1.3234 \\ 3.4171 \\ -0.0197 \end{bmatrix} \leq \begin{bmatrix} -1 \\ -1 \\ 5 \\ 1 \end{bmatrix}$$

Since $\mathbf{Ax}' \leq \mathbf{b}$, the current solution $\mathbf{x}' = (0.5728, 0.9481)$ is feasible for the given inequalities.

It is interesting to note that the measure of improvement for the above algorithm is the volume of the ellipse. Also, the ellipsoid algorithm as stated is not necessarily finite, much less polynomial. In order to obtain the desired polynomial result, it is necessary to make three assumptions.

1. Each element of $\mathbf{A}$ and each component of $\mathbf{b}$ is an integer.
2. If $F = \{\mathbf{x} : \mathbf{Ax} \leq \mathbf{b}\}$ is not empty, then its "volume" should be positive. Geometrically, this means that F should be a solid, full-dimensional set and not flat. Algebraically, for F to have positive volume means that there is a vector $\mathbf{x}^*$ with $\mathbf{Ax}^* \leq \mathbf{b}$ and a real number $t > 0$ such that $\{\mathbf{x} : \text{for all } i \text{ with } 1 \leq i \leq n,\ |x_i^* - x_i| \leq t\}$ is a subset of F.
3. It is somehow possible to find a positive definite matrix $\mathbf{B}$ such that the initial ellipse contains a vector $\mathbf{x}$ satisfying $\mathbf{Ax} \leq \mathbf{b}$, if such a vector exists.

With these three assumptions it is possible to prove the following theorem:

Theorem. *In a polynomial number of iterations, either the algorithm will produce a vector* $\mathbf{x}$ *satisfying* $\mathbf{Ax} \leq \mathbf{b}$, *or else the volume of the final ellipse will have shrunk to a size such that it is possible to conclude that there is no vector* $\mathbf{x}$ *satisfying* $\mathbf{Ax} \leq \mathbf{b}$.

Before the ellipsoid algorithm can be used to solve an LP, it is necessary to verify that the resulting system of inequalities satisfies the three assumptions stated above. The first assumption is certainly true if all of the data in the original LP consist of integers. Even if this is not the case, one can store on a computer only a finite number of digits of precision. Consequently, by an appropriate scaling (see Section 7.4), the elements of the resulting **A** matrix and **b** vector can be made into integers.

The second assumption—that the set F, if not empty, has positive volume—is a problem when solving an LP. The reason is that if the LP and its dual have optimal solutions, then the constraint that the primal and dual objective function be equal forces the set F to be flat and, consequently, to have zero volume. In this case, it has been shown that one can create a new system of inequalities that define a set F' such that

1. F' is not empty if and only if F is not empty;
2. if F' is not empty, then F' has positive volume;
3. if a vector is found that satisfies the inequalities in F', then, with a polynomial amount of computational effort, a solution to the original system of inequalities that define F can be found.

Thus it is possible to apply the ellipsoid algorithm to the new set of inequalities.

Finally, when applied to an LP, the phase 1 issue of how to find the initial ellipse satisfying the third assumption has been resolved by constructing $\mathbf{B} = s\mathbf{I}$, where s is a sufficiently large real number. Further detail can be found in the references at the end of this appendix.

In order for the ellipsoid algorithm to obtain numerically accurate solutions, a special computational implementation is required. In fact, if great care is not taken in the implementation, then the solution to an LP will not be obtained. The details of such an implementation, however, will not be discussed here.

DISCUSSION

Exponential Examples

For each pivot rule used in the simplex algorithm, there seems to be a specific numerical example that forces the algorithm to test an exponential number of basic feasible solutions before reaching optimality. The example presented in this appendix is from Klee and Minty [10]. For the rule in which one moves to the bfs that provides the greatest overall improvement in the objective function, an exponential example is presented by Jeroslow [8]. Finally, for Bland's rule, Avis and Chvâtal [1] give an exponential example (although that rule does avoid cycling).

Computational Complexity

One major aspect of computational complexity is the study of predicting how much effort will be required to solve a problem of a given size. This field took a big step forward after the work of Cook [4]. Subsequently, much work has been done in the area of grouping all problems that seem to be hard to solve—"hard" in the sense that there seems to be no polynomial algorithm. The hardest of these problems are referred to as NP-complete, and a discussion of these and related topics has been presented by Garey and Johnson [7].

Historically, computational complexity has been studied from the "worst case" point of view; i.e., what happens in the worst case. More recently, work is being done with the "average case" analysis (see Dantzig [5] and Borgwardt [3]). Currently, an attempt is being made to categorize a problem as being easy or hard, depending upon whether or not there exists a finite improvement algorithm for solving the problem (see Franco, Solow, and Emmons [6]).

Polynomial Algorithms

The ellipsoid algorithm presented in this appendix was based on the work of Shor [13]. The proof that this algorithm can be made to solve linear programming problems in a polynomial number of steps was established by Khachian [9], and these proofs can also be found in the book by Papadimitriou and Steiglitz [12]. Subsequently, there has been a great surge in attempts to improve the computational efficiency (and expand the uses) of the ellipsoid algorithm. However, current results indicate that, in linear programming, the ellipsoid method is not likely to replace the simplex algorithm (see McCall [11]). An extensive list of references on the ellipsoid algorithm can be found in the survey article by Bland, Goldfarb, and Todd [2].

REFERENCES

1. D. Avis and V. Chvâtal, "Notes on Bland's Pivoting Rule," *Mathematical Programming Study 8* (*Polyhedral Combinatorics*), North-Holland, Amsterdam, 1978, pp. 24–34.
2. R. G. Bland, D. Goldfarb, and M. J. Todd, "The Ellipsoid Method: A Survey," *Operations Research* 29(6), 1039–1091 (1981).
3. K. Borgwardt, "Some Distribution-Independent Results About the Asymptotic Order of the Average Number of Pivot Steps of the Simplex Method," *Mathematics of Operations Research* 7(3), 441–462 (1982).
4. S. A. Cook, "The Complexity of Theorem Proving Procedures," *Proceedings of the Third Annual ACM Symposium on Theory of Computing*, Association for Computing Machinery, New York, 1971, pp. 151–158.
5. G. B. Dantzig, "Expected Number of Steps of the Simplex Method for a Linear Program with a Convexity Constraint," Technical Report, SOL 80-3R, Department of Operations Research, Stanford University, Stanford, CA, 1980.
6. J. Franco, D. Solow, and H. Emmons, "Duality, Finite Improvement, and Efficiently Solved Problems," submitted to *Mathematics of Operations Research* (1984).

7. M. R. Garey and D. S. Johnson, *Computers and Intractability: A Guide to the Theory of NP-Completeness*, Freeman, San Francisco, CA, 1979.
8. R. G. Jeroslow, "The Simplex Algorithm with the Pivot Rule of Maximizing Criterion Improvement," *Discrete Mathematics* 4, 367–378 (1973).
9. L. G. Khachian, "A Polynomial Algorithm for Linear Programming," *Doklady Akademiia Nauk USSR*, **224**(5), 1093–1096, (1979); English translation in *Soviet Mathematics Doklady* 20, 191–194.
10. V. Klee and G. J. Minty, "How Good Is the Simplex Method?" in *Inequalities-III* (O. Shisha, ed.), Academic Press, New York, 1972, pp. 159–175.
11. E. H. McCall, "A Study of the Khachian Algorithm for Real-World Linear Programming," Sperry Univac, P.O. Box 43942, St. Paul, MN, 1980.
12. C. H. Papadimitriou and K. Steiglitz, *Combinatorial Optimization: Algorithms and Complexity*, Prentice-Hall, Englewood Cliffs, NJ, 1982.
13. N. Z. Shor, "Cut-off Method with Space Extension in Convex Programming Problems," *Kibernetika*, **13** (1), 94–95 (1977); English translation in *Cybernetics* 13, 94–96.

Appendix C

Spreadsheet Modeling Using Excel

In Chapter 1 of the book *Linear Programming: An Introduction to Finite Improvement Algorithms*, you learned to formulate a linear programming model (LP) by using given data to identify variables, an objective function and constraints. The next step is to *solve* the model, that is, to find values for the variables that simultaneously satisfy all of the constraints and provide the best possible value of the objective function or determine that the LP is *infeasible* or *unbounded*. In Chapters 5 and 6 of the book you learned the algebraic steps of the simplex algorithm for doing so. A computational implementation of this algorithm is available in Excel with an add-in package called *Solver*. In this appendix, you will learn how to use Solver to solve a linear programming problem and how to interpret the output reports that Solver provides.

C.1 GETTING STARTED WITH SOLVER

There are several versions of Solver, depending on which version of Office you have on your computer. Here, you will learn how to use Solver in Excel for Office 2010, which may be slightly different from your version but the basic approach will be the same.

The first step is to determine if Solver is installed in your computer. To do so, open Excel and click on the Data tab in the menu and see if Solver appears in the top right Analysis menu, as shown in Figure C.1. If Solver appears in this menu, then you are all set and can skip to Section C.2. Otherwise, you need to install Solver in Excel by performing the following instructions:

(1) In Excel, click on the File tab in the upper left corner and then on Options on the lower left of the screen (see Figure C.2).

(2) Click on Add-Ins on the left side of the resulting menu (see Figure C.3).

(3) In the drop-down box at the bottom of the screen, select Excel Add-Ins and click GO (see Figure C.3).

(4) In the resulting menu, check the boxes labeled Solver Add-In and click OK (see Figure C.4).

If everything has worked correctly, on the Data tab of Excel you should see the Solver add-in in the top right Analysis menu from now on, as seen in Figure C.1.

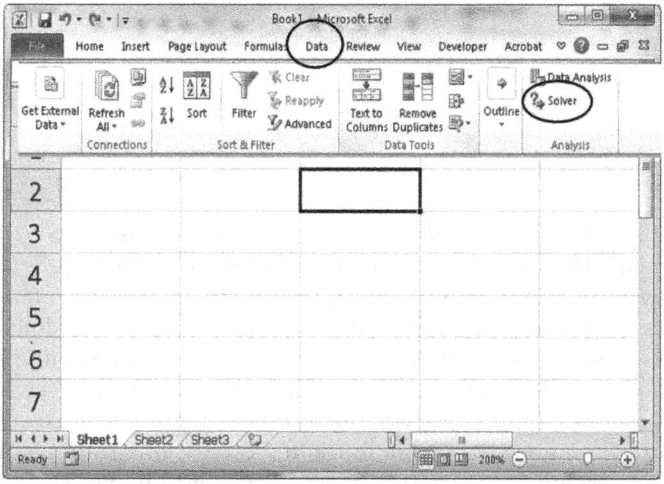

Fig. C.1 Checking if Solver is Installed in Excel

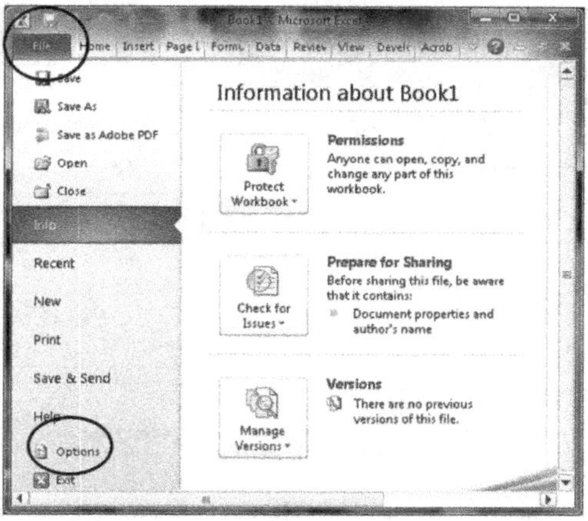

Fig. C.2 Getting to the Options Menu in Excel

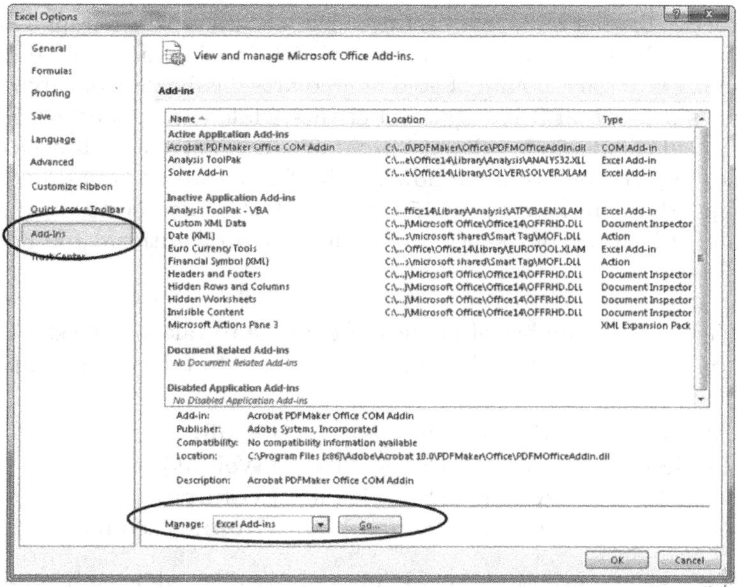

Fig. C.3 Getting to the Add-Ins Menu in Excel

C.2 CREATING A SPREADSHEET MODEL

From here on, it is assumed that Solver is properly installed in Excel. You will now learn how to build a linear programming (LP) model in an Excel

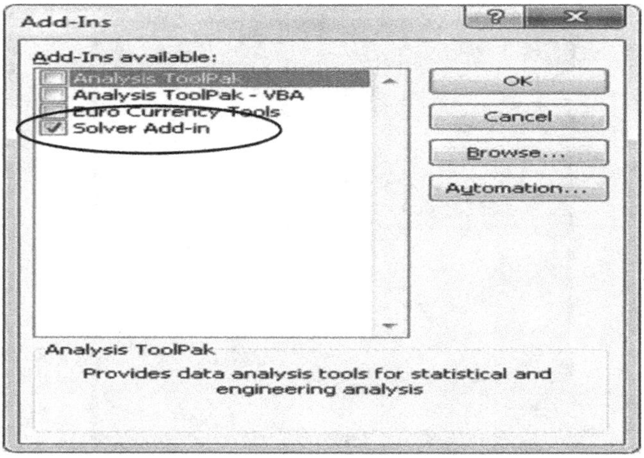

Fig. C.4 Installing the Solver Add-In

spreadsheet. To that end, recall the problem of Midas Miner from Chapter 1 of the book. Midas landed on an asteroid rich in silver and gold ore. After determining that each pound of gold ore requires 2 cubic feet of cargo volume and brings a profit of 3 intergalactic credits while each pound of silver ore requires 1 cubic foot and results in a profit of 2 intergalactic credits, he created the following LP to determine how much of each ore he could load on his spaceship—that can hold up to 150 cubic feet of cargo—so to maximize his profit and still lift off of the asteroid with up to an additional 100 pounds of cargo:

$$\begin{array}{lcl} G & = & \text{the number of pounds of gold ore to take on the ship} \\ S & = & \text{the number of pounds of silver ore to take on the ship} \end{array}$$

$$\begin{array}{lrcrcrl} \text{Maximize} & 3G & + & 2S & & & \textbf{(Profit)} \\ \text{Subject to} & 1G & + & 1S & \leq & 100 & \textbf{(Weight)} \\ & 2G & + & 1S & \leq & 150 & \textbf{(Volume)} \\ & G & , & S & \geq & 0 & \textbf{(Nonnegativity)} \end{array} \tag{C.1}$$

For discussion purposes, the LP in (C.1) is referred to subsequently as the *paper model.* To solve such a model with Excel, you must enter that information according to the following:

Golden Rule of Spreadsheet Modeling: The Excel spreadsheet must contain 100% of the information in the paper model.

For reasons that will soon be explained, it is best to separate the spreadsheet into two parts: one part for the *data* and one part for the *model.*

Entering the Data

The data for an LP consists of

(1) The objective function coefficients in the vector **c** [3 and 2 for the LP in (C.1)].

(2) The numerical values in the matrix A associated with the variables on the left-hand sides of the constraints [1 and 1 for the weight constraint in (C.1) and 2 and 1 for the volume constraint in (C.1)].

(3) The numerical values in the vector **b** associated with the right-hand side of each constraint [100 and 150 for the two constraints in (C.1)].

You are free to organize how this information is arranged in the spreadsheet. The most important considerations are to organize these data in a systematic and readable form so that

- You, or someone else, can understand the meaning of the values in the context of the model.
- You can easily find and modify specific values, if necessary.
- You can use these values easily in building the subsequent spreadsheet model.

These considerations are particularly important when the models are large and must be entered in more than one screen. One convenient way to enter the data for the problem of Midas Miner is shown in the top portion of the spreadsheet in Figure C.5.

Entering the Model in the Spreadsheet

Having entered the data in the spreadsheet, you must now enter the model itself, which consists of variables, an objective function, and constraints. The spreadsheet model must therefore contain cells of your choosing for the corresponding information, that is:

1. A collection of cells for the values of the decision variables.
2. One cell that contains the value of the objective function.
3. One cell that contains the value of the left-hand side (LHS) and one cell that contains the value of the right-hand side (RHS) of each constraint.

A spreadsheet model for the paper model of Midas Miner in (C.1) is shown in the bottom portion of Figure C.5 and explained in detail in what follows.

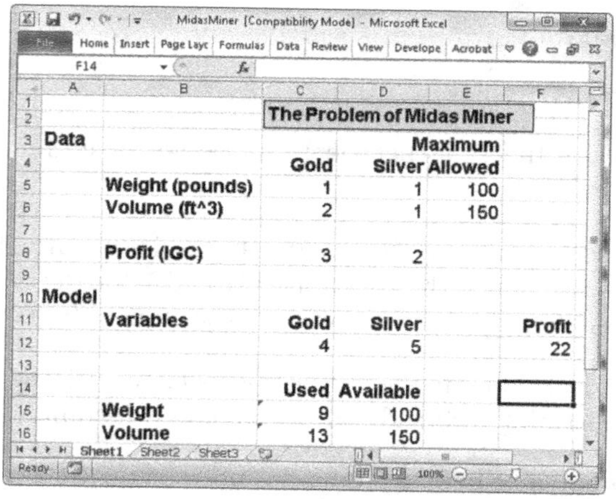

Fig. C.5 An Excel Spreadsheet for the Problem of Midas Miner

Selecting Cells for the Values of the Variables In Figure C.5, you can see that cells C12 and D12 are chosen to contain the values of the gold and silver variables, respectively. (Observe the use of labels **Gold** and **Silver** in cells C11 and D11 to indicate that the contents of the cells directly below contain information pertaining to the variable G and S in the paper model.) The numerical values of 4 and 5 in cells C12 and D12 are chosen arbitrarily so that you can verify subsequent formulas that use the values in these cells. You should choose contiguous cells in a single row, column, or rectangular block—hereafter referred to as a **range of cells**—for the values of the decision variables. This is because doing so simplifies the completion of the model, as you will see below. The cells you choose for the decision variables will eventually contain their optimal values, as determined by Solver.

Entering the Objective Function You must now enter a formula in a cell of the spreadsheet for computing the objective function $3G + 2S$ from the paper model in (C.1). For example, using the cells C12 and D12 that contain the respective values of the variables G and S, you could write the following formula in a cell for computing the objective function value:

=3 * C12 + 2 * D12

Consider, however, what would happen if the coefficient of 3 is changed in the future to 4, for example. In this case, you would have to change the foregoing formula. A better approach is to realize that the expression for the objective function depends on the coefficients 3 and 2, which are stored in cells C8 and D8 in the data portion of the spreadsheet (see the top portion of Figure C.5). You can use these cells to write the following formula for computing the objective function value:

$$=C8 * C12 + D8 * D12$$

The advantage of using the cells C8 and D8 to write the formula for computing the objective function value is that if the data for the objective function coefficients change in the future, then you need only change those values in the cells containing those data—you need not change the formula for computing the objective function value.

Note that the foregoing formula computes the sum of the products of the numbers in two lists, each of which is a range of cells: the list of numbers in cells C8 and D8 and the list of numbers in cells C12 and D12. This operation arises in so many types of problems that Excel has a built-in function called *SUMPRODUCT* that performs this computation. To use *SUMPRODUCT*, you must provide the range of cells that contain the two lists (both of the same vertical, horizontal, or square dimensions), as follows:

$$=\text{SUMPRODUCT(C8:D8, C12:D12)}$$

While you can type C8:D8 and C12:D12, for each of the two lists, you can instead simply left click and drag over the range of cells in the spreadsheet that you want. In any case, the foregoing *SUMPRODUCT* formula is entered in cell F12 in Figure C.6; however, note that what you see in cell F12 is the value 22 that results from computing this formula, namely, 3 * 4 + 2 * 5. Observe also the use of the label **Profit** in cell F11 to indicate the meaning of the contents in the cell immediately below.

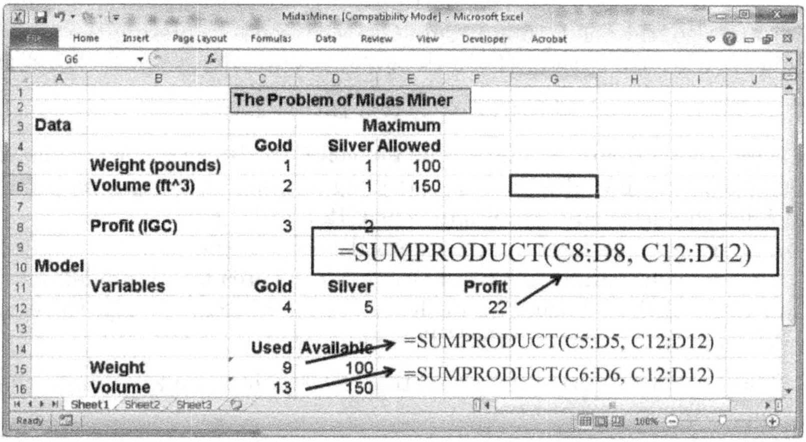

Fig. C.6 The SUMPRODUCT Formula for Computing the Objective Function for the Problem of Midas Miner

Entering the Constraints Recall that each constraint of a model is represented in the spreadsheet by choosing two cells: one to contain the value of the LHS and one for the RHS value of the constraint. To illustrate, consider the weight constraint in the problem of Midas Miner:

$$\underbrace{1G + 1S}_{LHS} \leq \underbrace{100}_{RHS}$$

The RHS value of 100 is entered in cell D15 in Figure C.6 using the formula =E5 which, as you can see, is the cell in the data portion of the spreadsheet that contains the value 100. Also, the cell C15 in Figure C.6 is chosen for entering the formula for computing the LHS value of the weight constraint, namely, $1G + 1S$. As with the objective function, observe that the value of $1G + 1S$ depends on the values of the decision variables in cells C12 and D12 and the data in cells C5 and D5. Using the built-in function *SUMPRODUCT*, the following is the appropriate formula for the LHS of this constraint:

=SUMPRODUCT(C5:D5, C12:D12)

The foregoing formula is entered in cell C15 but you do not see that formula in that cell in Figure C.6. Rather, you see the value 9 that results from evaluating the foregoing formula at the current values of 4 and 5 for the decision variables in cells C12 and D12.

Two cells are also needed for the LHS and RHS of the volume constraint:

$$\underbrace{2G + 1S}_{LHS} \leq \underbrace{150}_{RHS}$$

The RHS value of 150 is entered in cell D16 in Figure C.6 using the formula =E6, which is the cell in the data portion of the spreadsheet that contains the value 150. Also, the cell C16 in Figure C.6 is chosen for entering the following formula for computing the LHS of the weight constraint, namely, $2G + 1S$:

=SUMPRODUCT(C6:D6, C12:D12)

Once again, you do not see this formula in cell C16. Rather, you see the value 13 that results from evaluating the foregoing formula at the current values of 4 and 5 for the decision variables in cells C12 and D12.

You have now used cells in the spreadsheet for the variables, objective function, and constraints. However, the current spreadsheet model does *not* contain 100% of the information in the paper model. Specifically, the current spreadsheet model is missing the following items:

(1) The *type* of objective function (that is, whether you are minimizing or maximizing).

(2) The *relationship* of the LHS and RHS of each constraint, that is, $\leq$, $=$, or $\geq$.

(3) The nonnegativity constraints $G \geq 0$ and $S \geq 0$.

You will now learn how to specify these three remaining items.

Completing the Spreadsheet Model with Solver After entering the data and basic form of a paper model in an Excel spreadsheet, as just described, you can use the add-in Solver to obtain the optimal values of the variables. To do so, you first need to provide Solver with the remaining information to complete the model, as described now and illustrated with the problem of Midas Miner.

The first step is to click on the Data Tab of Excel and then on Solver in the far right (if you do not see Solver there, then you need to install Solver by following the instructions given in Section C.1). Doing so results in the Solver Parameters screen shown in Figure C.7. The next step is to enter all the necessary information in this screen to complete the model, as shown for the problem of Midas Miner in Figure C.8 and explained in the following discussion. In that regard, after clicking in a specific box in the Solver Parameters screen,

- To enter a single cell from the Excel worksheet, you can either type the address of that cell (for example, F12) or, more simply, left click on the cell in the spreadsheet and the appropriate cell will appear automatically in the Solver Parameters screen.

- To enter a range of cells from the Excel worksheet, you can either type the addresses of the first cell in the range, followed by a colon (:), followed by the last cells in the range (for example, C12:D12). Alternatively, you can left click on the first cell in the range and drag the cursor to the last cell in that range.

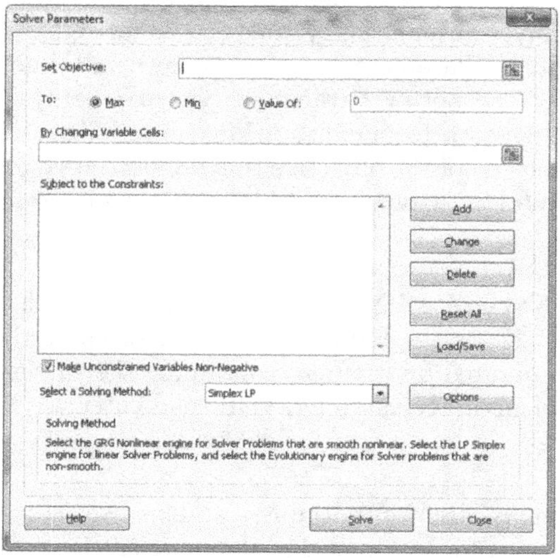

Fig. C.7 The Solver Parameters Screen

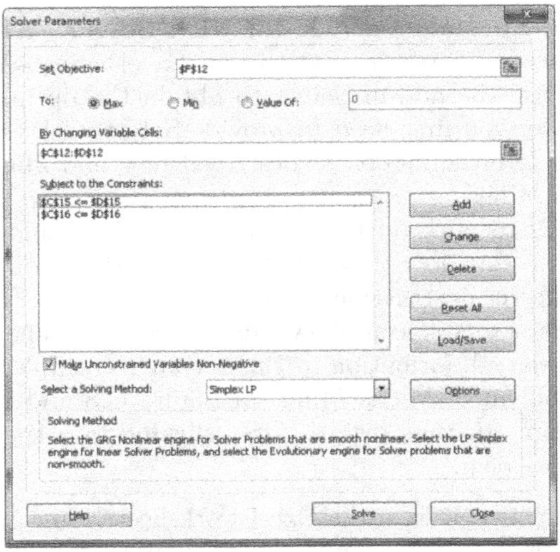

Fig. C.8 The Completed Solver Parameters Screen for the Problem of Midas Miner

Specifying the Objective Function To fill in the box labeled "Set Objective" at the top of the Solver Parameters screen, first click in the empty box next to "Set Objective" and then enter the cell of your spreadsheet that contains the formula for computing the value of the objective function. Referring to the spreadsheet in Figure C.6 for the problem of Midas Miner, you could click on cell F12. On so doing, you will see F12 appear in the "Set Objective" box (Solver automatically adds the dollar signs whenever any cell is entered in the Solver Parameters screen), as seen in Figure C.8.

The next line of the Solver Parameters screen labeled "To:" is used to indicate whether you want to maximize (Max), minimize (Min), or achieve a specific value (Value Of) of the objective function in the target cell. Select the appropriate option for your problem, which is Max for the problem of Midas Miner.

Specifying the Variables Next, click in the box labeled "By Changing Variable Cells" in the Solver Parameters screen to enter the list of cells in the spreadsheet, separated by commas, containing the values of the variables. Thus, for the problem of Midas Miner, you could enter the following:

C12, D12

When there are many decision variables, and hence many cells containing their values, you can enter the list using the more efficient colon (:) notation, provided those cells are in contiguous locations. Herein lies one of the advantages of selecting a group of contiguous cells for the values of the decision variables. You can enter the range of cells from C12 to D12 in Figure C.8 as follows by dragging the mouse over this range (to which Solver will add dollar signs):

C12:D12

If the cells containing the values of the variables form a rectangle, then you need only provide Solver with the upper-left and lower-right corners of the rectangle. Thus, for example, to indicate a rectangle of cells whose upper-left corner is cell C6 and whose lower-right corner is E10, use the following range:

C6:E10

Any combination of the comma and colon notation is allowed.

Specifying the Constraints You next enter all of the constraints in the box labeled "Subject to the Constraints" in the Solver Parameters screen shown in Figure C.8. Specifically, you must provide Solver with the cells that contain the LHS and RHS value of each constraint, as well as the relationship ($\leq, =,$ or $\geq$). To do so, first click the Add button in the Solvers Parameters screen to obtain the Add Constraint screen shown in Figure C.9. To enter the weight constraint

$$1G \; + \; 1S \; \leq \; 100$$

for the problem of Midas Miner, first click in the box labeled "Cell Reference" of the Add Constraint screen and then click on the cell in your spreadsheet that contains the formula for the LHS of this constraint, namely, cell C15 for the spreadsheet in Figure C.6.

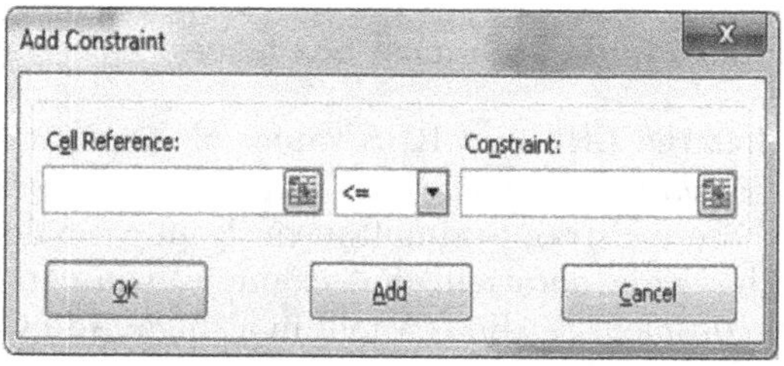

Fig. C.9 The Add Constraint Screen of Solver

Next, use the drop-down box in the middle of the Add Constraint screen to select the relationship of the constraint: $\leq, =,$ or $\geq$. For the foregoing weight constraint, $\leq$ is selected. (It is also here that you can specify that the value of a variable is to be integer.)

Then enter the cell of the spreadsheet containing the RHS value of this constraint in the rightmost box (labeled Constraint:) of the Add Constraint screen. For the weight constraint in the problem of Midas Miner, D15 is entered and the completed Add Constraint screen for this constraint is shown in Figure C.10.

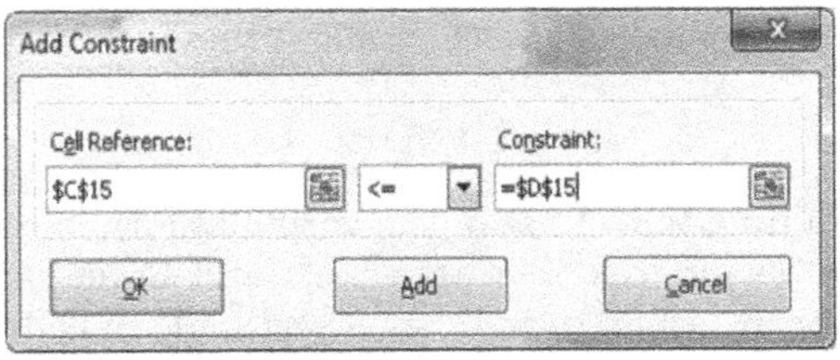

Fig. C.10 The Completed Add Constraint Screen for the Weight Constraint of Midas Miner

This completes the description of the first constraint for the problem of Midas Miner. To enter the next constraint, click Add in the Add Constraint screen and repeat the foregoing process of specifying the cells containing the LHS, RHS, and relationship of this constraint. Thus, for the volume constraint of Midas Miner, you would enter the following after clicking Add in the Add Constraint screen:

$$\text{C16} <= \text{D16}$$

Nonnegativity constraints are also included here by following these steps:

1. Enter the cell of the spreadsheet containing the value of the variable that is to be nonnegative in the box labeled "Cell Reference."
2. Select >= for the relationship of the constraint.
3. Type the value 0 in the rightmost box labeled Constraint:.

In the event that the LHS and RHS values of a collection of constraints *of the same relationship* ($\leq, =$, or $\geq$) are in a range of contiguous cells, you can enter all of those constraints simultaneously in a single Add Constraint screen by using the colon notation. For example, you can enter the two $\leq$ constraints for the problem of Midas Miner in a single Add Constraint screen, as follows:

$$\text{C15:C16} <= \text{D15:D16}$$

Likewise, you could enter the two nonnegativity constraints as follows:

$$\text{C12:D12} >= 0$$

When using the colon notation, the relationship of the constraint ($\leq, =$, or $\geq$) must be the same for all constraints in the specified range. It is therefore advisable to group all $\leq$ constraints in contiguous cells of the spreadsheet, all $=$ constraints in contiguous cells, and all $\geq$ constraints in contiguous cells, whenever doing so does not detract from the readability and logical organization of the spreadsheet.

After all constraints are added correctly, click OK in the Add Constraint creen. On so doing, the Solver Parameters screen reappears and the box la-eled "Subject to Constraints" now contains all of the constraints you entered, s shown in Figure C.11.

Notice the box labeled Make Unconstrained Variables Non-Negative. When ou click in that box, you will see a check symbol (√) appear, and this means hat the values of all variables in your problem are to be ≥ 0. Therefore, if ou check this box, you do not need to include the nonnegativity constraints n the Subject to Constraints box.

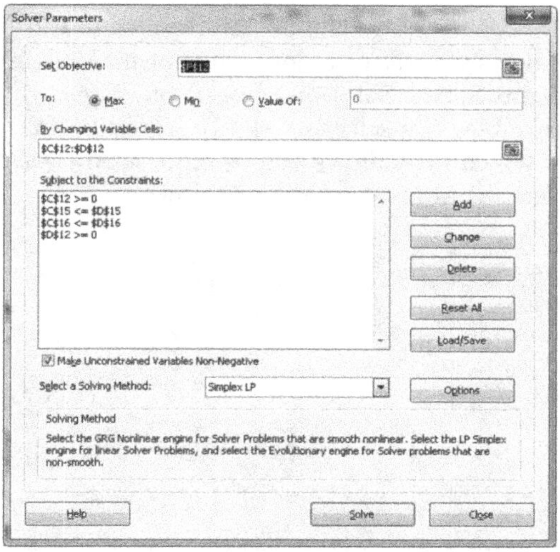

Fig. C.11 The Completed Solver Parameters Screen

The final step is to tell Solver that your problem is an LP (because Solver an also solve integer and nonlinear programming problems). This is accom-lished by using the drop-down box labeled Select a Solving Method that is ocated below the Subject to Constraint box in the Solver Parameters screen o choose Simplex LP, as has been done in Figure C.11.

Reviewing and Modifying the Solver Information You are now ready o ask Solver to use the simplex algorithm to solve your LP. However, before oing so, it is advisable to review the information you provide to Solver for ccuracy. Any errors should be corrected. For example, if you forget to in-lude a constraint, click the Add button next to the box labeled "Subject to Constraints" in the Solver Parameters screen shown in Figure C.11. Alter-atively, if you want to change one of the existing constraints, click on that onstraint in the Subject to Constraints box to highlight it and then click he Change button and make appropriate modifications. Finally, to delete a onstraint, highlight the constraint and then click the Delete button.

C.3 USING SOLVER TO SOLVE THE PROBLEM

After completing the information in the Solver Parameters screen and selecting Simplex LP, you instruct Solver to obtain the solution by clicking the Solve button in the lower-right portion of the Solver Parameters screen shown in Figure C.11. If Solver is able to obtain the solution to your problem, then the optimal values of the variables are stored in the cells of the spreadsheet specified in the "By Changing Variables Cells" box of the Solver Parameters screen. Otherwise Solver displays a message indicating either that your LP is infeasible or unbounded.

When finished, Solver displays the Solver Results screen shown in Figure C.12. This screen allows you either to keep the optimal values of the variables (by selecting the Keep Solver Solution option) or to restore all of the original values of the variables (by selecting the Restore Original Values option). The Solver Results screen also allows you to request Answer, Sensitivity, and Limits reports, described in greater detail in Section C.3.

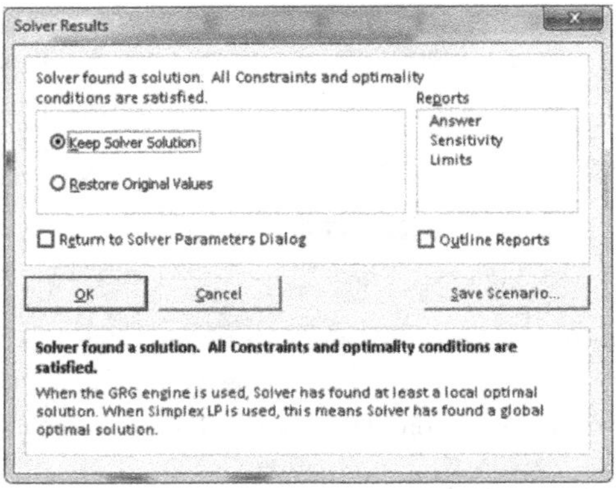

Fig. C.12 The Solver Results Screen

When done, click OK to close the Solver Results screen. On doing so for the problem of Midas Miner, you can see from the optimal values in cells C12 and D12 in Figure C.13 that Midas should load 50 pounds of gold ore and 50 pounds of silver ore on his spaceship. Doing so results in a profit of 250 intergalactic credits, shown in cell F12. This solution uses all 100 pounds of cargo weight capacity, as shown in cell C15 in Figure C.13. Likewise, all 150 cubic feet of cargo volume are used (see cell C16).

In this section, you have seen the steps involved in using Excel to solve a model that has already been developed. In Section C.4, you will see how to interpret the information in the Answer Report and Sensitivity Report that Solver provides.

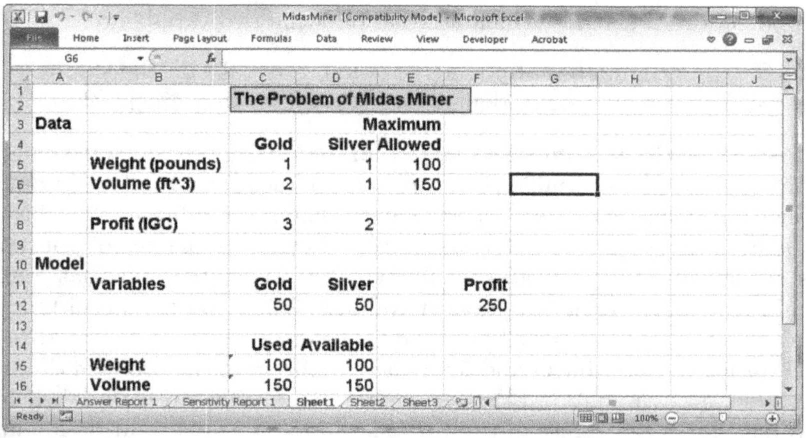

	A	B	C	D	E	F
1			The Problem of Midas Miner			
2						
3	Data				Maximum	
4			Gold	Silver	Allowed	
5		Weight (pounds)	1	1	100	
6		Volume (ft^3)	2	1	150	
7						
8		Profit (IGC)	3	2		
9						
10	Model					
11		Variables	Gold	Silver		Profit
12			50	50		250
13						
14			Used	Available		
15		Weight	100	100		
16		Volume	150	150		

Fig. C.13 The Optimal Solution Obtained by Solver for the Problem of Midas Miner

C.4 INTERPRETING THE ANSWER AND SENSITIVITY REPORTS FROM SOLVER

When you obtain the optimal solution to an LP, Solver also provides two reports that contain useful economic information about the solution: the *Answer Report* and the *Sensitivity Report.* In this section you will learn how to obtain and interpret the information in those two reports.

You obtain the Answer and Sensitivity Report by clicking on both of these two reports in the Solver Results screen that is displayed immediately after Solver finds the optimal solution to your LP (see Figure C.12) and then clicking OK. As seen at the bottom of Figure C.13, Solver stores these reports in two new Excel worksheets whose names are *Answer Report 1* and *Sensitivity Report 1* which you will find in two new tabs at the bottom of the spreadsheet that contains your data and model.

Interpreting the Answer Report

By clicking on the appropriate tab, you obtain the Answer Report, as shown for the problem of Midas Miner in Figure C.14. As you can see, the Answer Report contains three sections corresponding to the objective function, the variables, and the constraints each of which are discussed now, starting with the Variable Cells section in the middle that pertains to the values of the variables.

Interpreting the Variable Cells Section This section of the report contains one row of information pertaining to each variable in your problem. Recall that the problem of Midas Miner includes the following two variables:

$$\begin{aligned} G &= \text{the number of pounds of gold ore to take on the ship} \\ S &= \text{the number of pounds of silver ore to take on the ship.} \end{aligned}$$

The first line of the Variable Cells section in Figure C.14 contains information pertaining to the gold variable G and the second line contains information pertaining to the silver variable S, as you can see under the column labeled Name (this is another reason why it is advisable to enter the names of your variables in the cells of the spreadsheet immediately above the cells that contain the values of the variables). The next column of the report, labeled Original Value, displays the values of the variables before you called Solver to solve the LP (4 for the variable G and 5 for the variable S in the problem of Midas Miner, as seen in Figure C.14). The next column of the report, labeled Final Value, shows the values of the variables in the optimal solution found by Solver (namely, 50 for the variable G and 50 for the variable S in the problem of Midas Miner). Note that Solver knows nothing about the units of those variables and so you must interpret the values according to the appropriate units (pounds, for the problem of Midas Miner). The final column, labeled Integer indicates whether the variable is integer or continuous (and should *always* be continuous for an LP).

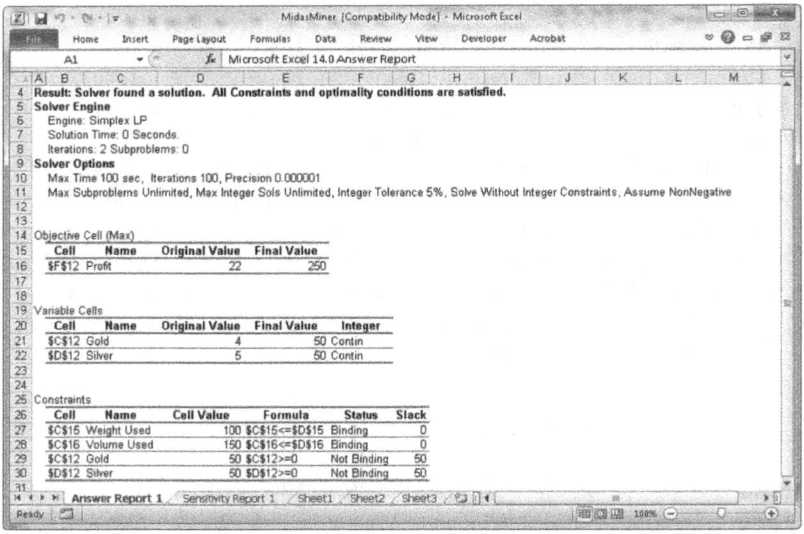

Fig. C.14 The Answer Report for the Problem of Midas Miner

Interpreting the Objective Cell Section The Objective Cell section of the Answer Report contains information pertaining to the objective function. As you can see in Figure C.14, for the problem of Midas Miner, the Original Value of the objective function $3G + 2S$ prior to calling Solver was 22 (corresponding to the original values in the spreadsheet of $G = 4$ and $S = 5$) and the Final Value after calling Solver is 250 (corresponding to the final values in the spreadsheet of $G = 50$ and $S = 50$ that Solver obtained). Note again that Solver does not know the units of the objective function and so you must interpret this value yourself (intergalactic credits, for the problem of Midas Miner).

Interpreting the Constraints Section The Constraints section of the Answer Report contains one row of information for each constraint that is listed in the Add Constraints portion of the Solver Parameters screen (see Figure C.11). The column labeled Cell in the Constraints section of the Answer Report shows the cell of the spreadsheet that contains the LHS of the constraint. The second column shows the names that appear immediately to the left and immediately above the cell containing the LHS value of the constraint. The third column, labeled Cell Value, contains the value of the LHS of the constraint at the optimal values of the variables. Thus, for example, for the weight constraint $1G + 1S \leq 100$ for the problem of Midas Miner, the value of 100 displayed in the column labeled Cell Value in Figure C.14 is the value of $1G + 1S$ at the optimal values of $G = 50$ and $S = 50$. The fourth column, labeled Formula, shows the cells with the LHS and RHS together with the relationship (<=, =, or >=) between the two sides of the constraint.

Of particular importance is the next two columns, labeled Status and Slack. Recall from Section 4.6 of the book that, prior to finding an optimal solution, the simplex algorithm adds a nonnegative slack variable to the LHS of each $\leq$ constraint and subtracts a nonnegative slack variable from the LHS of each $\geq$ constraint. In particular, for the problem of Midas Miner, a separate nonnegative slack variable is added to the LHS of each of the weight and volume ($\leq$) constraints. The final column labeled Slack in this section of the Answer Report shows the values of these slack variables. As seen in Figure C.14, the last column labeled Slack indicates that the values of the two slack variables associated with the weight and volume constraint are both 0. In the context of this problem, this means that the optimal solution of $G = 50$ and $S = 50$ uses all 100 pounds of available cargo weight and all 150 cubic feet of available cargo space—that is, there are 0 pounds of cargo weight and 0 cubic feet of cargo space that are unused by the optimal solution. When the slack variable associated with an inequality constraint has value 0—as in the weight and volume constraints for the problem of Midas Miner—the constraint is said to be **binding** at the optimal solution, as you can see in the next-to-last column, labeled Status, of the Constraints section of the Answer Report in Figure C.14 for these two constraints.

Likewise, because the nonnegativity constraints $G \geq 0$ and $S \geq 0$ for the problem of Midas Miner are included in the Add Constraints portion of the Solver Parameters screen (see Figure C.11), two more nonnegative slack variables are subtracted from the LHS of each of these constraints. As you can see from the final column of the Constraints section of the Answer Report in Figure C.14, for these two nonnegativity constraints, the values of the slack variables are both 50. In the context of this problem, this means that the optimal values of $G = 50$ and $S = 50$ are each 50 pounds above the allowable value of 0 on the RHS of the nonnegativity constraints. Because the values of these two slack variables are not 0, the statuses of these two nonnegativity constraints are said to be **not binding** at the optimal solution, as you can see in the next-to-last column, labeled Status, of the Constraints section of the Answer Report in Figure C.14 for these two constraints.

Now that you know how to interpret the Answer Report from Solver, you will learn how to do the same for the Sensitivity Report.

Interpreting the Sensitivity Report From Solver

Recall from Chapter 9 of the book that *sensitivity analysis* is the process of asking what happens to the optimal solution and optimal objective function value of an LP if some of the data in the A-matrix, **c**-vector, or **b**-vector change. While you can always change the appropriate values in the data portion of the Excel spreadsheet and then resolve the LP, in some cases, you can answer such questions more easily—and without having to use a computer—by working with the **Sensitivity Report** produced by Solver.

You obtain this report in the same way as the Answer Report described previously in this section, namely, from the Solver Results screen produced after obtaining the optimal solution (see Figure C.12). The Sensitivity Report for the problem of Midas Miner—which is stored in a separate Excel worksheet—is shown in Figure C.15 where you can see that there are two separate portions: one called **Variable Cells** and one called **Constraints**. As described below, you can use these two portions to answer the following two types of sensitivity questions, respectively:

1. What happens to the optimal solution and optimal objective function value if one (and only one) of the objective function coefficients in the **c**-vector changes, assuming that all other data remain the same?

2. What happens to the optimal solution and optimal objective function value if one (and only one) of the RHS values of a constraint in the **b**-vector changes, assuming that all other data remain the same?

Answer Sensitivity Questions About the Objective Function Coefficients The top portion of the Sensitivity Report, labeled Variable Cells, contains one row of information for each variable. For example, the first row in this portion of the Sensitivity Report for the problem of Midas Miner in Figure C.15 refers to the gold variable G and the second row to the silver variable S (as indicated in the second column labeled Name). The third column provides the **reduced cost** of the variables—described in Section 5.2 of the book—which

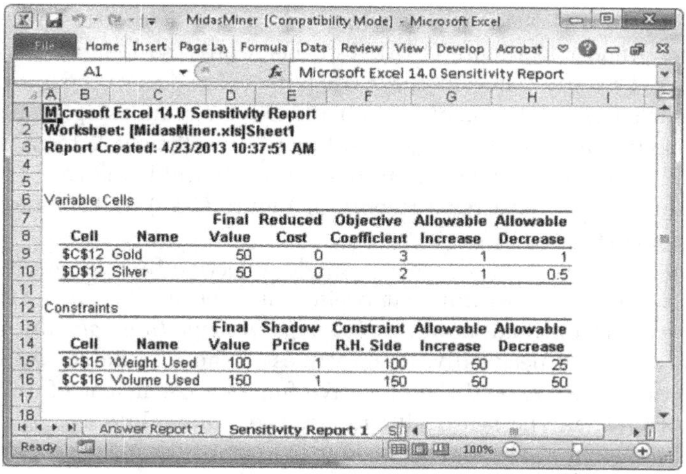

Cell	Name	Final Value	Reduced Cost	Objective Coefficient	Allowable Increase	Allowable Decrease
C12	Gold	50	0	3	1	1
D12	Silver	50	0	2	1	0.5

Cell	Name	Final Value	Shadow Price	Constraint R.H. Side	Allowable Increase	Allowable Decrease
C15	Weight Used	100	1	100	50	25
C16	Volume Used	150	1	150	50	50

Fig. C.15 The Sensitivity Reports for the Problem of Midas Miner

are 0 for both G and S, as seen in Figure C.15. The next column, labeled Objective Coefficient, shows the value of the objective function coefficient for the corresponding variable. For the problem of Midas Miner, the objective function is to maximize $3G + 2S$ and so the values of 3 and 2 are shown in the Objective Coefficient column of the report in Figure C.15.

The final two columns in this portion of the report, labeled Allowable Increase and Allowable Decrease, are used to answer the following sensitivity question about the objective function coefficient of a variable:

> **Question:** By how much can the objective function coefficient c_j for the variable x_j change without changing the current optimal solution?

Using the Allowable Decrease and the Allowable Increase from this portion of the Sensitivity Report associated with the variable x_j, the answer is the following:

> **Answer:** So long as the objective function coefficient c_j for the variable x_j is changed to any value between $c_j -$Allowable Decrease to $c_j +$Allowable Increase—hereafter called the **sensitivity range of** c_j—the current optimal solution remains optimal. However, if c_j is changed to a value outside of the sensitivity range, then the optimal solution will change in an unpredictable way and you will need to solve the modified LP to obtain the new optimal solution.

For example, for the objective function $3G + 2S$ in the problem of Midas Miner, the sensitivity range of the coefficient 3 is 3 − the Allowable Decrease of 1 = 2 up to 3 + the Allowable Increase of 1 = 4 (see this portion of the Sensitivity Report in Figure C.15). In other words, so long as the coefficient 3 is changed to any value in the sensitivity range [2, 4] and no other data change, the current solution of $G = 50$ and $S = 50$ remains optimal. Likewise the sensitivity range of the objective function coefficient 2 for S is 2 − the Allowable Decrease of 0.5 = 1.5 up to 3 + the Allowable Increase of 1 = 4 (see this portion of the Sensitivity Report in Figure C.15) and so the sensitivity range for 2 is [1.5, 4].

As you have just learned, when a single objective function coefficient c_j is changed to any value within that coefficient's sensitivity range, the current solution remains optimal; however, *the objective function value at the optimal solution changes.* Specifically, if the values of the n variables in an optimal solution are $x_1, \ldots, x_n$ and the objective function coefficient of c_j changes to a value, say c'_j, in the sensitivity range, then the values $x_1, \ldots, x_n$ remain optimal but the objective function value of $c_1x_1 + \cdots + c_jx_j + \cdots + c_nx_n = \mathbf{cx}$ changes to the following:

$$\begin{aligned}\text{New Obj. Func. Value} &= c_1x_1 + \cdots + c'_ix_j + \cdots + c_nx_n \\ &= \mathbf{cx} - c_jx_j + c'_ix_j \\ &= \mathbf{cx} + (c'_i - c_j)x_j\end{aligned}$$

For example, if the coefficient of 3 in the objective function $3G + 2S$ in the problem of Midas Miner is changed to 3.5, which is in the sensitivity range of [2, 4] for this coefficient, then the solution of $G = 50$ and $S = 50$ remains optimal but the objective function value changes from $3G + 2S = 3(50) + 2(50) = 250$ to $3.5G + 2S = 3.5(50) + 2(50) = 275$.

You have just learned how to use the Variable Cells portion of the Sensitivity Report to answer a sensitivity question about a change in a single objective function coefficient. You will now learn how to use the Constraints portion of the Sensitivity Report to answer a sensitivity question that pertains to a change in a RHS value of a single constraint.

Answer Sensitivity Questions About the RHS Value of a Constraint Unlike a change in an objective function coefficient in an LP, when a value on the RHS of a constraint changes, *the optimal solution changes in an unpredictable way* and so you would have to solve the modified LP to find the new optimal solution. However, when you change the value on the RHS of a constraint, in some cases, it is possible to predict how the *optimal objective function value changes,* as you will now learn to do using the bottom portion of the Sensitivity Report, labeled Constraints.

Specifically, this portion of the Sensitivity Report contains one row of information for each constraint (other than the logical constraints that involve only one variable, such as nonnegativity constraints). For example, the first row in this portion of the Sensitivity Report for the problem of Midas Miner in Figure C.15 refers to the weight constraint and the second row to the volume constraint (as indicated in the second column labeled Name). The third column, labeled Final Value, shows the value of the LHS of the associated constraint at the optimal solution and the fifth column, labeled Constraint R.H. Side, shows the RHS value of the constraint.

The fourth column provides the **shadow price** of each constraint which, for the problem of Midas Miner, is 1 for both the weight and volume constraints, as seen in Figure C.15. Recall, from Section 8.2 of the textbook, that the shadow price of a constraint is the optimal value of the dual variable associated with the constraint and has the following meaning:

shadow price = the change in the optimal objective function value per unit of increase in the RHS of the associated constraint, assuming all other data remain the same.

For the problem of Midas Miner in which it is optimal to load $G = 50$ pounds of gold ore and $S = 50$ pounds of silver ore on the spaceship, the shadow price of 1 for the weight constraint $1G + 1S \leq 100$ means that for each additional unit above 100 on the RHS of the foregoing constraint, the optimal objective function value of $3G+2S = 3(50)+2(50) = 250$ changes by the shadow price of 1. For example, if the RHS value of 100 for the weight constraint increases by 8 to 108, then the optimal objective function value of 250 changes by 8 times the shadow price of $1 = 8$ and thus becomes 258. Likewise, the shadow price of 1 for the volume constraint $2G + 1S \leq 150$ means that for each additional unit above 150 on the RHS of the foregoing constraint, the optimal objective function value of 250 changes by the shadow price of 1. For example, if the RHS value of 150 for the volume constraint decreases by 5 to 145, then the optimal objective function value of 250 changes by -5 times the shadow price of $1 = -5$ and thus becomes 245.

In the context of sensitivity analysis, the shadow price is used to answer the following question about the RHS value of a constraint:

> **Question:** What happens to the optimal objective function value of an LP if the RHS value of a single constraint i is changed from its original value of b_i to a new value, say b'_i, all other data being the same?

To answer such a question, you should think about using the shadow price; however, before doing so, it is important to keep in mind that *you can only use the shadow price if the new value b'_i of the RHS is in a certain* **sensitivity range** *around the original RHS value b_i.*

You can obtain the sensitivity range for the RHS of a constraint by using the Allowable Decrease and the Allowable Increase from the Constraints portion of the Sensitivity Report to provide the following answer to the foregoing sensitivity question:

> **Answer:** So long as the RHS b_i for constraint i is changed to a value b_i' in the sensitivity range from b_i − Allowable Decrease up to b_i + Allowable Increase, the optimal objective function value changes at the rate of the shadow price for that constraint. However, if b_i' is changed to a value outside of the sensitivity range, then the optimal objective function value will change in an unpredictable way and you will need to solve the modified LP to obtain the new optimal objective function value.

For example, using the Allowable Decrease of 25 and the Allowable Increase of 50 from the first row of the Constraints portion of the Sensitivity Report for the problem of Midas Miner in Figure C.15, the sensitivity range for the value 100 on the RHS of the weight constraint is:

$$[100 - 25, 100 + 50] = [75, 150].$$

In other words, so long as the value of 100 on the RHS of the weight constraint is changed to any value in the sensitivity range [75, 150] and no other data change, the optimal objective function value changes at the rate of the shadow price of 1. Likewise, the sensitivity range for the value of 150 on the RHS of the volume constraint is from 150 − the Allowable Decrease = 150 − 50 = 100 up to 150 + the Allowable Increase = 150 + 50 = 200, that is, [100, 200] (see the second row in this portion of the Sensitivity Report in Figure C.15).

You now have the knowledge of how to create a spreadsheet model of a linear programming problem and how to use Solver in Excel to obtain an optimal solution. You have learned how to request and interpret the Answer and Sensitivity Reports that Solver produces and to use them to answer certain types of sensitivity questions pertaining to how the optimal solution and optimal objective function value change when you change a single objective function coefficient or the RHS value of a constraint.

Index

D